Stem Cell Therapy for Diabetes

Stem Cell Therapy for Diabetes

Editor: Kobe Harding

AMERICAN
MEDICAL PUBLISHERS
www.americanmedicalpublishers.com

AMERICAN
MEDICAL PUBLISHERS
www.americanmedicalpublishers.com

Cataloging-in-Publication Data

Stem cell therapy for diabetes / edited by Kobe Harding.
p. cm.
Includes bibliographical references and index.
ISBN 979-8-88740-470-7
1. Diabetes--Treatment. 2. Stem cells--Therapeutic use. 3. Stem cells.
4. Cellular therapy. I. Harding, Kobe.
RC660 .S74 2023
616.462--dc23

American Medical Publishers,
41 Flatbush Avenue,
1st Floor, New York,
NY 11217, USA

ISBN 979-8-88740-470-7 (Hardback)

Contents

Preface VII

Chapter 1 **Modeling Type 1 Diabetes Using Pluripotent Stem Cell Technology** 1
Kriti Joshi, Fergus Cameron, Swasti Tiwari, Stuart I. Mannering, Andrew G. Elefanty and Edouard G. Stanley

Chapter 2 **Modeling Maturity Onset Diabetes of the Young in Pluripotent Stem Cells: Challenges and Achievements** 17
Carmel Braverman-Gross and Nissim Benvenisty

Chapter 3 **Genome Editing Human Pluripotent Stem Cells to Model β-Cell Disease and Unmask Novel Genetic Modifiers** 26
Matthew N. George, Karla F. Leavens and Paul Gadue

Chapter 4 **Stem Cell-Based Clinical Trials for Diabetes Mellitus** 41
Eleonora de Klerk and Matthias Hebrok

Chapter 5 **Use of Induced Pluripotent Stem Cells to Build Isogenic Systems and Investigate Type 1 Diabetes** 54
Lucas H. Armitage, Scott E. Stimpson, Katherine E. Santostefano, Lina Sui, Similoluwa Ogundare, Brittney N. Newby, Roberto Castro-Gutierrez, Mollie K. Huber, Jared P. Taylor, Prerana Sharma, Ilian A. Radichev, Daniel J. Perry, Natalie C. Fredette, Alexei Y. Savinov, Mark A. Wallet, Naohiro Terada, Todd M. Brusko, Holger A. Russ, Jing Chen, Dieter Egli and Clayton E. Mathews

Chapter 6 **Selection for $CD26^-$ and $CD49A^+$ Cells from Pluripotent Stem Cells-Derived Islet-Like Clusters Improves Therapeutic Activity in Diabetic Mice** 68
Kfir Molakandov, Denise A. Berti, Avital Beck, Ofer Elhanani, Michael D. Walker, Yoav Soen, Karina Yavriyants, Michal Zimerman, Ella Volman, Itzik Toledo, Anna Erukhimovich, Alon M. Levy, Arik Hasson, Joseph Itskovitz-Eldor, Judith Chebath and Michel Revel

Chapter 7 **Human Pluripotent Stem Cells: A Unique Tool for Toxicity Testing in Pancreatic Progenitor and Endocrine Cells** 82
Erin M. MacFarlane and Jennifer E. Bruin

Chapter 8 **Monogenic Diabetes Modeling: *In Vitro* Pancreatic Differentiation from Human Pluripotent Stem Cells Gains Momentum** 95
Juan Ignacio Burgos, Ludovic Vallier and Santiago A Rodríguez-Seguí

Chapter 9 **Human Pluripotent Stem Cells to Model Islet Defects in Diabetes** 114
Diego Balboa, Diepiriye G. Iworima and Timothy J. Kieffer

Chapter 10 **Lessons from Human Islet Transplantation Inform Stem Cell-Based Approaches in the Treatment of Diabetes** 133
Taylor M. Triolo and Melena D. Bellin

Chapter 11 **Epigenetic Memory: Lessons from iPS Cells Derived from Human β Cells** 140
Shimon Efrat

Chapter 12 **Human Pluripotent Stem Cells go Diabetic: A Glimpse on Monogenic Variants** 145
Sandra Heller, Michael Karl Melzer, Ninel Azoitei, Cécile Julier and Alexander Kleger

Chapter 13 **Harnessing Proliferation for the Expansion of Stem Cell-Derived Pancreatic Cells: Advantages and Limitations** 161
Amanda Oakie and Maria Cristina Nostro

Chapter 14 **Protecting Stem Cell Derived Pancreatic Beta-Like Cells from Diabetogenic T Cell Recognition** 168
Roberto Castro-Gutierrez, Aimon Alkanani, Clayton E. Mathews, Aaron Michels and Holger A. Russ

Chapter 15 **Multipotent Mesenchymal Stromal Cells Interact and Support Islet of Langerhans Viability and Function** 180
Naomi Koehler, Leo Buhler, Bernhard Egger and Carmen Gonelle-Gispert

Chapter 16 **Microvascular Complications in Type 1 Diabetes: A Comparative Analysis of Patients Treated with Autologous Nonmyeloablative Hematopoietic Stem-Cell Transplantation and Conventional Medical Therapy** 188
Jaquellyne G. Penaforte-Saboia, Renan M. Montenegro Jr., Carlos E. Couri, Livia A. Batista, Ana Paula D. R. Montenegro, Virginia O. Fernandes, Hussain Akhtar, Carlos A. Negrato, Kelen Cristina Ribeiro Malmegrim, Daniela Aparecida Moraes, Juliana B. E. Dias, Belinda P. Simões, Marilia Brito Gomes and Maria Carolina Oliveira

Chapter 17 **Lower Insulin-Dose Adjusted A1c (IDAA1c) is Associated with Less Complications in Individuals with Type 1 Diabetes Treated with Hematopoetic Stem-Cell Transplantation and Conventional Therapy** 196
Jaquellyne Gurgel Penaforte-Saboia, Carlos Eduardo Barra Couri, Virginia Oliveira Fernandes, Ana Paula Dias Rangel Montenegro, Lívia Aline De Araújo Batista, Lenita Zajdenverg, Carlos Antonio Negrato, Kelen Cristina Ribeiro Malmegrim, Daniela Aparecida Moraes, Juliana Bernardes Elias Dias, Maria Carolina Oliveira, Akhtar Hussain, Marilia Brito Gomes and Renan Magalhães Montenegro Jr.

Chapter 18 **Stem Cell Transplantation in the Treatment of Type 1 Diabetes Mellitus: From Insulin Replacement to Beta-Cell Replacement** 203
Xin-Xing Wan, Dan-Yi Zhang, Md. Asaduzzaman Khan, Sheng-Yuan Zheng, Xi-Min Hu, Qi Zhang, Rong-Hua Yang and Kun Xiong

Chapter 19 **Mesenchymal Stem Cell Exosomes as a New Strategy for the Treatment of Diabetes Complications** 216
Jiachao Xiong, Hao Hu, Rong Guo, Hui Wang and Hua Jiang

Permissions

List of Contributors

Index

Preface

Every book is a source of knowledge and this one is no exception. The idea that led to the conceptualization of this book was the fact that the world is advancing rapidly; which makes it crucial to document the progress in every field. I am aware that a lot of data is already available, yet, there is a lot more to learn. Hence, I accepted the responsibility of editing this book and contributing my knowledge to the community.

Stem cell therapy represents the use of stem cells for the treatment and prevention of various diseases. The researches in the field of stem cells can aid in an increased understanding of disease mechanisms. It can be used to produce healthy tissues to replace the diseased ones. Stem cell therapy holds immense opportunity for the management of neurodegenerative diseases and conditions such as diabetes and heart disease. A patient suffering from type 1 diabetes has an impaired function of insulin-producing beta cells that are present in the pancreas. In laboratory experiments, scientists have been able to coax embryonic stem cells into beta cells. The successful transplantation of beta cells can provide a substitute for damaged cells in a diabetic patient. This book explores the potential of stem cell therapy in the treatment of diabetes. A number of latest researches have been included to keep the readers up-to-date with the global concepts in this area of study. This book aims to serve as a resource guide for students and experts alike.

While editing this book, I had multiple visions for it. Then I finally narrowed down to make every chapter a sole standing text explaining a particular topic, so that they can be used independently. However, the umbrella subject sinews them into a common theme. This makes the book a unique platform of knowledge.

I would like to give the major credit of this book to the experts from every corner of the world, who took the time to share their expertise with us. Also, I owe the completion of this book to the never-ending support of my family, who supported me throughout the project.

Editor

1

Modeling Type 1 Diabetes Using Pluripotent Stem Cell Technology

*Kriti Joshi[1,2,3], Fergus Cameron[3,4,5], Swasti Tiwari[2], Stuart I. Mannering[6], Andrew G. Elefanty[3,5,7] and Edouard G. Stanley[3,5,7]**

[1] *Department of Endocrinology and Metabolism, All India Institute of Medical Sciences Rishikesh, Uttarakhand, India,* [2] *Department of Molecular Medicine & Biotechnology, Sanjay Gandhi Post Graduate Institute of Medical Sciences, Lucknow, India,* [3] *Department of Cell Biology, Murdoch Children's Research Institute, Parkville, Vic, Australia,* [4] *Department of Endocrinology and Diabetes, The Royal Children's Hospital, Parkville, Vic, Australia,* [5] *Department of Paediatrics, University of Melbourne, Parkville, Vic, Australia,* [6] *Immunology and Diabetes Unit, St. Vincent's Institute of Medical Research, Fitzroy, Vic, Australia,* [7] *Department of Anatomy and Developmental Biology, Monash University, Clayton, Vic, Australia*

***Correspondence:**
Edouard G. Stanley
ed.stanley@mcri.edu.au

Induced pluripotent stem cell (iPSC) technology is increasingly being used to create *in vitro* models of monogenic human disorders. This is possible because, by and large, the phenotypic consequences of such genetic variants are often confined to a specific and known cell type, and the genetic variants themselves can be clearly identified and controlled for using a standardized genetic background. In contrast, complex conditions such as autoimmune Type 1 diabetes (T1D) have a polygenic inheritance and are subject to diverse environmental influences. Moreover, the potential cell types thought to contribute to disease progression are many and varied. Furthermore, as HLA matching is critical for cell-cell interactions in disease pathogenesis, any model that seeks to test the involvement of particular cell types must take this restriction into account. As such, creation of an *in vitro* model of T1D will require a system that is cognizant of genetic background and enables the interaction of cells representing multiple lineages to be examined in the context of the relevant environmental disease triggers. In addition, as many of the lineages critical to the development of T1D cannot be easily generated from iPSCs, such models will likely require combinations of cell types derived from *in vitro* and *in vivo* sources. In this review we imagine what an ideal *in vitro* model of T1D might look like and discuss how the required elements could be feasibly assembled using existing technologies. We also examine recent advances towards this goal and discuss potential uses of this technology in contributing to our understanding of the mechanisms underlying this autoimmune condition.

Keywords: induced pluripotent stem cells, type 1 diabetes, macrophages, antigen presenting cells, T cells, T cell receptor

INTRODUCTION

Type 1 diabetes mellitus (T1D) is an autoimmune disorder disease involving the specific destruction of insulin-producing pancreatic beta cells (1). Beta cell loss leads to primary insulin deficiency and subsequent hyperglycemia, which presents as clinical diabetes. A complex interplay of genetic and environmental factors is thought to trigger beta cell specific autoimmunity. The disease predominantly affects children and young adults and current estimates suggest that more than a million children around the world are affected by T1D, with the prevalence rising by almost 3% each year (2) Our knowledge of how the disorder develops remains imperfect and therefore attempts at preventing or curing the disease have largely not met with success (3).

A major deficit in understanding human T1D has been the lack of appropriate models. While multiple therapeutic interventions have been found effective in the Non obese diabetic (NOD) mouse T1D disease model, none have been translatable to humans (4). This has led investigators to question rodent models as a platform for testing disease therapeutics and has also resulted in a quest for human T1D models (5).

An idealized *in vitro* model of T1D would necessarily enable the incorporation of the large number of variables and cell types that have been implicated in disease development. Indeed, understanding how different cell types and environmental factors interact to contribute to the pathogenesis of T1D will be critical for development of new models.

DISEASE PATHOGENESIS

Current understanding is that T1D is precipitated in genetically susceptible individuals by environmental triggers such as infections, diet, toxins or stress, which initiate the autoimmune response against beta cells. Failure of immune tolerance results in the expansion of autoreactive CD4+ and CD8+ T cells, autoantibody-producing B cells, and activation of the innate immune system, which then collude to lead to beta cell destruction (6, 7).

Most of our understanding of disease pathogenesis has been deduced from rodent models such as the NOD mouse (4). However, emerging data from human biobanks such as the Diabetes Virus Detection (DiViD) study (8) and the JDRF Network for Pancreatic Organ Donors (nPOD) (9) have yielded important details of the human disease pathology and highlighted the differences in rodent and human disease patterns.

The histological hallmark of the disease is the presence of insulitis, i.e., an infiltration of inflammatory cells consisting of T and B lymphocytes and macrophages around and within islets (10, 11). Although variable between subjects, CD8+ T cells have been found to be the predominant immune cell type in the insulitic lesion, followed by CD68+ macrophages, CD4+ T cells and CD20+ B cells (11, 12).

OVERVIEW OF PATHOGENESIS

There are numerous hypotheses regarding the events initiating the processes that eventually lead to T1D. For example, a triggering event, such as a viral infection, may lead to an initial phase of beta cell death causing release of beta cell autoantigens (13). MHC Class I hyper-expression has also been noted on beta cells from T1D tissue samples, potentially making these cells prone to attack from self-reactive cytotoxic CD8+ T cells and further antigen release (14). Islet autoantigens are phagocytosed by antigen presenting cells in the islets, and carried to the draining pancreatic lymph nodes where they are presented to CD4+ and CD8+ T cells (15). Due to loss of central and peripheral tolerance, these self-antigens are recognized by autoreactive CD4+ T cells leading to their activation and proliferation. B cell activation leads to formation of plasma cells and the appearance of autoantibodies against islet proteins (16). These immune cells then infiltrate the islets leading to insulitis and progressive beta cell death (17).

With the initiation of the autoimmune attack, inflammatory cytokines are released which amplify the immune response. These include IL1 and 6, IFNγ, and TNFα. It has been suggested that some of these cytokines directly precipitate beta cell destruction, diminishing insulin secretion from the beta cells and activating cytotoxic T cells (18). They also enhance the expression of HLA class 1 molecules on the beta cells (19). Production of superoxide radicals and high concentrations of nitric oxide increase the damage to the beta cells (18). Inevitably, this damage leads to the further release of beta cell antigens, which may serve to create a feedback loop that reinforces ongoing beta cell destruction. This process is summarized in **Figure 1**.

THE CASE FOR ESTABLISHING HUMAN STEM CELL-BASED MODELS OF T1D

Studying the disease pathogenesis of human T1D has been challenging for many reasons. The retroperitoneal location of the organ and the inherent risk of pancreatitis make pancreatic biopsies a risky procedure (20, 21), leading to an understandable scarcity of pancreatic tissue samples from affected individuals. The scattered and sparse nature of the insulitic lesions means that multiple tissue samples from one organ are needed for a comprehensive analysis. The tissue itself is difficult to handle due to the high content of pancreatic enzymes, predisposing it to autolysis. In addition to this, T1D has a long pre-symptomatic phase which means that affected individuals only present with established disease where most of the beta cell mass has been lost, making the study of the early disease pathogenesis difficult (5). Therefore, mouse models have been widely used as surrogates of the human disease. However, recent studies on human pancreata have brought to light important differences between human and rodent disease patterns (8, 9, 22).

Abbreviations: iPSCs, Induced pluripotent stem cells; T1D, Type 1 Diabetes Mellitus; APC, Antigen presenting cell; TCR, T cell receptor.

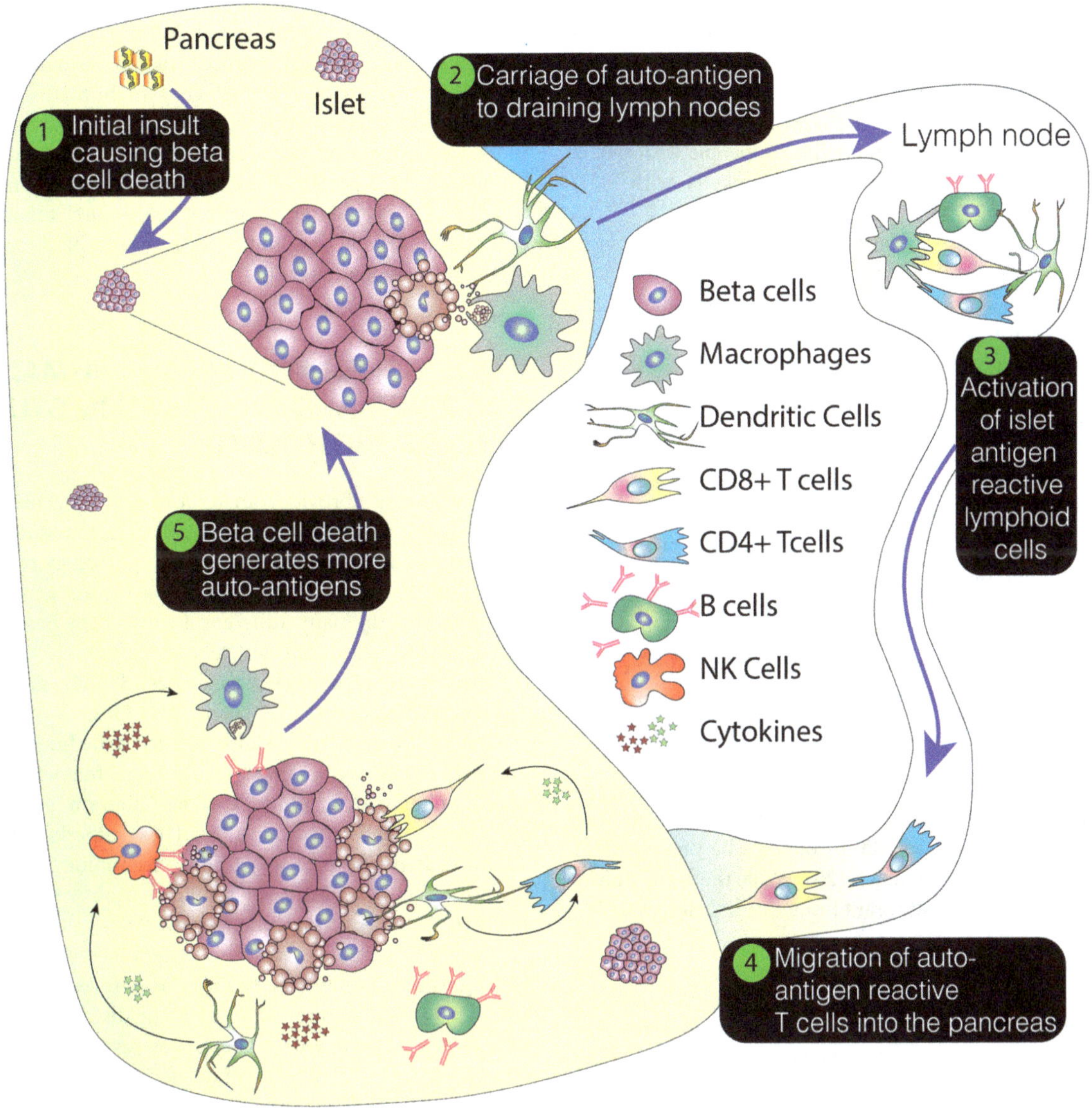

FIGURE 1 | Overview of the pathogenesis of T1D. This process envisages an initial insult that creates beta cell stress or death. The former potentially leading to the production of neoantigens and the latter resulting in the release of beta cell proteins. This damage results in the attraction of immune cells, with emigrant antigen presenting cells picking up and processing the proteins and conveying them to the pancreatic lymph nodes. Here, autoreactive or neoantigen specific T cells are recruited and these then migrate back to the pancreas, potentially promoting further inflammation, stress and cell death. This positive feedback loop ultimately results in the loss of beta cells.

ANIMAL MODELS—THE NON-OBESE DIABETIC (NOD) MOUSE

NOD mice are prone to spontaneously developing autoimmune diabetes, which mimics many features of human disease such as, islet infiltration by immune cells and the development of autoantibodies (23). However, in distinction to humans, affected mice display an intense insulitis including a peri-insulitic pattern of heavy infiltration, with clusters of lymphocytes often resembling tertiary lymphoid organs. A marked decrease in islet insulin content is seen after week 12 and, after a median period of 18 weeks, diabetes develops in most female mice (22).

HUMAN DATA AND HUMAN BIOBANKS

In recent years, biobanks have been established to collect pancreatic tissue specimens from donors with type 1 diabetes, autoantibody positive individuals, type 2 diabetics and healthy controls. The Diabetes Virus Detection study (DiViD) is one such biobank. DiViD was setup in Norway with the unique premise of collecting pancreatic biopsies from live adult patients newly diagnosed with type 1 diabetes (20). While this study provided valuable information, it also demonstrated the dangers associated with this investigative approach, as serious procedural complications in three out of the six enrolled patients (postoperative leak and bleeding) led to early termination of the study (24). In contrast to the small scale focused DiViD study described above, the Juvenile Diabetes Research Foundation established a multicenter collaborative effort, the Network for Pancreatic Organ Donors with Diabetes (nPOD) in 2007. This tissue biobank supports collection of tissue samples from donors with T1D, autoantibody positive individuals, people with type 2 diabetes (T2DM), and pancreas transplant recipients (with T1D) (25, 26).

What have human data shown and what are the differences when compared to animal models?

Analysis of human data suggests that the presence of insulitis in human T1D is much lower than in the NOD mouse with only 10-30% of islets affected on average, however, the variation between patients is large with the DiViD study showing that insulitis varied between 5 and 58% in patients (8, 27). Insulitis was mainly seen to affect the insulin positive islets (33%) with low levels of inflammation found in insulin negative islets (2%) indicating that inflammatory cell influx is predominantly seen in the early phase of the disease with efflux of cells following beta cell destruction and depletion of autoantigen targets (9).

Another important finding from human studies has challenged the classical dogma that >90% of beta cell mass is lost at the time of disease presentation. In fact, it was found that individuals who develop diabetes beyond their teenage years may retain as many as 40% of their insulin positive islets (28), confirming previous knowledge that the disease process is more fulminant in younger cases.

The distribution of insulitis and the degree of immune cell infiltration in human samples is also found to be very different when compared to the NOD mouse (29). While the NOD mouse has been a useful model to study autoimmune diabetes, available human data is bringing to light important differences in pathology between human and rodent disease patterns. It is possible that these differences go some way to explaining why interventions that have been successful in either preventing or reversing the disease process in the NOD mouse have not yielded similar outcomes in human clinical trials (4, 30).

Although studies from human biobanks have gone some way towards filling the gaps in our knowledge, our understanding of the human disease is still incomplete. Issues of limited tissue availability still remain valid, particularly in the current era of enhanced modern diabetes management where death from diabetes related complications is rare. Therefore, there is an urgent need for alternative human models of the disease, which can address species specific aspects of human physiology and allow the study of interventions for disease prevention. This need has therefore paved the way for stem cell derived *in vitro* human disease models.

NEWER WAYS OF MODELING DISEASE: PLURIPOTENT STEM CELLS

Disease models for T1D are necessary for understanding disease pathogenesis, as a platform for testing potential immune-modulatory therapies and for designing beta cell preservation strategies. *In vivo* studies on immune modulatory therapies have mainly been carried out in NOD mice, and whilst some of these therapies showed success in mice, this success has generally not translated to provide equal efficacy in human disease. This highlights the importance of having species specific disease models, which reflect the complexity and heterogeneity of the human disease process. A human *in vitro* disease model could provide a complementary experimental resource for studying the pathophysiology of the human disease and also for designing potential treatment strategies.

IPSCS GENERATION FROM HUMAN T1D SUBJECTS AND THEIR USE IN STUDYING DISEASE PATHOLOGY

iPSC technology provides an opportunity to generate patient specific cell lines that can be differentiated into tissues of interest and then be used for modeling disease pathology or potentially for cell replacement therapy. iPSCs have been generated from individuals with many different forms of diabetes including T2DM (31), cystic fibrosis related diabetes (32), neonatal diabetes, forms of monogenic diabetes (33, 34), maturity onset diabetes of the young (35–37) and T1D (38).

iPSCs have been successfully used to create human models of diabetes caused by monogenic disorders that effect beta cell development and function such as Wolfram syndrome (33) and insulin gene mutations (34). These experiments not only demonstrated the success of iPSC technology for modeling disease phenotypes but also provided proof of principle data for correcting the disease phenotype.

However, the potential of this system to investigate acquired forms of diabetes has only recently been investigated. T1D is a complex disease to model *in vitro* as the disease has a polygenic inheritance pattern with a heterogenous presentation and a strong influence of environmental factors as potential triggers of autoimmunity. Therefore, simply generating beta cells *in vitro* will likely be insufficient to reproduce the conditions that reflect the *in vivo* disease. Immune cells that have been identified in insulitis lesions from human pancreas, such as CD8+ T cells, CD4+ T cells, macrophages, dendritic cells and B cells (11) would also need to be generated *in vitro* and then co-cultured together with beta cells so as to mimic the pathologic process in the pancreas (39, 40). However, because of the complexity of T cell development, it is unlikely that methods for the generation of autoreactive T cells from iPSCs will be straightforward. As such, it is likely that, in the first instance, autoreactive T cells will need to be obtained from T1D tissue/cell donors. In such a scenario, immune interactions in the disease process could potentially be modeled by recreating key components such as antigen presenting cells and beta cells *in vitro* whilst obtaining autoreactive T-cells from *in vivo* sources.

A number of groups have examined iPSC derived beta cells in the context of T1D tissue donors (38, 41, 42) (**Table 1**). Maehr et al. were one of the first to report the generation of iPSC from

TABLE 1 | Generation of iPSCs and relevant cell types from T1D individuals.

Cell type	Starting cell	Results	Reference
Beta like cells	iPSCs generation from fibroblasts of 2 T1D individuals	Insulin & c-peptide positive beta like cells which were glucose responsive	Maehr et al. (38)
Islet like cells	iPSC from skin fibroblasts from 3 T1D individuals and 1-ND	Significant intra-individual variability found with only 1 of 3 iPSC clones from each donor being able to generate INS-positive cells	Thatava et al. (41)
Beta like cells	iPSC from skin fibroblasts from 3 T1D individuals and 3-ND	Generation of C-peptide+/NKX6-1 + glucose responsive beta cells with the ability to ameliorate alloxan induced diabetes in mice. No differences in morphology, marker profile, gene profile, functionality and propensity to cytokine induced stress seen in T1D versus ND iPSC-beta cells.	Millman et al. (42)
Beta like cells	iPSC from peripheral blood from 3 T1D individuals and 1-ND	iPSC beta cells from both, on undergoing ER stress elicit an immune activation response from autologous T cells from both T1D and non-diabetic individuals. T cell activation is specific to beta cells and exposure to iPSC- alpha cells elicits minimal immune activation.	Leite et al. (43)
Macrophages	iPSC from PBMC of 1 T1D individual & 1-ND	iPSC Macs displayed mature morphology and surface marker profile with ability of phagocytosis and capacity to process complex protein mixtures and present relevant epitopes derived from proinsulin C-peptide to TCRs derived from autologous islet infiltrating T cells leading to their activation.	Joshi et al. (44)

T1D, Type 1 Diabetes; ND, Non diabetic; PBMC, Peripheral blood mononuclear cells; ER, endoplasmic reticulum.

individuals with T1D and to differentiate these into beta-like insulin producing cells that were glucose responsive (38). More recently, Millman et al. reported on the generation of iPSCs from three T1D donors and compared their differentiation potential to iPSCs from non-diabetic individuals. Their study showed that beta cells derived from T1D iPSCs were similar to those from non-diabetic individuals in terms of their surface marker expression profile, morphology and *in vitro/in vivo* insulin secretion capacity. They were able to demonstrate that post transplantation into mice, T1D derived iPSC beta cells were equally efficient at rescuing the phenotype of alloxan induced diabetes (42). These experiments reinforce the view that T1D arises from factors that are not intrinsic to the beta cells, whether they are immune cells or environmental triggers. Consistent with this, Millman et al. also examined how the cytokine environment may contribute to beta cell death in the context of type 1 diabetes, hypothesizing that T1D beta cells may be more prone to cytokine induced damage. However, they found that beta cells derived from both diabetic and non-diabetic individuals were equally sensitive to cytokine induced stress, with both showing loss in expression of beta cell markers post exposure to inflammatory cytokines (42). This study therefore further underlines the fact that for studying T1D pathogenesis analyzing islet biology in isolation will not be sufficient and interactions with the immune system will be the key to understanding the complex mechanisms of autoimmune beta cell destruction.

A recent study by Hosokawa et al. described the generation of iPSCs derived from patients with fulminant diabetes, a subgroup of the Type 1b non-autoimmune or idiopathic T1D (45). The pathogenesis of this kind of diabetes, which has almost exclusively been reported from Japan, is not well understood. It differs from classical T1D by the rapidity of onset of symptoms, the degree of hyperglycemia and severity of ketoacidosis, almost complete loss of beta cells along with variable alpha cell loss, absence of islet autoantibodies and elevated levels of pancreatic enzymes (46). In distinction to work above related to classical T1D, Hosokawa and colleagues found beta cells generated from fulminant diabetes individuals had an increased sensitivity to proinflammatory cytokine induced damage (45). However since the pathogenesis of this kind of diabetes is believed to be different to classical Type 1A autoimmune diabetes, these results may not be extrapolatable to the classic human T1D disease pathology (46).

In a recent study, Leite et al. simulated T1D relevant immune interactions in an *in vitro* system. They generated iPSC derived beta cells and exposed these cells to ER stress, attempting to replicate the pro-inflammatory islet environment in T1D (**Table 1**). Interestingly, their experiments suggested that stressed iPSC derived beta cells elicit an immune activation response from autologous T cells from both T1D and non-diabetic individuals. They thus concluded that beta cells from T1D individuals are healthy to begin with and the process of islet inflammation makes them stressed and vulnerable to T cell mediated autoimmune destruction (43). This sequence of events is consistent with the hypothesis that an initial assault on beta cells, such as viral infections or an environmental toxin, trigger beta cell damage or stress, which subsequently leads to immune attack and widespread beta cell destruction.

The above experiments have focused on beta cells and their interaction with effector cells of the immune system, particularly CD8+ cytotoxic T cells. However, most current hypotheses regarding the genesis of T1D implicate CD4+ helper cells as underlying drivers of disease because of the very strong genetic association with HLA-class II (47). *In vivo*, the involvement of CD4+ requires the presence of antigen presenting cells: cells that could potentially take up and process antigens from damaged beta cells. For this reason, we investigated the generation of iPSC derived macrophages from individuals with T1D. We were able to demonstrate that these antigen presenting cells had mature functionality and were able to process and present islet lysate and purified synthetic C-peptide to autologous islet infiltrating CD4+ T-cells (44) (**Table 1**). Clearly, the major benefit of the iPSC system is the capacity to exactly match the HLA alleles of antigen presenting cells (APCs) to donor derived T cells. This important feature will be the key advantage of future iPSC-based models of T1D.

AN IDEALIZED MODEL OF T1D

The complexity of T1D pathogenesis presents a number of challenges for efforts to create an *in vitro* model that can be used to interrogate the many factors contributing to ongoing islet cell destruction. Key variables that would need to be addressed by a potential stem cell model of T1D include the different cellular insults used to trigger beta cell death or stress, the presence or absence of particular cell types, and, because T cells are an important player, HLA class matching. In order to visualize how such a model could be constructed, it is first necessary to briefly outline what is known about the role that putative environmental triggers and incriminated cell types play in disease onset and progress (**Figure 2**).

Model Inputs

Initiating Insults

Genetic Predisposition

T1D has a complex inheritance pattern with genome-wide association studies identifying more than 60 disease susceptibility loci. The HLA complex has been shown to have the strongest association with T1D and more than half the genetic susceptibility is attributed to this region. Other important loci are the cytotoxic T lymphocyte antigen 4 (CTLA4, IDDM12 locus), PTPN22 (gene encoding lymphoid tyrosine phosphatase) and the IL2RA (interleukin 2 receptor A) locus (48). Most of these loci are associated with immune regulation. Therefore, using iPSCs from T1D individuals carrying high-risk genetic susceptibility alleles would allow the factoring in of underlying genetic factors, which play a role in modulating autoimmunity and mechanisms of impaired immune tolerance in T1D, which lead to disease predisposition.

Environmental Factors

There is a complex interplay between genetic disease susceptibility and environmental triggers of autoimmunity. Factors such as viral infections, diet and toxins are believed to be potentiators of islet autoimmunity (49, 50). While an association has been reported with most of these factors, a direct causal relationship has yet to be found. Enteroviruses have been most commonly implicated in disease pathogenesis and are believed to either lead to direct beta cell destruction or to initiate autoimmunity because of molecular mimicry between enteroviral proteins and beta cell antigens (51). Similarly, various dietary factors such as cow's milk protein, gluten exposure and Vitamin D have long been implicated in the pathogenesis of T1D, however, their roles are still not very well defined. The beta cell stress hypothesis postulates that a combination of any of these factors leads to a state of beta cell endoplasmic reticulum stress promoting generation of neoantigens *via* post-translational modification of islet proteins (52–55).

Any *in vitro* model that seeks to study the entire autoimmune pathogenesis of T1D will likely have to incorporate these triggering factors which link beta cell stress to autoimmunity. It can be envisaged that certain factors such as viruses and their role in beta cell infection and stress could easily be studied in an iPSC derived *in vitro* model of T1D.

The Role of Individual Cell Types

Beta Cells

Most protocols designed to promote the *in vitro* differentiation of beta-cells from PSCs rely on recapitulating the key developmental steps, which occur during embryogenesis (56). Therefore, an understanding of the steps of early pancreatic organogenesis is essential. Most of the knowledge of

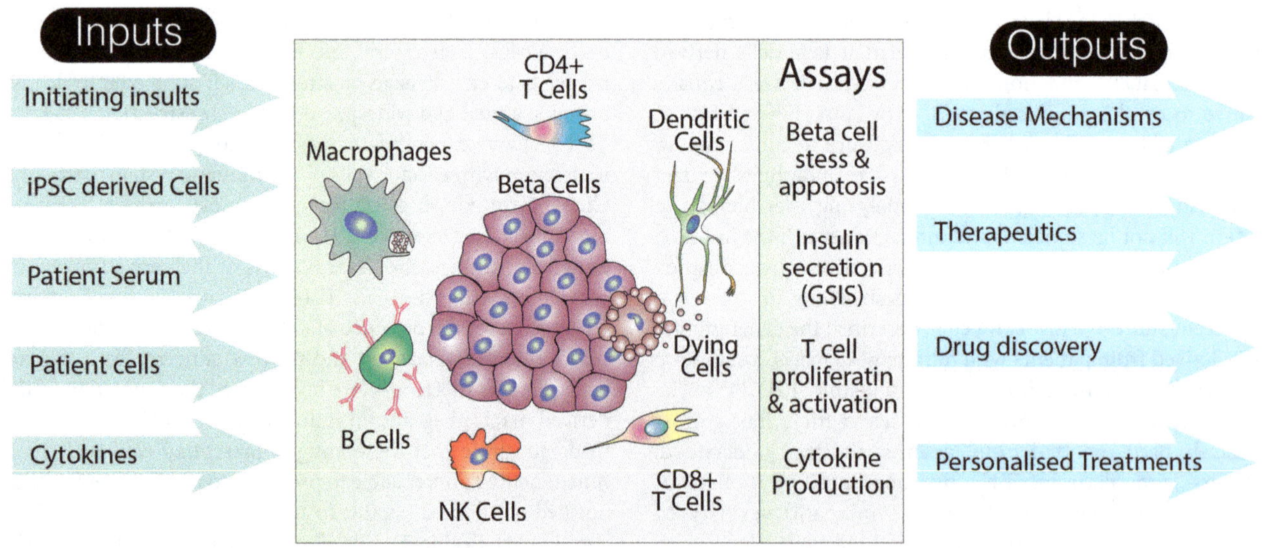

FIGURE 2 | Utility of an idealized model of T1D. The schematic suggests various inputs to the model, how the effects of these inputs could be assayed, as well as long term potential outputs. Note that this representation of the model does not specifically include a host of non-immune and non-endocrine cells which may also impact on the disease pathogenesis. Additionally, although a limited number of assays are shown, analyses specific to particular cell types are likely to increase the breadth and depth of data that could collected from such a system.

developmental biology and signaling pathways involved in pancreatic lineage specification has come from studying rodents (40, 57).

The development of protocols for generating functional endocrine cells has been hampered by a lack of knowledge of pathways involved in final stages of beta cell differentiation. The initial protocols generated cells which though insulin positive, were bi-hormonal and failed to secrete insulin in response to glucose in a glucose-stimulated insulin secretion (GSIS) assay (58–60). However, more recent protocols have described the generation of more mature cells, with development of insulin positive mono-hormonal cells which, show function both *in vitro* and *in vivo*, are glucose responsive, ameliorate diabetes in mice models and are transcriptionally more similar to native beta cells (61–65). Protocol development is an ongoing process, with recent improvements enabling the generation of beta like cells which have a more physiological glucose secretion profile and show appropriate dynamic insulin secretion to high and low glucose challenges (66, 67). However, in spite of these advances, *in vitro* derived beta cells are still ontologically and functionally immature when compared to adult beta cells, with a lower insulin secretion per cell at high glucose, lower glucose stimulation, slower first-phase insulin release and persisting differences in gene expression profiles (68). Studies using single cell RNA-sequencing techniques for transcriptomic profiling of *in vitro* derived beta cells will contribute to a more refined understanding of beta cell maturation pathways and help in development of more evolved beta cell generation protocols (69).

The maturity of the beta cells may indeed be an important factor determining their susceptibility to T cell mediated cell death. Similarly, death induced by cytokines may also be affected by beta cell maturity. It is also conceivable that the degree of functional and transcriptional maturity might also affect the cell's susceptibility to the initial triggering insult which sets of the process of beta cell death and autoimmunity, potentially a viral insult. With advances in the development beta cell differentiation protocols, generation of mature, functional beta cells that are transcriptionally similar to adult beta cells may become possible. Development of functional adult like beta cells is likely to be an important aspect of *in vitro* models attempting to study T1D *in vitro*.

Antigen Presenting Cells

Following an initial insult, macrophages and dendritic cells within the islet and/or draining lymph node are the first responders, phagocytosing cellular debris and processing it for antigen presentation. This is believed the to be the key step in the initiation of autoimmunity and involves the presentation of beta cell autoantigens by professional antigen-presenting cells to CD4+ T cells *via* HLA class II, leading to subsequent T cell activation (15). In studies on pancreata from human subjects with T1D, macrophages have been found to be an important part of the islet infiltrate, thereby underlining their role in the immune events, which precipitate autoimmune diabetes (11, 70). Macrophage depletion and functional inhibition have been shown to reduce the development of autoimmune diabetes in rodent models (71–73). Similarly, NOD mice deficient in CD103+ DCs were found to have a reduced islet infiltration of autoreactive T cells and a corresponding reduction in diabetes incidence (74). Finally, macrophages and DCs may also be involved in causing direct beta cell death by the secretion of proinflammatory cytokines such as interleukin 1-beta (IL1β), tumor necrosis factor alpha (TNFα) and ROS (70, 75). These experiments clearly indicate that antigen presenting cells are likely to play a part in initiating and propagating the T cell mediated autoimmune attack against beta cells.

In an idealized model of T1D, both macrophages and dendritic cells could be generated from iPSCs using established protocols (**Table 2**). In many protocols, PSCs are guided through the sequential stages of hematopoietic development using by stage specific growth factors and cytokines. Induction of mesoderm is achieved by the use of BMP4, activin and FGF2 followed by the addition of VEGF, SCF and FGF to generate hematopoietic precursors. Developing myeloid cells, which are shed from the cultures, are harvested from the supernatant and matured using M-CSF with or without IL3. These cells then pass through an intermediate monocyte stage where CD14+ monocytes can be collected using flow cytometric sorting. Macrophages can then be matured in adherent cultures using high concentrations M-CSF and activated using either Lipopolysaccharide (LPS)/IFNg (classic activation) or IL4 (alternate) activation (86).

In vitro PSC derived macrophages have been shown to have similar phenotypic, functional, and transcriptomic characteristics to peripheral blood monocyte derived macrophages (78, 87, 88), suggesting they could be used to provide the antigen processing and presentation functions thought to be key steps in T1D initiation and maintenance. Ideally, iPSCs would be derived from T1D donors from whom islet antigen specific T cells were also available, enabling the interactions between these two cell types to be assessed in a fully autologous HLA setting. However, as noted below, creating or isolating such T cells is likely to be major obstacle to the generation of an authentic *in vitro* model of T1D.

The developmental identity of *in vitro* derived macrophages however remains to be resolved. Most protocols promoting the *in vitro* hematopoietic differentiation of pluripotent stem cells are believed to create cells resembling those generated from embryonic primitive hematopoiesis rather than those derived from adult definitive hematopoiesis. Indeed, there is evidence that PSC-derived macrophages have a primitive embryonic-type macrophage phenotype, predominantly because they expand in the absence of cMyb, a transcription factor required for definitive hematopoiesis (89). However, differences, if any exist, between embryonic and adult origin macrophages in terms of function remains to be elucidated. Moreover, macrophages of fetal origin continue to exist in adult humans in the form of tissue resident macrophages, which self renew with a minimal contribution from adult blood monocytes (90). Therefore *in vitro* derived macrophages may have a role in modeling the functions of these specialized tissue resident macrophages as well (91).

Protocols for the production of dendritic cells from PSCs are limited and most rely on derivation of cells of the myeloid lineage

TABLE 2 | Generation of T1D relevant immune cells from PSCs.

Cell type	Starting cell	Method	Cell functionality and maturity	Reference
Dendritic cells (DCs), monocytes and macrophages	hiPSC-	OP9 mouse stromal cell co-culture for generation of hematopoietic precursors, followed by further differentiation along the macrophage/DC pathway with the use of GM-CSF/M-CSF, Flt3L, SCF and IL1β	DCs capable of cytokine secretion and antigen presentation and activation of allogenic and autologous HLA matched T cells. Macs showed ability of phagocytosis and antitumor activity	Senju et al. (76)
Monocytes and Macrophages	hiPSC-	Embryoid body formation followed by hematopoietic specification by IL-3/M-CSF combination with high dose M-CSF for terminal differentiation	iPSC Macs capable of phagocytosis pro inflammatory cytokine release on LPS stimulation	Lachmann et al. (77)
	hiPSC-	Monolayer culture on a layer of matrigel, an extra- cellular matrix component by using stage specific hematopoietic cytokines to generate monocytes followed by differentiation to macrophages using high dose M-CSF	iPSC Macs showed capacity of bacterial and tumor cell phagocytosis along with relevant cytokine and chemokine release. Phenotypic, functional, and transcriptomic characteristics to peripheral blood monocyte derived macrophage	Cao et al. (78)
	hiPSC-	Embryoid body formation using rotational cultures followed by hematopoietic differentiation using IL-3, M-CSF,GM-CSF and FLT3-L combination with high dose M-CSF for terminal differentiation to mature macrophages. Activation using IFN-γ	iPSC Macs displayed mature morphology and surface marker profile with ability of phagocytosis and capacity to process complex protein mixtures and present relevant epitopes derived from proinsulin C-peptide to TCRs derived from autologous islet infiltrating T cells leading to their activation.	Joshi et al. (44)
Dendritic cells	hiPSC-	Embryoid body formation followed by guided differentiation using hematopoietic growth factors and final DC differentiation using GM-CSF and IL-4	Generation of CD141+ DCs with ability of phagocytosis and features reminiscent of tolerogenic DCs as evinced by capacity of IL-10 secretion, reduced capacity of immunostimulation and polarization of naïve CD4 cells to Tregs.	Sachamitr et al. (79)
	iPSCs derived from an individual with Sjögren's syndrome	Co-cultured on C3H10T1/2 mouse mesenchymal cells to differentiate into hematopoietic cells	Generation of CD141+ myeloid DCs with ability of phagocytosis. Mature functionality as demonstrated by capacity to stimulate allogenic T cells and present antigen to and activate autoreactive CD4+ T cells.	Iizuka-Koga et al. (80)
Antigen specific T cells	iPSC and hESCs	Directed differentiation and artificial thymic organoids (containing DLL expressing mouse stromal cells). Lentiviral vector introduction of transgenes encoding antigen specific TCRs	Antigen recognition and antigen specific activation CD8+ T cells	Montel-Hagen et al. (81)
	iPSCs from tumor infiltrating CTLs	Mesoderm differentiation followed by co-culture on DLL1 expressing OP9 stromal cells	Antigen recognition and antigen specific activation of CD8+ T cells	Nishimura et al. (82)
	iPSCs from antigen specific lymphocytes	Mesoderm differentiation followed by co-culture on DLL1 expressing OP9 stromal cells	Antigen recognition and antigen specific activation of CD8+ T cells.	Nagano et al. (83)
NK cells	iPSCs from umbilical cord blood	Embryoid body formation, hematopoiesis induction and expansion of NK cells with IL7	*In vitro* and *in vivo* killing of ovarian cancer cell lines.	Hermanson et al. (84)
	iPSCs from peripheral blood	Directed differentiation	Target cell killing and antibody mediated cytotoxicity	Zeng et al. (85)
Antigen specific B cells		N.D		N.D

M-CSF, Macrophage colony stimulating factor; GM-CSF, Granulocyte Macrophage colony stimulating factor; FLT3L, fms like tyrosine kinase 3 receptor ligand; IL-3, Interleukin 3; IL1β, Interleukin 1β; LPS, lipopolysaccharide; IFN-γ, interferon gamma; hESCs, human embryonic stem cells; CTL, cytotoxic T lymphocytes; IL-7, interleukin 7; ND, not done.

as described above, followed by the further differentiation towards a dendritic cell lineage with the use of GM-CSF and IL4, mirroring methods for deriving DCs from peripheral blood. Most protocols describe an intermediate monocyte stage (92–94) following which blood cells in suspension are harvested and matured in media containing GM-CSF and IL4. These immature DCs then undergo a final maturation step with the use of proinflammatory stimuli like LPS, TNFα (94, 95) or IFNγ, IL1β, PGE2 (79, 93).

Functionally, these DCs have been found to have a cytokine profile, chemotaxis ability and capacity for allogenic T cell stimulation, which is reminiscent of peripheral blood derived myeloid DCs (92, 93, 95). Finally, the antigen presenting functions of iPSC-derived DCs have been used to characterize T cell responses in Sjogrens syndrome, thereby demonstrating that the antigen presenting functions of iPSC derived APCs such as dendritic cells can be used to study the repertoire of pathogenic T cells in autoimmune disorders (80).

Lymphocytes in Type 1 Diabetes

T cells are thought to be the primary mediators of beta cell loss in T1D, with cytotoxic CD8+ T lymphocytes mainly responsible for causing beta cell death. The evidence supporting their key role in pathogenesis includes the ability of beta cell specific CD8+ and CD4+ T cell clones to transfer T1D to immunocompromised hosts (96). Furthermore, the use of an anti-CD3 antibody has been shown to reverse T1D in the NOD mouse model (97); a result that has translated to humans with anti-CD3 antibodies shown to preserve beta cell function in recent onset T1D (98) and to delay diabetes progression in secondary prevention trials (99).

CD8+ Cytotoxic T Cells

Cytotoxic T lymphocytes (CTLs) are the most common immune cell found in insulitic lesions in human T1D pancreatic specimens (100). They recognize beta cell autoantigens presented by HLA class I expressed on the beta cell surface. CTLs can lead to beta cell death by a variety of mechanisms including the induction of molecules involved in the granule exocytosis pathway such as perforin, granzyme, or granulysin, as well as increased surface expression of death inducing molecules such as Fas ligand and TNF-related apoptosis inducing ligand (101, 102). Autoreactive CD8+ T cells from insulitic lesions in human T1D pancreatic specimens have been found to react to known islet autoantigens such as insulin, islet amyloid polypeptide (IAPP) and islet- specific glucose-6-phosphatase catalytic subunit-related protein (IGRP), pre-proinsulin, GAD65, pre-proislet amyloid protein, and IA-2, providing direct evidence for involvement of these cells in autoimmune beta cell destruction (14).

CD4+ Helper T Cells

CD4+ T helper cells are a key player in the pathogenesis of Type 1 diabetes mellitus and have been consistently identified in the inflammatory infiltrate of islets from T1D patients (27). The strong association of HLA class II molecules with the genetic disease susceptibility risk also underscores the important role of CD4+ T cells in the pathogenesis of the disease (48). While CD4+ T cells do not lead to direct beta cell killing, they are important effector cells involved in pro-inflammatory cytokine secretion, which amplify and propagate the immune response and lead to activation of immune cells such as macrophages and CD8+ cytotoxic T cells (102).

Characterizing the antigenic repertoire of these autoreactive T cells has been difficult as the frequency of islet antigen-specific T cells is very low in the most readily available tissue sample, blood, and access to T cells in the islets is limited by the availability of pancreatic tissue samples (103). However, the increasing availability of pancreatic specimens from T1D tissue donors has made possible the study of islet infiltrating autoreactive T cells (104–106). Another approach has been to express T cell receptors (TCRs) from islet infiltrating T cells in an immortalized T cell line to provide a readily available and expandable source of T cells for antigenic testing (107). These studies have played a crucial role in understanding the antigenic targets of CD4+ T cells in T1D, which has not only provided novel insights into disease pathogenesis but may also be useful for testing antigen specific therapies for disease prevention (108).

Any model that seeks to include these cell types needs to be cognizant of the importance of specific T cell receptors that recognize islet antigens. Although a number of methods for making T cells from iPSCs have been published, only a handful of these have addressed the production of T cells bearing specific TCRs (**Table 2**). In short, T cells with specific TCRs can be created by "rejuvenating" T cells isolated from tissue donors or by providing a known TCR in the form of a transgene. In the case of the "rejuvenated" T cells, which can be made from iPSCs generated from T cells, care has to be taken to ensure that the TCR expressed by the iPSC derived T cell is identical to that possessed by the starting cell that was reprogrammed to generate the iPSC. Similarly, iPSC derived T cells expressing a TCR encoded by transgenes may also express endogenously encoded TCRs, potentially complicating the interpretation of antigen specific activation studies.

Although the technical issues surrounding the fidelity of TCRs expressed by *in vitro* derived T cells can be addressed, the work required to generate T cells with specific TCRs from iPSC is still significant. For this reason, direct isolation of T cells from T1D donors may provide a more accessible route for examining this aspect of the autoimmune reactions. However, this path also presents its own challenges, including the phenomenon of T cell exhaustion (82, 109, 110) and the issue of whether the TCR repertoire of T cells in circulation reflects that of autoreactive T cells present with the islets (105, 108). Ideally, an *in vitro* model would incorporate T cells isolated from islets of T1D individuals (44, 104, 107), a scenario that would limit the scope of such a model to deceased tissue donors.

B Cells

CD20+ B cell infiltrates have been described in insulitis in human T1D pancreatic specimens (11, 28), however, their exact role in the pathogenesis remains unclear. B cell activation leads to production of autoantibodies against key islet autoantigens that are used as markers for disease onset and as entry points for enrollment in secondary prevention trials (16, 111). However, the conventional understanding is that antibodies by themselves are not pathogenic in T1D (112). There is conflicting data on the need of B cells in initiating autoimmunity as B lymphocyte depletion with the anti CD20 antibody (rituximab) has been associated with reversal of diabetes in the NOD mouse (113) and preservation of beta cell function in newly diagnosed T1D subjects (114). However, in contrast, a report of T1D development in a child with X linked agammaglobulinemia (115) suggests that B cells are not an absolute necessity for disease causation. Nevertheless, B cells can also function as antigen presenting cells and thus may play a role in disease pathogenesis by activating and diversifying the responses of autoreactive T cells in T1D (112).

In an idealized model of T1D, the dual role for B cells as antigen presenting cells and antibody producers could be addressed separately. Specifically, the potential effects of circulating antibodies directed against islet specific antigens could be examined by including patient serum or purified immunoglobulin fractions as input into the *in vitro* model. On the other hand, B cells themselves could be included as APCs. Currently, reports describing protocols for generating B cells from iPSCs have been scant (116, 117) and the robustness of methods for *in vitro* B cell maturation limited. As such, if B cells are to be incorporated into and *in vitro* model of T1D it is likely these will also need to be initially sourced directly from donors. In a similar scenario to that described above for T cells, this approach is likely to exclude the use of B cells producing antibodies with a known specificity.

As a final point, although generation of all the required cells types for a complete model of T1D is onerous, the fact that blood cells can be effectively cryopreserved means that experiments in which cells are recombined into a single culture can be separated from the process of cell generation.

Natural Killer Cells

Natural Killer (NK) cells are an innate immune cell that plays a critical role in identifying and killing abnormal cells, particular those that are the target of viral infection or have undergone tumorigenic transformation (reviewed in (118). Historically, NK cells have been classified as lineage negative cells that express CD16 in conjunction with either high or low levels of CD56. The designation "Natural killer" is indicative of this class of lymphocyte's capacity to kill cells without the requirement for activation by specific HLA antigen complexes, distinguishing them from conventional cytotoxic CD8+ T cells. Indeed, an important characteristic of NK cells is their ability to recognize and kill cells that fail to display self-HLAs, a property important for their role in tumor surveillance. Similarly, NK cells lack of requirement for antigen specific activation means they are first responders to viral infections, recognizing and destroying cells under stress. In addition to the lysis of abnormal cells, another key characteristic of NK cells is their ability to rapidly produce high levels of numerous cytokines and chemokines, putting them in a position to orchestrate immune attacks, as well as serving as an active participate (118).

Only a limited number of studies have examined the role of NK cells in human T1DM (119). Analysis of peripheral blood samples from individuals with T1DM suggested that those with long-standing disease had an NK population that showed reduced levels of activation [for example, reduced production of IFNg (120) and potentially decreased lytic activity (121). Rodacki et al. suggested this reduced NK activity was more likely a consequence than a cause of T1DM, given its association with long-standing disease. Nevertheless, others have suggested that reduced NK activity might make individuals more susceptible viral insults that may precipitate T1DM in the first instance (119)].

Consistent with their role as first responders to viral infection Dotta et al. (122), identified NK cell infiltrates within the islets of 3 T1DM individuals who also showed evidence of Coxsackie B4 enteroviral infection. The presence of NK cells coincided with non-destructive islet inflammation, suggesting these cells could represent a stepping-stone between and initial insult and an expanding inflammatory reaction.

Current protocols for generating NK cells from iPSCs have focused on those representing the myeloid lineages, characterized by expression of CD56 and CD16 (see **Table 2**). These methods have been primarily developed with a view to using iPSC derived NK cells as anti-tumor therapies (84, 85). Given the role of NK cells in detecting cellular stress, inclusion of this cell type in an *in vitro* model of T1DM could provide information related to beta cell stress, whether that be induced by exogenous stimuli or by the presence of the NK cells themselves.

Patient Derived Serum—Role of Autoantibodies in T1D

The release of autoantigens following the initiation of autoimmune attack on the pancreatic beta cells leads to the formation of antibodies against key islet proteins such as insulin (micro IAA or mIAA), glutamic acid decarboxylase (GAD), islet antigen 2 (IA-2), and zinc transporter 8 (ZnT8) (13, 123). As noted above the role of autoantibodies in T1D is not well defined. It is believed that the antibodies themselves are not pathogenic and do not cause disease by forming immune complexes as has been described for other autoimmune diseases (124). However, they do predict the risk of development of disease and rate of progression of disease (115). One potential important role played by autoantibodies in the T1D disease process is their effect on autoantigen processing and presentation by class II major histocompatibility complexes (125).

Several effector mechanisms render autoantibodies potentially harmful. These include antibody-dependent, cell-mediated cytotoxicity; release of inflammatory mediators through stimulation of Fc receptors on natural killer cells, macrophages, or mast cells; opsonization of islet autoantigen, which promotes phagocytosis by macrophages; and complement activation with subsequent assembly of the membrane-attack complex (126, 127).

Earlier studies demonstrated that sera from patients with T1D can have a cytotoxic effects on cultured rat beta cells (128, 129). Increased complement activation has been seen in serum of patients with recent onset T1D and similarly was found to cause apoptosis in rat islets (130). A study analyzing human pancreatic tissue specimens from the Network for Pancreatic Organ Donors

Diabetes (nPOD) program has also demonstrated evidence of complement activation in the pancreas (131). However, there have been few studies that examine the direct interaction of these antibodies with live beta cells and data on effect of these antibodies on human beta cells is sparse. In the same vein, few experiments have directly addressed the possibility that other serum bound factors may influence beta cell viability or function. An iPSC based model would be a useful platform to study the functional effects of islet autoantibodies/patient serum derived factors on beta cell function and the propagation of the autoimmune process. Indeed, such platforms have been successfully used for modeling autoantibody mediated neuromuscular diseases such as Myasthenia gravis (132).

Model Outputs

Disease Mechanisms

One of the most important uses of this system would be to study disease pathogenesis by *in vitro* assays that could examine beta cell function and immune cell activation

Assessment of Beta Cell Mass and Function

- Beta cell apoptosis assays
- Insulin content of beta cells and Glucose stimulated insulin assay (GSIS)

Immune Cell Activation

- Cytokine production assays
- T cell activation assays
- T cell proliferation assays

In addition to these cell specific assays, single cell RNAseq analysis could be employed to examine how the complex collections of cells respond to changes in their environment or to the presence of other cell types.

Potential Therapeutic Outputs

One application of an *in vitro* immune model of T1D would be to explore interventions that might modify the autoimmune response.

Drug Discovery and Screening

With the availability of patient specific iPSCs it is possible to recapitulate disease pathogenesis *in vitro* and to use this knowledge to guide development of patient specific targeted therapies (133).This can be particularly useful in disorders with a long preclinical phase such as T1D where, at the time of clinical disease onset, a significant proportion of tissue function is already lost (134). Disorders for which iPSCs have been used for drug discovery include spinal muscular atrophy (135), familial-dysautonomia (136) and amyotrophic lateral sclerosis (137). The failure of translation of most therapies, which are found successful in rodent models in human trials, has highlighted specific issues that are crucial to consider when designing future intervention trials. These issues include the significant knowledge gaps that exist in the understanding of human disease and the realization that rodent disease patterns and key physiological responses are significantly different from humans. Indeed, emerging knowledge suggests the disease process itself is very heterogenous in humans and therefore personalized strategies for immune intervention may be needed (138). A human iPSC derived *in vitro* model could account for interindividual variations in disease pattern and also circumvent the problems relating to pathophysiological differences between rodent and human disease.

In most autoimmune diseases the therapeutic interventions can be tested even when the disease state is well established. On the contrary, in T1D, intervention strategies would ideally be instituted at the pre-symptomatic phase where significant residual beta cell mass and function still remain. Therefore, iPSC-based models, which recreate the early disease milieu of T1D, are fertile ground for testing strategies for secondary and tertiary preservation in Type 1 Diabetes. Recent trials have focused on the use of immunomodulatory agents which inhibit T cell activation, cytokine action and promote Treg formation such as the use of anti CD3 antibody, CTLA-4 Ig, Anti thymocyte globulin (ATG), anti TNF alpha and IL-2 (139, 140). Cell based therapies such as tolerogenic DCs, Tregs and cord blood cells are also being studied for induction of immune tolerance in T1D. Hematopoietic stem cells have been used to reset the immune system in human trials and trials utilizing mesenchymal stem cells for immunomodulation are also underway (141). An iPSC model would be an ideal testing platform for pre-clinical trials of these therapeutic agents and provide output data relevant to human disease.

Drug repositioning, that is, uncovering new applications for existing drugs, is another application of iPSC technology. This approach has been investigated for conditions such as skeletal dysplasia's, Alzheimer's disease and amyotrophic lateral sclerosis (142). Drugs such as hydroxychloroquine (an anti-malarial with immunomodulatory activity) and imatinib mesylate (a tyrosine kinase inhibitor used in chronic myeloid leukemia) are currently being tested in beta cell preservation trials (143) after promising results in preclinical studies. iPSC based models provide a unique human platform for drug discovery and testing of novel therapeutic agents and also for validating the efficacy of these therapeutic agents in pre-clinical studies before translation to clinical trials.

POTENTIAL LIMITATIONS

Any model that seeks to recapitulate pathogenic events of human T1D may need to examine all of the cell types that have been implicated in disease causality and progression. At present, this is one limitation of trying to create such a model as robust protocols for generating cells with the correct characteristics are currently not available. As such, rather than trying to incorporate the multitude of environmental, genetic and cellular factors that could potentially affect disease pathogenesis, modeling will need to focus on specific aspects of the disease process that are experimentally tractable. Thus, combinations of iPSC and patient derived native cells will need to be used until more robust protocols for generation of immune cells are available.

In this review, we have not discussed the inclusion of other cell types such as endothelial, epithelial and mesenchymal cells, which may also play a role in disease pathogenesis. Similarly, studies have identified a role for the autonomic nervous system in the control of both insulin and glucagon release, as well as a regulation of islet mass (144). It would be clearly very challenging to construct an islet-like organoid *in vitro* that could fully account for such modulatory neural inputs.

In addition to the non-endocrine cells mentioned above, the endocrine components of the islet itself constitute a complex mixture of multihormonal cell types. Moreover, some of these cell types, such as alpha cells, may have a role in T1D pathogenesis. The *in vivo* islet environment and cross talk between various endocrine cells is believed to be important for normal islet function and hormone release (145). Although many beta cell differentiation protocols generate other islet cell types including alpha cells (59, 60), the relative proportions of cell types generated are often difficult to control. In this respect, the islet-like milieu recreated *in vitro* will, at best, be an approximation to the rich interconnected environment of the native human islet.

Finally, such a model would also find it difficult to take into account the effects of non-islet derived hormones and growth factors within circulation that may collectively contribute to inflammation and beta cell stress.

Our omission of specific cell types and circulating factors is a reminder that any *in vitro* model cannot fully mimic the subtle multi-systemic interactions that occur *in vivo*. Future models which incorporate multi-lineage organoid cultures may circumvent some of these issues. Similarly, our current conception of an *in vitro* model of T1D does not address the initial loss of tolerance, which portends the onset of autoimmunity. Models for examining this question might require improvements in T cell and thymic epithelium differentiation protocols that will allow the study of T cell selection.

CONCLUSION

The advent of iPSC technology, which brings the possibility of creating diverse human cell types *in vitro*, has provided the opportunity to construct a fully humanized model of T1D that recapitulates human disease pathology. A major impetus for this work has come from recent improvements in protocols for generating iPSC derived beta cells with mature functionality, allowing investigators to generate the cellular target of autoimmunity in the context of specific HLA haplotypes.

The successful creation of a human iPSC based T1D model will allow a more nuanced understanding of the disease process and help investigators design better beta cell preservation strategies. Such models will also capture the heterogeneity of the human disease process and provide a landscape for testing patient tailored therapies. Finally, and most importantly, stem cell models of T1D will lessen our dependence on rodent disease models and pave the way for better translation of preclinical therapeutic strategies to the clinical arena.

AUTHOR CONTRIBUTIONS

KJ reviewed the literature and wrote the paper. FC, ST, SM, AE, and ES were involved with the writing, editing, as well as the key concepts behind the paper. All authors contributed to the article, approved the submitted version, and agree to be accountable for the content of the work.

FUNDING

The work quoted in this manuscript was supported by grants from the National Health and Medical Research Council (Australia) (GNT1068866, GNT1079004, GNT1117596, GNT1129861, GNT1138717), the Juvenile Diabetes research foundation (3-SRA-2018-603-M-B), Diabetes Australia (Y19G-STAE), the Government of India: Department of Science and Technology (DST/INT/ISR/P-23/2017), Human resource division (HRD-F.5-6/2013-TS. VII), and from Sanjay Gandhi Post Graduate Institute of Medical Sciences India (A PGI/DIR/RC/844/2018).

REFERENCES

1. Mannering SI, Pathiraja V, Kay TW. The case for an autoimmune aetiology of type 1 diabetes. *Clin Exp Immunol* (2016) 1831(1):8–15. doi: 10.1111/cei.12699
2. Federation ID. International Diabetes Federation. IDF Diabetes Atlas, 9th edn.: Brussels, Belgium: International Diabetes Federation. (2019) 2019.
3. Staeva TP, Chatenoud L, Insel R, Atkinson MA. Recent lessons learned from prevention and recent-onset type 1 diabetes immunotherapy trials. *Diabetes* (2013) 621(5):9–17. doi: 10.2337/db12-0562
4. Reed JC, Herold KC. Thinking bedside at the bench: the NOD mouse model of T1DM. *Nat Rev Endocrinol* (2015) 115(5):308–14. doi: 10.1038/nrendo.2014.236
5. Roep BO, Atkinson M, von Herrath M. Satisfaction (not) guaranteed: re-evaluating the use of animal models of type 1 diabetes. *Nat Rev Immunol* (2004) 412(12):989–97. doi: 10.1038/nri1502
6. Bluestone JA, Herold K, Eisenbarth G. Genetics, pathogenesis and clinical interventions in type 1 diabetes. *Nature* (2010) 464(7923):1293–300. doi: 10.1038/nature08933
7. Atkinson MA. The pathogenesis and natural history of type 1 diabetes. *Cold Spring Harb Perspect Med* (2012) 2(11):1–18. doi: 10.1101/cshperspect.a007641
8. Krogvold L, Wiberg A, Edwin B, Buanes T, Jahnsen FL, Hanssen KF, et al. Insulitis and characterisation of infiltrating T cells in surgical pancreatic tail resections from patients at onset of type 1 diabetes. *Diabetologia* (2016) 59 (3):492–501. doi: 10.1007/s00125-015-3820-4
9. Campbell-Thompson M, Fu A, Kaddis JS, Wasserfall C, Schatz DA, Pugliese A, et al. Insulitis and beta-Cell Mass in the Natural History of Type 1 Diabetes. *Diabetes* (2016) 65(3):719–31. doi: 10.2337/db15-0779
10. Foulis AK, Liddle CN, Farquharson MA, Richmond JA, Weir RS. The histopathology of the pancreas in type 1 (insulin-dependent) diabetes mellitus: a 25-year review of deaths in patients under 20 years of age in the United Kingdom. *Diabetologia* (1986) 29(5):267–74. doi: 10.1007/BF00452061
11. Willcox A, Richardson SJ, Bone AJ, Foulis AK, Morgan NG. Analysis of islet inflammation in human type 1 diabetes. *Clin Exp Immunol* (2009) 155 (2):173–81. doi: 10.1111/j.1365-2249.2008.03860.x
12. In't Veld P. Insulitis in human type 1 diabetes: The quest for an elusive lesion. *Islets* (2011) 3(4):131–8. doi: 10.4161/isl.3.4.15728

13. van Belle TL, Coppieters KT, von Herrath MG. Type 1 diabetes: etiology, immunology, and therapeutic strategies. *Physiol Rev* (2011) 91(1):79–118. doi: 10.1152/physrev.00003.2010
14. Coppieters KT, Dotta F, Amirian N, Campbell PD, Kay TW, Atkinson MA, et al. Demonstration of islet-autoreactive CD8 T cells in insulitic lesions from recent onset and long-term type 1 diabetes patients. *J Exp Med* (2012) 209(1):51–60. doi: 10.1084/jem.20111187
15. Carrero JA, Ferris ST, Unanue ER. Macrophages and dendritic cells in islets of Langerhans in diabetic autoimmunity: a lesson on cell interactions in a mini-organ. *Curr Opin Immunol* (2016) 43:54–9. doi: 10.1016/j.coi.2016.09.004
16. Silveira PA, Grey ST. B cells in the spotlight: innocent bystanders or major players in the pathogenesis of type 1 diabetes. *Trends Endocrinol Metab* (2006) 17(4):128–35. doi: 10.1016/j.tem.2006.03.006
17. DiMeglio LA, Evans-Molina C, Oram RA. Type 1 diabetes. *Lancet* (2018) 391(10138):2449–62. doi: 10.1016/S0140-6736(18)31320-5
18. Padgett LE, Broniowska KA, Hansen PA, Corbett JA, Tse HM. The role of reactive oxygen species and proinflammatory cytokines in type 1 diabetes pathogenesis. *Ann N Y Acad Sci* (2013) 1281:16–35. doi: 10.1111/j.1749-6632.2012.06826.x
19. Coomans de Brachene A, Dos Santos RS, Marroqui L, Colli ML, Marselli L, Mirmira RG, et al. IFN-alpha induces a preferential long-lasting expression of MHC class I in human pancreatic beta cells. *Diabetologia* (2018) 61 (3):636–40. doi: 10.1007/s00125-017-4536-4
20. Krogvold L, Edwin B, Buanes T, Ludvigsson J, Korsgren O, Hyoty H, et al. Pancreatic biopsy by minimal tail resection in live adult patients at the onset of type 1 diabetes: experiences from the DiViD study. *Diabetologia* (2014) 57 (4):841–3. doi: 10.1007/s00125-013-3155-y
21. Morgan NG, Richardson SJ. Fifty years of pancreatic islet pathology in human type 1 diabetes: insights gained and progress made. *Diabetologia* (2018) 61(12):2499–506. doi: 10.1007/s00125-018-4731-y
22. In't Veld P. Insulitis in human type 1 diabetes: a comparison between patients and animal models. *Semin Immunopathol* (2014) 36(5):569–79. doi: 10.1007/s00281-014-0438-4
23. Atkinson MA, Leiter EH. The NOD mouse model of type 1 diabetes: as good as it gets? *Nat Med* (1999) 5(6):601–4. doi: 10.1038/9442
24. Atkinson MA. Pancreatic biopsies in type 1 diabetes: revisiting the myth of Pandora's box. *Diabetologia* (2014) 57(4):656–9. doi: 10.1007/s00125-013-3159-7
25. Pugliese A, Yang M, Kusmarteva I, Heiple T, Vendrame F, Wasserfall C, et al. The Juvenile Diabetes Research Foundation Network for Pancreatic Organ Donors with Diabetes (nPOD) Program: goals, operational model and emerging findings. *Pediatr Diabetes* (2014) 15(1):1–9. doi: 10.1111/pedi.12097
26. Campbell-Thompson M. Organ donor specimens: What can they tell us about type 1 diabetes? *Pediatr Diabetes* (2015) 16(5):320–30. doi: 10.1111/pedi.12286
27. Pugliese A. Insulitis in the pathogenesis of type 1 diabetes. *Pediatr Diabetes* (2016) 17 Suppl 22:31–6. doi: 10.1111/pedi.12388
28. Leete P, Willcox A, Krogvold L, Dahl-Jorgensen K, Foulis AK, Richardson SJ, et al. Differential Insulitic Profiles Determine the Extent of beta-Cell Destruction and the Age at Onset of Type 1 Diabetes. *Diabetes* (2016) 65 (5):1362–9. doi: 10.2337/db15-1615
29. Eisenbarth GS. Banting Lecture 2009: An unfinished journey: molecular pathogenesis to prevention of type 1A diabetes. *Diabetes* (2010) 59(4):759–74. doi: 10.2337/db09-1855
30. Roep BO. Are insights gained from NOD mice sufficient to guide clinical translation? Another inconvenient truth. *Ann N Y Acad Sci* (2007) 1103:1–10. doi: 10.1196/annals.1394.018
31. Kudva YC, Ohmine S, Greder LV, Dutton JR, Armstrong A, De Lamo JG, et al. Transgene-free disease-specific induced pluripotent stem cells from patients with type 1 and type 2 diabetes. *Stem Cells Transl Med* (2012) 1 (6):451–61. doi: 10.5966/sctm.2011-0044
32. Simsek S, Zhou T, Robinson CL, Tsai SY, Crespo M, Amin S, et al. Modeling Cystic Fibrosis Using Pluripotent Stem Cell-Derived Human Pancreatic Ductal Epithelial Cells. *Stem Cells Transl Med* (2016) 5(5):572–9. doi: 10.5966/sctm.2015-0276
33. Shang L, Hua H, Foo K, Martinez H, Watanabe K, Zimmer M, et al. beta-cell dysfunction due to increased ER stress in a stem cell model of Wolfram syndrome. *Diabetes* (2014) 63(3):923–33. doi: 10.2337/db13-0717
34. Ma S, Viola R, Sui L, Cherubini V, Barbetti F, Egli D. beta Cell Replacement after Gene Editing of a Neonatal Diabetes-Causing Mutation at the Insulin Locus. *Stem Cell Rep* (2018) 11(6):1407–15. doi: 10.1016/j.stemcr.2018.11.006
35. Teo AK, Windmueller R, Johansson BB, Dirice E, Njolstad PR, Tjora E, et al. Derivation of human induced pluripotent stem cells from patients with maturity onset diabetes of the young. *J Biol Chem* (2013) 288(8):5353–6. doi: 10.1074/jbc.C112.428979
36. Stepniewski J, Kachamakova-Trojanowska N, Ogrocki D, Szopa M, Matlok M, Beilharz M, et al. Induced pluripotent stem cells as a model for diabetes investigation. *Sci Rep* (2015) 5:8597. doi: 10.1038/srep08597
37. Braverman-Gross C, Nudel N, Ronen D, Beer NL, McCarthy MI, Benvenisty N. Derivation and molecular characterization of pancreatic differentiated MODY1-iPSCs. *Stem Cell Res* (2018) 31:16–26. doi: 10.1016/j.scr.2018.06.013
38. Maehr R, Chen S, Snitow M, Ludwig T, Yagasaki L, Goland R, et al. Generation of pluripotent stem cells from patients with type 1 diabetes. *Proc Natl Acad Sci U S A* (2009) 106(37):15768–73. doi: 10.1073/pnas.0906894106
39. Teo AK, Gupta MK, Doria A, Kulkarni RN. Dissecting diabetes/metabolic disease mechanisms using pluripotent stem cells and genome editing tools. *Mol Metab* (2015) 4(9):593–604. doi: 10.1016/j.molmet.2015.06.006
40. Balboa D, Saarimaki-Vire J, Otonkoski T. Concise Review: Human Pluripotent Stem Cells for the Modeling of Pancreatic beta-Cell Pathology. *Stem Cells* (2019) 37(1):33–41. doi: 10.1002/stem.2913
41. Thatava T, Kudva YC, Edukulla R, Squillace K, De Lamo JG, Khan YK, et al. Intrapatient variations in type 1 diabetes-specific iPS cell differentiation into insulin-producing cells. *Mol Ther* (2013) 21(1):228–39. doi: 10.1038/mt.2012.245
42. Millman JR, Xie C, Van Dervort A, Gürtler M, Pagliuca FW, Melton DA. Generation of stem cell-derived β-cells from patients with type 1 diabetes. *Nat Commun* (2016) 7(1):1–8. doi: 10.1038/ncomms11463
43. Leite NC, Sintov E, Meissner TB, Brehm MA, Greiner DL, Harlan DM, et al. Modeling Type 1 Diabetes In Vitro Using Human Pluripotent Stem Cells. *Cell Rep* (2020) 32(2):107894. doi: 10.1016/j.celrep.2020.107894
44. Joshi K, Elso C, Motazedian A, Labonne T, Schiesser JV, Cameron F, et al. Induced pluripotent stem cell macrophages present antigen to proinsulin-specific T cell receptors from donor-matched islet-infiltrating T cells in type 1 diabetes. *Diabetologia* (2019) 62(12):2245–51. doi: 10.1007/s00125-019-04988-6
45. Hosokawa Y, Toyoda T, Fukui K, Baden MY, Funato M, Kondo Y, et al. Insulin-producing cells derived from 'induced pluripotent stem cells' of patients with fulminant type 1 diabetes: Vulnerability to cytokine insults and increased expression of apoptosis-related genes. *J Diabetes Investig* (2017) 9 (3):481–93. doi: 10.1111/jdi.12727
46. Imagawa A, Hanafusa T. Pathogenesis of fulminant type 1 diabetes. *Rev Diabetes Stud* (2006) 3(4):169–77. doi: 10.1900/RDS.2006.3.169
47. Todd JA, Bell JI, McDevitt HO. HLA-DQ beta gene contributes to susceptibility and resistance to insulin-dependent diabetes mellitus. *Nature* (1987) 329(6140):599–604. doi: 10.1038/329599a0
48. Concannon P, Rich SS, Nepom GT. Genetics of type 1A diabetes. *N Engl J Med* (2009) 360(16):1646–54. doi: 10.1056/NEJMra0808284
49. Knip M, Simell O. Environmental triggers of type 1 diabetes. *Cold Spring Harb Perspect Med* (2012) 2(7):a007690. doi: 10.1101/cshperspect.a007690
50. Eringsmark Regnell S, Lernmark A. The environment and the origins of islet autoimmunity and Type 1 diabetes. *Diabetes Med* (2013) 30(2):155–60. doi: 10.1111/dme.12099
51. Hober D, Sauter P. Pathogenesis of type 1 diabetes mellitus: interplay between enterovirus and host. *Nat Rev Endocrinol* (2010) 6(5):279–89. doi: 10.1038/nrendo.2010.27
52. Marre ML, Profozich JL, Coneybeer JT, Geng X, Bertera S, Ford MJ, et al. Inherent ER stress in pancreatic islet beta cells causes self-recognition by autoreactive T cells in type 1 diabetes. *J Autoimmun* (2016) 72:33–46. doi: 10.1016/j.jaut.2016.04.009

53. Rewers M, Ludvigsson J. Environmental risk factors for type 1 diabetes. *Lancet* (2016) 387(10035):2340–8. doi: 10.1016/S0140-6736(16)30507-4
54. James EA, Pietropaolo M, Mamula MJ. Immune Recognition of beta-Cells: Neoepitopes as Key Players in the Loss of Tolerance. *Diabetes* (2018) 67 (6):1035–42. doi: 10.2337/dbi17-0030
55. Marre ML, McGinty JW, Chow IT, DeNicola ME, Beck NW, Kent SC, et al. Modifying Enzymes Are Elicited by ER Stress, Generating Epitopes That Are Selectively Recognized by CD4(+) T Cells in Patients With Type 1 Diabetes. *Diabetes* (2018) 67(7):1356–68. doi: 10.2337/db17-1166
56. Balboa D, Otonkoski T. Human pluripotent stem cell based islet models for diabetes research. *Best Pract Res Clin Endocrinol Metab* (2015) 29(6):899–909. doi: 10.1016/j.beem.2015.10.012
57. Tan G, Elefanty AG, Stanley EG. beta-cell regeneration and differentiation: how close are we to the 'holy grail'? *J Mol Endocrinol* (2014) 53(3):R119–29. doi: 10.1530/JME-14-0188
58. D'Amour KA, Bang AG, Eliazer S, Kelly OG, Agulnick AD, Smart NG, et al. Production of pancreatic hormone-expressing endocrine cells from human embryonic stem cells. *Nat Biotechnol* (2006) 24(11):1392–401. doi: 10.1038/nbt1259
59. Micallef SJ, Li X, Schiesser JV, Hirst CE, Yu QC, Lim SM, et al. INS(GFP/w) human embryonic stem cells facilitate isolation of in vitro derived insulin-producing cells. *Diabetologia* (2012) 55(3):694–706. doi: 10.1007/s00125-011-2379-y
60. Rezania A, Bruin JE, Riedel MJ, Mojibian M, Asadi A, Xu J, et al. Maturation of human embryonic stem cell-derived pancreatic progenitors into functional islets capable of treating pre-existing diabetes in mice. *Diabetes* (2012) 61(8):2016–29. doi: 10.2337/db11-1711
61. Pagliuca FW, Millman JR, Gurtler M, Segel M, Van Dervort A, Ryu JH, et al. Generation of functional human pancreatic beta cells in vitro. *Cell* (2014) 1592(2):428–39. doi: 10.1016/j.cell.2014.09.040
62. Rezania A, Bruin JE, Arora P, Rubin A, Batushansky I, Asadi A, et al. Reversal of diabetes with insulin-producing cells derived in vitro from human pluripotent stem cells. *Nat Biotechnol* (2014) 32(11):1121–33. doi: 10.1038/nbt.3033
63. Russ HA, Parent AV, Ringler JJ, Hennings TG, Nair GG, Shveygert M, et al. Controlled induction of human pancreatic progenitors produces functional beta-like cells in vitro. *EMBO J* (2015) 34(13):1759–72. doi: 10.15252/embj.201591058
64. Vegas AJ, Veiseh O, Gurtler M, Millman JR, Pagliuca FW, Bader AR, et al. Long-term glycemic control using polymer-encapsulated human stem cell-derived beta cells in immune-competent mice. *Nat Med* (2016) 22(3):306–11. doi: 10.1038/nm.4030
65. Zhu S, Russ HA, Wang X, Zhang M, Ma T, Xu T, et al. Human pancreatic beta-like cells converted from fibroblasts. *Nat Commun* (2016) 7:10080. doi: 10.1038/ncomms10080
66. Velazco-Cruz L, Song J, Maxwell KG, Goedegebuure MM, Augsornworawat P, Hogrebe NJ, et al. Acquisition of Dynamic Function in Human Stem Cell-Derived beta Cells. *Stem Cell Rep* (2019) 12(2):351–65. doi: 10.1016/j.stemcr.2018.12.012
67. Hogrebe NJ, Augsornworawat P, Maxwell KG, Velazco-Cruz L, Millman JR. Targeting the cytoskeleton to direct pancreatic differentiation of human pluripotent stem cells. *Nat Biotechnol* (2020) 38(4):460–70. doi: 10.1038/s41587-020-0430-6
68. Velazco-Cruz L, Goedegebuure MM, Millman JR. Advances Toward Engineering Functionally Mature Human Pluripotent Stem Cell-Derived beta Cells. *Front Bioeng Biotechnol* (2020) 8:786. doi: 10.3389/fbioe.2020.00786
69. Augsornworawat P, Maxwell KG, Velazco-Cruz L, Millman JR. Single-Cell Transcriptome Profiling Reveals beta Cell Maturation in Stem Cell-Derived Islets after Transplantation. *Cell Rep* (2020) 32(8):108067. doi: 10.1016/j.celrep.2020.108067
70. Uno S, Imagawa A, Okita K, Sayama K, Moriwaki M, Iwahashi H, et al. Macrophages and dendritic cells infiltrating islets with or without beta cells produce tumour necrosis factor-alpha in patients with recent-onset type 1 diabetes. *Diabetologia* (2007) 50(3):596–601. doi: 10.1007/s00125-006-0569-9
71. Jun HS, Yoon CS, Zbytnuik L, van Rooijen N, Yoon JW. The role of macrophages in T cell-mediated autoimmune diabetes in nonobese diabetic mice. *J Exp Med* (1999) 189(2):347–58. doi: 10.1084/jem.189.2.347
72. Carrero JA, McCarthy DP, Ferris ST, Wan X, Hu H, Zinselmeyer BH, et al. Resident macrophages of pancreatic islets have a seminal role in the initiation of autoimmune diabetes of NOD mice. *Proc Natl Acad Sci U S A* (2017) 114(48):E10418–E27. doi: 10.1073/pnas.1713543114
73. Tahvili S, Torngren M, Holmberg D, Leanderson T, Ivars F. Paquinimod prevents development of diabetes in the non-obese diabetic (NOD) mouse. *PLoS One* (2018) 13(5):e0196598. doi: 10.1371/journal.pone.0196598
74. Ferris ST, Carrero JA, Mohan JF, Calderon B, Murphy KM, Unanue ER. A minor subset of Batf3-dependent antigen-presenting cells in islets of Langerhans is essential for the development of autoimmune diabetes. *Immunity* (2014) 41(4):657–69. doi: 10.1016/j.immuni.2014.09.012
75. Lehuen A, Diana J, Zaccone P, Cooke A. Immune cell crosstalk in type 1 diabetes. *Nat Rev Immunol* (2010) 10(7):501–13. doi: 10.1038/nri2787
76. Senju S, Haruta M, Matsumura K, Matsunaga Y, Fukushima S, Ikeda T, et al. Generation of dendritic cells and macrophages from human induced pluripotent stem cells aiming at cell therapy. *Gene Ther* (2011) 18(9):874–83. doi: 10.1038/gt.2011.22
77. Lachmann N, Ackermann M, Frenzel E, Liebhaber S, Brennig S, Happle C, et al. Large-scale hematopoietic differentiation of human induced pluripotent stem cells provides granulocytes or macrophages for cell replacement therapies. *Stem Cell Rep* (2015) 4(2):282–96. doi: 10.1016/j.stemcr.2015.01.005
78. Cao X, Yakala GK, van den Hil FE, Cochrane A, Mummery CL, Orlova VV. Differentiation and Functional Comparison of Monocytes and Macrophages from hiPSCs with Peripheral Blood Derivatives. *Stem Cell Rep* (2019) 12 (6):1282–97. doi: 10.1016/j.stemcr.2019.05.003
79. Sachamitr P, Leishman AJ, Davies TJ, Fairchild PJ. Directed Differentiation of Human Induced Pluripotent Stem Cells into Dendritic Cells Displaying Tolerogenic Properties and Resembling the CD141(+) Subset. *Front Immunol* (2017) 8:1935. doi: 10.3389/fimmu.2017.01935
80. Iizuka-Koga M, Asashima H, Ando M, Lai CY, Mochizuki S, Nakanishi M, et al. Functional Analysis of Dendritic Cells Generated from T-iPSCs from CD4+ T Cell Clones of Sjogren's Syndrome. *Stem Cell Rep* (2017) 8(5):1155–63. doi: 10.1016/j.stemcr.2017.04.010
81. Montel-Hagen A, Seet CS, Li S, Chick B, Zhu Y, Chang P, et al. Organoid-Induced Differentiation of Conventional T Cells from Human Pluripotent Stem Cells. *Cell Stem Cell* (2019) 24(3):376–89.e8. doi: 10.1016/j.stem.2018.12.011
82. Nishimura T, Kaneko S, Kawana-Tachikawa A, Tajima Y, Goto H, Zhu D, et al. Generation of rejuvenated antigen-specific T cells by reprogramming to pluripotency and redifferentiation. *Cell Stem Cell* (2013) 12(1):114–26. doi: 10.1016/j.stem.2012.11.002
83. Nagano S, Maeda T, Ichise H, Kashima S, Ohtaka M, Nakanishi M, et al. High Frequency Production of T Cell-Derived iPSC Clones Capable of Generating Potent Cytotoxic T Cells. *Mol Ther Methods Clin Dev* (2020) 16:126–35. doi: 10.1016/j.omtm.2019.12.006
84. Hermanson DL, Bendzick L, Pribyl L, McCullar V, Vogel RI, Miller JS, et al. Induced Pluripotent Stem Cell-Derived Natural Killer Cells for Treatment of Ovarian Cancer. *Stem Cells* (2016) 34(1):93–101. doi: 10.1002/stem.2230
85. Zeng J, Tang SY, Toh LL, Wang S. Generation of "Off-the-Shelf" Natural Killer Cells from Peripheral Blood Cell-Derived Induced Pluripotent Stem Cells. *Stem Cell Rep* (2017) 9(6):1796–812. doi: 10.1016/j.stemcr.2017.10.020
86. Shi J, Xue C, Liu W, Zhang H. Differentiation of Human-Induced Pluripotent Stem Cells to Macrophages for Disease Modeling and Functional Genomics. *Curr Protoc Stem Cell Biol* (2019) 48(1):e74. doi: 10.1002/cpsc.74
87. Zhang H, Xue C, Shah R, Bermingham K, Hinkle CC, Li W, et al. Functional analysis and transcriptomic profiling of iPSC-derived macrophages and their application in modeling Mendelian disease. *Circ Res* (2015) 117(1):17–28. doi: 10.1161/CIRCRESAHA.117.305860
88. Zhang H, Reilly MP. Human Induced Pluripotent Stem Cell-Derived Macrophages for Unraveling Human Macrophage Biology. *Arterioscler Thromb Vasc Biol* (2017) 37(11):2000–6. doi: 10.1161/ATVBAHA.117.309195
89. Buchrieser J, James W, Moore MD. Human Induced Pluripotent Stem Cell-Derived Macrophages Share Ontogeny with MYB-Independent Tissue-Resident Macrophages. *Stem Cell Rep* (2017) 8(2):334–45. doi: 10.1016/j.stemcr.2016.12.020
90. Haake K, Ackermann M, Lachmann N. Concise Review: Towards the Clinical Translation of Induced Pluripotent Stem Cell-Derived Blood

Cells-Ready for Take-Off. *Stem Cells Transl Med* (2019) 8(4):332–9. doi: 10.1002/sctm.18-0134

91. Takata K, Kozaki T, Lee CZW, Thion MS, Otsuka M, Lim S, et al. Induced-Pluripotent-Stem-Cell-Derived Primitive Macrophages Provide a Platform for Modeling Tissue-Resident Macrophage Differentiation and Function. *Immunity* (2017) 47(1):183–98.e6. doi: 10.1016/j.immuni.2017.06.017
92. Su Z, Frye C, Bae KM, Kelley V, Vieweg J. Differentiation of human embryonic stem cells into immunostimulatory dendritic cells under feeder-free culture conditions. *Clin Cancer Res* (2008) 14(19):6207–17. doi: 10.1158/1078-0432.CCR-08-0309
93. Tseng SY, Nishimoto KP, Silk KM, Majumdar AS, Dawes GN, Waldmann H, et al. Generation of immunogenic dendritic cells from human embryonic stem cells without serum and feeder cells. *Regener Med* (2009) 4(4):513–26. doi: 10.2217/rme.09.25
94. Yanagimachi MD, Niwa A, Tanaka T, Honda-Ozaki F, Nishimoto S, Murata Y, et al. Robust and highly-efficient differentiation of functional monocytic cells from human pluripotent stem cells under serum- and feeder cell-free conditions. *PLoS One* (2013) 8(4):e59243. doi: 10.1371/journal.pone.0059243
95. Slukvin II, Vodyanik MA, Thomson JA, Gumenyuk ME, Choi KD. Directed differentiation of human embryonic stem cells into functional dendritic cells through the myeloid pathway. *J Immunol* (2006) 176(5):2924–32. doi: 10.4049/jimmunol.176.5.2924
96. Phillips JM, Parish NM, Raine T, Bland C, Sawyer Y, De La Pena H, et al. Type 1 diabetes development requires both CD4+ and CD8+ T cells and can be reversed by non-depleting antibodies targeting both T cell populations. *Rev Diabetes Stud* (2009) 6(2):97–103. doi: 10.1900/RDS.2009.6.97
97. Chatenoud L, Thervet E, Primo J, Bach JF. Anti-CD3 antibody induces long-term remission of overt autoimmunity in nonobese diabetic mice. *Proc Natl Acad Sci U S A* (1994) 91(1):123–7. doi: 10.1073/pnas.91.1.123
98. Herold KC, Gitelman SE, Masharani U, Hagopian W, Bisikirska B, Donaldson D, et al. A single course of anti-CD3 monoclonal antibody hOKT3gamma1(Ala-Ala) results in improvement in C-peptide responses and clinical parameters for at least 2 years after onset of type 1 diabetes. *Diabetes* (2005) 54(6):1763–9. doi: 10.2337/diabetes.54.6.1763
99. Herold KC, Bundy BN, Long SA, Bluestone JA, DiMeglio LA, Dufort MJ, et al. An Anti-CD3 Antibody, Teplizumab, in Relatives at Risk for Type 1 Diabetes. *N Engl J Med* (2019) 381(7):603–13. doi: 10.1056/NEJMoa1902226
100. Morgan NG, Leete P, Foulis AK, Richardson SJ. Islet inflammation in human type 1 diabetes mellitus. *IUBMB Life* (2014) 66(11):723–34. doi: 10.1002/iub.1330
101. Barry M, Bleackley RC. Cytotoxic T lymphocytes: all roads lead to death. *Nat Rev Immunol* (2002) 2(6):401–9. doi: 10.1038/nri819
102. Graham KL, Sutherland RM, Mannering SI, Zhao Y, Chee J, Krishnamurthy B, et al. Pathogenic mechanisms in type 1 diabetes: the islet is both target and driver of disease. *Rev Diabetes Stud* (2012) 9(4):148–68. doi: 10.1900/RDS.2012.9.148
103. Mannering SI, Wong FS, Durinovic-Bello I, Brooks-Worrell B, Tree TI, Cilio CM, et al. Current approaches to measuring human islet-antigen specific T cell function in type 1 diabetes. *Clin Exp Immunol* (2010) 162(2):197–209. doi: 10.1111/j.1365-2249.2010.04237.x
104. Pathiraja V, Kuehlich JP, Campbell PD, Krishnamurthy B, Loudovaris T, Coates PT, et al. Proinsulin-specific, HLA-DQ8, and HLA-DQ8-transdimer-restricted CD4+ T cells infiltrate islets in type 1 diabetes. *Diabetes* (2015) 64 (1):172–82. doi: 10.2337/db14-0858
105. Babon JA, DeNicola ME, Blodgett DM, Crevecoeur I, Buttrick TS, Maehr R, et al. Analysis of self-antigen specificity of islet-infiltrating T cells from human donors with type 1 diabetes. *Nat Med* (2016) 22(12):1482–7. doi: 10.1038/nm.4203
106. Delong T, Wiles TA, Baker RL, Bradley B, Barbour G, Reisdorph R, et al. Pathogenic CD4 T cells in type 1 diabetes recognize epitopes formed by peptide fusion. *Science* (2016) 3516274):711–4. doi: 10.1126/science.aad2791
107. Michels AW, Landry LG, McDaniel KA, Yu L, Campbell-Thompson M, Kwok WW, et al. Islet-Derived CD4 T Cells Targeting Proinsulin in Human Autoimmune Diabetes. *Diabetes* (2017) 66(3):722–34. doi: 10.2337/db16-1025
108. Nakayama M, Michels AW. Determining Antigen Specificity of Human Islet Infiltrating T Cells in Type 1 Diabetes. *Front Immunol* (2019) 10:365. doi: 10.3389/fimmu.2019.00365
109. Schietinger A, Greenberg PD. Tolerance and exhaustion: defining mechanisms of T cell dysfunction. *Trends Immunol* (2014) 35(2):51–60. doi: 10.1016/j.it.2013.10.001
110. Wiedeman AE, Muir VS, Rosasco MG, DeBerg HA, Presnell S, Haas B, et al. Autoreactive CD8+ T cell exhaustion distinguishes subjects with slow type 1 diabetes progression. *J Clin Invest* (2020) 130(1):480–90. doi: 10.1172/JCI126595
111. Knip M, Siljander H, Ilonen J, Simell O, Veijola R. Role of humoral beta-cell autoimmunity in type 1 diabetes. *Pediatr Diabetes* (2016) 17 Suppl 22:17–24. doi: 10.1111/pedi.12386
112. Wong FS, Hu C, Xiang Y, Wen L. To B or not to B–pathogenic and regulatory B cells in autoimmune diabetes. *Curr Opin Immunol* (2010) 22 (6):723–31. doi: 10.1016/j.coi.2010.10.002
113. Hu CY, Rodriguez-Pinto D, Du W, Ahuja A, Henegariu O, Wong FS, et al. Treatment with CD20-specific antibody prevents and reverses autoimmune diabetes in mice. *J Clin Invest* (2007) 117(12):3857–67. doi: 10.1172/JCI32405
114. Pescovitz MD, Greenbaum CJ, Krause-Steinrauf H, Becker DJ, Gitelman SE, Goland R, et al. Rituximab, B-lymphocyte depletion, and preservation of beta-cell function. *N Engl J Med* (2009) 361(22):2143–52. doi: 10.1056/NEJMoa0904452
115. Martin S, Wolf-Eichbaum D, Duinkerken G, Scherbaum WA, Kolb H, Noordzij JG, et al. Development of type 1 diabetes despite severe hereditary B-cell deficiency. *N Engl J Med* (2001) 345(14):1036–40. doi: 10.1056/NEJMoa010465
116. Carpenter L, Malladi R, Yang CT, French A, Pilkington KJ, Forsey RW, et al. Human induced pluripotent stem cells are capable of B-cell lymphopoiesis. *Blood* (2011) 117(15):4008–11. doi: 10.1182/blood-2010-08-299941
117. French A, Yang CT, Taylor S, Watt SM, Carpenter L. Human induced pluripotent stem cell-derived B lymphocytes express sIgM and can be generated via a hemogenic endothelium intermediate. *Stem Cells Dev* (2015) 24(9):1082–95. doi: 10.1089/scd.2014.0318
118. Freud AG, Mundy-Bosse BL, Yu J, Caligiuri MA. The Broad Spectrum of Human Natural Killer Cell Diversity. *Immunity* (2017) 47(5):820–33. doi: 10.1016/j.immuni.2017.10.008
119. Fraker C, Bayer AL. The Expanding Role of Natural Killer Cells in Type 1 Diabetes and Immunotherapy. *Curr Diabetes Rep* (2016) 16(11):109. doi: 10.1007/s11892-016-0806-7
120. Rodacki M, Svoren B, Butty V, Besse W, Laffel L, Benoist C, et al. Altered natural killer cells in type 1 diabetic patients. *Diabetes* (2007) 56(1):177–85. doi: 10.2337/db06-0493
121. Lorini R, Moretta A, Valtorta A, d'Annunzio G, Cortona L, Vitali L, et al. Cytotoxic activity in children with insulin-dependent diabetes mellitus. *Diabetes Res Clin Pract* (1994) 23(1):37–42. doi: 10.1016/0168-8227(94)90125-2
122. Dotta F, Censini S, van Halteren AG, Marselli L, Masini M, Dionisi S, et al. Coxsackie B4 virus infection of beta cells and natural killer cell insulitis in recent-onset type 1 diabetic patients. *Proc Natl Acad Sci U S A* (2007) 104 (12):5115–20. doi: 10.1073/pnas.0700442104
123. Wenzlau JM, Juhl K, Yu L, Moua O, Sarkar SA, Gottlieb P, et al. The cation efflux transporter ZnT8 (Slc30A8) is a major autoantigen in human type 1 diabetes. *Proc Natl Acad Sci U S A* (2007) 104(43):17040–5. doi: 10.1073/pnas.0705894104
124. Wong FS, Wen L. B cells in autoimmune diabetes. *Rev Diabetes Stud* (2005) 2 (3):121–35. doi: 10.1900/RDS.2005.2.121
125. Pihoker C, Gilliam LK, Hampe CS, Lernmark A. Autoantibodies in diabetes. *Diabetes* (2005) 54 Suppl 2:S52–61. doi: 10.2337/diabetes.54.suppl_2.S52
126. Elkon K, Casali P. Nature and functions of autoantibodies. *Nat Clin Pract Rheumatol* (2008) 4(9):491–8. doi: 10.1038/ncprheum0895
127. Ludwig RJ, Vanhoorelbeke K, Leypoldt F, Kaya Z, Bieber K, McLachlan SM, et al. Mechanisms of Autoantibody-Induced Pathology. *Front Immunol* (2017) 8:603. doi: 10.3389/fimmu.2017.00603
128. Dobersen MJ, Scharff JE. Preferential lysis of pancreatic B-cells by islet cell surface antibodies. *Diabetes* (1982) 31(5 Pt 1):459–62. doi: 10.2337/diabetes.31.5.459
129. Dobersen MJ, Scharff JE, Ginsberg-Fellner F, Notkins AL. Cytotoxic autoantibodies to beta cells in the serum of patients with insulin-dependent diabetes mellitus. *N Engl J Med* (1980) 303(26):1493–8. doi: 10.1056/NEJM198012253032601

130. Caraher EM, Conroy SJ, Newsholme P. Evidence for enhanced rates of complement activation in serum from patients with newly diagnosed insulin-dependent diabetes mellitus exposed to rat islet cells and complement-dependent induction of islet cell apoptosis. *J Endocrinol* (1999) 162 (1):143–53. doi: 10.1677/joe.0.1620143
131. Rowe P, Wasserfall C, Croker B, Campbell-Thompson M, Pugliese A, Atkinson M, et al. Increased complement activation in human type 1 diabetes pancreata. *Diabetes Care* (2013) 36(11):3815–7. doi: 10.2337/dc13-0203
132. Steinbeck JA, Jaiswal MK, Calder EL, Kishinevsky S, Weishaupt A, Toyka KV, et al. Functional Connectivity under Optogenetic Control Allows Modeling of Human Neuromuscular Disease. *Cell Stem Cell* (2016) 18 (1):134–43. doi: 10.1016/j.stem.2015.10.002
133. Wu SM, Hochedlinger K. Harnessing the potential of induced pluripotent stem cells for regenerative medicine. *Nat Cell Biol* (2011) 13(5):497–505. doi: 10.1038/ncb0511-497
134. Inoue H, Nagata N, Kurokawa H, Yamanaka S. iPS cells: a game changer for future medicine. *EMBO J* (2014) 33(5):409–17. doi: 10.1002/embj.201387098
135. Ebert AD, Yu J, Rose FFJr., Mattis VB, Lorson CL, Thomson JA, et al. Induced pluripotent stem cells from a spinal muscular atrophy patient. *Nature* (2009) 457(7227):277–80. doi: 10.1038/nature07677
136. Lee G, Papapetrou EP, Kim H, Chambers SM, Tomishima MJ, Fasano CA, et al. Modelling pathogenesis and treatment of familial dysautonomia using patient-specific iPSCs. *Nature* (2009) 461(7262):402–6. doi: 10.1038/nature08320
137. Burkhardt MF, Martinez FJ, Wright S, Ramos C, Volfson D, Mason M, et al. A cellular model for sporadic ALS using patient-derived induced pluripotent stem cells. *Mol Cell Neurosci* (2013) 56:355–64. doi: 10.1016/j.mcn.2013.07.007
138. Atkinson MA, Roep BO, Posgai A, Wheeler DCS, Peakman M. The challenge of modulating β-cell autoimmunity in type 1 diabetes. *Lancet Diabetes Endocrinol* (2019) 7(1):52–64. doi: 10.1016/S2213-8587(18)30112-8
139. Jacobsen L, Schatz D. Current and future efforts toward the prevention of type 1 diabetes. *Pediatr Diabetes* (2016) 17 Suppl 22:78–86. doi: 10.1111/pedi.12333
140. Jacobsen LM, Haller MJ, Schatz DA. Understanding Pre-Type 1 Diabetes: The Key to Prevention. *Front Endocrinol (Lausanne)* (2018) 9:70. doi: 10.3389/fendo.2018.00070
141. Barcala Tabarrozzi AE, Castro CN, Dewey RA, Sogayar MC, Labriola L, Perone MJ. Cell-based interventions to halt autoimmunity in type 1 diabetes mellitus. *Clin Exp Immunol* (2013) 171(2):135–46. doi: 10.1111/cei.12019
142. Karagiannis P, Takahashi K, Saito M, Yoshida Y, Okita K, Watanabe A, et al. Induced Pluripotent Stem Cells and Their Use in Human Models of Disease and Development. *Physiol Rev* (2019) 99(1):79–114. doi: 10.1152/physrev.00039.2017
143. Greenbaum C, VanBuecken D, Lord S. Disease-Modifying Therapies in Type 1 Diabetes: A Look into the Future of Diabetes Practice. *Drugs* (2019) 79 (1):43–61. doi: 10.1007/s40265-018-1035-y
144. Thorens B. Neural regulation of pancreatic islet cell mass and function. *Diabetes Obes Metab* (2014) 16 Suppl 1:87–95. doi: 10.1111/dom.12346
145. Gaisano HY, Macdonald PE, Vranic M. Glucagon secretion and signaling in the development of diabetes. *Front Physiol* (2012) 3:349. doi: 10.3389/fphys.2012.00349

2

Modeling Maturity Onset Diabetes of the Young in Pluripotent Stem Cells: Challenges and Achievements

Carmel Braverman-Gross and Nissim Benvenisty**

The Azrieli Center for Stem Cells and Genetic Research, Department of Genetics, The Alexander Silberman Institute of Life Sciences, The Hebrew University of Jerusalem, Jerusalem, Israel

***Correspondence:**
Carmel Braverman-Gross
carmel.braverman@mail.huji.ac.il
Nissim Benvenisty
nissimb@mail.huji.ac.il

Maturity onset diabetes of the young (MODY), is a group of monogenic diabetes disorders. Rodent models for MODY do not fully recapitulate the human phenotypes, calling for models generated in human cells. Human pluripotent stem cells (hPSCs), capable of differentiation towards pancreatic cells, possess a great opportunity to model MODY disorders *in vitro*. Here, we review the models for MODY diseases in hPSCs to date and the molecular lessons learnt from them. We also discuss the limitations and challenges that these types of models are still facing.

Keywords: maturity onset diabetes of the young, human embryonic stem cells, induced pluripotent stem cells, disease modeling, genetic and epigenetic aberrations

INTRODUCTION

Monogenic diabetes refers to a group of disorders caused by mutations in a single gene resulting in diabetes. To date, more than 40 subtypes of monogenic diabetes have been identified, most of them results in β cell loss or function impairment. In rare cases, diabetes is caused by mutations leading to insulin resistance or associated with other features of genetic syndromes affecting multiple organs. The typical classification of monogenic diabetes includes two main subgroups; neonatal diabetes, usually presenting before 6 months of age, and maturity onset diabetes of the young (MODY), usually presenting in youth and adults. While neonatal diabetic cases are rare (1 in 1,000,000 birth), MODY disorders are more common, accounting for 1–5% of all diabetic cases (1–4). The acronym MODY was first used in 1975 by Fajans and Tattersall, to distinguish a hereditary form of diabetes presented in juvenile patients from classical type 1 diabetes patients (5). Today, the term MODY is used to describe a group of clinically heterogeneous metabolic disorders that are characterized by pancreatic β cell functional impairment. The clinical features of MODY are varied and depend on the causal gene. Some of the common features of MODY include hyperglycemia, diagnosed usually in childhood or adolescence (under 25), family history (autosomal dominant inheritance) and lack of pancreatic auto-antibodies (2, 6).

MATURITY ONSET DIABETES OF THE YOUNG GENETICS AND PATHOGENESIS

Fourteen distinct subtypes of MODY have been identified to date, all caused by mutations in genes important for pancreatic *β* cell development, regulation, and function. Most of these genes encode for transcription factors (TFs), *i.e.*, MODY1, MODY3, MODY4, MODY5, MODY6, MODY7, and MODY9 (caused by mutations in the genes *HNF4A*, *HNF1A*, *PDX1*, *HNF1B*, *NEOROD1*, *KLF11*, and *PAX4*, respectively). Some of them encode for enzymes, *i.e.*, MODY2, MODY8, and MODY11 (caused by mutation in the genes *GCK*, *CEL*, and *BLK*, respectively), and some involve other pancreatic genes, *i.e.*, MODY10, MODY12, MODY13 and MODY14 (caused by mutations in *INS*, *SUR1*, *KCNJ11*, and *APPL1* genes, respectively) (7–9).

Clinical diagnosis of MODY is still suboptimal, mainly due to the variability of clinical presentations and their similarity to symptoms of other types of diabetes, leading to misdiagnosis of MODY as type 1 or type 2 diabetes (10, 11). However, with the increasing availability and price reduction of genetic tests, MODY diagnosis is rising. An accurate and timely diagnosis of MODY can dramatically affect the medical treatment given as treatment is tailored to the specific mutation. This treatment is often dramatically different from that of type 1 or type 2 diabetes (12). Precise diagnosis is also important for early identification of asymptomatic or undiagnosed family members, in order to minimize the disorder's impact on multiple organs (1).

Some of the MODY genes are specifically involved in *β* cells' function, while others are related to different stages of the endocrine pancreatic development. Studying the effect of perturbation in these latter genes may also help understand the developmental processes and pathogenesis of other pancreatic diseases. Furthermore, understanding the mechanisms underlying *β* cell formation could improve *in vitro* differentiation protocols of *β* cells from human stem cells, enhancing the feasibility of pancreatic islet transplantation in type 1 diabetes patients and other pancreatic disorders.

MODELING MATURITY ONSET DIABETES OF THE YOUNG

A great part of the current understanding of pancreatic *β* cell development and function was achieved using rodent models. However, as pancreatic development and architecture, as well as glucose response, differ substantially between mice and humans, rodent models do not always accurately represent the human phenotypes. Such are the cases of heterozygous mutations in the genes *ABCC8*, *HNF1B*, *HNF4A*, *HNF1A*, *GATA4*, and *GATA6* that cause neonatal diabetes or MODY in humans, but do not present any diabetic pathophysiology in mice (13–15) (**Figure 1**).

Human studies of diabetes mechanisms can also be done by using cadaveric islets, which are human primary islets harvested *post mortem* from pancreatic donors (15, 16),. Recent studies that used human islets from donors diagnosed with type 1 diabetes revealed mutations in genes causing monogenic diabetes, including MODY, that are the primary cause of diabetic symptoms (17, 18). This approach is limited due to high variability between islets, the short life span of the cells composing them and mainly due to low donor accessibility. The latter is especially challenging when modeling MODY, given the low prevalence and diagnosis of this disease (17, 18). Immortalized *β* cells, or *β* cell lines, are another human cell type that are used in the field of diabetes research. To date, no cell lines were established from MODY patients, calling for gene editing to model these diseases. Although CRISPR/Cas9 editing has been previously used in EndoC-BH lines (19), these lines have low clonal efficiency which makes editing at clonal level challenging (15, 20). In addition, both cadaveric islets and immortalized *β* cells enable the study of mature pancreatic cells and are less suitable for studying genes that have a role during pancreatic development (**Figure 1**).

This calls for suitable monogenic diabetes models that can be fulfilled by human pluripotent stem cells (hPSCs). Both induced pluripotent stem cells (iPSCs) and embryonic stem cells (ESCs) can be differentiated from their pluripotent state to pancreatic *β*-like cells. Over the last decade, such differentiation protocols were developed and improved, enabling the generation of functional human *β* cells (21–25). Glucose-responsive and insulin secreting *β* cells can be generated *in vitro*, and were even proved to reduce blood glucose levels when transplanted in Streptozotocin-induced diabetic mice. Differentiation protocols are performed in a stepwise manner, typically mimicking *in vivo* development of *β* cells (26). By using PSC-derived cells carrying a mutation in MODY-causing gene, one can follow the natural course of development and elucidate the relevant stage or stages that are interrupted by the mutation (**Figure 1**).

Since MODY genes vary in their function and in the developmental stages in which they are expressed and act, there may be different mechanisms underlying each of the MODY disorders. These mechanisms can be studied using PSCs.

iPSC MODELS FOR MATURITY ONSET DIABETES OF THE YOUNG

hPSC models for several MODY types were established in the past decade (**Table 1**), mainly for the manifestations caused by mutant transcription factors. The vast majority of these models were based on iPSCs. iPSCs are generated by reprogramming patients' somatic cells, harboring the mutated gene. These reprogrammed cells are capable of self-renewal and can be maintained *in vitro*. They can be further differentiated into pancreatic cells, enabling the study of the effects of particular mutations.

Such models were generated for MODY1 (27–30), MODY2 (27), MODY3 (27, 31–33), MODY4 (34, 35), MODY5 (27, 37, 38), MODY8 (27), and MODY13 (39) (**Table 1**).

These studies modeled different mutations and used different differentiation protocols, leading to complex conclusions in some cases. Nevertheless, these studies contribute to our understanding of

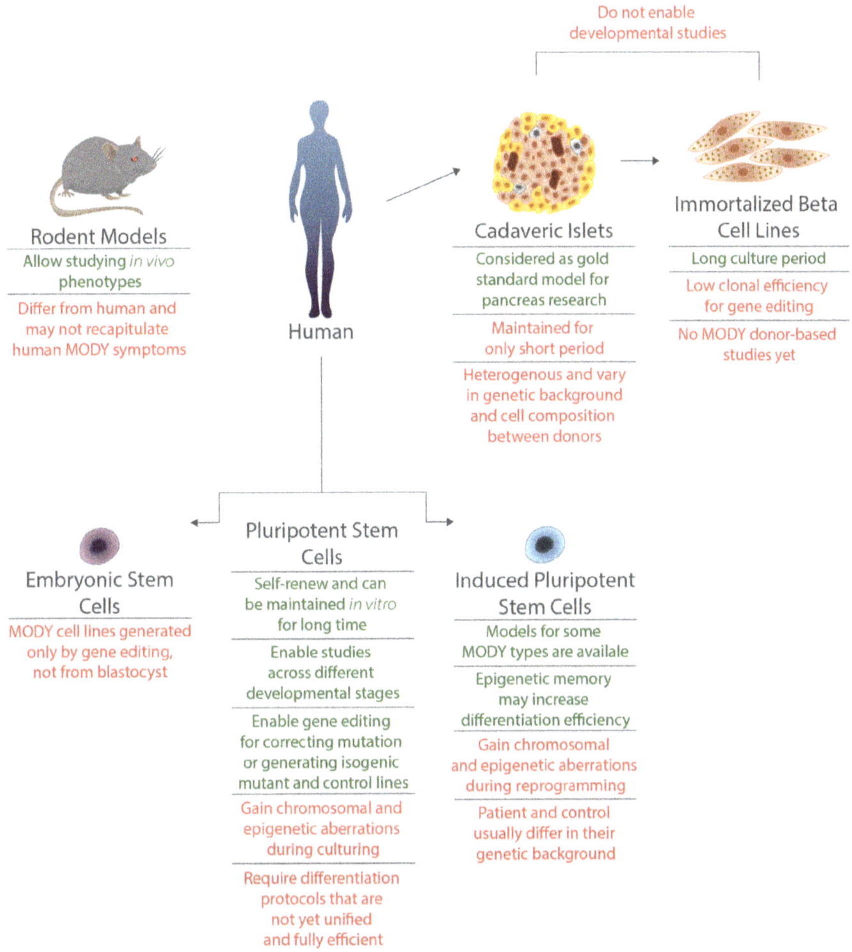

FIGURE 1 | Advantages and limitations of various models for MODY disorders. Models for MODY disorders can be generated using different *in vivo* and *in vitro* systems, such as rodents, human cadaveric islets, immortalized beta cell lines, hiPSCs and hESCs. Advantages (appear in green) and disadvantages (appear in red) for each type of model are presented.

the relations between TFs and their gene targets throughout pancreatic development and of the mechanisms underlying MODY symptoms. We will therefore focus on the analyses of MODY disorders that involve TFs.

Maturity Onset Diabetes of the Young 1

MODY1 is characterized by progressive β cell dysfunction, macrosomia, and neonatal hyperinsulinemic hypoglycemia in some of the cases. In 2017, Vethe et al. generated iPSCs from patients carrying I271fs mutation in *HNF4A* gene and differentiated them towards β-like cells (28). Comparing MODY1 iPSCs to iPSCs from a healthy member of the same family, they reported that the mutation did not alter the differentiation capacity of the cells. Mutated and control cells expressed similar levels of insulin, as well as other pancreatic proteins such as PDX1, NEUROD1 and MAFB. Comparing the proteomic landscape and function (by glucose stimulated insulin secretion (GSIS) test) of mutant and control β-like cells to adult human islet, they reported an immature state of the differentiated iPSCs.

A recent study from the same group by Ghila et al. used the same MODY1 iPSCs differentiated towards hormone producing islet-like cells, to investigate their miRNA expression profile (40). Their bioinformatic analysis revealed miRNAs that are differentially expressed between mutant and control cells. These miRNAs are composed of two subgroups, one that showed different expression during early differentiation stages (during the formation of posterior foregut) and the other was differentially expressed in later stages (after pancreatic endoderm formation). They observed alterations in miRNA networks related to *TP53* regulation in both stages. A combination of the miRNA data with transcriptomic data in MODY1 cells further highlighted an activation of *TP53* in the MODY1 cells, suggesting that mutations in *HNF4A* affect cell cycle arrest during late differentiation.

Ng et al. used iPSCs generated from a MODY1 patient carrying the same I271fs mutation. They showed that this

TABLE 1 | MODY models in hPSCs.

Type	Gene Name	Mutation Type	Type of cells	Reference
MODY1	HNF4A	p.I271fs	hiPSC	(27)
		p.I271fs	hiPSC	(28)
		p.I271fs	hiPSC	(29)
		p.Q268X	hiPSC	(30)
MODY2	GCK	V62A	hiPSC	(27)
MODY3	HNF1A	p.P291fsinsC	hiPSC	(27)
		p.R271W and p.P379fs	hiPSC	(31)
		p.S142F	hiPSC	(32)
		19/16bp δ-exon 1	hESC	(19)
		indel-exon 1	human β cell lines	(19)
		p.P291fsinsC	hiPSC	(33)
MODY4	PDX1	p.P33T	hiPSC	(34)
		p.C18R	hiPSC	(35)
		p.L36fs and p.A34fs	hESC	(36)
MODY5	HNF1B	g.1-1671del	hiPSC	(27)
		p.R177X	hiPSC	(37)
		p.S148L	hiPSC	(38)
MODY8	CEL	p.C563fsX673	hiPSC	(27)
MODY13	KCNJ11	p.E227K	hiPSC	(39)

mutation has haploinsufficiency, rather than a dominant negative mode of action, and that it leads to cytoplasmic mislocalization of the HNF4A TF (29). In order to assess the mutation effect on the developmental process, they focused on the hepatopancreatic foregut endoderm stage, an early stage of pancreatic differentiation. RNA sequencing (RNAseq) data of MODY1 cells showed an upregulation of hindgut HOX genes in parallel to downregulation of a set of pancreatic and liver genes including *GATA4*, *HNF1B*, and *PDX1*. Together, these results suggested that HNF4A has a role in hindgut repression. Downregulation of hepatic genes was also detected in further differentiated cells towards hepatocytes, while a significant reduction in *HNF1A* was detected in differentiated pancreatic β cells.

Braverman-Gross et al. generated MODY1 iPSCs from patients carrying a Q268X mutation in the *HNF4A* gene [previously shown to cause haploinsufficiency (41),]. The authors differentiated the cells towards pancreatic progenitors and exhibited an upregulation of some pancreatic markers, such as *PAX6*, *NEUROD1* and *NEOUROG3* in the MODY1 cells, indicating a possible partial compensatory mechanism for the mutation (30). RNAseq data of MODY1 cells from the primitive gut tube stage of differentiation revealed subgroups of putative HNF4A gene targets. Targets with less HNF4A binding sites and more sites for other TFs in their promotors were less affected by HNF4A haploinsufficiency, further suggesting a redundancy mechanism for the mutation.

Maturity Onset Diabetes of the Young 3

MODY3, caused by mutations in another TF, HNF1A, is characterized by gradual β cell dysfunction and progressive hyperglycemia. In 2015, Stepniewsky et al. generated two MODY3 iPSC lines derived from patients carrying either R271W or P379fs mutations in *HNF1A* (31). They then differentiated the cells towards hormone producing cells. Even though the differentiation capacity was rather poor and resulted in immature polyhormonal cells with low percentage of insulin-positive cells, they showed that the MODY3 cells had a similar capacity to differentiate as iPSCs generated from healthy individuals (31).

Yabe et al. generated MODY3 iPSCs carrying the most common mutation in *HNF1A* gene, P291fsinsC (33). They differentiated the iPSCs to pancreatic cells, and showed that the mutated RNA is abolished by nonsense mediate decay (NMD), consequently leading to minimal expression of the mutant protein, indicating haploinsufficiency mechanism for this MODY3 mutation as well.

Maturity Onset Diabetes of the Young 4

MODY4 is caused by mutated PDX1 TF, which usually results in defective insulin secretion, while homozygous mutations in this gene also lead to pancreatic agenesis. In 2016, Wang et al. generated two iPSC lines carrying heterozygous P33T and C18R mutations in *PDX1* gene (34, 35). By differentiating the cells towards pancreatic endoderm, they exhibited no difference in PDX1 expression and early pancreatic differentiation between mutant and control cells (42). At later stages, β-like cells derived from MODY4 iPSCs had low insulin expression and reduced GSIS. Using CRISPR/Cas9 gene editing, they also induced the same mutations in homozygous state, as well as heterozygous frame shift mutation in control iPSCs. The mutated cells showed impairment in differentiation towards pancreatic precursors by reduced C-peptide expression and reduced GSIS. In addition, they showed reduction of transcription levels in pancreatic genes including *NEUROD1*, *ISL1*, and *INS*, which were previously shown to be direct PDX1 targets (43).

Maturity Onset Diabetes of the Young 5

MODY5, caused by mutations in HNF1B TF, is characterized by reduced insulin secretion and renal cysts in some patients. Yabe et al. generated MODY5 iPSCs from a patient carrying R177X mutation in *HNF1B* (37). As in the case of their MODY3 study, they showed that NMD destruction of the mutant RNA is taking place, hinting for haploinsufficiency of HNF1B as the underlying cause of MODY5. Teo et al. generated iPSCs from MODY5 patients, carrying a different mutation, S148L (38). Differentiated MODY5 cells showed upregulation of definitive endoderm and early pancreatic genes, including *SOX17*, *FOXA2*, *GATA4*, *GATA6* and *PDX1*, as well as *HNF1B* itself, suggesting a compensatory gene expression circuit. Using luciferase assays they also showed that the increase in *PDX1* expression was directly related to the S148L mutant allele. The only pancreatic gene that was downregulated in the MODY5 cells was *PAX6*. It was hypothesized that this occurred by an indirect regulation, as *PAX6* promoter was not found to bind HNF1B. This downregulation could explain some of the diabetic symptoms of MODY5 patients.

EMBRYONIC STEM CELL MODELS FOR MATURITY ONSET DIABETES OF THE YOUNG

The scarcity of MODY patients and tissues donors, as well as inherent disadvantages of iPSC-based disease models, as discussed below, call for additional *in vitro* MODY models. One of the promising models currently being used are based on targeted gene editing, which became much simpler and prevalent with the introduction of the CRISPR/Cas9 methodology. Some groups recently combined gene editing of human embryonic stem cells together with pancreatic differentiation, to study the effects of MODY-related gene disruption (**Table 1**).

In 2016, Zhu et al. used the HUES8 ESC line to generate 62 ESC sublines carrying mutations in eight different pancreatic TFs, in order to study their role in poly hormonal β-like cell differentiation. One of these genes was *PDX1*, the gene causing MODY4 disease, for which they created two homozygous lines carrying biallelic frame shift mutation, L36fs and A34fs (36). ESCs carrying the same heterozygous monoallelic mutations showed a reduction in PDX1 expression compared to WT cells, indicating haploinsufficiency mechanism for these mutants, leading to decrease in endocrine hormone gene expression (*INS, SST, GCG, GHRL*) in differentiated cells.

Cardenas-Diaz et al. generated HNF1A homozygous and heterozygous deletion mutations in MEL1 and H1 ESC lines (19). They differentiated the cells and showed effect of the mutation in the β-like cells' stage, where they observed a reduction in the expression of pancreatic TFs including *PDX1, RFX6, HNF4A* and *PAX4.* Mutant cells showed an increase in *ARX* gene levels, accompanied by higher GCG levels and lower *INS* expression in the mutant cells. These results indicated a role for *HNF1*A in inhibiting α cell development. Mutations in *HNF1A* also impaired the β-like cells GSIS, and cell with heterozygous mutations showed decreased mitochondrial respiration. In addition, they revealed a downregulation in the heterozygous cells of *LINKA*, a primate-specific lncRNA, and suggested it has a role as a mediator of *HNF1A* regulation for subset of *HNF1A* targets, specifically related to mitochondrial respiration and pancreatic genes expression.

LESSONS LEARNED FROM MATURITY ONSET DIABETES OF THE YOUNG MODELS IN PSCs

The diverse studies discussed above shed light on roles of TFs during the different stages of pancreatic development. Some of these TFs may have roles in a specific time point of the differentiation. Focusing on that exact step may be crucial to understanding the effects of a gene mutation. A mutation in *PDX1*, for example, was shown to affect the cells only at the terminal stage of differentiation (42). In contrast, mutations in *HNF4A* were found to alter foregut formation (29) and affect posterior foregut genes (30). Although insulin expression was reduced in mutant *HNF4A* pancreatic β-like cells in one study (29), in another, similar cells did not functionally differ from control cells (28). It is possible that compensatory mechanisms overcome the mutation and enable the mutant cells to fully or partially differentiate towards β-like cells. Such compensatory mechanism was shown for cells carrying mutations in *HNF1B*, where the mutant protein caused a possible direct upregulation of *HNF1B* itself and of *HLXB9*, and indirect upregulation of *PDX1* and other early pancreatic genes (38). HNF genes, including *HNF4A, HNF1A* and *HNF1B* are known to cross-regulate each other during pancreas development (44), and may cause the upregulation of shared targets when one of them is mutated. It is also likely that the variability of differentiation protocols used in different studies influence the final outcome (45)—highlighting the need for a uniform, efficient differentiation protocol, from the pluripotent stage to β cells.

Within each of the 14 known genes causing MODY, various mutations have been identified. The vast majority of MODY mutations are heterozygous, however, in some of the MODY genes a biallelic mutation may occur, usually leading to a more severe phenotype. Homozygous mutations in GCG, PDX1, ABCC8, and KCNJ11 cause permanent or transient neonatal diabetes, as well as pancreatic agenesis (as is the case in PDX1) or neurological abnormalities (in NEUROD1) (6, 9). MODY models in iPSCs were only generated from heterozygous patients, but gene editing using CRISPR/Cas9 studies generated homozygous cells that showed some differences from their heterozygous parallels. For example, cells homozygous for the P33T mutation in *PDX1* gene showed impairment in pancreatic differentiation to a much higher extent than the heterozygous cells (42). HNF1A homozygous mutants showed lower insulin levels than the heterozygous upon differentiation while heterozygous mutation affected the cellular respiration more than the homozygous knock-out mutation (19).

Since MODY disorders are typically characterized by heterozygous mutations, and symptoms of patients with homozygous mutations are distinct, it is hypothesized that a dosage dependent mechanism underlies MODY genes' mode of action. One such dosage dependent model is haploinsufficiency (29, 33, 36, 37, 41). Thus, in a patient with heterozygous mutation, only one allele forms the gene product which is not enough for the normal function of the protein. The mechanisms underlying haploinsufficiency are varied and include threshold requirement of protein amount (*i.e.* in the case of TF that has to bind multiple sites of the DNA), or imbalanced stoichiometry of multi subunits complexes (46). Understanding the exact mechanism of action is important for future therapeutic purposes. While some haploinsufficiency genes are dosage sensitive and their upregulation may be toxic (46, 47), in other cases, activation of the normal allele transcription may rescue the haploinsufficiency disease symptoms (48).

MODY symptoms are affected not only from the mono or biallelic expression of a mutation, but also from the specific type of mutation, even within the same gene. Hence, specific mutations in a gene can cause MODY, while other mutations in the same gene can cause other types of disease. Permanent neonatal diabetes can be caused by mutations in *INS* and *KCNJ11* genes, while other mutations in these genes will cause MODY10 and MODY13, respectively, with symptoms presenting only later in life; congenital hyperinsulinism can be caused by mutations in *ABCC8*, *KCNJ11*, *GCK*, *HNF1A*, and *HNF4A* genes, the same genes that, when harboring different mutations, cause MODY12, MODY13, MODY2, MODY3, and MODY1, respectively (9, 49, 50). Even within patients with the same type of MODY, symptoms and cellular gene profiling may be different due to different types of mutations. For example, cells carrying P33T mutation in *PDX1* show lower levels of PDX1, NKX6.1 and C-peptide than cells carrying C18R mutation in the same gene (42).

Although generating a specific heterozygous mutation may be technically challenging (51), these observations highlight the need to model the exact type and location of a mutation, especially when using gene editing.

CHALLENGES OF PSC BASED MATURITY ONSET DIABETES OF THE YOUNG MODELING

Using hPSCs for disease modeling, specifically monogenic developmental diseases such as MODY, have great benefits (52). In the case of MODY, one of the most important reasons to use hPSCs is the lack of appropriate animal models that recapitulate the human diabetic phenotypes (13). Furthermore, using hPSCs enables studying the complex pancreatic developmental process and assessing the effects of different mutations on its progress. This cannot be achieved using adult, differentiated tissues. However, when modeling a disease using hPSC, there are several issues one should consider (**Figure 1**).

Genomic Aberrations

Since hPSCs are cultured and passaged *in vitro*, and iPSCs are going through the clonal process of reprogramming; they are prone to gain genomic aberrations. Such aberrations include large chromosomal aberrations, sub-chromosomal aberrations, copy number variations (CNVs), and point mutations (53, 54). These changes in genome content may affect the differentiation performance and global gene expression (55).

Although a normal karyotype is validated in all MODY iPSC models, usually by Giemsa band staining, in order to avoid clones containing large chromosomal gains or deletions, smaller mutations are more difficult to identify. Of note, it was found that hPSCs bear CNV hot spots, mainly on chromosomes 1, 12, 17q, 20q, and X (56). This finding may be crucial when modeling genes located within these genomic loci. These genes include *HNF4A*, located on chromosome 20q13, *HNF1A* on chromosome 12q24, and *HNF1B* on chromosome 17q12.

Recent studies showed that some hPSC lines harbor point mutations mainly in *TP53*, coding for the tumor suppressor p53 (57). p53 is a master regulator of apoptosis, cell cycle, and proliferation and was recently suggested to be involved in cell cycle arrest of MODY4 cells (40). Since *TP53* mutations probably provide the cells with selective advantages and since these mutations remain and expand in the culture during pancreatic differentiation (57), they could affect the phenotypes observed in the models they are derived from and their interpretation. These mutations should thus be tightly monitored when using hPSC for disease modeling.

Epigenetic Aberrations

Another type of aberrations affecting hPSCs is epigenetic aberrations, including DNA methylation alterations, loss of parental imprinting, and variation in X chromosome inactivation (58). Specific genomic regions are prone to acquire hypermethylation leading to gene silencing during *in vitro* culturing and during reprogramming. These may affect the differentiation capacity of the cells, as was shown for *TSPYL5*, which was recently found to cause aberrant differentiation of cells upon its silencing (59). Various methylation aberrations were shown to originate from the reprogramming process of iPSCs, altering several genes in different cell lines (60). These aberrations, however, were suggested to diminish over time and passaging (61). Other methylation aberrations, usually corelated with cancerous mutations, provide growth advantage and are selected for during culturing (58).

Parental imprinted genes are genes that are expressed from a single allele, while the other allele is methylated and silenced depending on the parental origin. Wang et al. (43) found a decrease in PDX1 targets upon its haploinsufficiency in pancreatic progenitors differentiated cells (43). Two of these genes were *NNAT* and *MEG3*, paternally and maternally imprinted genes, respectively, which are related to insulin synthesis and secretion. This finding is based on comparing isogenic control iPSCs and mutant generated *via* CRISPR editing (43). However, many of the MODY studies are based on mutant and control iPSCs from different genetic and epigenetic backgrounds, such as diseased and healthy family members. In these cases, imprinted genes may be influenced by loss of imprinting aberrations, that consequently cause an expression upregulation through the biallelic expression (62).

Another epigenetic aspect relates to MODY, and a diabetes model is the epigenetic memory affecting differentiation capacity. It was previously shown that hESC and iPSC can be differentiated towards β-like cells in similar efficiencies (21) and that the majority of differences between lines in that context were related to the genetic background of the hPSC donor (63). iPSCs derived from specific somatic cells were thought to differentiate better towards the same cell-type of origin. Specifically, iPSCs generated from human pancreatic β cells were shown to differentiate better than ESC or iPSC derived from fibroblasts towards pancreatic progenitor cells (64). It was proposed that this observation originates from the open chromatin sites in the β cell-derived iPSCs, that were found in genomic regulatory regions related to endoderm and pancreatic islets' development, indicating an epigenetic memory of iPSCs (65, 66). This feature of iPSCs can be beneficial when modeling MODY, since β cells are usually formed and function in patients (6).

Gene Editing

The high rates of point mutations and genetic aberration occurrence in hPSCs and their clonal mode of enrichment, together with differences in differentiation capacity between cell lines, highlight the need for carefully supervised controls.

Many of the MODY studies discussed above used non-isogenic cells as control to the mutated cells, which may affect their conclusions. The introduction of gene editing and, specifically, the CRISPR/Cas9 methodology enables the formation of isogenic lines, with the same genetic background, enabling the study of the exact effect of a mutant gene. This can be done by either introducing a mutation to a healthy control background, or by correcting a mutation in iPSCs generated from patients. MODY models that utilized CRISPR editing used the former option, and moreover, created both monoallelic and biallelic mutations to study dosage dependence effect of the mutated gene (19, 36, 42).

Though the CRISPR system enables many promising applications (20), it has some limitations that should be considered. While the immune response and obstacles in vector delivery are the main pitfalls for its *in vivo* use in regenerative medicine, the off-target and low efficiency are the main disadvantages for *in vitro* disease modeling approaches (67). Off-targets are genomic regions that may be mutated, undesirably, when using Cas9, especially when it is constitutively expressed. It was also suggested that CRISPR editing in hPSCs is sub-optimal in the sense of efficiency, possibly due to TP53-mediated cell cycle arrest and apoptosis which are induced by the Cas9. These obstacles may be tackled by predicting the off-targets, restricting the Cas9 expression and using improved CRISPR protocols (67).

β Cell Differentiation

MODY modeling requires the differentiation of mutant ESCs or iPSCs towards pancreatic cells. As mentioned above, differentiation protocols are diverse and the differentiation capacity varies between cell lines because of genetic and epigenetic background. Since different studies are based on different differentiation protocols, and use different types of cells (and in the case of iPSCs—different reprogramming protocols), the final conclusions may be affected and even cause discrepancies between studies (28, 29). Current protocols generate β-like cells, which are not fully mature *in vitro*, thus limiting the study of MODY mutations' effect on the final stages of maturation and function of mature β cells. During recent years, several groups generated hESCs and iPSCs derived islet-like organoid. Such organoids containing MODY mutations are yet to be done. They could contribute to the study of these mutations' effect on pancreatic cells other than β cells, for example the impact of *HNF1A* mutation on α cells (19) or even on the development of the islet as a whole (13, 26, 45, 49).

CONCLUSIONS

In this review, we discussed the great power, as well as the limitations, of using hPSCs for modeling the monogenic diabetes MODY group of disorders. MODY are relatively rare diseases, underdiagnosed with multiple existing subtypes—modeling all MODY types is an ongoing quest. Indeed, nowadays, hPSCs models were generated only for a few MODY types (**Table 1**). Models generated so far contributed to our knowledge of essential pancreatic TF mode of action, the relation between them, and the mechanisms that cause the disease when those are mutated. These models are essential to studying human pancreatic development.

The field of pancreatic developmental disease modeling in hPSCs is still facing challenges, including the requirement for an improved robust and unified differentiation protocol for the generation of mature functional β-like cells *in vitro*. Modeling disease using hPSCs requires tight monitoring of genetic and epigenetic aberrations that may be acquired during reprogramming, culturing, and gene editing (**Figure 1**). However, the great improvement of gene editing technology as well as the increase in clinical identification of MODY patients will further promote MODY research thus, helping better understanding of human pancreatic development and offering new treatment options for patients.

AUTHOR CONTRIBUTIONS

CB-G wrote the manuscript with support and supervision from NB. All authors contributed to the article and approved the submitted version.

REFERENCES

1. De Franco E. From Biology to Genes and Back Again : Gene Discovery for Monogenic Forms of Beta-Cell Dysfunction in Diabetes. *J Mol Biol* (2020) 432:1535–50. doi: 10.1016/j.jmb.2019.08.016
2. Hattersley AT, Greeley SAW, Rubio-cabezas MPO, Njølstad PR, Mlynarski W, Castano L, et al. ISPAD Clinical Practice Consensus Guidelines 2018: The diagnosis and management of monogenic diabetes in children and adolescents. *Pediatr Diabetes* (2018) 19:47–63. doi: 10.1111/pedi.12772
3. Misra S, Owen KR, Owen KR. Genetics of Monogenic Diabetes : Present Clinical Challenges. *Curr Diabetes Rep* (2018) 18:1–11. doi: 10.1007/s11892-018-1111-4
4. Gaál Z, Balogh I. Monogenic Forms of Diabetes Mellitus. *Experientia Supplementum* (2012) 111:385–416. doi: 10.1007/978-3-030-25905-1_18
5. Fajans SS, Bell GI. MODY: History, genetics, pathophysiology, and clinical decision making. *Diabetes Care* (2011) 34:1878–84. doi: 10.2337/dc11-0035
6. Urakami T. Maturity-onset diabetes of the young (MODY): current perspectives on diagnosis and treatment. *Diabetes Metab Syndr Obes Targets Ther* (2019) 12:1047–56. doi: 10.2147/DMSO.S179793
7. Ufuoma SB. Genetics and Pathophysiology of Maturity-onset Diabetes of the Young (MODY): A Review of Current Trends. *Oman Med J* (2020) 35:e126. doi: 10.5001/omj.2020.44
8. Nkonge KM, Nkonge DK, Nkonge TN. The epidemiology, molecular pathogenesis,diagnosis,and treatment of maturity-onset diabetes of the

young (MODY). *Clin Diabetes Endocrinol* (2020) 5:1–10. doi: 10.1186/s40842-020-00112-5
9. Jang KM. Maturity-onset diabetes of the young : update and perspectives on diagnosis and treatment. *Yeungnam Univ J Med* (2020) 37:13–21. doi: 10.12701/yujm.2019.00409
10. Barbosa P, Metab D, Barbosa RP, Reis AF, Giuffrida FMA. Update on clinical screening of maturity – onset diabetes of the young (MODY). *Diabetol Metab Syndr* (2020) 9:1–14. doi: 10.1186/s13098-020-00557-9
11. Shields BM, Hicks S, Shepherd MH, Colclough K, Hattersley AT, Ellard S. Maturity-onset diabetes of the young (MODY): How many cases are we missing? *Diabetologia* (2010) 53:2504–8. doi: 10.1007/s00125-010-1799-4
12. Hattersley AT, Patel KA. Precision diabetes : learning from monogenic diabetes. *Diabetologia* (2017) 60:769–77. doi: 10.1007/s00125-017-4226-2
13. Shabrina N, Su B, Low J, Onn K, Tai ES, Kee A, et al. New insights into human beta cell biology using human pluripotent stem cells. *Semin Cell Dev Biol* (2020) 103:31–40. doi: 10.1016/j.semcdb.2019.11.004
14. Santosa MM, Low BSJ, Pek NMQ, Teo AKK. Knowledge gaps in rodent pancreas biology: Taking human pluripotent stem cell-derived pancreatic beta cells into our own hands. *Front Endocrinol* (2016) 6:194. doi: 10.3389/fendo.2015.00194
15. Mattis KK, Gloyn AL. From Genetic Association to Molecular Mechanisms for Islet-cell Dysfunction in Type 2 Diabetes. *J Mol Biol* (2020) 432:1551–78. doi: 10.1016/j.jmb.2019.12.045
16. Hart NJ, Powers AC. Use of human islets to understand islet biology and diabetes: progress, challenges and suggestions. *Diabetologia* (2019) 62:212–22. doi: 10.1007/s00125-018-4772-2
17. Haliyur R, Tong X, Sanyoura M, Shrestha S, Lindner J, Saunders DC, et al. Human islets expressing HNF1A variant have defective β cell transcriptional regulatory networks. *J Clin Invest* (2019) 129:246–51. doi: 10.1172/JCI121994
18. Sanyoura M, Jacobsen L, Carmody D, del Gaudio D, Alkorta-Aranburu G, Arndt K, et al. Pancreatic Histopathology of Human Monogenic Diabetes Due to Causal Variants in KCNJ11, HNF1A, GATA6, and LMNA. *J Clin Endocrinol Metab* (2018) 103:35–45. doi: 10.1210/jc.2017-01159
19. Cardenas-Diaz FL, Osorio-Quintero C, Diaz-Miranda MA, Kishore S, Leavens K, Jobaliya C, et al. Modeling Monogenic Diabetes using Human ESCs Reveals Developmental and Metabolic Deficiencies Caused by Mutations in HNF1A. *Cell Stem Cell* (2019) 25:273–289.e5. doi: 10.1016/j.stem.2019.07.007
20. Balboa D, Prasad RB, Groop L, Otonkoski T. Genome editing of human pancreatic beta cell models: problems, possibilities and outlook. *Diabetologia* (2019) 62:1329–36. doi: 10.1007/s00125-019-4908-z
21. Pagliuca FW, Millman JR, Gürtler M, Segel M, Van Dervort A, Ryu JH, et al. Generation of functional human pancreatic beta cells in vitro. *Cell* (2014) 159:428–39. doi: 10.1016/j.cell.2014.09.040
22. Rezania A, Bruin JE, Arora P, Rubin A, Batushansky I, Asadi A, et al. Reversal of diabetes with insulin-producing cells derived in vitro from human pluripotent stem cells. *Nat Biotechnol* (2014) 32:1121–33. doi: 10.1038/nbt.3033
23. Russ HA, Parent AV, Ringler JJ, Hennings TG, Nair GG, Shveygert M, et al. Controlled induction of human pancreatic progenitors produces functional beta-like cells in vitro. *EMBO J* (2015) 34:1759–72. doi: 10.15252/embj.201591058
24. Nair GG, Liu JS, Russ HA, Tran S, Saxton MS, Chen R, et al. Recapitulating endocrine cell clustering in culture promotes maturation of human stem-cell-derived β cells. *Nat Cell Biol* (2019) 21:263–74. doi: 10.1038/s41556-018-0271-4
25. Veres A, Faust AL, Bushnell HL, Engquist EN, Kenty H, Harb G, et al. Charting cellular identity during human in vitro β-cell differentiation. *Nature* (2019) 569:368–73. doi: 10.1038/s41586-019-1168-5
26. Shahjalal H, Dayem AA, Lim KM, Jeon T, Cho S. Generation of pancreatic β cells for treatment of diabetes: advances and challenges. *Stem Cell Res Ther* (2018) 9:355. doi: 10.1186/s13287-018-1099-3
27. Teo AKK, Windmueller R, Johansson BB, Dirice E, Njolstad PR, Tjora E, et al. Derivation of human induced pluripotent stem cells from patients with maturity onset diabetes of the young. *J Biol Chem* (2013) 288:5353–6. doi: 10.1074/jbc.C112.428979
28. Vethe H, Bjørlykke Y, Ghila LM, Paulo JA, Scholz H, Gygi SP, et al. Probing the missing mature β-cell proteomic landscape in differentiating patient iPSC-derived cells. *Sci Rep* (2017) 7:4780. doi: 10.1038/s41598-017-04979-w
29. Ng NHJ, Vallier L, Hoon S, Jasmen JB, Lim CS, Lau HH, et al. HNF4A haploinsufficiency in MODY1 abrogates liver and pancreas differentiation from patient- derived induced pluripotent stem cells. *ISCIENCE* (2019) 16:192–205. doi: 10.1016/j.isci.2019.05.032
30. Braverman-Gross C, Nudel N, Ronen D, Beer NL, Mccarthy MI, Benvenisty N. Derivation and molecular characterization of pancreatic differentiated. *Stem Cell Res* (2018) 31:16–26. doi: 10.1016/j.scr.2018.06.013
31. Stepniewski J, Kachamakova-Trojanowska N, Ogrocki D, Szopa M, Matlok M, Beilharz M, et al. Induced pluripotent stem cells as a model for diabetes investigation. *Sci Rep* (2015) 5:8597. doi: 10.1038/srep08597
32. Griscelli F, Ezanno H, Soubeyrand M, Feraud O, Oudrhiri N, Bonnefond A, et al. Generation of an induced pluripotent stem cell (iPSC) line from a patient with maturity-onset diabetes of the young type 3 (MODY3) carrying a hepatocyte nuclear factor 1-alpha (HNF1A) mutation. *Stem Cell Res* (2018) 29:56–9. doi: 10.1016/j.scr.2018.02.017
33. Yabe SG, Nishida J, Fukuda S, Takeda F, Nasiro K, Yasuda K, et al. Expression of mutant mRNA and protein in pancreatic cells derived from MODY3- iPS cells. *PloS One* (2019) 14:e0217110. doi: 10.1371/journal.pone.0217110
34. Wang X, Chen S, Burtscher I, Sterr M, Hieronimus A, Machicao F, et al. Generation of a human induced pluripotent stem cell (iPSC) line from a patient carrying a P33T mutation in the PDX1 gene. *Stem Cell Res* (2016) 17:273–6. doi: 10.1016/j.scr.2016.08.004
35. Wang X, Chen S, Burtscher I, Sterr M, Hieronimus A, Machicao F, et al. Generation of a human induced pluripotent stem cell (iPSC) line from a patient with family history of diabetes carrying a C18R mutation in the PDX1 gene. *Stem Cell Res* (2016) 17:292–5. doi: 10.1016/j.scr.2016.08.005
36. Zhu Z, Li QV, Lee K, Rosen BP, Lez FG, Soh CL, et al. Genome Editing of Lineage Determinants in Human Pluripotent Stem Cells Reveals Mechanisms of Pancreatic Development and Diabetes. *Cell Stem Cell* (2016) 18:755–68. doi: 10.1016/j.stem.2016.03.015
37. Yabe SG, Iwasaki N, Yasuda K, Hamazaki TS, Konno M, Fukuda S, et al. Establishment of maturity-onset diabetes of the young-induced pluripotent stem cells from a Japanese patient. *J Diabetes Investig* (2015) 6:543–7. doi: 10.1111/jdi.12334
38. Teo AKK, Lau HH, Valdez IA, Dirice E, Tjora E, Raeder H, et al. Early developmental perturbations in a human stem cell model of MODY5/HNF1B pancreatic hypoplasia. *Stem Cell Rep* (2016) 6:357–67. doi: 10.1016/j.stemcr.2016.01.007
39. Griscelli F, Feraud O, Ernault T, Turhan AG. Generation of an induced pluripotent stem cell (iPSC) line from a patient with maturity-onset diabetes of the young type 13 (MODY13) with a the potassium inwardly-rectifying channel, subfamily J, member 11 (KCNJ11) mutation. *Stem Cell Res* (2017) 23:178–81. doi: 10.1016/j.scr.2017.07.023
40. Ghila L, Bjørlykke Y, Legøy TA, Vethe H. Bioinformatic Analyses of miRNA–mRNA Signature during hiPSC Differentiation towards Insulin-Producing Cells upon HNF4α Mutation. *Biomedicines* (2020) 8:179. doi: 10.3390/biomedicines8070179
41. Stoffel M, Duncan SA. The maturity-onset diabetes of the young (MODY1) transcription factor HNF4α regulates expression of genes required for glucose transport and metabolism. *Proc Natl Acad Sci U.S.A.* (1997) 94:13209–14. doi: 10.1073/pnas.94.24.13209
42. Wang X, Sterr M, Burtscher I, Böttcher A, Beckenbauer J, Siehler J, et al. Point mutations in the PDX1 transactivation domain impair human b -cell development and function. *Mol Metab* (2019) 24:80–97. doi: 10.1016/j.molmet.2019.03.006
43. Wang X, Sterr M, Burtscher I, Chen S, Hieronimus A, Cernilogar FM, et al. Genome-wide analysis of PDX1 target genes in human pancreatic progenitors. *Mol Metab* (2018) 9:57–68. doi: 10.1016/j.molmet.2018.01.011
44. Lau HH, Hui N, Ng J, Sai L, Loo W, Jasmen JB, et al. The molecular functions of hepatocyte nuclear factors – In and beyond the liver. *J Hepatol* (2018) 68:1033–48. doi: 10.1016/j.jhep.2017.11.026
45. Gaertner B, Carrano AC, Sander M. Human stem cell models: lessons for pancreatic development and disease. *Dev Dis* (2019) 33:1475–90. doi: 10.1101/gad.331397.119
46. Johnson AF, Nguyen HT, Veitia RA. Causes and effects of haploinsufficiency. *Biol Rev Camb Philos Soc* (2020) 94:1774–85. doi: 10.1111/brv.12527
47. Morrill SA, Amon A. Why haploinsufficiency persists. *Pnas* (2019) 116:11866–71. doi: 10.1073/pnas.1900437116
48. Matharu N, Rattanasopha S, Tamura S, Maliskova L, Wang Y, Bernard A, et al. CRISPR-mediated activation of a promoter or enhancer rescues obesity

caused by haploinsufficiency. *Science* (2019) 363:eaau0629. doi: 10.1126/science.aau0629
49. Balboa D, Saarimaki-Vire J, Otonkoski T. Human Pluripotent Stem Cells for the Modeling of Pancreatic β -Cell Pathology. *Stem Cells* (2019) 37:33–41. doi: 10.1002/stem.2913
50. Galcheva S, Demirbilek H, Al-khawaga S, Hussain K. The Genetic and Molecular Mechanisms of Congenital Hyperinsulinism. *Front Endocrinol* (2019) 10:1–18. doi: 10.3389/fendo.2019.00111
51. Paquet D, Kwart D, Chen A, Sproul A, Jacob S, Teo S, et al. Efficient introduction of specific homozygous and heterozygous mutations using CRISPR/Cas9. *Nature* (2016) 533:125–9. doi: 10.1038/nature17664
52. Avior Y, Sagi I, Benvenisty N. Pluripotent stem cells in disease modelling and drug discovery. *Nat Rev Mol Cell Biol* (2016) 17:170–82. doi: 10.1038/nrm.2015.27
53. Weissbein U, Benvenisty N, Ben-david U. Genome maintenance in pluripotent stem cells. *J Cell Bilogy* (2014) 204:153–63. doi: 10.1083/jcb.201310135
54. Lund RJ, Närvä E, Lahesmaa R. Genetic and epigenetic stability of human pluripotent stem cells. *Nat Rev Genet* (2012) 13:732–44. doi: 10.1038/nrg3271
55. Keller A, Dziedzicka D, Zambelli F, Markouli C, Sermon K, Spits C. Genetic and epigenetic factors which modulate differentiation propensity in human pluripotent stem cells. *Hum Reprod Update* (2018) 24:162–75. doi: 10.1093/humupd/dmx042
56. Assou S, Girault N, Plinet M, Bouckenheimer J, Sansac C, Combe M, et al. Recurrent genetic abnormalities in human pluripotent stem cells: definition and routine detection in culture supernatant by targeted droplet digital PCR. *Stem Cell Rep* (2020) 14:1–8. doi: 10.1016/j.stemcr.2019.12.004
57. Merkle FT, Ghosh S, Kamitaki N, Mitchell J, Avior Y, Mello C, et al. Human pluripotent stem cells recurrently acquire and expand dominant negative P53 mutations. *Nature* (2017) 545:229–33. doi: 10.1038/nature22312
58. Bar S, Benvenisty N. Epigenetic aberrations in human pluripotent stem cells. *EMBO J* (2019) 38:e101033. doi: 10.15252/embj.2018101033
59. Weissbein U, Plotnik O, Vershkov D, Benvenisty N. Culture-induced recurrent epigenetic aberrations in human pluripotent stem cells. *PloS Genet* (2017) 13:e1006979. doi: 10.1371/journal.pgen.1006979
60. Lister R, Pelizzola M, Kida YS, Hawkins RD, Nery JR, Hon G, et al. Hotspots of aberrant epigenomic reprogramming in human induced pluripotent stem cells. *Nature* (2011) 471:68–73. doi: 10.1038/nature09798
61. Tesarova L, Simara P, Stejskal S, Koutna I. The aberrant DNA methylation profile of human induced pluripotent stem cells is connected to the reprogramming process and is normalized during in vitro culture. *PloS One* (2016) 11:e0157974. doi: 10.1371/journal.pone.0157974
62. Bar S, Schachter M, Eldar-Geva T, Benvenisty N. Large-scale analysis of loss of imprinting in human pluripotent stem cells. *Cell Rep* (2017) 19:957–68. doi: 10.1016/j.celrep.2017.04.020
63. Jacobson EF, Tzanakakis ES. Human pluripotent stem cell differentiation to functional pancreatic cells for diabetes therapies: Innovations, challenges and future directions. *J Biol Eng* (2017) 11:21. doi: 10.1186/s13036-017-0066-3
64. Bar-nur O, Russ HA, Efrat S, Benvenisty N. Epigenetic memory and preferential lineage-specific differentiation in induced pluripotent stem cells derived from human pancreatic islet beta cells. *Cell Stem Cell* (2011) 9:17–23. doi: 10.1016/j.stem.2011.06.007
65. Jacobson EF, Tzanakakis ES. Who will win: induced pluripotent stem cells versus embryonic stem cells for β cell replacement and diabetes disease modeling ? *Curr Diabetes Rep* (2018) 18:133. doi: 10.1007/s11892-018-1109-y
66. Thurner M, Shenhav L, Wesolowska-Andersen A, Bennett AJ, Barrett A, Gloyn AL, et al. Genes associated with pancreas development and function maintain open chromatin in iPSCs generated from human pancreatic beta cells. *Stem Cell Rep* (2017) 9:1395–405. doi: 10.1016/j.stemcr.2017.09.020
67. Hsu MN, Chang YH, Truong VA, Lai PL, Nguyen TKN, Hu YC. CRISPR technologies for stem cell engineering and regenerative medicine. *Biotechnol Adv* (2019) 37:8. doi: 10.1016/j.biotechadv.2019.107447

3

Genome Editing Human Pluripotent Stem Cells to Model β-Cell Disease and Unmask Novel Genetic Modifiers

Matthew N. George[1], Karla F. Leavens[1,2,3] and Paul Gadue[1,4]*

[1] Center for Cellular and Molecular Therapeutics, The Children's Hospital of Philadelphia, Philadelphia, PA, United States,
[2] Department of Pediatrics, Perelman School of Medicine, University of Pennsylvania, Philadelphia, PA, United States,
[3] Division of Endocrinology and Diabetes, The Children's Hospital of Philadelphia, Philadelphia, PA, United States,
[4] Department of Pathology and Laboratory Medicine, The Children's Hospital of Philadelphia, Philadelphia, PA, United States

***Correspondence:**
Paul Gadue
gaduep@chop.edu

A mechanistic understanding of the genetic basis of complex diseases such as diabetes mellitus remain elusive due in large part to the activity of genetic disease modifiers that impact the penetrance and/or presentation of disease phenotypes. In the face of such complexity, rare forms of diabetes that result from single-gene mutations (monogenic diabetes) can be used to model the contribution of individual genetic factors to pancreatic β-cell dysfunction and the breakdown of glucose homeostasis. Here we review the contribution of protein coding and non-protein coding genetic disease modifiers to the pathogenesis of diabetes subtypes, as well as how recent technological advances in the generation, differentiation, and genome editing of human pluripotent stem cells (hPSC) enable the development of cell-based disease models. Finally, we describe a disease modifier discovery platform that utilizes these technologies to identify novel genetic modifiers using induced pluripotent stem cells (iPSC) derived from patients with monogenic diabetes caused by heterozygous mutations.

Keywords: IPS (induced pluripotent stem) cell, pluripotent stem cell (PSC), beta cell (β cell), diabetes, disease modifier, MODY (mature onset diabetes of the young), candidate gene approach, GWAS (genome-wide association study)

INTRODUCTION

Diabetes mellitus is a worldwide healthcare problem that is rapidly increasing in prevalence. In the United States alone, over 10% of the population is affected, with approximately 1.5 million Americans newly diagnosed with diabetes each year (1). Particularly troubling is the dramatic increase in the incidence of diabetes in children, the consequences of which are expected to lead to increased complications and comorbidity as adults (2, 3). With obesity rates projected to increase in the United States over the upcoming decades, there is little chance that the trend of increasing diabetes prevalence will reverse itself (4). In addition to decreasing quality of life, diabetes is associated with significant morbidity and mortality, including retinopathy, neuropathy, cardiovascular and kidney disease (5). Diabetes also puts individuals at risk of having more complicated courses of common illnesses, with recent studies documenting increased morbidity and mortality in patients with diabetes who contracted COVID-19 (6–8).

Though often referred to as a single condition, diabetes is likely many overlapping diseases, with most stemming from pancreatic β-cell dysfunction and the disruption of glucose homeostasis due to abnormal insulin secretion and/or responsiveness (9). The two most common forms of diabetes, type I (T1D) and type II (T2D), are associated with the eventual loss of insulin-secreting pancreatic β-cells, which can occur either early (T1D) or late (T2D) in disease progression. In the case of T1D, autoimmune destruction results in β-cell death and subsequent insulinopenia, although there is increasing support for the role of β-cell stress in T1D onset (10, 11); T2D is characterized by a combination of peripheral insulin resistance and inadequate β-cell compensation, resulting in a metabolic syndrome that leads to eventual β-cell exhaustion and loss of β-cell mass (12, 13). T1D and T2D display a multifactorial etiology on both a population and individual level, likely motivated by a complex combination of genetic, epigenetic, and environmental factors (14). Furthermore, there is substantial heterogeneity within the underlying β-cell pathophysiology of each of these disorders (9).

Apart from the two major types, there are also 14 known types of monogenic diabetes, historically called MODYs (mature onset diabetes of the young), which are caused by single gene mutations that result in β-cell dysfunction (15–18). Monogenic diabetes is typically underrecognized and underdiagnosed, with identified subtypes likely making up 2-5% of all diabetes cases while additional causative genes undoubtably remain to be discovered (17). To complicate matters further, a number of MODY genes have been associated with the development of T1D and T2D (19–21), suggesting that the pathophysiology of the diabetes subtypes can often overlap. For example, there are additional genes that can cause neonatal diabetes through downstream impacts on pancreatic development, such as GATA6 and NKX2.2, which are not traditionally included as MODYs, although overlap in disease pathology occurs depending on the timing of their presentation (22). There are also numerous monogenic syndromes that have diabetes as a component in all or some affected individuals, including cystic fibrosis and Friedreich's ataxia. While the underlying pathogenesis of monogenic and neonatal diabetes predominantly involves intrinsic defects of the β-cell, syndrome-associated forms of diabetes may result from both peripheral and β-cell defects, with the latter being understudied in many cases (15, 16). Better studies of all forms of diabetes are necessary to understanding the underlying pathophysiology of this complex disease.

The Need for Human β-Cell Models in Diabetes Research

For decades, studies using mouse models greatly advanced our knowledge of diabetes (23). Mice are inexpensive relative to larger animal models (i.e., non-human primates), recapitulate human disease more closely than more basal organismal models, and were genetically-manipulatable even before the invention of modern genome editing techniques. As a result, mouse models continue to be incredibly valuable for the study of whole-body physiology, capable of providing complex metabolic readouts of multiorgan systems, as is the case in oral glucose tolerance tests and hyperinsulinemic euglycemic clamp studies. However, while mouse models play an essential role in the study of diabetes, important differences in rodent versus human physiology have sometimes limited the translatability of rodent datasets to the complex presentation of the human diabetes subtypes. Therefore, the development of *in vitro* human β-cell models will provide an important adjunct to *in vivo* rodent models in the study of diabetes.

While a variety of T1D and T2D mouse models exist, none have been able to comprehensively mimic human diabetes (23–25). For T1D, both genetic and chemically-induced models are used, with the latter resulting in β-cell destruction and insulinopenia, but neither method allowing for the study of the autoimmune processes that drive disease pathophysiology (24, 25). Even mouse models with an autoimmune component do not exactly resemble T1D due to interspecies differences, including well described mechanisms for the regulation of immune cell activation, homing, and target cell interactions (26). As a result, diabetes manifests differently in the two species: for example, the commonly-used non-obese diabetic (NOD) mouse strain exhibits pronounced insulitis and rapid β-cell destruction, while β-cell loss in human T1D is associated with gradual and milder infiltration of islets (27). In the case of T2D, there are a myriad of diet-induced obesity and genetic models that can be used (23, 25). However, the complex polygenic nature of human obesity can be difficult to model using inbred mice strains, and observed effects of sex, age, and epigenetic factors on diet-induced obesity may not be the same between species, though these differences may provide some insight into genetic loci that result in phenotypic variation (23).

While mouse models have yielded significant insights into monogenic causes of diabetes, they do not always fully recapitulate the human disease. One example involves GATA6 and GATA4, members of the GATA family of transcription factors, which are the most highly expressed isoforms in the pancreas. Heterozygous, largely *de novo*, mutations in GATA6 are the most common cause of pancreatic agenesis, resulting in neonatal diabetes as described by multiple groups (28–37). Though less common, GATA4 haploinsufficiency can also result in neonatal and early childhood-onset diabetes (38, 39). However, in mice, GATA4 and GATA6 appear to be completely redundant isoforms, as the loss of a single allele of either *GATA4* or *GATA6* does not appear to impact pancreatic development or glucose homeostasis, and the loss of three of the four Gata4/6 alleles is needed to recapitulate the human phenotype (40, 41). Another example is the most common form of monogenic diabetes, MODY3, caused by heterozygous mutations in the transcription factor HNF1α. Mice with heterozygous mutations in *HNF1A* are healthy (42) and mice with *HNF1A* null mutations can have a diabetic phenotype, but with significant variability dependent on genetic background (43). These results suggest that there are complex, human-specific genome-phenotype interactions that must be additionally investigated using human models. Therefore, the combined and complementary use of both *in vitro* human β-cell models and *in vivo* rodent models will greatly advance our knowledge of the pathophysiology of diabetes.

The Emergence of the Stem Cell-Derived β-Cell as a Model

While mouse models have and certainly will continue to advance our knowledge of diabetes, β-cell-centric research in diabetes has unfortunately been hindered by the lack of human models. Most immortalized β-cell lines are rodent-derived, though several human lines, including the EndoC-βH1 (and subsequent βH2 and βH3) line, are becoming more widely used (44, 45). However, β-cell lines can exhibit differences from *in vivo* human β-cells due to their immortalized status, can be difficult to genetically manipulate, and cannot be used to study β-cell development. While the use of cadaveric human islets in research has greatly expanded due to the success of programs such as the Network for Pancreatic Organ Donors with Diabetes (https://www.jdrfnpod.org) and the Integrated Islet Distribution Program (https://iidp.coh.org), these resources unfortunately remain scarce and precocious. Furthermore, the genetic, epigenetic, and environmental characteristics of donors are largely unknown, while islets themselves cannot be genetically manipulated efficiently.

To address these limitations, great strides have been made over the last two decades in the development of human pluripotent stem cells (hPSC), a term which includes both embryonic stem cells (ESC) and induced pluripotent stem cells (iPSC) [reviewed in (46, 47)]. Efficient methods for the production of hPSCs have completely changed the face of biomedical research and have opened new avenues of therapeutic development for a multitude of diseases. The subsequent development of techniques to differentiate hPSCs into pancreatic β-cells have enormous potential to contribute to the study of diabetes. Built upon foundational research within mouse developmental biology [reviewed in (48)], modern techniques of stem cell differentiation leverage known inductive signals that drive development *in vivo* by replicating these signals temporally and spatially to drive development *in vitro* [reviewed in (49)]. The first lab-guided hPSC differentiation protocols were developed to generate definitive endoderm (50), followed quickly by protocols capable of driving hPSCs towards pancreatic progenitors and endocrine cells (51). While these protocols initially required that pancreatic progenitors be transplanted into mice to mature into functional β-cells, current protocols can achieve functionality *in vitro* without transplantation (52–54) This field has become robust with technical advancements being published regularly by laboratories around the world, resulting in the generation of β-like cells that are closer and closer to their natural counterparts, though continued optimization of functionality is required (55–62).

With advances in directed *in vitro* differentiation, stem cell-derived β-cells provide a tremendous opportunity to study pancreatic development and endocrine diseases in a human model system, particularly when combined with recent advances in genome editing technology. Using clustered regularly interspaced short palindromic repeats (CRISPR)-Cas9 technology (63, 64) (see *Section 2*), targeted mutations in hESC lines can be made, generating mutant and control isogenic lines that avoid confounding results due to differing genetic backgrounds. In addition, iPSCs can be generated from reprogramed patient donor blood or skin fibroblasts (46, 47), resulting in a unique platform within which to study the specific contribution of single mutations to β-cell development and/or function. While these systems certainly have caveats, including expense, labor-intensiveness, and lab-to-lab variability, the expanding use of stem cell-derived β-cells stands to drive our knowledge of diabetes pathophysiology forward beyond the prior limitations of mouse and cell line models. In this review, we will review the current methods of genome editing in hPSCs, discuss how this can be applied to the evaluation of candidate genes in the study of diabetes, and examine the use of stem cell-derived β-cells as a platform for the identification of novel genetic modifiers of diabetes.

GENOME EDITING IN hPSCs

The development of genome editing technologies capable of selectively targeting sites within the human genome has revolutionized our ability to investigate the genetic underpinnings of disease. In the case of diabetes, ESCs and patient-derived iPSCs from multiple genetic backgrounds can now be genetically edited and paired with lab-guided differentiation protocols to build powerful and scalable cell-based models of multiple diabetes subtypes (65, 66). Genome modifications in each system are achieved through nuclease localization with a target sequence, the induction of a double stranded DNA break (DSB), and the activation of endogenous cellular DNA repair mechanisms such as homologous recombination (HR) and non-homologous end-joining (NHEJ) (67, 68). Several types of gene modification can be accomplished through these mechanisms, including (1) 'gene disruption' through the addition/subtraction of short nucleotide sequences and frame shift mutation induction (2), 'gene correction' through targeted base substitutions that restore gene function using a homologous donor DNA construct as a template, and (3) 'gene addition' through the introduction of a complete transgene into a specific locus. Here we briefly review several of the most popular methods for the selective editing of hPSCs, each of which exhibit advantages and disadvantages when editing specific cell types (69).

Zinc Finger Nuclease (ZFN) and Transcription Activator-Like Effector Nuclease (TALEN)

Zinc finger nuclease (ZFN) and transcription activator-like effector nuclease (TALEN) are structurally similar in that they both rely on the C-terminal *Fok*I endonuclease domain to generate DSBs within a targeted sequence (70–72). ZFN architecture combines multiple zinc finger protein DNA-binding domains (motifs) (73) with the nuclease domain of the *Fok*I restriction enzyme that performs optimally when targeting long (12-18 bp) and unique sequences within the eukaryotic genome (74). In contrast, TALENs employ multiple transcription activator-like effector (TALE) DNA binding domains – a class of proteins isolated from the *Xanthomonas* bacteria that have evolved to alter the transcription of host plants (75). In either case, the two distinct regions of the nuclease each perform a unique function, with zinc finger motifs or TALEs binding to DNA while the *FokI* nuclease domain induces a DSB within a target sequence upon dimerization (76, 77).

While structurally similar, there are distinct advantages and disadvantages of each system. Typically, ZFN-based platforms afford greater flexibility in targeting, while also allowing for independent optimization of the two subunits for simplified retargeting (78). Drawbacks of ZFNs include the cost of application, a suite of complex design constraints that must account for context-dependent interactions between fingers (79), and a higher prevalence off-target effects and translocations than other methods (80). In contrast, the highly conserved stretches of 33-35 amino acids (AA) that TALEN-based approaches employ addresses many of the design complexity concerns of ZFNs, while maintaining high cleavage activity rates, a broad targeting range specificity, and improved cytotoxicity (81). However, TALEN-based approaches have been shown to produce off-target effects and suffer from dramatically lower efficiencies when targeting sequences that are methylated or do not include thymidine (82, 83). Comprehensive reviews that provide specific recommendations for the design and reproducible integration of ZFN (78, 84) and TALEN-based (85) genome editing approaching in hESCs and iPSCs are available elsewhere.

Clustered Regularly Interspaced-Short Palindromic Repeats (CRISPR)/Cas

(CRISPR)/Cas-based gene editing platforms have become incredibly popular tools to modify the genome of hPSCs since the introduction of the technology in 2012 (86). Based on the adaptive immune system of bacteria and archaea, CRISPR was made possible by the discovery of DNA fragments within the *E. coli* genome from past viral and bacteriophage invaders known as clustered regularly interspaced short palindromic repeats (CRISPRs). Unlike ZFNs and TALENs which use proteins, CRISPR loci are transcribed during viral infections to produce an RNA-guided DNA endonuclease that selectively binds and cuts invading viral DNA (87). CRISPR sequences exhibit a repetitive pattern, wherein short DNA sequences (24-48 bp) are followed by their reverse complement and a protospacer that matches part of the viral genome. Through coordination with RNase III, CRISPR-associated (Cas) proteins, and trans-activating CRISPR RNA (tracrRNA), long RNA transcribed from CRISPR loci are cleaved into short, spacer-derived RNA (crRNA) (88). TracrRNA and crRNA then act together to guide the Cas9 protein to a target cut site located within the genome of an invading virus, causing a DSB [for a review, see (89)].

When used in genome-editing platforms, tracrRNA and crRNA can now be combined into a single "guide RNA" (gRNA) molecule (86) and administered with the Cas9 protein to selectively cut target DNA sequences (90). Multiple CRISPR/Cas9 systems have been developed specifically for hPSCs that are capable of editing or replacing genome sequences (91, 92) and are quickly replacing TALEN-based systems due to their ease of generation, efficiency, and cytocompatibility (93, 94). Drawbacks of CRISPR-based methods include the re-cutting of target regions after DSB repair and the prevalence of erroneous insertion or deletions (indels) on the non-targeted allele, making the generation of single allele edits difficult. Recent protocols, including one from our laboratory (95), address these issues through the use of two single stranded oligonucleotide repair templates, with one expressing the desired sequence change and the other maintaining the normal sequence. These repair templates also contain silent mutations that prevent gRNA recognition and re-cutting, facilitating the selective editing of a single allele with an average efficiency of close to 10%. In addition, off target cutting of CRISPR/Cas9 at other sites in the genome is also an issue but it can be mitigated. First, in the hPSC system, off target cutting is less prevalent than in somatic cells, mostly likely due to the fact the pluripotent stem cells are very sensitive to DNA damage and cells that have undergone multiple DNA cuts are less likely to survive (96). Second, careful design and testing of potential off target sites can also be used to minimize the impact of off target cutting. Overall, genome editing technologies in hPSCs are advancing to the stage where virtually all coding mutations can be repaired or introduced in a single allele manner, an important advance given the majority of monogenic genetic diseases of the pancreas are caused by heterozygous mutations.

THE ROLE OF CANDIDATE GENE AND GWAS APPROACHES IN THE STUDY OF DIABETES

The availability of standardized laboratory protocols for the generation of glucose responsive β-cells from hPSCs that have undergone selective genome editing has the potential to dramatically increase our knowledge of genes that contribute to diabetes. Traditionally, the functional contribution of genes to disease states has come from the deletion or mutation of a single gene target. The use of candidate genes, chosen as they are known clinically to cause disease, has been employed for the study of monogenic diabetes. However, given the increasing ease of genome and exome sequencing, comparisons of genetic variants between diabetic and non-diabetic populations through GWAS analysis using genome sequencing has generated a large list of variants associated with all forms of diabetes, most of which are in non-coding regions of the genome (97–100). Newly developed stem cell platforms can be used to target these variants, initially by targeted mutation of the gene thought to be regulated by a given variant (101). The direct targeting of a non-coding variant has been studied in neonatal diabetes caused by GATA6 (102), and similar approaches could also be used for variants associated with more common causes of diabetes. Care will need to be taken as it is possible that most non-coding variants may have a small impact on gene expression and disease penetrance on their own. While we are still in the early stages of utilizing these approaches in stem cell-derived β-cells to interrogate the roles of specific genes in β-cell development and function, we predict that this will become more commonplace and contribute to our understanding of β-cell physiology and disease.

Studying Known Causes of Monogenic Diabetes

There are dozens of types of diabetes caused by single gene mutations, including monogenic, neonatal and syndrome-associated diabetes (15–17, 103–105). Monogenic forms of diabetes are often caused by heterozygous coding mutations within genes that influence β-cell function [reviewed in (106)]. Many forms of monogenic diabetes have been recognized for decades and more gene causes are likely to be found in the upcoming years with the increasing use of exome and genome sequencing (17, 18). In addition, many of the genes associated with monogenic diabetes have numerous reported mutations with potentially different consequences on protein function and therefore on β-cell dysfunction (107–109). Stem cell-derived β-cells may provide a platform in which to investigate some of these polymorphisms, and may help to provide some insight into genotype-phenotype correlations.

The modeling of monogenic diabetes using stem cell-derived β-cells has been extensively described. To date, hPSC lines have been made with mutations in HNF1A (110–114), HNF1B (112, 115, 116), HNF4A (112, 117–119), PDX1 (107, 120, 121), KCNJ11 (122), GCK (112), and CEL (112). Of the studies listed, only two used genome-editing to make mutant hESC lines (113, 121), with the remainder generating patient-derived iPSC lines. While some studies simply described the derivation of iPSC lines from patient samples, most publications included lab-guided differentiations to β-cells and subsequent studies on β-cell gene expression and biology (111, 113–115, 117–119, 121).

MODY3, caused by heterozygous mutations in the transcription factor HNF1α, is the most common form of monogenic diabetes (15, 103) and is currently the most extensively-researched monogenic form of diabetes using stem cell-derived β-cells. HNF1α has been of particular interest because of its additional association with T1D and T2D in several large population studies (19–21, 123). Several studies on the role of HNF1A has been performed in mice, but mouse models with heterozygous deletion of *HNF1A* do not have diabetes and thus have provided limited insight into MODY3 (124). To date, three publications from different groups have modeled HNF1A-deficiency in stem cell derived β-cells, with two employing patient-derived iPSC lines (111, 114) and the other using hESC lines (113). These studies have yielded significant insights into the complex role HNF1α plays in controlling β-cell development, metabolism and function and have discovered new downstream targets of this transcription factor that had not previously been identified in mouse studies.

While the underlying pathogenesis of monogenic diabetes results from intrinsic β-cell defects, the role of β-cell dysfunction in many syndrome-associated forms of diabetes, such as cystic fibrosis (CF)-related diabetes, is largely unknown (105). There is significant interest in these fields to generate syndrome-related iPSC lines for use in multiple areas, but this will ultimately aid in the study of rare causes of diabetes by providing accessible resource lines. Groups have already generated stem cell derived β-cells to model β-cell dysfunction in Wolfram syndrome (125, 126) and Friedreich's ataxia (127). In addition, CF iPSC lines have been made and differentiated in the pancreatic ductal endothelium to study the effects of CF-related pancreatic exocrine function (128). These lines and others generated for non-diabetes study can always be used to produce stem cell derived β-cells and advance our knowledge of these understudied forms of diabetes.

Another avenue of study using stem cell derived β-cells as a model is to focus on genes that are thought to play a role in β-cell development, identity, or function but that may not have been described as monogenic causes of disease. In an impressive paper by Zhu and colleagues, researchers used genome editing to generate hPSCs knockout lines to further probe the role of 8 known pancreatic transcription factors, including PDX1, RFX6, PTF1A, GLIS3, MNX1, NGN3, HES1 and ARX (121). Many of these factors had only been studied previously in rodent models and, through lab-directed differentiation, their role in human β-cell development and function could be interrogated for the first time. This reverse candidate approach using stem cell derived β-cells will provide a significant basis for future advances.

As genome editing techniques improve and become more well-established, the field is turning more to the use of isogenic lines in which to study the contribution of a specific genes on β-cell physiology. This involves making targeted mutations in hESC lines or correcting mutations in patient-derived iPSC lines, generating mutant and control isogenic lines that avoid confounding results due to differing genetic backgrounds. In all studies above using patient-derived iPSC lines to study monogenic diabetes, the mutant stem cell-derived β-cells were compared to unaffected family members or unrelated wild type iPSC lines and not isogenic controls (111, 114, 115, 117–119). With the increasing use of CRISPR/Cas9 technology, we advocate the use of at least 2 pairs of isogenic lines, examining a single clone for each, for interrogating the impact of a given genome alteration. Alternatively, if using a single stem cell line, the examination of several genome edited clones has been suggested by leading stem cell journals such as *Stem Cell Reports* (Information for authors). We would argue that 2 isogenic pairs is superior because it controls for both artifact due to an acquired mutation in a single genome edited clone as well as confirm any phenotype is general enough to be seen in 2 independent genetic backgrounds. The use of a single edited clone per isogenic pair we feel is a good tradeoff between the effort required to differentiate and analyze these clones while still minimizing the impacts of clonal artifacts.

Leveraging GWAS to Identify and Validate Candidate Genes in Diabetes

The cause of T1D and T2D is likely a complex combination of genetic, epigenetic and environmental factors (14). In addition, there is substantial heterogeneity within each of these disorders, so that the disease-causing combination in each affected individual is slightly different (9). Therefore, a single gene-to-disease strategy is not always effective in the study of T1D and T2D. Technological advancements in next generation sequencing, combined with the targeted efforts of several consortia, continue to expand the size and scope of available genomic datasets from diabetic patients (129–131). Previously, the identification of diabetes-linked genes was the product of quantitative trait mapping (QTL), obtained through the cross of genetically engineered mice (132–134). Over the past decade as the cost and availability of sequencing technology has improved, GWAS have identified more than 60 loci for T1D (135) and more than 240 associated with T2D (136), with the hereditability of each now explaining approximately 15% (137) and 25-80% (138) of the disease-risk for each subtype, respectively.

The explosion of available GWAS datasets for both T1D and T2D can be leveraged by using stem cell derived β-cells. Using this technique, novel genes that are revealed by GWAS, individual exome, or genome sequencing associated with diabetic populations can be targeted for study either in isolation or as part of co-cultures where interactions between adipocytes, immune cells, critical biological components and β-cells are replicated *in vitro* (139–141). Through the use of stem cell derived β-cells, these systems can ascertain whether a specific locus causes β-cell-intrinsic dysfunction, while also probing the contributions of the surrounding environment. For example, polymorphisms in the human leukocyte antigen (HLA) DR and DQ alleles increase T1D-risk by altering T-cell binding [reviewed in (142)], which is predictably a β-cell-extrinsic effect that can be observed in cell culture studies. Alternatively, some polymorphisms in the insulin (INS) gene, a β-cell-specific gene, have been described to influence T1D risk due to changing insulin mRNA production in the thymus altering immune tolerance (143, 144). However, other mutations in INS lead to neonatal diabetes, thought to be caused by β-cell death due to increased cell stress from misfolded insulin protein [reviewed in (145)]. While the difference between polymorphisms and mutations may be determined by their prevalence in the population, modeling these gene differences in stem cell derived β-cells may prove useful for understanding their significance.

While GWAS studies can yield a potential target gene which could be directly involved in disease, sometimes they identify an associated region of unclear significance. GWAS comparing the islets of diabetic and non-diabetic individuals suggest that most T2D-associated variants do not reside in coding regions (146, 147). In order to understand the role of these variants, iPSC banks from T1D, T2D and non-diabetic patients can be used to probe these differences on a multigene scale. Multiple iPSCs from T1D and T2D patients have been made (148–151), and there are consortia and foundations that are focused on making larger banks of available diabetes and non-diabetes iPSC lines, including the Human Islet Research Network (HIRN, https://hirnetwork.org/hpap-overview) and the New York Stem Cell Foundation (NYSCF, https://nyscf.org/research-institute/repository-stem-cell-search/). Several groups have also recently performed lab-directed differentiations on patient-derived iPSC lines to generate stem cell derived β-cells to examine broad molecular and functional differences (150, 151). Unlike the need for the generation of isogenic lines in the study of monogenic diabetes, making banks of T1D, T2D, and non-diabetic stem cell derived β-cells can be used to study many factors contributing to β-cell pathophysiology in diabetes. One caveat of the use of non-isogeneic lines is that differences in genetic background amongst disease and control lines leads to tremendous variability in phenotype and necessitates large sample numbers to dissect the underlying biology.

GENETIC DISEASE MODIFIERS AND DIABETES

The way genetic factors interact with disease can be highly variable. Even in canonical examples of monogenic Mendelian diseases such as cystic fibrosis and sickle-cell anemia where a disease endophenotype is linked to a single mutation (152), fraternal twins that reside within the same household can present vastly discordant phenotypes (153). The results of longitudinal twin studies add to a growing body of clinical evidence that underscores the importance of 'disease modifiers' that alter the penetrance, expressivity, rate of progression, and/or presence of disease endophenotypes through the modification of disease-linked genes (154, 155). While the terminology used to describe the mechanisms of oligogenic inheritance continues to evolve, for the purposes of this review we have chosen to classify genetic disease modifiers within two groups based on their mode of action, or as either 'protein coding' or 'non-protein coding'.

Protein coding disease modifiers typically affect the phenotypic expression of a disease through mutations in the coding sequence of intact genes, leading to changes in protein function (156). These changes can be either sufficient to elicit a diseased state (i.e., a frame shift mutation within a coding sequence of an important functional protein), or can affect the molecular expression of other disease-linked genes through alterations in regulatory DNA such as promoters and enhancers (e.g. modifier genes) (157). In contrast, non-protein coding disease modifiers include non-coding regulatory elements and non-coding RNAs (ncRNA) (158). In either case, disease modifiers can act to either enhance, silence, or modify the expression of genes that can modify the activity of important proteins whose dysregulation result in changes in the penetrance, expressivity, and/or presence of a disease endophenotype.

Diabetes is a complex disease wherein patients express significant heterogeneity in the progression, clinical phenotype, and response to treatment. T1D and T2D show clear evidence of a genetic component and familial reoccurrence (159, 160), with observed associations with lifestyle, obesity, and cancer playing a particularly significant role in T2D (161). In the face of this variability, it is important to note that the direct influence of specific mutations within protein-coding regions on the etiology of diabetes have been described (162, 163). However, while allelic variants have been shown to confer an increased risk of

T1D/T2D, other subtypes of diabetes, such as monogenic diabetes, are causally linked to single mutations, as described above. Apart from changes in coding sequences, there is substantial evidence that disease progression and severity of all forms of diabetes results from the interaction of multiple non-coding genetic, epigenetic, and environmental factors, which act in concert to cause β-cell dysregulation and islet dysfunction (164, 165).

The influence of protein coding and non-protein coding disease modifiers on each of the diabetes subtypes remain active areas of research. In the case of monogenic diabetes, modifier genes have been shown to explain some degree of clinical variability (166), while several studies suggest that non-coding disease modifiers influence the development of gestational diabetes (167, 168). In this section, we provide a brief overview of the two classes of disease modifiers, discussing known associations with the diabetes subtypes when available. To facilitate the further discovery of such mechanisms, we then outline a disease modifier discovery platform that leverages recent advancements in RNA sequencing, genome editing, and laboratory-guided stem cell differentiation to identify genetic disease modifiers of genetic disease caused by heterozygous mutations, using monogenic diabetes as a model. Given the limited availability of research on this topic, it is our hope that the platform outlined here will support the discovery of novel protein coding and non-protein coding disease modifiers that can help explain the heterogeneity observed in diabetes subtypes.

Protein Coding Disease Modifiers of Diabetes

The influence of modifier genes and allelic heterogeneity on human disorders has been the subject of an ongoing discussion within medicine for over a century (169), with multiple parallel avenues of investigation within genetics (e.g. epistasis, oligogenic inheritance) dedicated to understanding the effect of one gene/allele on the phenotypic expression of a second gene/allele (154, 155). In the case of diabetes, there is a growing body of evidence that some subtypes may be the result of oligogenic inheritance, wherein the underlying etiology of the disorder is primarily genetic, but still requires the synergistic action of several genetic modifiers at disparate disease-linked loci (156, 170, 171). In this continuum between classical Mendelian and complex traits, possible protein coding disease modifiers include allelic heterogeneity that results from mutations within disease-linked loci, the activity of modifier genes that regulate others with important roles in glucose homeostasis, and the presence/absence of single nucleotide polymorphisms (SNPs) that are either necessary or sufficient to change the presence, penetrance, expressivity/heritability, or rate of progression of a disease.

As GWAS datasets expand to include sampling of diabetic patients from varied ethnic backgrounds that present different endophenotypes, the technology is poised to assist in the identification of candidate disease-linked genes and SNPs that reside within 'modifier loci' (172). While hundreds of candidate genes that are linked with increased diabetes-risk have been identified, the mechanisms underlying their action often remain unclear. One of the first identified and best understood examples of how genetic protein coding disease modifiers modulate the phenotypic expression of diabetes are the multiple identified polymorphisms within the base pair sequence of the HLA region of chromosome 6p21.3 on T1D (173). In this case, variation within the coding sequence of the HLA DR and DQ alleles produce changes in the amino acid sequence of cell surface receptors, altering their binding affinity to T-cells and increasing T1D-risk [reviewed in (142)].

Apart from polymorphisms in the HLA region, coding mutations within genes that encode important pancreatic transcription factors (TFs) have also been shown to modify the phenotypic expression of diabetes (162, 163). Coding mutations within the TCF7L2 (174), PDX1 (107), HNF1A (108), HNF4A (175), and TM6SF2 (176) genes can result in the dysregulation of blood glucose homeostasis by altering TF expression or imparting direct functional consequences on β-cell or islet function through alterations in a TF's amino acid sequence. For example, hundreds of missense mutations within PDX1 coding regions have been identified, with mutations within the transactivation domain reducing gene activation and impairing both β-cell development and function (107). In the case of HNF1A, 11 rare coding variants have been identified that result in a >40% reduction in transcription and are strongly associated with monogenic diabetes (MODY3) in the general population (108, 109). Similarly, a number of coding SNPs can impart T1D and T2D susceptibility within groups with a shared ancestral heritage, including SNPs in the SUMO4 (177) and MGEA5 (178) genes, identified within Japanese and Mexican populations, respectively.

Non-Protein Coding Disease Modifiers of Diabetes

Recent advances in targeted RNA sequencing technology (RNA Seq, RNA CaptureSeq) have greatly expanded our understanding of transcriptomics (179), underscoring the potential importance of regulatory elements in the control of disease-linked genes (180–182). Rather than coding for a protein directly, non-coding RNA (ncRNA) regulate the transcriptional or post-transcriptional production and modification of proteins. ncRNA make up 98% of the transcripts in the human genome, can be either trans- or cis-acting, and are classified into groups according to their length, morphology, and function (179, 183). 'Short' ncRNAs are less than 200 bp in length and perform a diversity of functions during gene regulation, protein synthesis, and the post-translational modification of proteins. Short ncRNA include nuclear RNAs (snRNAs), small nucleolar RNAs (snoRNAs), micro-RNAs (miRNAs), and transfer RNA (tRNA), to name a only a few (184). In contrast, long non-coding RNAs (lncRNAs) are 200 bps or longer and are generally only involved in the regulation of gene transcription and epigenetic regulation, although in some rare occasions they may produce peptides (185).

To date, the influence of non-protein coding disease modifiers on the pathogenesis of diabetes remains underdefined, providing an exciting avenue for future research. GWAS comparing the islets of diabetic and non-diabetic individuals suggest that most T2D-associated variants do not reside in coding regions (146, 147), adding to a growing, yet sparse, body of evidence that glucose homeostasis is heavily controlled by the activity of non-coding regulatory elements (186). For example, thousands of miRNAs and lncRNAs have been isolated from islets (136), with preliminary evidence suggesting that some miRNA are required for islet development in mice (187) and β-cell function (188, 189). LncRNA in particular have been linked to several important processes in diabetes (190), with overexpression resulting in enhanced cell proliferation and fibrosis in the early state of diabetic nephropathy [LncRNA CYP4B1-PS1-001 (191)].

Non-coding single nucleotide polymorphisms (ncSNPs), or variations in a single DNA base pair that code for non-coding regulatory elements, can also act as disease modifiers of diabetes (182). More than 90% of disease-associated SNPs are located within non-coding regions, resulting in possible functional variants of promoters, enhancers, and ncRNA genes (192). Through this mechanism, ncSNPs within important regulatory regions can alter the splicing, binding, degradation, or sequence of a ncRNA, which in turn can modulate the activity of multiple regulatory elements that act to control other cellular processes, such as transcription factor binding and chromatin states (193, 194). As an example, a recent study from our laboratory that used genome editing to knock-out *HNF1A* in hESCs identified a human-specific lncRNA (*LINKA*) that is a target of HNF1A and is necessary for normal mitochondrial respiration within stem cell-derived β-cells (113). Given that there is recent evidence that islet-specific lncRNA and transcription factors co-regulate genes associated with enhancer clusters (195, 196), we expect that additional studies that expand upon the functional consequences on ncSNPs and the potential targets of lncRNA in human islets have great potential to explain some of the clinical heterogeneity each diabetes subtype (197, 198).

Patient iPSCs as a Diabetes Disease Modifier Discovery Platform for Monogenic Diabetes

The discovery of genetic modifiers of diabetes have been slowed by the complex presentation of the diabetes subtypes, with the cause of each existing on a multi-dimensional continuum of genetic, epigenetic, and environmental factors (14). However, while progress has been hindered in some areas, success has been achieved over the past two decades within monogenic diabetes, with advancements in molecular genetics enabling the definition of discrete etiological disease subtypes that can inform preventative treatments through precision medicine (199). As discussed in section 3, monogenic forms of diabetes result from coding mutations in single identified genes which cause β-cell-intrinsic dysfunction. This has allowed for targeted studies focused on elucidating the role of the specific causative gene in β-cell development and function. However, disease penetrance and presentation can vary among individuals with the same underlying pathogenic mutation, suggesting that additional factors can influence the genotype-phenotype association (15–18).

The multifaceted nature of monogenic diabetes provides a unique opportunity to model gene-phenotype relationships that contribute to endophenotypes seen in the more common forms of the diabetes and aid in the discovery of novel disease modifiers (200, 201). Recent technological advancements in sequencing technology, genome editing (see *Section 2*), and the generation and guided-differentiation of iPSCs (see *Section 1.2*) now provide the foundation for an iPSC-based discovery platform that can identify novel protein coding and non-protein coding genetic elements that modify the presentation and penetrance of endophenotypes. Presented in **Figure 1**, genetic disease modifier discovery begins with the identification of a heterozygous coding mutation that results in monogenic diabetes. Coding variant identification can be done on demand through partial or whole genome sequencing given available information regarding candidate genes (97), through exome sequencing (98), or by leveraging publicly available GWAS datasets that compare non-diabetic and diabetic patients (99) (see *Section 3*). A useful database for monogenic diabetes includes the products of the DIAbetes Genetics Replication And Meta-analysis (DIAGRAM) Consortium (https://www.diagram-consortium.org/).

Upon candidate allele or gene selection, the next step within the disease modifier discovery platform is the production of iPSC lines from diabetic patients with the desired coding mutation (step 2, **Figure 1**). Due to recent technological advancements, iPSCs can easily be generated from adult cells that are harvested from blood or skin tissue (112, 202, 203). Once generated, endodermal cells with the desired coding mutation can be produced from iPSCs through exposure to the inductive signals that drive *in vivo* development (48, 204). To this end, several stepwise protocols that move cells through the multiple stages of pancreatic development *in vitro* over a few weeks have been developed (52–54). This process represents a relatively efficient method for the generation of pancreatic β-cells using tissue from multiple donors that share the identified coding mutation but have varied genetic backgrounds that result in different genetic modifiers, a distinct advantage when addressing observed heterogeneity in phenotypic expression.

After the generation of stable iPSC lines, the next step within the disease modifier discovery platform is to selectively correct the identified coding mutation and compare the resulting gene mRNA and protein expression before and after correction (step 3, **Figure 1**). As described in section 2, there are a number of genome editing technologies available to accomplish line correction, although CRISPR/cas9-mediated systems are becoming the most frequency used within stem cell models (63, 64). The goal of model generation is to compare corrected and non-corrected lines to non-diabetic controls, to which the mRNA and protein expression of the corrected gene can be measured (step 4, **Figure 1**). Throughout this process, one of two outcomes may be observed. *Outcome 1: the correction of the observed coding mutation can result in the complete normalization of gene/protein expression.* This result implies that the coding mutation was completely responsible for the decrease in gene expression and/or function (step 5A, **Figure 1**).

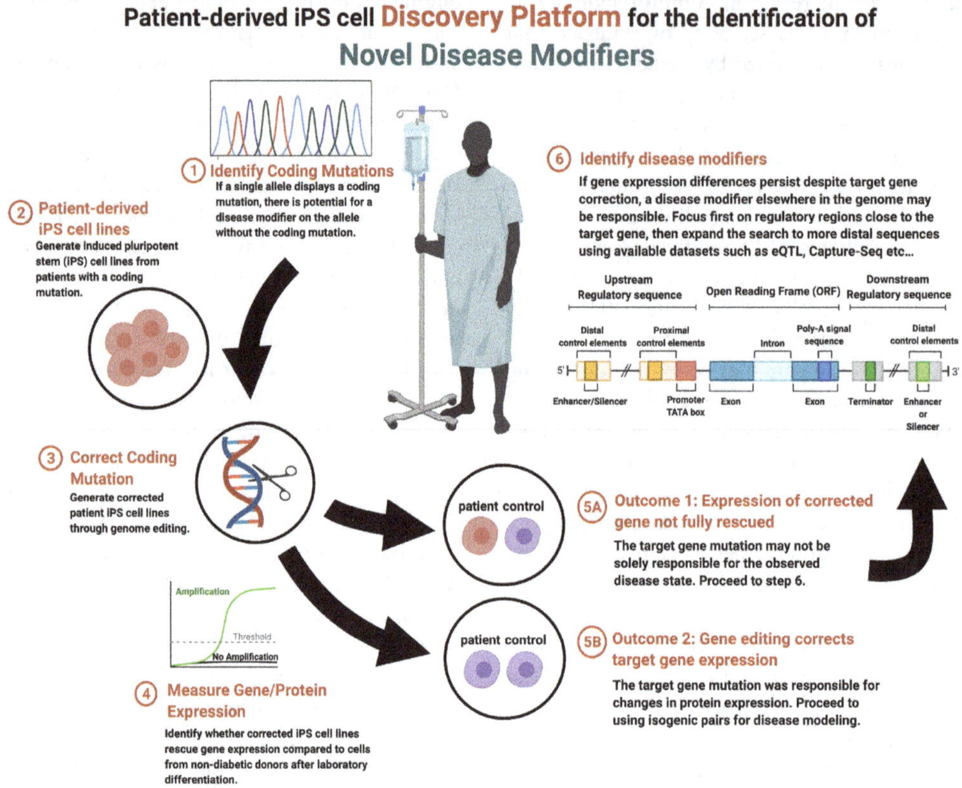

FIGURE 1 | Stepwise flow diagram of the process of genetic disease modifier discovery.

Outcome 2: the expression of the monogenic diabetes gene is not rescued to the levels observed in non-diabetic control cells. In this case, expression is possibly being regulated by a modifier elsewhere in the genome (step 5B, **Figure 1**). It is possible that certain coding mutations may disrupt protein function while not impacting mRNA or protein expression. Such a mutation can still be studied with this platform as regulatory region variants which decrease gene expression may still be detectable when comparing the patient iPSC cell line to control lines.

In the event that the candidate gene expression is not normalized by selective correction of the coding mutation, the target gene may be under the regulatory control of one or more disease modifiers (step 6, **Figure 1**). Disease modifiers can reside either proximally or distally with respect to the coding mutation, as well as either upstream or downstream from the affected gene, making their location difficult to determine. An effective search strategy can be to focus on proximal regulatory regions near the gene of interest first, although if this approach doesn't prove fruitful then there are a number of computational approaches that are specifically designed for the identification of regulatory elements [reviewed in (205)]. Similarly, the sequencing and chromatin mapping efforts of the ENCODE (https://www.encodeproject.org/) (206), Epigenome Roadmap Consortia (https://egg2.wustl.edu/roadmap/web_portal/) (207), and Common Metabolic Diseases Knowledge Portal (https://hugeamp.org) have provided extensive annotation of coding and non-coding regions within the human genome, as well as the likelihood of variants to impact metabolic disease.

Through the use of public databases, it is now possible to determine likely regulatory regions of the target gene of interest that can then be interrogated by targeted sequencing of patient iPS cell lines that could not be completely rescued by correction of the coding mutation. If variants are discovered, they can be studied by genome editing in the context of coding mutations or in isolation to determine the impact on expression of the target gene. For example, this strategy has been successfully used to study pancreas agenesis caused by mutations in GATA6 within our laboratory (102). A non-coding SNP was discovered in a patient iPS cell line that regulated expression of GATA6 during pancreas development *in vitro* and when interrogated in a cohort of patients with the disorder was confirmed to be a disease modifier. This strategy is especially useful in studying variants that impact genetic disease caused by heterozygous coding mutations. Variants that may have only a small influence on gene expression and no impact on disease alone can synergize with heterozygous coding mutations to bring target gene expression below a critical threshold needed for function. This platform does have some limitations including the requirement to focus of monogenic heterozygous disorders and may not be scalable to examine large numbers of genes with current differentiation technologies. We suggest that this methodology could be applied secondarily to any heterozygous iPSC disease modeling project that entails the creation of isogenic corrected lines with minimal additional effort.

CONCLUSION AND FUTURE OUTLOOK

The recent development of techniques to differentiate hPSCs into pancreatic β-cells has opened up new pathways to study the pathogenesis of diabetes. These human-centric models, combined with rapidly advancing genome editing techniques, provide incredibly powerful and scalable platforms in which to study the contribution of genetic elements to β-cell function, while also addressing the limitation of mouse models. Furthermore, the use of hPSCs provide unique opportunities in which to accomplish the targeted study of β-cell dysfunction as well as provide a platform to discover protein coding and non-protein coding genetic modifiers. Given recent evidence that large numbers of disease-linked variants do not reside in coding regions and the presence of variants can be population-specific, iPSC platforms that use patient-derived tissue hold great promise for the discovery of novel genetic disease modifiers that may help to explain the variability seen across and within diabetes subtypes.

While hPSC-based platforms represent a great leap forward in our ability to study β-cell function, there are caveats to their use that must be taken into account. hPSC derived β-cell generation and culture is labor-intensive, requiring approximately 40 days of differentiation and maturation. Additionally, though they do have some degree of insulin secretion in response to glucose and other secretagogues, significant uncertainty regarding their functionality and maturity still exists (66, 208–211). Ideally, these protocols need to be optimized to support the efficiency and accuracy of discovery platforms utilizing stem cell-derived β-cells. However, further fine-tuning of the established differentiation protocols will drive us closer to an ideal *ex vivo* human model of pancreatic β-cells. In addition to improving β-cell function, protocols need to be improved so that they are more universally successful, as certain hPSC lines can more easily differentiated into β-cells than others using current protocols.

The development of more universally-applicable protocols is required as the use of patient iPSC lines expands. There has been a recent flurry of publications that promise improved protocols with better function and wider applicability, and advances will continue to build on those already made (57, 212). Finally, generation of islet cells in platforms combining different stem cell-derived cell types will allow for more complex modeling of the genetic and environmental factors driving all forms of diabetes. Improving our knowledge of pancreatic β-cells function and development in humans is essential for the development of treatments for the millions of people affected by diabetes.

AUTHOR CONTRIBUTIONS

MG and KL conducted the literature review, designed figures, and wrote the manuscript. MG, KL, and PG conceived of the topic and edited the text. All authors contributed to the article and approved the submitted version.

ACKNOWLEDGMENTS

Figure 1 was created using BioRender.com.

REFERENCES

1. Prevention CfDCa. *National Diabetes Statistics Report, 2020*. Atlanta, GA: Centers for Disease Control and Prevention, U.S. Dept of Health and Human Services (2020).
2. Divers J, Mayer-Davis E, Lawrence L, Isom S, Dabelea D, Dolan L, et al. Trends in Incidence of Type 1 and Type 2 Diabetes Among Youths — Selected Counties and Indian Reservations, United States, 2002-2015. *MMWR Morb Mortal Wkly Rep* (2020) 69:161–5. doi: 10.15585/mmwr.mm6906a3
3. Mayer-Davis EJ, Lawrence JM, Dabelea D, Divers J, Isom S, Dolan L, et al. Incidence Trends of Type 1 and Type 2 Diabetes Among Youths, 2002-2012. *N Engl J Med* (2017) 376:1419–29. doi: 10.1056/NEJMoa1610187
4. Ward ZJ, Bleich SN, Cradock AL, Barrett JL, Giles CM, Flax C, et al. State-Level Prevalence of Adult Obesity and Severe Obesity. *N Engl J Med* (2019) 381:2440–50. doi: 10.1056/NEJMsa1909301
5. Huang ES, Brown SE, Ewigman BG, Foley EC, Meltzer DO. Patient Perceptions of Quality of Life With Diabetes-Related Complications and Treatments. *Diabetes Care* (2007) 30:2478–83. doi: 10.2337/dc07-0499
6. Barron E, Bakhai C, Kar P, Weaver A, Bradley D, Ismail H, et al. Associations of Type 1 and Type 2 Diabetes With COVID-19-Related Mortality in England: A Whole-Population Study. *Lancet Diabetes Endocrinol* (2020) 8:813–22. doi: 10.1016/S2213-8587(20)30272-2
7. Holman N, Knighton P, Kar P, O'Keefe J, Curley M, Weaver A, et al. Risk Factors for COVID-19-Related Mortality in People With Type 1 and Type 2 Diabetes in England: A Population-Based Cohort Study. *Lancet Diabetes Endocrinol* (2020) 8:823–33. doi: 10.1016/S2213-8587(20)30271-0
8. Riddle MC, Buse JB, Franks PW, Knowler WC, Ratner RE, Selvin E, et al. Covid-19 in People With Diabetes: Urgently Needed Lessons From Early Reports. *Diabetes Care* (2020) 43:1378–81. doi: 10.2337/dci20-0024
9. Polonsky KS. The Past 200 Years in Diabetes. *N Engl J Med* (2012) 367:1332–40. doi: 10.1056/NEJMra1110560
10. von Scholten BJ, Kreiner FF, Gough SC, von Herrath M. Current and Future Therapies for Type 1 Diabetes. *Diabetologia* (2021) 64:1037–48. doi: 10.1007/s00125-021-05398-3
11. Roep BO, Thomaidou S, van Tienhoven R, Zaldumbide A. Type 1 Diabetes Mellitus as a Disease of the β-Cell (Do Not Blame the Immune System)? *Nat Rev Endocrinol* (2020) 17:150–61. doi: 10.1038/s41574-020-00443-4
12. DeFronzo RA, Ferrannini E, Groop L, Henry RR, Herman WH, Holst JJ, et al. Type 2 Diabetes Mellitus. *Nat Rev Dis Primers* (2015) 1:1–22. doi: 10.1038/nrdp.2015.19
13. Katsarou A, Gudbjörnsdottir S, Rawshani A, Dabelea D, Bonifacio E, Anderson BJ, et al. Type 1 Diabetes Mellitus. *Nat Rev Dis Primers* (2017) 3:1–17. doi: 10.1038/nrdp.2017.16
14. Katsanis N. The Continuum of Causality in Human Genetic Disorders. *Genome Biol* (2016) 17:233. doi: 10.1186/s13059-016-1107-9

15. Fajans SS, Bell GI, Polonsky KS. Molecular Mechanisms and Clinical Pathophysiology of Maturity-Onset Diabetes of the Young. *N Engl J Med* (2001) 345:971–80. doi: 10.1056/NEJMra002168
16. Nyunt O, Wu JY, McGown IN, Harris M, Huynh T, Leong GM, et al. Investigating Maturity Onset Diabetes of the Young. *Clin Biochemist Rev* (2009) 30:67.
17. Shields BM, Shepherd M, Hudson M, McDonald TJ, Colclough K, Peters J, et al. Population-Based Assessment of a Biomarker-Based Screening Pathway to Aid Diagnosis of Monogenic Diabetes in Young-Onset Patients. *Diabetes Care* (2017) 40:1017–25. doi: 10.2337/dc17-0224
18. Naylor R, Knight JA, del Gaudio D. *Maturity-Onset Diabetes of the Young Overview.* In: MP Adam, RA Pagon, SE Wallace, LJH Bean, G Mirzaa and and A Amemiya, editors. Seattle: University of Washington (2018).
19. Fuchsberger C, Flannick J, Teslovich TM, Mahajan A, Agarwala V, Gaulton KJ, et al. The Genetic Architecture of Type 2 Diabetes. *Nature* (2016) 536:41–7. doi: 10.1038/nature18642
20. Grarup N, Sandholt CH, Hansen T, Pedersen O. Genetic Susceptibility to Type 2 Diabetes and Obesity: From Genome-Wide Association Studies to Rare Variants and Beyond. *Diabetologia* (2014) 57:1528–41. doi: 10.1007/s00125-014-3270-4
21. Mishra R, Chesi A, Cousminer DL, Hawa MI, Bradfield JP, Hodge KM, et al. Relative Contribution of Type 1 and Type 2 Diabetes Loci to the Genetic Etiology of Adult-Onset, Non-Insulin-Requiring Autoimmune Diabetes. *BMC Med* (2017) 15:88. doi: 10.1186/s12916-017-0846-0
22. Lemelman MB, Letourneau L, Greeley SAW. Neonatal Diabetes Mellitus: An Update on Diagnosis and Management. *Clin Perinatol* (2018) 45:41–59. doi: 10.1016/j.clp.2017.10.006
23. Kleinert M, Clemmensen C, Hofmann SM, Moore MC, Renner S, Woods SC, et al. Animal Models of Obesity and Diabetes Mellitus. *Nat Rev Endocrinol* (2018) 14:140. doi: 10.1038/nrendo.2017.161
24. Al-Awar A, Kupai K, Veszelka M, Szucs G, Attieh Z, Murlasits Z, et al. Experimental Diabetes Mellitus in Different Animal Models. *J Diabetes Res* (2016) 2016:9051426. doi: 10.1155/2016/9051426
25. Rodrigues R. A Comprehensive Review: The Use of Animal Models in Diabetes Research. *J Analytical Pharm Res* (2016) 3. doi: 10.15406/japlr.2016.03.00071
26. Kitada M, Ogura Y, Koya D. Rodent Models of Diabetic Nephropathy: Their Utility and Limitations. *Int J Nephrol Renovascular Dis* (2016) 9:279. doi: 10.2147/IJNRD.S103784
27. Driver JP, Serreze DV, Chen Y-G. Mouse Models for the Study of Autoimmune Type 1 Diabetes: A NOD to Similarities and Differences to Human Disease. *Semin Immunopathol Springer* (2011) 33(1):67–87. doi: 10.1007/s00281-010-0204-1
28. Bonnefond A, Clément N, Fawcett K, Yengo L, Vaillant E, Guillaume J-L, et al. Rare MTNR1B Variants Impairing Melatonin Receptor 1B Function Contribute to Type 2 Diabetes. *Nat Genet* (2012) 44:297–301. doi: 10.1038/ng.1053
29. Catli G, Abaci A, Flanagan S, De Franco E, Ellard S, Hattersley A, et al. A Novel GATA6 Mutation Leading to Congenital Heart Defects and Permanent Neonatal Diabetes: A Case Report. *Diabetes Metab* (2013) 39:370–4. doi: 10.1016/j.diabet.2013.01.005
30. Chao A, Whittemore R, Minges KE, Murphy KM, Grey M. Self-Management in Early Adolescence and Differences by Age at Diagnosis and Duration of Type 1 Diabetes. *Diabetes Educator* (2014) 40:167–77. doi: 10.1177/0145721713520567
31. Eifes S, Chudasama KK, Molnes J, Wagner K, Hoang T, Schierloh U, et al. A Novel GATA6 Mutation in a Child With Congenital Heart Malformation and Neonatal Diabetes. *Clin Case Rep* (2013) 1:86. doi: 10.1002/ccr3.33
32. Gong H, Zhang X, Cheng B, Sun Y, Li C, Li T, et al. Bisphenol A Accelerates Toxic Amyloid Formation of Human Islet Amyloid Polypeptide: A Possible Link Between Bisphenol A Exposure and Type 2 Diabetes. *PLoS One* (2013) 8:e54198. doi: 10.1371/journal.pone.0054198
33. Allen HL, Flanagan SE, Shaw-Smith C, De Franco E, Akerman I, Caswell R, et al. GATA6 Haploinsufficiency Causes Pancreatic Agenesis in Humans. *Nat Genet* (2012) 44:20–2. doi: 10.1038/ng.1035
34. Stanescu DE, Hughes N, Patel P, De León DD. A Novel Mutation in GATA6 Causes Pancreatic Agenesis. *Pediatr Diabetes* (2015) 16:67–70. doi: 10.1111/pedi.12111
35. Bonnefond A, Sand O, Guerin B, Durand E, De Graeve F, Huyvaert M, et al. GATA6 Inactivating Mutations Are Associated With Heart Defects and, Inconsistently, With Pancreatic Agenesis and Diabetes. *Diabetologia* (2012) 55:2845–7. doi: 10.1007/s00125-012-2645-7
36. Gong M, Simaite D, Kuhnen P, Heldmann M, Spagnoli F, Blankenstein O, et al. Two Novel GATA6 Mutations Cause Childhood-Onset Diabetes Mellitus, Pancreas Malformation and Congenital Heart Disease. *Horm Res Paediatr* (2013) 79:250–6. doi: 10.1159/000348844
37. Chao CS, McKnight KD, Cox KL, Chang AL, Kim SK, Feldman BJ. Novel GATA6 Mutations in Patients With Pancreatic Agenesis and Congenital Heart Malformations. *PLoS One* (2015) 10:e0118449. doi: 10.1371/journal.pone.0118449
38. D'Amato E, Giacopelli F, Giannattasio A, d'Annunzio G, Bocciardi R, Musso M, et al. Genetic Investigation in an Italian Child With an Unusual Association of Atrial Septal Defect, Attributable to a New Familial GATA4 Gene Mutation, and Neonatal Diabetes Due to Pancreatic Agenesis. *Diabetic Med* (2010) 27:1195–200. doi: 10.1111/j.1464-5491.2010.03046.x
39. Shaw-Smith C, De Franco E, Allen HL, Batlle M, Flanagan SE, Borowiec M, et al. GATA4 Mutations Are a Cause of Neonatal and Childhood-Onset Diabetes. *Diabetes* (2014) 63:2888–94. doi: 10.2337/db14-0061
40. Carrasco M, Delgado I, Soria B, Martín F, Rojas A. GATA4 and GATA6 Control Mouse Pancreas Organogenesis. *J Clin Invest* (2012) 122:3504–15. doi: 10.1172/JCI63240
41. Xuan S, Borok MJ, Decker KJ, Battle MA, Duncan SA, Hale MA, et al. Pancreas-Specific Deletion of Mouse Gata4 and Gata6 Causes Pancreatic Agenesis. *J Clin Invest* (2012) 122:3516–28. doi: 10.1172/JCI63352
42. Dukes ID, Sreenan S, Roe MW, Levisetti M, Zhou Y-P, Ostrega D, et al. Defective Pancreatic β-Cell Glycolytic Signaling in Hepatocyte Nuclear Factor-1α-Deficient Mice. *J Biol Chem* (1998) 273:24457–64. doi: 10.1074/jbc.273.38.24457
43. Garcia-Gonzalez MA, Carette C, Bagattin A, Chiral M, Makinistoglu MP, Garbay S, et al. A Suppressor Locus for MODY3-Diabetes. *Sci Rep* (2016) 6:33087. doi: 10.1038/srep35697
44. Skelin M, Rupnik M, Cencic A. Pancreatic Beta Cell Lines and Their Applications in Diabetes Mellitus Research. *ALTEX* (2011) 27:105–13. doi: 10.14573/altex.2010.2.105
45. Scharfmann R, Staels W, Albagli O. The Supply Chain of Human Pancreatic Beta Cell Lines. *J Clin Invest* (2019) 129:3511–20. doi: 10.1172/JCI129484
46. Takahashi K, Yamanaka S. A Decade of Transcription Factor-Mediated Reprogramming to Pluripotency. *Nat Rev Mol Cell Biol* (2016) 17:183–93. doi: 10.1038/nrm.2016.8
47. Wu J, Izpisua Belmonte JC. Stem Cells: A Renaissance in Human Biology Research. *Cell* (2016) 165:1572–85. doi: 10.1016/j.cell.2016.05.043
48. Zorn AM, Wells JM. Vertebrate Endoderm Development and Organ Formation. *Annu Rev Cell Dev* (2009) 25:221–51. doi: 10.1146/annurev.cellbio.042308.113344
49. Nair G, Hebrok M. Islet Formation in Mice and Men: Lessons for the Generation of Functional Insulin-Producing β-Cells From Human Pluripotent Stem Cells. *Curr Opin Genet Dev* (2015) 32:171–80. doi: 10.1016/j.gde.2015.03.004
50. D'Amour KA, Agulnick AD, Eliazer S, Kelly OG, Kroon E, Baetge EE. Efficient Differentiation of Human Embryonic Stem Cells to Definitive Endoderm. *Nat Biotechnol* (2005) 23:1534–41. doi: 10.1038/nbt1163
51. D'Amour KA, Bang AG, Eliazer S, Kelly OG, Agulnick AD, Smart NG, et al. Production of Pancreatic Hormone–Expressing Endocrine Cells From Human Embryonic Stem Cells. *Nat Biotechnol* (2006) 24:1392–401. doi: 10.1038/nbt1259
52. Kroon E, Martinson LA, Kadoya K, Bang AG, Kelly OG, Eliazer S, et al. Pancreatic Endoderm Derived From Human Embryonic Stem Cells Generates Glucose-Responsive Insulin-Secreting Cells In Vivo. *Nat Biotechnol* (2008) 26:443–52. doi: 10.1038/nbt1393
53. Pagliuca FW, Millman JR, Gürtler M, Segel M, Van Dervort A, Ryu JH, et al. Generation of Functional Human Pancreatic β Cells In Vitro. *Cell* (2014) 159:428–39. doi: 10.1016/j.cell.2014.09.040
54. Rezania A, Bruin JE, Arora P, Rubin A, Batushansky I, Asadi A, et al. Reversal of Diabetes With Insulin-Producing Cells Derived *In Vitro* From Human Pluripotent Stem Cells. *Nat Biotechnol* (2014) 32:1121. doi: 10.1038/nbt.3033

55. Alvarez-Dominguez JR, Donaghey J, Rasouli N, Kenty JH, Helman A, Charlton J, et al. Circadian Entrainment Triggers Maturation of Human *In Vitro* Islets. *Cell Stem Cell* (2020) 26:108–22:e10. doi: 10.1016/j.stem.2019.11.011
56. Davis JC, Alves TC, Helman A, Chen JC, Kenty JH, Cardone RL, et al. Glucose Response by Stem Cell-Derived β Cells *In Vitro* is Inhibited by a Bottleneck in Glycolysis. *Cell Rep* (2020) 31:107623. doi: 10.1016/j.celrep.2020.107623
57. Hogrebe NJ, Augsornworawat P, Maxwell KG, Velazco-Cruz L, Millman JR. Targeting the Cytoskeleton to Direct Pancreatic Differentiation of Human Pluripotent Stem Cells. *Nat Biotechnol* (2020) 38:460–70. doi: 10.1038/s41587-020-0430-6
58. Li X, Yang KY, Chan VW, Leung KT, Zhang X-B, Wong AS, et al. Single-Cell RNA-Seq Reveals That CD9 is a Negative Marker of Glucose-Responsive Pancreatic β-Like Cells Derived From Human Pluripotent Stem Cells. *Stem Cell Rep* (2020) 15:1111–26. doi: 10.1016/j.stemcr.2020.09.009
59. Mahaddalkar PU, Scheibner K, Pfluger S, Sterr M, Beckenbauer J, Irmler M, et al. Generation of Pancreatic β Cells From CD177+ Anterior Definitive Endoderm. *Nat Biotechnol* (2020) 28:1061–72. doi: 10.1038/s41587-020-0492-5
60. Nair GG, Liu JS, Russ HA, Tran S, Saxton MS, Chen R, et al. Recapitulating Endocrine Cell Clustering in Culture Promotes Maturation of Human Stem-Cell-Derived β Cells. *Nat Cell Biol* (2019) 21:263–74. doi: 10.1038/s41556-018-0271-4
61. Lithovius V, Saarimäki-Vire J, Balboa D, Ibrahim H, Montaser H, Barsby T, et al. SUR1-Mutant iPS Cell-Derived Islets Recapitulate the Pathophysiology of Congenital Hyperinsulinism. *Diabetologia* (2021) 64:630–40. doi: 10.1007/s00125-020-05346-7
62. Veres A, Faust AL, Bushnell HL, Engquist EN, Kenty JH-R, Harb G, et al. Charting Cellular Identity During Human *In Vitro* β-Cell Differentiation. *Nature* (2019) 569:368–73. doi: 10.1038/s41586-019-1168-5
63. Hsu PD, Lander ES, Zhang F. Development and Applications of CRISPR-Cas9 for Genome Engineering. *Cell* (2014) 157:1262–78. doi: 10.1016/j.cell.2014.05.010
64. Ran FA, Hsu PD, Wright J, Agarwala V, Scott DA, Zhang F. Genome Engineering Using the CRISPR-Cas9 System. *Nat Protoc* (2013) 8:2281–308. doi: 10.1038/nprot.2013.143
65. Millette K, Georgia S. Gene Editing and Human Pluripotent Stem Cells: Tools for Advancing Diabetes Disease Modeling and Beta-Cell Development. *Curr Diabetes Rep* (2017) 17:116. doi: 10.1007/s11892-017-0947-3
66. Balboa D, Saarimäki-Vire J, Otonkoski T. Concise Review: Human Pluripotent Stem Cells for the Modeling of Pancreatic β-Cell Pathology. *Stem Cells* (2019) 37:33–41. doi: 10.1002/stem.2913
67. Lieber MR. The Mechanism of Double-Strand DNA Break Repair by the Nonhomologous DNA End-Joining Pathway. *Annu Rev Biochem* (2010) 79:181–211. doi: 10.1146/annurev.biochem.052308.093131
68. Jackson SP, Bartek J. The DNA-Damage Response in Human Biology and Disease. *Nature* (2009) 461:1071–8. doi: 10.1038/nature08467
69. Waryah CB, Moses C, Arooj M, Blancafort P. *Zinc Fingers, TALEs, and CRISPR Systems: A Comparison of Tools for Epigenome Editing, Epigenome Editing.* Springer (2018). p. 19–63.
70. Miller JC, Tan S, Qiao G, Barlow KA, Wang J, Xia DF, et al. A TALE Nuclease Architecture for Efficient Genome Editing. *Nat Biotechnol* (2011) 29:143–8. doi: 10.1038/nbt.1755
71. Zhang F, Cong L, Lodato S, Kosuri S, Church GM, Arlotta P. Efficient Construction of Sequence-Specific TAL Effectors for Modulating Mammalian Transcription. *Nat Biotechnol* (2011) 29:149–53. doi: 10.1038/nbt.1775
72. Wright DA, Li T, Yang B, Spalding MH. TALEN-Mediated Genome Editing: Prospects and Perspectives. *Biochem J* (2014) 462:15–24. doi: 10.1042/BJ20140295
73. Segal DJ, Crotty JW, Bhakta MS, Barbas CFIII, Horton NC. Structure of Aart, A Designed Six-Finger Zinc Finger Peptide, Bound to DNA. *J Mol Biol* (2006) 363:405–21. doi: 10.1016/j.jmb.2006.08.016
74. Porteus MH, Carroll D. Gene Targeting Using Zinc Finger Nucleases. *Nat Biotechnol* (2005) 23:967–73. doi: 10.1038/nbt1125
75. Boch J, Bonas U. Xanthomonas AvrBs3 Family-Type III Effectors: Discovery and Function. *Annu Rev Phytopathol* (2010) 48. doi: 10.1146/annurev-phyto-080508-081936
76. Bibikova M, Carroll D, Segal DJ, Trautman JK, Smith J, Kim Y-G, et al. Stimulation of Homologous Recombination Through Targeted Cleavage by Chimeric Nucleases. *Mol Cell Biol* (2001) 21:289–97. doi: 10.1128/MCB.21.1.289-297.2001
77. Kim Y-G, Cha J, Chandrasegaran S. Hybrid Restriction Enzymes: Zinc Finger Fusions to Fok I Cleavage Domain. *Proc Natl Acad Sci* (1996) 93:1156–60. doi: 10.1073/pnas.93.3.1156
78. Urnov FD, Rebar EJ, Holmes MC, Zhang HS, Gregory PD. Genome Editing With Engineered Zinc Finger Nucleases. *Nat Rev Genet* (2010) 11:636–46. doi: 10.1038/nrg2842
79. Pabo CO, Peisach E, Grant RA. Design and Selection of Novel Cys2His2 Zinc Finger Proteins. *Annu Rev Biochem* (2001) 70:313–40. doi: 10.1146/annurev.biochem.70.1.313
80. Mussolino C, Cathomen T. On Target? Tracing Zinc-Finger-Nuclease Specificity. *Nat Methods* (2011) 8:725–6. doi: 10.1038/nmeth.1680
81. Joung JK, Sander JD. Talens: A Widely Applicable Technology for Targeted Genome Editing. *Nat Rev Mol Cell Biol* (2013) 14:49–55. doi: 10.1038/nrm3486
82. Cox DBT, Platt RJ, Zhang F. Therapeutic Genome Editing: Prospects and Challenges. *Nat Med* (2015) 21:121–31. doi: 10.1038/nm.3793
83. Bultmann S, Morbitzer R, Schmidt CS, Thanisch K, Spada F, Elsaesser J, et al. Targeted Transcriptional Activation of Silent Oct4 Pluripotency Gene by Combining Designer TALEs and Inhibition of Epigenetic Modifiers. *Nucleic Acids Res* (2012) 40:5368–77. doi: 10.1093/nar/gks199
84. Carroll D. Progress and Prospects: Zinc-Finger Nucleases as Gene Therapy Agents. *Gene Ther* (2008) 15:1463–8. doi: 10.1038/gt.2008.145
85. Ding Q, Lee Y-K, Schaefer EA, Peters DT, Veres A, Kim K, et al. A TALEN Genome-Editing System for Generating Human Stem Cell-Based Disease Models. *Cell Stem Cell* (2013) 12:238–51. doi: 10.1016/j.stem.2012.11.011
86. Jinek M, Chylinski K, Fonfara I, Hauer M, Doudna JA, Charpentier E. A Programmable Dual-RNA–Guided DNA Endonuclease in Adaptive Bacterial Immunity. *Science* (2012) 337:816–21. doi: 10.1126/science.1225829
87. Barrangou R, Fremaux C, Deveau H, Richards M, Boyaval P, Moineau S, et al. CRISPR Provides Acquired Resistance Against Viruses in Prokaryotes. *Science* (2007) 315:1709–12. doi: 10.1126/science.1138140
88. Deltcheva E, Chylinski K, Sharma CM, Gonzales K, Chao Y, Pirzada ZA, et al. Crispr RNA Maturation by Trans-Encoded Small RNA and Host Factor RNase III. *Nature* (2011) 471:602–7. doi: 10.1038/nature09886
89. Doudna JA, Charpentier E. The New Frontier of Genome Engineering With CRISPR-Cas9. *Science* (2014) 346. doi: 10.1126/science.1258096
90. Zhang Z, Zhang Y, Gao F, Han S, Cheah KS, Tse H-F, et al. CRISPR/Cas9 Genome-Editing System in Human Stem Cells: Current Status and Future Prospects. *Mol Therapy-Nucleic Acids* (2017) 9:230–41. doi: 10.1016/j.omtn.2017.09.009
91. González F, Zhu Z, Shi Z-D, Lelli K, Verma N, Li QV, et al. An iCRISPR Platform for Rapid, Multiplexable, and Inducible Genome Editing in Human Pluripotent Stem Cells. *Cell Stem Cell* (2014) 15:215–26. doi: 10.1016/j.stem.2014.05.018
92. Hockemeyer D, Jaenisch R. Induced Pluripotent Stem Cells Meet Genome Editing. *Cell Stem Cell* (2016) 18:573–86. doi: 10.1016/j.stem.2016.04.013
93. Ding Q, Regan SN, Xia Y, Oostrom LA, Cowan CA, Musunuru K. Enhanced Efficiency of Human Pluripotent Stem Cell Genome Editing Through Replacing TALENs With Crisprs. *Cell Stem Cell* (2013) 12:393. doi: 10.1016/j.stem.2013.03.006
94. Nerys-Junior A, Braga-Dias LP, Pezzuto P, Cotta-de-Almeida V, Tanuri A. Comparison of the Editing Patterns and Editing Efficiencies of TALEN and CRISPR-Cas9 When Targeting the Human CCR5 Gene. *Genet Mol Biol* (2018) 41:167–79. doi: 10.1590/1678-4685-gmb-2017-0065
95. Maguire JA, Cardenas-Diaz FL, Gadue P, French DL. Highly Efficient CRISPR-Cas9-Mediated Genome Editing in Human Pluripotent Stem Cells. *Curr Protoc Stem Cell Biol* (2019) 48:e64. doi: 10.1002/cpsc.64
96. Veres A, Gosis BS, Ding Q, Collins R, Ragavendran A, Brand H, et al. Low Incidence of Off-Target Mutations in Individual CRISPR-Cas9 and TALEN

Targeted Human Stem Cell Clones Detected by Whole-Genome Sequencing. *Cell Stem Cell* (2014) 15:27–30. doi: 10.1016/j.stem.2014.04.020
97. Ellard S, Allen HL, De Franco E, Flanagan S, Hysenaj G, Colclough K, et al. Improved Genetic Testing for Monogenic Diabetes Using Targeted Next-Generation Sequencing. *Diabetologia* (2013) 56:1958–63. doi: 10.1007/s00125-013-2962-5
98. Johansson S, Irgens H, Chudasama KK, Molnes J, Aerts J, Roque FS, et al. Exome Sequencing and Genetic Testing for MODY. *PLoS One* (2012) 7: e38050. doi: 10.1371/journal.pone.0038050
99. Meyre D. Give GWAS a Chance. *Diabetes* (2017) 66:2741–2. doi: 10.2337/dbi17-0026
100. Cai L, Wheeler E, Kerrison ND, Ja L, Deloukas P, PW F, et al. Genome-Wide Association Analysis of Type 2 Diabetes in the EPIC-InterAct Study. *Sci Data* (2020) 7:1–6. doi: 10.1038/s41597-020-00716-7
101. Guo M, Zhang T, Dong X, Xiang JZ, Lei M, Evans T, et al. Using hESCs to Probe the Interaction of the Diabetes-Associated Genes CDKAL1 and MT1E. *Cell Rep* (2017) 19:1512–21. doi: 10.1016/j.celrep.2017.04.070
102. Kishore S, De Franco E, Cardenas-Diaz FL, Letourneau-Freiberg LR, Sanyoura M, Osorio-Quintero C, et al. A Non-Coding Disease Modifier of Pancreatic Agenesis Identified by Genetic Correction in a Patient-Derived iPSC Line. *Cell Stem Cell* (2020). doi: 10.1016/j.stem.2020.05.001
103. Colclough K, Bellanne-Chantelot C, Saint-Martin C, Flanagan SE, Ellard S. Mutations in the Genes Encoding the Transcription Factors Hepatocyte Nuclear Factor 1 Alpha and 4 Alpha in Maturity-Onset Diabetes of the Young and Hyperinsulinemic Hypoglycemia. *Hum Mutat* (2013) 34:669–85. doi: 10.1002/humu.22279
104. Fajans SS, Bell GI. MODY: History, Genetics, Pathophysiology, and Clinical Decision Making. *Diabetes Care* (2011) 34:1878–84. doi: 10.2337/dc11-0035
105. Tamaroff J, Kilberg M, Pinney SE, McCormack S. Overview of Atypical Diabetes. *Endocrinol Metab Clin North Am* (2020) 49:695–723. doi: 10.1016/j.ecl.2020.07.004
106. Steck AK, Winter WE. Review on Monogenic Diabetes. *Curr Opin Endocrinol Diabetes Obes* (2011) 18:252–8. doi: 10.1097/MED.0b013e3283488275
107. Wang X, Sterr M, Burtscher I, Böttcher A, Beckenbauer J, Siehler J, et al. Point Mutations in the PDX1 Transactivation Domain Impair Human β-Cell Development and Function. *Mol Metab* (2019) 24:80–97. doi: 10.1016/j.molmet.2019.03.006
108. Najmi LA, Aukrust I, Flannick J, Molnes J, Burtt N, Molven A, et al. Functional Investigations of HNF1A Identify Rare Variants as Risk Factors for Type 2 Diabetes in the General Population. *Diabetes* (2017) 66:335–46. doi: 10.2337/db16-0460
109. Haliyur R, Tong X, Sanyoura M, Shrestha S, Lindner J, Saunders DC, et al. Human Islets Expressing HNF1A Variant Have Defective β Cell Transcriptional Regulatory Networks. *J Clin Invest* (2019) 129:246–51. doi: 10.1172/JCI121994
110. Griscelli F, Ezanno H, Soubeyrand M, Feraud O, Oudrhiri N, Bonnefond A, et al. Generation of an Induced Pluripotent Stem Cell (iPSC) Line From a Patient With Maturity-Onset Diabetes of the Young Type 3 (MODY3) Carrying a Hepatocyte Nuclear Factor 1-Alpha (HNF1A) Mutation. *Stem Cell Res* (2018) 29:56–9. doi: 10.1016/j.scr.2018.02.017
111. Stepniewski J, Kachamakova-Trojanowska N, Ogrocki D, Szopa M, Matlok M, Beilharz M, et al. Induced Pluripotent Stem Cells as a Model for Diabetes Investigation. *Sci Rep* (2015) 5:8597. doi: 10.1038/srep08597
112. Teo AK, Windmueller R, Johansson BB, Dirice E, Njolstad PR, Tjora E, et al. Derivation of Human Induced Pluripotent Stem Cells From Patients With Maturity Onset Diabetes of the Young. *J Biol Chem* (2013) 288:5353–6. doi: 10.1074/jbc.C112.428979
113. Cardenas-Diaz FL, Osorio-Quintero C, Diaz-Miranda MA, Kishore S, Leavens K, Jobaliya C, et al. Modeling Monogenic Diabetes Using Human ESCs Reveals Developmental and Metabolic Deficiencies Caused by Mutations in HNF1A. *Cell Stem Cell* (2019) 25:273–89. doi: 10.1016/j.stem.2019.07.007
114. Yabe SG, Nishida J, Fukuda S, Takeda F, Nasiro K, Yasuda K, et al. Expression of Mutant mRNA and Protein in Pancreatic Cells Derived From MODY3- iPS Cells. *PLoS One* (2019) 14:e0217110. doi: 10.1371/journal.pone.0217110
115. Teo AK, Lau HH, Valdez IA, Dirice E, Tjora E, Raeder H, et al. Early Developmental Perturbations in a Human Stem Cell Model of MODY5/HNF1B Pancreatic Hypoplasia. *Stem Cell Rep* (2016) 6:357–67. doi: 10.1016/j.stemcr.2016.01.007
116. Yabe SG, Iwasaki N, Yasuda K, Hamazaki TS, Konno M, Fukuda S, et al. Establishment of Maturity-Onset Diabetes of the Young-Induced Pluripotent Stem Cells From a Japanese Patient. *J Diabetes Investig* (2015) 6:543–7. doi: 10.1111/jdi.12334
117. Braverman-Gross C, Nudel N, Ronen D, Beer NL, McCarthy MI, Benvenisty N. Derivation and Molecular Characterization of Pancreatic Differentiated MODY1-Ipscs. *Stem Cell Res* (2018) 31:16–26. doi: 10.1016/j.scr.2018.06.013
118. Ng NHJ, Jasmen JB, Lim CS, Lau HH, Krishnan VG, Kadiwala J, et al. Hnf4a Haploinsufficiency in MODY1 Abrogates Liver and Pancreas Differentiation From Patient-Derived Induced Pluripotent Stem Cells. *iScience* (2019) 16:192–205. doi: 10.1016/j.isci.2019.05.032
119. Vethe H, Bjorlykke Y, Ghila LM, Paulo JA, Scholz H, Gygi SP, et al. Probing the Missing Mature Beta-Cell Proteomic Landscape in Differentiating Patient iPSC-derived Cells. *Sci Rep* (2017) 7:4780. doi: 10.1038/s41598-017-04979-w
120. Wang X, Chen S, Burtscher I, Sterr M, Hieronimus A, Machicao F, et al. Generation of a Human Induced Pluripotent Stem Cell (iPSC) Line From a Patient Carrying a P33T Mutation in the PDX1 Gene. *Stem Cell Res* (2016) 17:273–6. doi: 10.1016/j.scr.2016.08.004
121. Zhu Z, Li QV, Lee K, Rosen BP, González F, Soh C-L, et al. Genome Editing of Lineage Determinants in Human Pluripotent Stem Cells Reveals Mechanisms of Pancreatic Development and Diabetes. *Cell Stem Cell* (2016) 18:755–68. doi: 10.1016/j.stem.2016.03.015
122. Griscelli F, Feraud O, Ernault T, Oudrihri N, Turhan AG, Bonnefond A, et al. Generation of an Induced Pluripotent Stem Cell (iPSC) Line From a Patient With Maturity-Onset Diabetes of the Young Type 13 (MODY13) With a the Potassium Inwardly-Rectifying Channel, Subfamily J, Member 11 (KCNJ11) Mutation. *Stem Cell Res* (2017) 23:178–81. doi: 10.1016/j.scr.2017.07.023
123. Voight BF, Scott LJ, Steinthorsdottir V, Morris AP, Dina C, Welch RP, et al. Twelve Type 2 Diabetes Susceptibility Loci Identified Through Large-Scale Association Analysis. *Nat Genet* (2010) 42:579. doi: 10.1038/ng.609
124. Pontoglio M, Sreenan S, Roe M, Pugh W, Ostrega D, Doyen A, et al. Defective Insulin Secretion in Hepatocyte Nuclear Factor 1alpha-Deficient Mice. *J Clin Invest* (1998) 101:2215–22. doi: 10.1172/JCI2548
125. Maxwell KG AP, Velazco-Cruz L, Kim MH, Asada R, Hogrebe NJ, Morikawa S, et al. Gene-Edited Human Stem Cell-Derived β Cells From a Patient With Monogenic Diabetes Reverse Preexisting Diabetes in Mice. *Sci Trans Med* (2020) 12:eaax9106. doi: 10.1126/scitranslmed.aax9106
126. Shang L HH, Foo K, Martinez H, Watanabe K, Zimmer M, Kahler DJ, et al. β-Cell Dysfunction Due to Increased ER Stress in a Stem Cell Model of Wolfram Syndrome. *Diabetes* (2014) 63:923–33. doi: 10.2337/db13-0717
127. Igoillo-Esteve M, Oliveira AF, Cosentino C, Fantuzzi F, Demarez C, Toivonen S, et al. Exenatide Induces Frataxin Expression and Improves Mitochondrial Function in Friedreich Ataxia. *JCI Insight* (2020) 5. doi: 10.1172/jci.insight.134221
128. Simsek S, Zhou T, Robinson CL, Tsai SY, Crespo M, Amin S, et al. Modeling Cystic Fibrosis Using Pluripotent Stem Cell-Derived Human Pancreatic Ductal Epithelial Cells. *Stem Cells Transl Med* (2016) 5:572–9. doi: 10.5966/sctm.2015-0276
129. Bennett SN, Caporaso N, Fitzpatrick AL, Agrawal A, Barnes K, Boyd HA, et al. Phenotype Harmonization and Cross-Study Collaboration in GWAS Consortia: The GENEVA Experience. *Genet Epidemiol* (2011) 35:159–73. doi: 10.1002/gepi.20564
130. Zeggini E, Scott LJ, Saxena R, Voight BF, Marchini JL, Hu T, et al. Meta-Analysis of Genome-Wide Association Data and Large-Scale Replication Identifies Additional Susceptibility Loci for Type 2 Diabetes. *Nat Genet* (2008) 40:638–45. doi: 10.1016/S0084-3741(08)79224-2
131. Rich S, Akolkar B, Concannon P, Erlich H, Hilner J, Julier C, et al. Overview of the Type I Diabetes Genetics Consortium. *Genes Immun* (2009) 10:S1–4. doi: 10.1038/gene.2009.84
132. Yaguchi H, Togawa K, Moritani M, Itakura M. Identification of Candidate Genes in the Type 2 Diabetes Modifier Locus Using Expression QTL. *Genomics* (2005) 85:591–9. doi: 10.1016/j.ygeno.2005.01.006

133. Laber S, Cox RD. Mouse Models of Human GWAS Hits for Obesity and Diabetes in the Post Genomic Era: Time for Reevaluation. *Front Endocrinol* (2017) 8:11. doi: 10.3389/fendo.2017.00011
134. Takeshita S, Suzuki T, Kitayama S, Moritani M, Inoue H, Itakura M. Bhlhe40, A Potential Diabetic Modifier Gene on Dbm1 Locus, Negatively Controls Myocyte Fatty Acid Oxidation. *Genes Genet Syst* (2012) 87:253–64. doi: 10.1266/ggs.87.253
135. Bakay M, Pandey R, Grant SF, Hakonarson H. The Genetic Contribution to Type 1 Diabetes. *Curr Diabetes Rep* (2019) 19:1–14. doi: 10.1007/s11892-019-1235-1
136. Mahajan A, Taliun D, Thurner M, Robertson NR, Torres JM, Rayner NW, et al. Fine-Mapping Type 2 Diabetes Loci to Single-Variant Resolution Using High-Density Imputation and Islet-Specific Epigenome Maps. *Nat Genet* (2018) 50:1505–13. doi: 10.1038/s41588-018-0241-6
137. Pociot F. Type 1 Diabetes Genome-Wide Association Studies: Not to be Lost in Translation. *Clin Trans Immunol* (2017) 6:e162. doi: 10.1038/cti.2017.51
138. Szabó M, Máté B, Csép K, Benedek T. Epigenetic Modifications Linked to T2D, the Heritability Gap, and Potential Therapeutic Targets. *Biochem Genet* (2018) 56:553–74. doi: 10.1007/s10528-018-9863-8
139. Green AD, Vasu S, Moffett RC, Flatt PR. Co-Culture of Clonal Beta Cells With GLP-1 and Glucagon-Secreting Cell Line Impacts on Beta Cell Insulin Secretion, Proliferation and Susceptibility to Cytotoxins. *Biochimie* (2016) 125:119–25. doi: 10.1016/j.biochi.2016.03.007
140. Seo H, Son J, Park J-K. Controlled 3D Co-Culture of Beta Cells and Endothelial Cells in a Micropatterned Collagen Sheet for Reproducible Construction of an Improved Pancreatic Pseudo-Tissue. *APL Bioeng* (2020) 4:046103. doi: 10.1063/5.0023873
141. Nguyen DTT, van Noort D, Jeong I-K, Park S. Endocrine System on Chip for a Diabetes Treatment Model. *Biofabrication* (2017) 9:015021. doi: 10.1088/1758-5090/aa5cc9
142. Pociot F, Akolkar B, Concannon P, Erlich HA, Julier C, Morahan G, et al. Genetics of Type 1 Diabetes: What's Next? *Diabetes* (2010) 59:1561–71. doi: 10.2337/db10-0076
143. Vafiadis P, Bennett ST, Todd JA, Nadeau J, Grabs R, Goodyer CG, et al. Insulin Expression in Human Thymus Is Modulated by INS VNTR Alleles at the IDDM2 Locus. *Nat Genet* (1997) 15:289–92. doi: 10.1038/ng0397-289
144. Pugliese A, Zeller M, Fernandez A, Zalcberg LJ, Bartlett RJ, Ricordi C, et al. The Insulin Gene is Transcribed in the Human Thymus and Transcription Levels Correlate With Allelic Variation at the INS Vntr-IDDM2 Susceptibility Locus for Type 1 Diabetes. *Nat Genet* (1997) 15:293–7. doi: 10.1038/ng0397-293
145. Stoy J, Steiner DF, Park SY, Ye H, Philipson LH, Bell GI. Clinical and Molecular Genetics of Neonatal Diabetes Due to Mutations in the Insulin Gene. *Rev Endocr Metab Disord* (2010) 11:205–15. doi: 10.1007/s11154-010-9151-3
146. Maurano MT, Humbert R, Rynes E, Thurman RE, Haugen E, Wang H, et al. Systematic Localization of Common Disease-Associated Variation in Regulatory DNA. *Science* (2012) 337:1190–5. doi: 10.1126/science.1222794
147. Gusev A, Lee SH, Trynka G, Finucane H, Vilhjálmsson BJ, Xu H, et al. Partitioning Heritability of Regulatory and Cell-Type-Specific Variants Across 11 Common Diseases. *Am J Hum Genet* (2014) 95:535–52. doi: 10.1016/j.ajhg.2014.10.004
148. Maehr R, Chen S, Snitow M, Ludwig T, Yagasaki L, Goland R, et al. Generation of Pluripotent Stem Cells From Patients With Type 1 Diabetes. *Proc Natl Acad Sci USA* (2009) 106:15768–73. doi: 10.1073/pnas.0906894106
149. Kudva YC, Ohmine S, Greder LV, Dutton JR, Armstrong A, De Lamo JG, et al. Transgene-Free Disease-Specific Induced Pluripotent Stem Cells From Patients With Type 1 and Type 2 Diabetes. *Stem Cells Transl Med* (2012) 1:451–61. doi: 10.5966/sctm.2011-0044
150. Millman JR, Xie C, Van Dervort A, Gurtler M, Pagliuca FW, Melton DA. Generation of Stem Cell-Derived Beta-Cells From Patients With Type 1 Diabetes. *Nat Commun* (2016) 7:11463. doi: 10.1038/ncomms11463
151. Kim MJ, Lee EY, You YH, Yang HK, Yoon KH, Kim JW. Generation of iPSC-derived Insulin-Producing Cells From Patients With Type 1 and Type 2 Diabetes Compared With Healthy Control. *Stem Cell Res* (2020) 48:101958. doi: 10.1016/j.scr.2020.101958
152. van Heyningen V, Yeyati PL. Mechanisms of Non-Mendelian Inheritance in Genetic Disease. *Hum Mol Genet* (2004) 13:R225–33. doi: 10.1093/hmg/ddh254
153. Walia S, Fishman GA, Swaroop A, Branham KE, Lindeman M, Othman M, et al. Discordant Phenotypes in Fraternal Twins Having an Identical Mutation in Exon ORF15 of the RPGR Gene. *Arch Ophthalmol* (2008) 126:379–84. doi: 10.1001/archophthalmol.2007.72
154. Dipple KM, McCabe ER. Phenotypes of Patients With "Simple" Mendelian Disorders Are Complex Traits: Thresholds, Modifiers, and Systems Dynamics. *Am J Hum Genet* (2000) 66:1729–35. doi: 10.1086/302938
155. Badano JL, Katsanis N. Beyond Mendel: An Evolving View of Human Genetic Disease Transmission. *Nat Rev Genet* (2002) 3:779–89. doi: 10.1038/nrg910
156. Kousi M, Katsanis N. Genetic Modifiers and Oligogenic Inheritance. *Cold Spring Harbor Perspect Med* (2015) 5:a017145. doi: 10.1101/cshperspect.a017145
157. Genin E, Feingold J, Clerget-Darpoux F. Identifying Modifier Genes of Monogenic Disease: Strategies and Difficulties. *Hum Genet* (2008) 124:357. doi: 10.1007/s00439-008-0560-2
158. Goyal N, Kesharwani D, Datta M. Lnc-Ing non-Coding RNAs With Metabolism and Diabetes: Roles of Lncrnas. *Cell Mol Life Sci* (2018) 75:1827–37. doi: 10.1007/s00018-018-2760-9
159. Franks PW, Pearson E, Florez JC. Gene-Environment and Gene-Treatment Interactions in Type 2 Diabetes: Progress, Pitfalls, and Prospects. *Diabetes Care* (2013) 36:1413–21. doi: 10.2337/dc12-2211
160. Knip M, Simell O. Environmental Triggers of Type 1 Diabetes. *Cold Spring Harbor Perspect Med* (2012) 2:a007690. doi: 10.1101/cshperspect.a007690
161. Lega IC, Lipscombe LL. Diabetes, Obesity, and Cancer—Pathophysiology and Clinical Implications. *Endocrine Rev* (2020) 41:33–52. doi: 10.1210/endrev/bnz014
162. Krentz NA, Gloyn AL. Insights Into Pancreatic Islet Cell Dysfunction From Type 2 Diabetes Mellitus Genetics. *Nat Rev Endocrinol* (2020) 16:202–12. doi: 10.1038/s41574-020-0325-0
163. Redondo MJ, Concannon P. Genetics of Type 1 Diabetes Comes of Age. *Diabetes Care* (2020) 43:16–8. doi: 10.2337/dci19-0049
164. Sheikhpour M, Abolfathi H, Khatami S, Meshkani R, Barghi TS. The Interaction Between Gene Profile and Obesity in Type 2 Diabetes: A Review. *Obes Med* (2020) 18:100197. doi: 10.1016/j.obmed.2020.100197
165. Redondo MJ, Steck AK, Pugliese A. Genetics of Type 1 Diabetes. *Pediatr Diabetes* (2018) 19:346–53. doi: 10.1111/pedi.12597
166. Nadeau JH. Modifier Genes in Mice and Humans. *Nat Rev Genet* (2001) 2:165–74. doi: 10.1038/35056009
167. Franzago M, Fraticelli F, Stuppia L, Vitacolonna E. Nutrigenetics, Epigenetics and Gestational Diabetes: Consequences in Mother and Child. *Epigenetics* (2019) 14:215–35. doi: 10.1080/15592294.2019.1582277
168. Singh R, Chandel S, Dey D, Ghosh A, Roy S, Ravichandiran V, et al. Epigenetic Modification and Therapeutic Targets of Diabetes Mellitus. *Bioscience Rep* (2020) 40. doi: 10.1042/BSR20202160
169. Haldane J. The Relative Importance of Principal and Modifying Genes in Determining Some Human Diseases. *J Genet* (1941) 41:149–57. doi: 10.1007/BF02983018
170. Verge CF, Vardi P, Babu S, Bao F, Erlich HA, Bugawan T, et al. Evidence for Oligogenic Inheritance of Type 1 Diabetes in a Large Bedouin Arab Family. *J Clin Invest* (1998) 102:1569–75. doi: 10.1172/JCI3379
171. Gioeva OA, Zubkova NA, Tikhonovich YV, Petrov VM, Vasilyev EV, Kiyaev AV, et al. Clinical and Molecular Genetic Characteristics of MODY Cases With Digenic and Oligogenic Inheritance as Defined by Targeted Next-Generation Sequencing. *Problems Endocrinol* (2016) 62:20–7. doi: 10.14341/probl201662620-27
172. Vardarli I, Baier LJ, Hanson RL, Akkoyun I, Fischer C, Rohmeiss P, et al. Gene for Susceptibility to Diabetic Nephropathy in Type 2 Diabetes Maps to 18q22. 3-23. *Kidney Int* (2002) 62:2176–83. doi: 10.1046/j.1523-1755.2002.00663.x
173. Nerup J, Platz P, Andersen OO, Christy M, Lyngsøe J, Poulsen J, et al. HL-a Antigens and Diabetes Mellitus. *Lancet* (1974) 304:864–6. doi: 10.1016/S0140-6736(74)91201-X
174. Grant SF, Thorleifsson G, Reynisdottir I, Benediktsson R, Manolescu A, Sainz J, et al. Variant of Transcription Factor 7-Like 2 (TCF7L2) Gene

Confers Risk of Type 2 Diabetes. *Nat Genet* (2006) 38:320–3. doi: 10.1038/ng1732
175. Mahajan A, Wessel J, Willems SM, Zhao W, Robertson NR, Chu AY, et al. Refining the Accuracy of Validated Target Identification Through Coding Variant Fine-Mapping in Type 2 Diabetes. *Nat Genet* (2018) 50:559–71. doi: 10.1038/s41588-018-0084-1
176. Mahdessian H, Taxiarchis A, Popov S, Silveira A, Franco-Cereceda A, Hamsten A, et al. TM6SF2 Is a Regulator of Liver Fat Metabolism Influencing Triglyceride Secretion and Hepatic Lipid Droplet Content. *Proc Natl Acad Sci* (2014) 111:8913–8. doi: 10.1073/pnas.1323785111
177. Guo D, Li M, Zhang Y, Yang P, Eckenrode S, Hopkins D, et al. A Functional Variant of SUMO4, a New Iκbα Modifier, Is Associated With Type 1 Diabetes. *Nat Genet* (2004) 36:837. doi: 10.1038/ng1391
178. Lehman DM, Fu D-J, Freeman AB, Hunt KJ, Leach RJ, Johnson-Pais T, et al. A Single Nucleotide Polymorphism in Mgea5 Encoding O-Glcnac–Selective N-Acetyl-β-D Glucosaminidase Is Associated With Type 2 Diabetes in Mexican Americans. *Diabetes* (2005) 54:1214–21. doi: 10.2337/diabetes.54.4.1214
179. Hon C-C, Ramilowski JA, Harshbarger J, Bertin N, Rackham OJ, Gough J, et al. An Atlas of Human Long Non-Coding RNAs With Accurate 5′ Ends. *Nature* (2017) 543:199–204. doi: 10.1038/nature21374
180. Gordon CT, Lyonnet S. Enhancer Mutations and Phenotype Modularity. *Nat Genet* (2014) 46:3–4. doi: 10.1038/ng.2861
181. Thurman RE, Rynes E, Humbert R, Vierstra J, Maurano MT, Haugen E, et al. The Accessible Chromatin Landscape of the Human Genome. *Nature* (2012) 489:75–82. doi: 10.1038/nature11232
182. Zhang F, Lupski JR. Non-Coding Genetic Variants in Human Disease. *Hum Mol Genet* (2015) 24:R102–10. doi: 10.1093/hmg/ddv259
183. Maston GA, Evans SK, Green MR. Transcriptional Regulatory Elements in the Human Genome. *Annu Rev Genomics Hum Genet* (2006) 7:29–59. doi: 10.1146/annurev.genom.7.080505.115623
184. Djebali S, Davis CA, Merkel A, Dobin A, Lassmann T, Mortazavi A, et al. Landscape of Transcription in Human Cells. *Nature* (2012) 489:101–8. doi: 10.1038/nature11233
185. Cabili MN, Trapnell C, Goff L, Koziol M, Tazon-Vega B, Regev A, et al. Integrative Annotation of Human Large Intergenic Noncoding RNAs Reveals Global Properties and Specific Subclasses. *Genes Dev* (2011) 25:1915–27. doi: 10.1101/gad.17446611
186. Ruan X. Long Non-Coding RNA Central of Glucose Homeostasis. *J Cell Biochem* (2016) 117:1061–5. doi: 10.1002/jcb.25427
187. Lynn FC, Skewes-Cox P, Kosaka Y, McManus MT, Harfe BD, German MS. MicroRNA Expression is Required for Pancreatic Islet Cell Genesis in the Mouse. *Diabetes* (2007) 56:2938–45. doi: 10.2337/db07-0175
188. Kameswaran V, Bramswig NC, McKenna LB, Penn M, Schug J, Hand NJ, et al. Epigenetic Regulation of the DLK1-MEG3 microRNA Cluster in Human Type 2 Diabetic Islets. *Cell Metab* (2014) 19:135–45. doi: 10.1016/j.cmet.2013.11.016
189. Morán I, Akerman I, Van De Bunt M, Xie R, Benazra M, Nammo T, et al. Human β Cell Transcriptome Analysis Uncovers lncRNAs That Are Tissue-Specific, Dynamically Regulated, and Abnormally Expressed in Type 2 Diabetes. *Cell Metab* (2012) 16:435–48. doi: 10.1016/j.cmet.2012.08.010
190. He X, Ou C, Xiao Y, Han Q, Li H, Zhou S. LncRNAs: Key Players and Novel Insights Into Diabetes Mellitus. *Oncotarget* (2017) 8:71325. doi: 10.18632/oncotarget.19921
191. Wang M, Wang S, Yao D, Yan Q, Lu W. A Novel Long Non-Coding RNA Cyp4b1-PS1-001 Regulates Proliferation and Fibrosis in Diabetic Nephropathy. *Mol Cell Endocrinol* (2016) 426:136–45. doi: 10.1016/j.mce.2016.02.020
192. Ricaño-Ponce I, Wijmenga C. Mapping of Immune-Mediated Disease Genes. *Annu Rev Genomics Hum Genet* (2013) 14:325–53. doi: 10.1146/annurev-genom-091212-153450
193. Sakabe NJ, Savic D, Nobrega MA. Transcriptional Enhancers in Development and Disease. *Genome Biol* (2012) 13:1–11. doi: 10.1186/gb-2012-13-1-238
194. Sanyal A, Lajoie BR, Jain G, Dekker J. The Long-Range Interaction Landscape of Gene Promoters. *Nature* (2012) 489:109–13. doi: 10.1038/nature11279
195. Akerman I, Tu Z, Beucher A, Rolando DM, Sauty-Colace C, Benazra M, et al. Human Pancreatic β Cell lncRNAs Control Cell-Specific Regulatory Networks. *Cell Metab* (2017) 25:400–11. doi: 10.1016/j.cmet.2016.11.016
196. Pasquali L, Gaulton KJ, Rodríguez-Seguí SA, Mularoni L, Miguel-Escalada I, Akerman I, et al. Pancreatic Islet Enhancer Clusters Enriched in Type 2 Diabetes Risk-Associated Variants. *Nat Genet* (2014) 46:136–43. doi: 10.1038/ng.2870
197. Mirza AH, Kaur S, Pociot F. Long Non-Coding RNAs as Novel Players in β Cell Function and Type 1 Diabetes. *Hum Genomics* (2017) 11:17. doi: 10.1186/s40246-017-0113-7
198. Leung A, Natarajan R. Long Noncoding RNAs in Diabetes and Diabetic Complications. *Antioxidants Redox Signaling* (2018) 29:1064–73. doi: 10.1089/ars.2017.7315
199. Hattersley AT, Patel KA. Precision Diabetes: Learning From Monogenic Diabetes. *Diabetologia* (2017) 60:769–77. doi: 10.1007/s00125-017-4226-2
200. Vaxillaire M, Bonnefond A, Froguel P. The Lessons of Early-Onset Monogenic Diabetes for the Understanding of Diabetes Pathogenesis. *Best Pract Res Clin Endocrinol Metab* (2012) 26:171–87. doi: 10.1016/j.beem.2011.12.001
201. Yang Y, Chan L. Monogenic Diabetes: What it Teaches Us on the Common Forms of Type 1 and Type 2 Diabetes. *Endocrine Rev* (2016) 37:190–222. doi: 10.1210/er.2015-1116
202. Maguire JA, Gagne AL, Jobaliya CD, Gandre-Babbe S, Gadue P, French DL. Generation of Human Control iPS Cell Line CHOPWT10 From Healthy Adult Peripheral Blood Mononuclear Cells. *Stem Cell Res* (2016) 16:338–41. doi: 10.1016/j.scr.2016.02.017
203. Yang W, Mills JA, Sullivan S, Liu Y, French DL, Gadue P. iPSC Reprogramming From Human Peripheral Blood Using Sendai Virus Mediated Gene Transfer. *Stem Book* (2008) 1–6. doi: 10.3824/stembook.1.73.1
204. Tsonkova VG, Sand FW, Wolf XA, Grunnet LG, Ringgaard AK, Ingvorsen C, et al. The Endoc-βh1 Cell Line Is a Valid Model of Human Beta Cells and Applicable for Screenings to Identify Novel Drug Target Candidates. *Mol Metab* (2018) 8:144–57. doi: 10.1016/j.molmet.2017.12.007
205. Doane AS, Elemento O. Regulatory Elements in Molecular Networks. *Wiley Interdiscip Rev Syst Biol Med* (2017) 9:e1374. doi: 10.1002/wsbm.1374
206. Dunham I, Birney E, Lajoie BR, Sanyal A, Dong X, Greven M, et al. An Integrated Encyclopedia of DNA Elements in the Human Genome. *Nature* (2012) 489(7414):57–74. doi: 10.1038/nature11247
207. Kundaje A, Meuleman W, Ernst J, Bilenky M, Yen A, Heravi-Moussavi A, et al. Integrative Analysis of 111 Reference Human Epigenomes. *Nature* (2015) 518:317–30. doi: 10.1038/nature14248
208. Johnson JD. The Quest to Make Fully Functional Human Pancreatic Beta Cells From Embryonic Stem Cells: Climbing a Mountain in the Clouds. *Diabetologia* (2016) 59:2047–57. doi: 10.1007/s00125-016-4059-4
209. Tremmel DM, Mitchell SA, Sackett SD, Odorico JS. Mimicking Nature-Made Beta Cells: Recent Advances Towards Stem Cell-Derived Islets. *Curr Opin Organ Transplant* (2019) 24:574. doi: 10.1097/MOT.0000000000000687
210. Melton D. The Promise of Stem Cell-Derived Islet Replacement Therapy. *Diabetologia* (2021) 64:1030–36. doi: 10.1007/s00125-020-05367-2
211. Velazco-Cruz L, Goedegebuure MM, Millman JR. Advances Toward Engineering Functionally Mature Human Pluripotent Stem Cell-Derived β Cells. *Front Bioeng Biotechnol* (2020) 8:786. doi: 10.3389/fbioe.2020.00786
212. Augsornworawat P, Maxwell KG, Velazco-Cruz L, Millman JR. Single-Cell Transcriptome Profiling Reveals β Cell Maturation in Stem Cell-Derived Islets After Transplantation. *Cell Rep* (2020) 32(8):108067. doi: 10.1016/j.celrep.2020.108067

4

Stem Cell-Based Clinical Trials for Diabetes Mellitus

Eleonora de Klerk and Matthias Hebrok[*]

Diabetes Center, University of California San Francisco, San Francisco, CA, United States

Correspondence:
Matthias Hebrok
Matthias.Hebrok@ucsf.edu

Since its introduction more than twenty years ago, intraportal allogeneic cadaveric islet transplantation has been shown to be a promising therapy for patients with Type I Diabetes (T1D). Despite its positive outcome, the impact of islet transplantation has been limited due to a number of confounding issues, including the limited availability of cadaveric islets, the typically lifelong dependence of immunosuppressive drugs, and the lack of coverage of transplant costs by health insurance companies in some countries. Despite improvements in the immunosuppressive regimen, the number of required islets remains high, with two or more donors per patient often needed. Insulin independence is typically achieved upon islet transplantation, but on average just 25% of patients do not require exogenous insulin injections five years after. For these reasons, implementation of islet transplantation has been restricted almost exclusively to patients with brittle T1D who cannot avoid hypoglycemic events despite optimized insulin therapy. To improve C-peptide levels in patients with both T1 and T2 Diabetes, numerous clinical trials have explored the efficacy of mesenchymal stem cells (MSCs), both as supporting cells to protect existing β cells, and as source for newly generated β cells. Transplantation of MSCs is found to be effective for T2D patients, but its efficacy in T1D is controversial, as the ability of MSCs to differentiate into functional β cells *in vitro* is poor, and transdifferentiation *in vivo* does not seem to occur. Instead, to address limitations related to supply, human embryonic stem cell (hESC)-derived β cells are being explored as surrogates for cadaveric islets. Transplantation of allogeneic hESC-derived insulin-producing organoids has recently entered Phase I and Phase II clinical trials. Stem cell replacement therapies overcome the barrier of finite availability, but they still face immune rejection. Immune protective strategies, including coupling hESC-derived insulin-producing organoids with macroencapsulation devices and microencapsulation technologies, are being tested to balance the necessity of immune protection with the need for vascularization. Here, we compare the diverse human stem cell approaches and outcomes of recently completed and ongoing clinical trials, and discuss innovative strategies developed to overcome the most significant challenges remaining for transplanting stem cell-derived β cells.

Keywords: stem cells, type 1 diabetes (T1D), type 2 diabetes (T2D), clinical trial (CT), transplantation, encapsulation, islets

WHY THE NEED FOR STEM-CELL BASED THERAPY IN DIABETES?

Intraportal allogeneic cadaveric islet transplantation is considered the best available treatment for patients with Type 1 Diabetes who cannot control their blood glucose levels with exogenous insulin, despite optimal intensive medical management. It consists of the isolation of pancreatic islets from deceased donors, and their infusion into the liver through the portal vein, which results in engraftment in the hepatic parenchyma (1).

Over two decades have passed since the first seven Type 1 diabetic patients were treated with what is known as the Edmonton Protocol (2), a procedure developed by Dr. James Shapiro and his team. Before the introduction of this ground-breaking protocol, the success rate of islet transplantation (measured as percentage of patients able to remain insulin independent for more than one year) was only 8% (3). Modifications to the standard protocol led to an unprecedent 100% success rate in the first seven patients (2). These modifications included a steroid-free immunosuppressive regimen, the use of xeno-protein-free media during islet isolation, and the immediate transplantation of the purified islets from multiple donors (mean islet mass of 11,547 ± 1,604 islet equivalents per kilogram of body weight), to reach an adequate islet mass capable of restoring normoglycemia. Despite the short-term success in maintaining insulin independence during the first year after transplantation, only 11% of those patients remained insulin-independent after five years. Further improvements to the Edmonton protocol over the last twenty years have markedly increased the safety of islet transplantation, with regard to the rate of adverse events related to the infusion procedure, and to the immunosuppression regimen.

Adverse events categorized as "possibly or definitely related" to the infusion procedure include peritoneal hemorrhage, hepatic hematoma or hemorrhage, portal vein thrombosis, and abnormal liver function, while those related to immunosuppression include leukopenia, mucosal inflammation, graft vs host disease, pneumonia, increased blood creatinine, renal disorder, skin disorder, and hypertension. Data from annual reports released by the Collaborative Islet Transplant Registry (CITR, https://citregistry.org/), which includes clinical trial data from 37 islet transplantation centers (28 in North America, seven in Europe, three in Australia) collected between 1999–2016, indicate that the rate of adverse events in the first 30 days following transplantation dropped from 66 to 22%. Reports from the most recent clinical trial data being collected are not yet publicly available. Although islet transplantation has become one of the safest and least invasive transplant procedures, it currently still requires life-long immunosuppression. In addition, long-term insulin independence, which is often reached right after transplantation, declines over time. According to the latest CITR annual report (10th Annual Report, released in January 2017), the percentage of patients remaining insulin-independent after one year is approximately 50%, and a drop to 25% is observed after five years. A drop in insulin-independency was also reported by the more recent phase 3 CIT trial of human islet-after-kidney transplantation (4), where only 16.7% of the patients retained insulin independence three years after transplantation. These percentages are much lower than what was reported for patients undergoing total pancreas transplantation. Based on a report summarizing data from 2005 to 2016 (from one single transplantation center), 75% of the patients who obtained pancreas transplantation following total pancreatectomy remained insulin-independent [until their time of death, or until the present day (5)].

While these numbers indicate that certain aspects of cadaveric islet transplantation need to be further optimized to achieve long lasting relief from exogenous insulin injections, it is important to note that many clinical goals are still achieved with this procedure, especially with regards to the restoration of hypoglycemia awareness and protection from severe hypoglycemic events (6–8). Hypoglycemia unawareness, a state in which a person is unaware of inappropriately low blood glucose levels, is a severe, relatively frequent and potentially life threatening complication occurring in approximately 40% of patients with type 1 diabetes mellitus (T1DM) (9) (10).

Cell survival and graft rejection are the two key unresolved challenges for increasing insulin independence rates in islet transplantation. Currently, most standard islet transplantations are performed through infusion into the portal vein. Despite encouraging results, the liver might not be the optimal place for transplanted islets as entrapment in the hepatic vasculature results in hypoxia (11), and the revascularization process can take up to 14 days to be fully established (12). In addition to hypoxia that impairs β cell function and survival, acute graft loss is caused by instant blood-mediated inflammatory reaction (IBMIR), which results in activation of the complement cascade, clot formation, and lymphocyte recruitment (13). Together, hypoxia and IBMIR lead to destruction of more than 50% of transplanted islets in the first 48 h following infusion into the portal vein (14). Only a few alternative transplantation sites have been tested in clinical trials so far: the bone marrow (15), the muscle of the forearm (16, 17), and the omentum (18–20). Despite positive outcomes for autologous islet transplantation, clinical trials have shown that the bone marrow site is not a suitable alternative site for pancreatic islet allotransplantation in T1D patients, due to recurrence of autoimmunity (21). Survival of alloislets in the intramuscular site has also been shown to be limited so far (17). The omentum represents a promising site, but the protocols utilizing this site need further optimizing to ensure better vascularization and improved management of immunosuppression for long term success (19).

Risks associated with life-long immune suppressive drugs, together with the limited availability of cadaveric islets, are the two prominent current obstacles to a broader use of islet transplantation for the treatment of diabetes. Immune suppression in general remains critical to prevent rejection of the graft. A combination of induction (administered only at the time of transplant) and maintenance (administered for long-term regime) immune suppressive agents are necessary for graft survival, and despite the advancements in the immunosuppressive regimen (22), the majority of the patients still require additional islet infusions (10th Annual Report, CITR).

Islet transplantation is a government-funded, standard-of care therapy in Canada, Europe, China, and other parts of Asia (23), but only for a minority of T1D patients suffering from glycemic lability, hypoglycemia unawareness, severe hypoglycemic episodes, and/or diabetic ketoacidosis, despite optimal intensive medical management. These extreme scenarios are often referred to as "brittle" diabetes. Given the limited supply of cadaveric islets, most transplant centers also limit enrollment only to T1D patients who have complete loss of C-peptide production (24). Even when islet transplantation is government-funded, access to this procedure may still be problematic. In Canada, while islet transplantation is available in the province of Alberta, access to it is more difficult in all the other provinces, where the procedure is not always recognized as a non-research therapy (25). Access to islet transplantation is highly restricted in the United States where the procedure is not approved by health insurance companies as it is still considered an experimental treatment and requires filing of an investigational new drug application (NDA) with the Food and Drug Administration (FDA). This is one of the possible reasons why the total number of centers performing islet transplantations and the total number of transplantations performed per year have declined in the US since 2014, according to the 10th Annual CITR Report. Interestingly, this decline is evident not only for centers in North America, but also in the European and Australian centers that are part of the CITR. The exact reasons for this reduced activity for these centers need to be determined, but American Centers have observed a reduction in pancreas donors since the mid-2000s (https://optn.transplant.hrsa.gov/data/view-data-reports/national-data/#). Unfortunately, among all organs isolated for transplantation the pancreas has become the organ with the lowest donation rate [approximately 11 donors per 100 eligible deaths were recorded in 2018 (26)].

A curated review of islet transplant trials registered on ClinicalTrials.gov (24) also revealed that although there are a number of newly registered trials focusing on testing alternative implant sites and innovative approaches to reduce graft rejection, including encapsulation devices and immune modulators, the overall number of clinical trials for cadaveric islet transplantation is not growing.

While it is highly likely that cell survival and graft rejection will continue to improve in the future, the low supply of cadaveric islets remains the critical limitation prohibiting wide-spread use of this therapy. In contrast, the prevalence of patients with T1D is increasing globally (27). Based on a recent diabetes forecasting model (28), by 2030 the total number of people with T1D and T2D in the United States alone will increase to 50 million, a 54% increase from 2015.

One possible pathway to a treatment, and perhaps a cure, for a broad number of diabetic patients, would be access to an alternate, unlimited source of insulin-producing cells that can reconstitute physiological glucose homeostasis, eliminating the reliance on organ donors.

Stem cell-derived β-cell therapy overcomes the barrier of limited donor availability, while also possibly representing a more cost-effective therapy compared to exogenous insulin. Although the future cost of stem cell-derived β-cell therapy is unknown, a speculative cost-effectiveness analysis from an early health technology assessment study (29) calculates that 8–9 years after transplantation both cadaveric islet transplantation and stem cell-derived β-cell transplantation would be more cost-effective than exogenous insulin therapy. This calculation assumes that the manufacturing costs of stem cell-derived β cells will be similar to the costs necessary to isolate cadaveric islets. In both cases, successful long-term engraftment is essential for the therapies to become profitable (30). The cost of stem cell-derived β-cell therapy will depend on a number of variables, including the requirement of immunosuppression, the duration of graft survival, and, most importantly, the optimization of the manufacturing process (31). While avoidance of autoimmune rejection will reduce the price, higher upfront costs are needed for the development of a scalable manufacturing process and the creation of stem cell banks for clinical use. Emerging technologies]such as stirred suspension bioreactor culture, wave bag bioreactor culture, multiplate culture, and roller bottle culture (32)] may eventually allow for mass production of stem cell-derived β cells and islet-like organoids. If a manufacturing expenses can be reduced, early health technology assessment studies demonstrated that stem cell-derived β-cell therapy will be a cost-effective (33, 34).

In addition to endogenous cell therapies, wearable computerized devices, such as insulin pumps and closed-loop systems, present alternative options for the delivery of exogenous insulin, and clinical trials are currently testing their efficacy. Insulin pumps (and sensor-augmented insulin pumps) can be programmed to continuously deliver a basal level of insulin as well as extra doses (bolus) during mealtimes. In contrast to these pumps that require manual adjustments and input from the patient, closed-loop systems (also known as artificial pancreas or automated insulin delivery systems) constitute a combined sensing-delivery system in which an external glucose sensor directs delivery of insulin from a sensor-responsive pump guided by real-time glucose sensor readings. A recent clinical trial on T1D patients reported that closed loop systems maintain control within the near normoglycemic range up to 71 ± 12% of the time. This represents a marked improvement over sensor-augmented insulin pumps, with which the normoglycemic range is maintained up to 59 ± 14% of the time (35). Moreover, closed loop systems are also able to improve glycated hemoglobin levels (HbA1c), an indicator of long-term systemic glucose levels. Similar improvements were also noticed when comparing sensor-augmented pumps against hybrid closed-loop system (where insulin is continuously administered, except during boosts at mealtime) (36), or bihormonal closed loop systems (37) that release both insulin and glucagon. The higher efficacy obtained with the bihormonal closed loop system represents an important achievement considering that glucagon's stability in solution is much lower than that of insulin, and that its remarkably dose-response relationship requires tight regulation of its release. A meta-analysis comparing 40 studies resulted in similar finding (38), asserting that closed loop systems are more efficient than any other insulin pump therapies, and represent an efficacious and safe approach for management of T1 diabetes. While closed loop systems clearly constitute a marked improvement in blood glucose control, there remains room for improvement of the algorithms controlling insulin release, for instance during physical activity. Variables such as the duration and intensity of physical exercise, and the proper timing of hormone release relative to food intake remain hard to control with a fully automated closed-loop system (39, 40). Other concerns regard the wearability, and the cost-effectiveness, especially if considering the constant rise in the price of insulin (a 200% increase from 2002 to 2013 (41), and a 14% annual increase from 2012 to 2018).

The technical advances described above have dramatically improved the lives of patients with diabetes. However, the limitations of these systems, including need for attaching external devices that penetrate the skin and thus raise the chance of infection and scarring over time, the dependence on fully functional pumps and sensors whose dysfunction can result in rapid and life threatening changes in glucose levels, and the complexities of algorithms tasked with anticipating ever changing aspects of patients metabolism, indicate that they present a powerful temporal solution but not a cure for diabetes.

STEM CELL-BASED APPROACHES: PROTECTION OR RESTORATION OF β CELLS MASS

A curated list of completed, active, recruiting, and suspended stem cell-based clinical trials for both T1D and T2D, registered at ClinicalTrials.gov within the last ten years, is presented in **Table 1**. The majority of the recently completed and active trials use adult mesenchymal stem cells (MSC) derived from different origins, hematopoietic stem cells, or a combination of both. Although initial studies might have suggested the possibility of generating insulin-producing cells from MSCs, clear evidence supporting this hypothesis is currently lacking. Thus, the purpose of these current trials is to understand the mechanisms of protection provided by MSCs and evaluate their efficacy, especially in modulating the immune response. In contrast, a number of publications have demonstrated the potential for human embryonic stem cells (hESCs) (42–46) and induced pluripotent stem cells (iPSCs) (43, 47) to form functional mature insulin-producing β cells, and trials with such cells address the issue of directly restoring β cell mass. Despite these efforts being very promising, only three trials have so far utilized hESCs to derive pancreatic progenitors for β cell replacement therapy. However, successful preclinical studies in non-human primates, led by Vertex Pharmaceutical and Sigilon Therapeutics, are paving the road towards more pluripotent stem cell-based clinical trials.

MESENCHYMAL STEM CELL-BASED THERAPY

Mesenchymal stem cells (MSCs), or stromal stem cells, are currently the most widely used stem cells in clinical trials (www.clinicaltrials.gov). MSCs are multipotent adult stem cells that can be derived from both adult and neonatal tissues. Although adult bone marrow is the most prevalent source, MSCs can be obtained from almost all tissues that include a perivascular area (48, 49). MSCs are derived from the mesodermal germ layer and have a trilineage differentiation potential, that is the ability to differentiate *in vitro* into osteoblasts (bone tissue), chondroblasts (cartilage), and adipocytes (fat tissue) 29. A number of studies have also shown neuronal crest-derived MSCs (50–52) and given the high heterogeneity of MSCs it remains to be determined if additional sources besides the paraxial mesoderm and the neural crest exist. From a regulatory perspective, MSCs have been classified as an advanced therapy medicinal product (https://www.ema.europa.eu/en/human-regulatory/overview/advanced-therapy-medicinal-products-overview).

Although MSCs are emerging as the most promising source for allogeneic cell therapy (53), the therapeutic use of MSCs in T1D clinical trials is highly controversial. Three different hypotheses have been explored in clinical settings: (a) the use of MSC-derived pancreatic progenitors that develop into functional β cells capable of restoring normoglycemia, (b) the use of undifferentiated MSCs to generate β cells through direct transdifferentiation *in vivo* upon transplantation, and (c) the use of undifferentiated MSCs to support islet health and survival without differentiating into pancreatic progenitors (**Figure 1**). As of yet, strong evidence to support the hypothesis that MSCs can differentiate into functional mature β cells or islet-like organoids, both *in vitro* and *in vivo*, is lacking.

MSCs' MECHANISMS OF ACTION

Early studies have investigated the hypothesis that MSCs differentiate into insulin-producing cells (54–61). This was in part based on the observation that expression of insulin and other pancreatic transcription factors increase in differentiating MSCs. However, the mere presence of such markers, including PDX1, NGN3, NEUROD1, NKX6.1, and ISL, is not proof of fully matured β cells, as some of these factors are found to be expressed also upon expansion of MSCs *in vitro* (58), and during development of other cell types, such as neurons (62). Furthermore, the presence of these proteins alone does not guarantee mature β cell activities, as expression of non-β cell factors could interfere with critical processes, whereas expression of other markers essential for mature function (including, but not limited to, K^+-channels, Ca^{2+}-channels, secretory vesicles) might still be missing. Functionality of MSC-derived insulin-producing cells has been tested by glucose stimulated insulin secretion *in vitro* and by glucose tolerance in mice *in vivo*. The claimed ability to respond to glucose by secretion of insulin may have been overestimated as insulin levels, rather than the more appropriate C-peptide levels (57, 60, 61), were measured. Insulin secretion may not indicate true insulin production as the hormone is often present in culture media (63). This is supported by the observation that when both insulin and C-peptide levels were determined, a significant increase in secretion was observed only for insulin (56) but not for C-peptide, a normal by-product generated during the maturation of the hormone. Nevertheless, one study showed a four-fold increase

TABLE 1 | Completed and active stem-cell based clinical trials for T1D and T2D.

Trial ID	Study start date	Sponsor and Collaborators	Cell type	Diabetes subtype	Status	Official title	Purpose of the study	Treatment method
NCT01068951	2010-06-01	Uppsala University Hospital	MSCs	T1D	Completed	Open Study to Evaluate the Safety and Efficacy of Autologous Mesenchymal Stem Cells in Treatment of Recently Diagnosed Patients With Type 1 Diabetes Mellitus	Test if development of autoimmune diabetes may be halted by the immune modulatory properties of mesenchymal stem cells	Intravenous injection of autologous mesenchymal stem cells
NCT02239354	2014-09-01	ViaCyte. California Institute for Regenerative Medicine (CIRM)	hESCs	T1D	Suspended	A Prospective, Multicenter, Open-Label, First-in-Human Phase 1/2 Study With Two Cohorts to Evaluate the Safety, Tolerability, and Efficacy of Various Doses of VC-01™ Combination Product in Subjects With Type 1 Diabetes Mellitus	Test if VC-01™ combination product can be implanted subcutaneously and maintained safely for two years. It will also test if VC-01 is an effective treatment	Subcutenous transplantation of combination product VC-01 (PEC-01 cells loaded into PEC-Encap)
NCT03920397	2015-03-01	Universidade Federal do Rio de Janeiro	MSCs	T1D	Active / Recruiting	Allogenic Adipose Derived Mesenchymal Stem Cells and Vitamin D Supplementation in Patients With Recent-onset Type 1 Diabetes Mellitus	Unspecified	Intravenous injection of adipose tissue-derived stem/stromal cells and oral Cholecalciferol supplementation
NCT04078308	2015-07-06	Royan Institute. Tehran University of Medical Sciences, Iranian Stem Cell Council	MSCs	T1D	Active / Recruiting	Phase I/II Clinical Trial to Examine the Safety and Efficacy of Transplantation of Mesenchymal Stem Cells in New-onset Type 1 Diabetes Patients	Modulate immune response and improve regeneration	Intravenous injection of autologous mesenchymal stem cells
NCT02940418	2017-02-19	Sophia Al-Adwan	MSCs	T1D	Active / Recruiting	The Use of Mesenchymal Stromal Cells (MSC) in Type 1 Diabetes Mellitus in Adult Humans: Phase I Clinical Trial	Unspecified	Intravenous injection of allogenic adipose-derived mesenchymal cells with autologous bone marrow mononuclear cells
NCT03162926	2017-07-05	ViaCyte	hESCs	T1D	Completed	An Open-Label Study Evaluating the Safety and Tolerability of VC-02™ Combination Product in Subjects With Type 1 Diabetes Mellitus	Test if VC-02™ combination product can be implanted subcutaneously and maintained safely for up to four months	Subcutenous transplantation of combination product VC-02 (PEC-01 cells loaded into PEC-Direct). Up to six VC-02-20 implants
NCT03163511	2017-07-06	ViaCyte. California Institute for Regenerative Medicine (CIRM), Horizon 2020 - European Commission	hESCs	T1D	Active / Recruiting	An Open-Label, First-In-Human Study Evaluating the Safety, Tolerability, and Efficacy of VC-02™ Combination Product in Subjects With Type 1 Diabetes Mellitus and Hypoglycemia Unawareness	Test if VC-02™ combination product can be implanted subcutaneously and maintained safely for up to two years. It will also test if VC-02 is an effective treatment	Subcutenous transplantation of combination product VC-02 (PEC-01 cells loaded into PEC-Direct). Cohort 1: up to six VC-02-20 implants and up to two VC-02-300 implants. Cohort 2: up to ten VC-02-300 and up to two VC-02-20.
NCT03406585	2017-11-28	NextCell Pharma Ab	MSCs	T1D	Active / Recruiting	A Double-blinded, Randomized, Placebo-controlled Trial With Wharton's Jelly Derived Allogeneic Mesenchymal Stromal Cells (WJMSCs) for Preserving Endogenous Insulin Production in Adult Patients Diagnosed for Type 1 Diabetes	Unspecified	Transplantation of cell suspension with expanded allogenic MSC's procured from donated Wharton's Jelly from umbilical cord tissue
NCT03912480	2019-01-05	CAR-T (Shanghai) Biotechnology Co., Ltd.	MSCs	T1D	Active / Recruiting	Study on the Efficacy and Safety of Stem Cells From Human Exfoliated Teeth in Treating Diabetic Patients With Significantly Reduced Islet	Unspecified	Intravenous drip of dental pulp mesenchymal stem cells

(Continued)

TABLE 1 | Continued

Trial ID	Study start date	Sponsor and Collaborators	Cell type	Diabetes subtype	Status	Official title	Purpose of the study	Treatment method
NCT03973827	2019-05-17	NextCell Pharma Ab	MSCs	T1D	Active / Recruiting	An Open Label, Parallel Single Center Trial of Wharton's Jelly Derived Allogeneic Mesenchymal Stromal Cells Repeatedly Treated to Preserve Endogenous Insulin Production in Adult Patients Diagnosed With Type 1 Diabetes	Investigate safety and tolerance after a repeated allogeneic infusion of WJMSCs intravenously after one year following the repeated treatment.	Transplantation of cell suspension with expanded allogenic MSC's procured from donated Wharton's Jelly from umbilical cord tissue
NCT04061746	2020-02-13	Medical University of South Carolina. National Institute of Diabetes and Digestive and Kidney Diseases (NIDDK)	MSCs	T1D	Active / Recruiting	Cellular Therapy for Type 1 Diabetes Using Mesenchymal Stem Cells	Determine the safety and efficacy of allogeneic umbilical cord-derived mesenchymal stromal cells for the treatment of new-onset T1D and to understand the mechanisms of protection	Intravenous injection of autologous mesenchymal stem cells
NCT01719640	2011-01-01	Fuzhou General Hospital	MSCs	T2D	Completed	Autologous Bone Marrow Mesenchymal Stem Cell and Bone Marrow Mononuclear Cell Infusion in Type 2 Diabetes Mellitus	Provide signals for regeneration and improve recovery from inflammation-induced lesion	Intra-arterial pancreatic infusion of autologous bone marrow mononuclear cells in combination with autologous bone marrow mesenchymal stem cells
NCT01576328	2012-04-01	Mesoblast, Ltd.	MSCs	T2D	Completed	A Randomized, Placebo-Controlled Dose-Escalation Study to Assess the Safety and Tolerability of a Single Intravenous Infusion of Allogeneic Mesenchymal Precursor Cells (MPCs) in Patients With Type 2 Diabetes Sub-optimally Controlled on Metformin	Assess safety and tolerability of a single intravenous infusion of three doses of Mesenchymal Precursor Cells	Single intravenous infusion of MPCs
NCT01759823	2012-12-01	Postgraduate Institute of Medical Education and Research	MSCs	T2D	Completed	Efficacy and Safety of Autologous Bone Marrow Derived Stem Cell Transplantation in Patients With Type 2 Diabetes Mellitus	Unspecified	mesenchymal stem cell will be injected into superior pancreatic duodenal artery
NCT03343782	2017-11-01	Vinmec Research Institute of Stem Cell and Gene Technology	MSCs	T2D	Completed	Outcomes of Expanded Autologous Bone Marrow-derived Mesenchymal Stem Cells Therapy in Type 2 Diabetes	Evaluate safety and effectiveness of autologous bone marrow-derived mesenchymal stem cells transplantation	Transplantation of autologous bone marrow-derived mesenchymal stem cells
NCT03943940	2019-04-24	Van Hanh General Hospital	MSCs	T2D	Active / Recruiting	A Preliminary Safety and Efficacy Evaluation of Bone Marrow Mononuclear Cells (BM-MNCs) and Umbilical Cord Tissue-derived Mesenchymal Stem Cells (UC-MSC) Infusion for Type 2 Diabetes Mellitus (T2DM) Patients	Unspecified	Intravenous injection of autologous bone marrow mononuclear cells and allogeneic umbilical cord tissue-derived mesenchymal stem cells

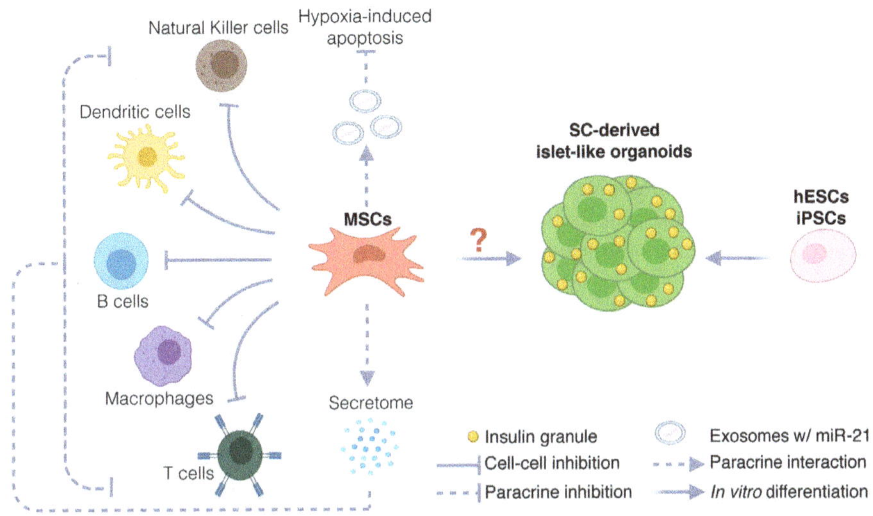

FIGURE 1 | Potential therapeutic mechanisms. Potential mechanisms include protection of endogenous islets and restoration of β cells mass. MSCs could protect endogenous β cells *via* immunomodulation and inhibition of hypoxia-induced apoptosis. Immunomodulation is exerted *via* two mechanisms: inhibition through direct cell-cell interaction with immune cells, and inhibition through paracrine activity, by secretion of chemokines, cytokines, and growth factors (secretome). Inhibition of hypoxia-induced apoptosis could be exerted through release of exosomes carrying miR21, targeting messenger RNAs involved in the hypoxia-mediated ER stress preceding apoptosis. The therapeutic use of MSCs as source for generating stem-cell derived β cells and islet-like organoids is uncertain. hESCs and iPSCs instead are used to generate functional islet-like organoids to restore β cell mass.

in C-peptide secretion upon glucose stimulation *in vitro* (58), suggesting that some MSCs may differentiate into glucose-responsive insulin secreting cells. Glucose tolerance tests have shown some ability in decreasing blood glucose levels upon transplantation into pancreactomized or Streptozotocin (STZ)-induced diabetic mice, although at suboptimal levels. Guo et al. (60) showed an initial maintenance of low blood glucose levels, followed by a drastic increase in blood sugar just 14 days after transplantation. Thus, in the absence of human C-peptide measurements it is difficult to conclude whether the initial low blood glucose levels in mice transplanted with MSC-derived insulin-producing cells were due to the presence of human C-peptide or because of persisting mouse C-peptide, especially without a proper control (STZ-treated mice without transplanted cells). The time-window was extended in the study by Dong-Qui et al. (57), but with a less favorable outcome. Although mice transplanted with MSC-derived insulin-producing cells showed an initial reduction in blood glucose levels by almost 50%, these mice developed tumors and became diabetic within 45 days. Tumor mass is most likely derived from undifferentiated MSCs as bone marrow-derived MSCs have shown to spontaneously transform into neoplastic cells during long-term culture *in vitro*. The neoplastic propensity is not observed in every study (58), possibly due to differences in the source of MSCs, the differentiation protocols employed, or the differentiation stage of the transplanted cells (64). Normoglycemia was also not determined in the study conducted by Kamalaveni et al. (58). Low levels of human C-peptide were detectable in response to glucose only 60 days after transplantation, and although still measurable 150 days post-transplantation, the range was highly variable and below therapeutic purposes (range 0.0–7.97 pmol/ml). Absence of pro-insulin transcript, suggesting the inefficiency in generating mature insulin-producing cells, was also observed in the study from Phadnis et al. (55). Furthermore, none of these studies has extensively investigated the transcriptome profile of the MSC-derived insulin-producing cells, their heterogeneity, and the percentages of poly-hormonal cells versus mono-hormonal cells. Neither were critical functional aspects of mature β cells analyzed, such as biphasic dynamic insulin secretion, proper calcium signaling, mitochondrial respiratory function, and induction of mitochondrial oxidative phosphorylation for glucose oxidation. Summarily, the evidence for differentiation of MSCs into β cells remains unconvincing. In addition, cell tracing studies in mice do not support direct transdifferentiation of multi-potent MSCs into pancreatic progenitors after transplantation *in vivo* (65–67).

Conversely, a number of pre-clinical studies performed over the last 15 years support the hypothesis that MSCs protect islet grafts (68–73) *via* at least two different mechanisms, improvement of cell survival, and immune-modulation. Back in 2012, Ezquer at al. (67) showed that intravenous administration of murine bone marrow MSCs in STZ-treated mice improves blood glucose levels, decreases glycated hemoglobin to levels similar to non-diabetic mice, and increases insulin total insulin levels. Fluorescence-tracing confirmed that MSCs do not differentiate into insulin producing cells, but engraft in lymphoid organs where they restore both the systemic and the local balance of regulatory T cells, increase anti-inflammatory markers such as IL13, and decrease proinflammatory markers such as IL1 beta, IL18, tumor necrosis factor alpha (TNF alpha), and MCP1. In addition, STZ-treated mice transplanted with MSCs also showed an increase in EGF, a trophic factor involved in cell survival. A recent study (74) has shown that highly proliferative bone marrow MSCs can promote autochthonous β cell regeneration *in vivo* in mice with partial pancreatomy. Higher levels of proliferation were reported based on an increase in number of bromodeoxyuridine positive cells, and a plausible mechanism points to the downregulation of the FoxO1 pathway. In addition to enhanced proliferation, MSC-treated mice also displayed an increase in EGF and total insulin content together with a decrease in interferon gamma and TNF alpha. These data suggest that at least some of the beneficial effects of MSC treatment are mediated *via* a reduction of inflammation.

MSCs-BASED CLINICAL TRIALS

Despite the lack of convincing evidence from pre-clinical studies, clinical trials have been performed testing the hypothesis that MSC-derived pancreatic progenitors generated *in vitro* maturate into β cells *in vivo*, either alone (75) or upon co-transplantation with bone marrow-derived stem cells (76). Although both trials reported positive outcomes in terms of improvement in HbA1c, in addition to an increase in serum C-peptide, a decrease in glutamic acid decarboxylase antibodies (GAD), and a decrease in exogenous insulin requirement, clear evidence for the presence of mature functioning MSC-derived β cells is still lacking. What is more likely is that the benefits of MSC transplantation derive from the immune-modulation and/or the protective role of these cells towards endogenous islets. In fact, co-infusion of MSCs-derived islet pancreatic progenitors with bone marrow-derived hematopoietic stem cells resulted in better long-term control of hyperglycemia as compared with MSC-pancreatic progenitors only (75, 76).

For the reasons listed above, the most recent trials (**Table 1**) focus on the third hypothesis: MSCs support islet health and survival *via* indirect means. Potential mechanisms of action include a paracrine effect through secretion of growth factors (77), modulation of extracellular matrix, ability to scavenge reactive oxygen species (ROS), ability to protect against hypoxia-induced apoptosis through micro RNAs (miRNAs) derived from exosomes (78), and the ability to modulate the immune system (79) (80, 81) through inhibition of T-cell proliferation and promotion of regulatory T-cells, or through interactions with other immune cell types, such as macrophages, B-cells, dendritic cells, and natural killer cells (53).

Meta-analyses of data from completed clinical trials suggest that MSCs can protect islets in T2D but the effect in T1D patients remains questionable. A clear interpretation of early MSC-based clinical trials for T1D has been challenged by the limited number of enrolled patients, and/or by the trial's design that does not allow a proper statistical analysis. A trial conducted by Mesples and his team in 2013 (80) to test the efficacy of autologous bone marrow stem cell transplant reported improvements especially in the reduction of anti-pancreatic islet antibodies. The follow up study at 12 months showed negative value in islet cell antibodies (ICA), GAD, and insulin antibody levels, followed by an increased level of C-peptide and decreased levels of blood glucose and HbA1c. However, a decrease in blood glucose and HbA1c was also seen in the only available control patient, and the levels of C-peptide were not fully maintained in one of the two enrolled patients after 12 months. The small number of treated patients and controls rendered the interpretation of the data inconclusive. In 2015, Carlsson and his team conducted a similar trial (81) on twenty adult patients with newly diagnosed T1D. Treated patients showed preservation or even increase in C-peptide levels in response to a mixed-meal tolerance test 12 months after transplantation.

Although each of these trials suggested that MSC-based therapy promotes β cell health and function in T1D patients, systematic reviews and meta-analysis studies of controlled clinical trials are still debating their positive outcomes. A meta-analysis performed in 2018 (82), comprising 9 randomized-controlled trials and 14 self-controlled trials, concluded that the pooled effect of hematopoietic stem cells therapy, MSC-based therapy, and co-infusion of hematopoietic and multipotent MSCs, resulted in an increased C-peptide level, compared with conventional insulin therapy, whereas trials based on umbilical cord blood-derived MSCs did not reach a significance. A separate meta-analysis of 6 controlled T1D trials published in 2019 (83) showed that there was no difference in the levels of stimulated C-peptide and fasting C-peptide. The reduction in HbA1c was the only difference observed between treated and control patients. Results from ongoing randomized-controlled trials, with larger number of enrolled patients and controls, are needed to elucidate the efficacy of MSC therapy for T1D.

Contrary to the questionable benefits of MSC-based therapies for T1D, MSC-based clinical trials for T2D have shown a constant and robust efficacy. T2D MSC-based trials make use of multipotent MSCs derived from different sources (**Table 1**). From the first trial in 2009 (84) to the most recent trials (85–87) improvements have been observed in C-pep levels, HbA1c values, and reductions in the required insulin dosage. A systematic review of 10 T2D MSC-based trials confirmed a significant increase in the levels of stimulated C-peptide and fasting C-peptide (83). However, despite the large number of *in vitro* studies and *in vivo* pre-clinical studies already conducted, the exact mechanism by which MSCs improve outcomes still remains to be elucidated. Whether the discrepancy between T1D and T2D trials outcomes is caused by technical limitations relative to how the trials for T1D patients have been designed, or by differences in the etiology, needs to be also elucidated.

PLURIPOTENT STEM CELL-BASED CLINICAL TRIALS

Human embryonic stem cells (hESCs) are pluripotent cells isolated from the inner cell mass (ICM) of the blastocyst (88). They possess self-renewal capacity, genomic stability, and can give rise to all three lineages (endoderm, mesoderm, and ectoderm). Induced pluripotent stem cells (iPSCs) (89) are generated from somatic cells by ectopic overexpression of specific transcription factors. iPSCs also have the capacity of self-renewal and differentiation potential, though their genomic stability is still questionable. hESCs and iPSCs maintain their pluripotency after expansion (90), thus fulfilling that need of unlimited supply required for therapeutic purposes. Although iPSCs are emerging as a potential alternative to hESCs, their ability to differentiate into mature pancreatic endocrine cells has not yet reached the same quality observed with hESC protocols (91).

In 2015, the first T1D patient was treated with a hESC-based pancreatic progenitor transplant in Edmonton. The study was driven by the regenerative medicine company ViaCyte, under the supervision of J. Shapiro's team. The purpose of the trial (NCT02239354, submitted in 2014) was to test a combination of hESC-derived pancreatic progenitor cells (PEC-01) (92, 93) expected to mature into functional insulin-producing cells upon transplantation based on prior studies with surrogate animals, within an encapsulation device called PEC-Encap (VC-01) (94). The first transplantation consisted of 40 million pancreatic progenitor cells, divided into two encapsulation devices implanted subcutaneously in the abdomen, along with six smaller encapsulation devises implanted subcutaneously in the arm, serving as sentinels to be removed at different time points to follow cell survival and maturation. This first encapsulation device was designed to protect the pancreatic progenitor cells from the immune system, preventing both allogeneic (foreign organ) reaction and autoimmune rejection, eliminating the necessity of immunosuppressive drugs. The device had a semipermeable membrane that allowed exchange of molecules but not cells. VC-01 (consisting of the combination of PEC-01 cells and PEC-Encap device) was meant to be evaluated in an open-label, dose-escalating Phase 1/2 study in T1D patients with minimal insulin-producing β-cell function. The trial was suspended due to inconsistencies in cell survival and poor cell engraftment, primarily caused by a foreign body response, similar to a wound healing which clogged the membrane and prevented vascularization. This first trial indicated the necessity for optimization of the encapsulation device.

In 2017 ViaCyte launched a second 12-months trial (NCT03162926) which introduced an alternative encapsulation device (PEC-Direct, VC-02), with a modified membrane that does not provide immune protection, but allows vascularization, and therefore requires the re-introduction of immunosuppressants. No changes were performed in the type of hESC-derived pancreatic progenitors used. Successful outcomes led to the currently ongoing 2-year trial (NCT03163511), aimed at testing safety and tolerability of VC-02 implanted subcutaneously in T1D subjects with hypoglycemia unawareness. The purpose of this trial is also to test whether VC-02 is an effective treatment. PEC-01 cells were able to engraft, survive, and produce measurable C-peptide levels (95). Preliminary results from a small subset of patients (six out of 18) showed substantial engraftment of sentinel devices containing insulin positive cells (9 months after transplantation), and production of C-peptide in all patients up to 12-months (with some patients already reaching 15, 18, or 21 months). Moreover, the immunosuppression regimen prevented allogeneic and autoimmune destruction of the cells, without causing a foreign body response. The intended islet mass transplanted was intentionally insufficient to normalize HbA1c levels, therefore no data regarding the efficacy is available. Although these positive outcomes are restricted to 30% of the transplanted patients, further optimization of microencapsulation device materials might improve future outcomes. In support of this notion, a press release from Viacyte from August 2020 (https://viacyte.com/news-events/) announced a clinical phase agreement with Gore, a materials science company, for the development of a modified version of the original PEC-Encap, which has the potential to eliminate the need for immunosuppression while still allowing vascularization.

In addition to Viacyte, two other companies, Vertex Pharmaceutical and Sigilon Therapeutics, are moving forward towards clinical trials with stem cell derived beta cells. These companies are taking a different approach in terms of cell type and immune protection. Viacyte's cells (PEC-01) (92, 93) consist of a mixture of hESC-derived multipotent pancreatic progenitors (which can differentiate into endocrine, exocrine, or ductal cells) and immature hormone-producing cells. This choice derives from the observation that immature progenitors can better overcome the inflammation initiated by the transplantation procedure (96). Vertex Pharmaceutical and Sigilon Therapeutics produce stem cell-derived islet-like organoids which lack the progenitor population. Islet-like organoids generated by Vertex have been tested in a pre-clinical study performed on non-human primates, mimicking cadaveric islet transplantation. Organoids were delivered through the portal vein, in combination with immunosuppressants. The transplanted organoids successfully engrafted in the liver and were functional over a period of 6 weeks. Although the transplanted amount did not lead to insulin independence, the study showed a 60% reduction in the required insulin dosage (International Society for Stem Cell Research (ISSCR) Annual Meeting 2019). Vertex is also developing its own macroencapsulation device, consisting of a porous membrane which allows immune protection. The first pre-clinical study in pigs showed that the device was able to confer immune protection, while still balancing cell survival and foreign body reaction (ISSCR Annual Meeting 2019). Alice Tomei and her team recently reported an alternative encapsulation strategy, termed conformal coating, consisting in a uniformly thin hydrogel layer that conforms to the islet shape. Preliminary data in mice revealed that the conformal-coated stem cell-derived islets could reverse diabetes and maintain euglycemia for more than 80 days (97). Positive results were also obtained from pre-clinical studies performed on macaques by Sigilon (98), using a different encapsulation strategy. Rather than generating islet-like organoids that aggregate solely by cell-cell interaction, the company utilize a microencapsulation technology consisting of gel-based spheres that can hold up to 30,000 cells (99). Chemical modifications on the surface of the spheres allows for immune protection. Endocrine cells within these clusters were shown to remain functional after transplantation for up to four months (98).

The timing for clinical trials has not been released, but both companies aim to bring their technology into clinical use in the near future.

It is worth noting that numerous companies are developing and clinically testing encapsulation devices that, although initially aimed at preserving human pancreatic islets, could be quickly applied to stem cell-derived therapies. An example of such a company is BetaO2 Technologies, which developed a bioartificial pancreas, called Beta-Air, designed to contain macro-encapsulated human islets together with an oxygen tank. Human islets encapsulated within an alginate-based hydrogel are protected from the immune system by a permselective membrane, and are continuously supplied with oxygen. The bioartificial pancreas was tested in patients with T1D in 2014 (NCT02064309), and the company is currently developing a second-generation device specifically adapted for stem cell-derived pancreatic clusters.

FUTURE DIRECTIONS

After decades of MSC-based clinical trials for both T1 and T2 diabetic patients, hESCs and iPSCs-based β cell replacement therapies are finally becoming a tangible reality with the first hESC-derived islet-like organoids transplanted in T1D patients in 2014. A tremendous amount of clinical testing is now necessary to investigate the many aspects involved in stem cell transplantation, including the long-term safety of each encapsulation device, the optimal implant size to reach a therapeutic effect, and the long-term viability of the transplanted cells. Improvements of cell survival in subcutaneous and/or intramuscular space would alleviate safety concerns and allow for easier transplant monitoring. A preclinical study on non-human primates has shown that long-term survival (over 800 days) of human islets transplanted subcutaneously can be achieved when islets are mixed with a matrix, termed islet viability matrix, consisting of human collagen 1, l-glutamine, fetal bovine serum, sodium bicarbonate and medium199 (100). Similarly, numerous groups are developing drug-eluting scaffolds to modulate the immune reaction. These biomaterials have been shown to not only reduce local inflammation following transplantation (101) but also maintain long-term graft survival (102). Islet viability matrixes and drug eluting matrices/scaffolds should be tested in clinical trials, in combination with encapsulation devices. Co-transplantation of autologous non-endocrine tissues which may help cell survival and engraftment may also be considered, such as co-transplantation with parathyroid gland tissue, a method currently tested with cadaveric islets in the intramuscular space (NCT03977662).

Efforts to compare the quality of the cell mixtures required to achieve optimal metabolic control are crucial. Currently, three different cell mixtures can be generated: pancreatic progenitors, islet-like organoids, and enriched β cell clusters; of those, only the first two are explored in clinical trials. Numerous studies have shown that cell-cell communication between different endocrine cells are critical for correct glucose responsiveness and electrical coupling of stimulus with insulin secretion (103). Current ongoing clinical trials are utilizing a mixture of pancreatic progenitor cells which can differentiate and mature into endocrine, exocrine, or ductal cells *in vivo*. Although the architecture of these organoids more likely resembles that of endogenous human islets, it is still not known whether immature polyhormonal cells remain in small quantities after transplantation, and whether they may disrupt proper function over time. The use of mature enriched β cell clusters has not been tested in clinical settings, but considering that enriched β cell clusters lack other islet cell types, such aggregates may not provide optimal metabolic control either. Furthermore, human islets contain specialized β cells, termed hub cells, which have reduced β cell identity but regulate efficient islet response to changes in glucose levels (104). Enrichment strategies that aim at targeting highly insulin expressing cells may therefore exclude β cell hubs. Previous studies have shown that enriched β cell clusters that have already reached maturity *in vitro* can still continue to mature *in vivo*, and generate mono-hormonal glucagon and somatostatin positive cells (105). Whether these *in vivo* matured β cell clusters contain β cell hubs remains to be determined. Transplantation of islet-like organoids, consisting of mature β, alpha, and delta cells, which have been individually differentiated, and subsequently clustered based on endogenous percentages, has also not yet been tested in pre-clinical nor clinical settings, although protocols for the *in vitro* generation of both β and alpha cells have been optimized (32, 106). The generation of somatostatin-producing cells is currently achieved in small percentages as a bio-product of the β cell differentiation protocols (105).

Another aspect to consider is the use of iPSCs over hESCs. iPSCs have some advantages in terms of safety, at least in the sphere of alloimmunity, but from a technical perspective the time-consuming generation of iPSC lines from each single patient may pose an insurmountable economic burden. This is one of the reasons why numerous studies are currently testing alternative methods to eliminate overall immune rejection. These methods can be classified into two categories: induction of immune tolerance, and gene editing to generate 'cloaked' cells invisible to the immune system. Immune tolerance can be induced with tolerogenic cytokines and immunomodulatory proteins such as CTLA-4, and PD-L1 (107), whereas the generation of 'cloaked' cells is attempted by removal of HLA proteins, mainly through genome editing (108–110). ViaCyte, in partnership with CRISPR Therapeutics, is currently developing immune-evasive stem cell lines that combine both strategies. Approaches aimed at inducing immune protection have the potential concern of creating cells that cannot be recognized and thus eliminated by the immune system if they should become infected by a virus or should form a teratoma, a major concern in stem cell therapy. A possible solution may be provided by the introduction of inducible suicide genes, such as the inducible Caspase-9 (iC9) (111), combined with its in-frame insertion into a locus transcriptionally active in undifferentiated stem cells, such as SOX2 (112). Overall, while there are challenges that still need to be addressed, generating immune "cloaked" cells would remove the need for immune-suppressive regimen, thus broadening the applicability of stem cell therapies to treat patients afflicted by both type 1 and type 2 diabetes.

AUTHOR CONTRIBUTIONS

EdK wrote the manuscript with support from MH. All authors contributed to the article and approved the submitted version.

ACKNOWLEDGMENTS

The authors would like to thank Dr. Sapna Puri and Dr. Audrey Parent for reviewing the manuscript and for providing helpful suggestions, and Greg Szot for providing insights into cadaveric islets transplantation.

REFERENCES

1. Barker M. Pancreatic islet transplantation. *Lancet* (1990) 336(8726):1323. doi: 10.1016/0140-6736(90)93009-E
2. Shapiro AM, Lakey JR, Ryan EA, Korbutt GS, Toth E, Warnock GL, et al. Islet transplantation in seven patients with type 1 diabetes mellitus using a glucocorticoid-free immunosuppressive regimen. *N Engl J Med* (2000) 343 (4):230–8. doi: 10.1056/NEJM200007273430401
3. Bretzel RG, Hering BJ, Schultz AO, Geier C, Federlin K. International islet transplant registry report. In: RP Lanza, WL Chick, editors. *Yearbook of Cell and Tissue Transplantation 1996–1997.* Dordrecht: Springer Netherlands (1996). p. 153–60. doi: 10.1007/978-94-009-0165-0_15
4. Markmann JF, Rickels MR, Eggerman TL, Bridges ND, Lafontant DE, Qidwai J, et al. Phase 3 Trial of Human Islet-after-Kidney Transplantation in Type 1 Diabetes. *Am J Transplant* (2020). doi: 10.1111/ajt.16174
5. Cerise A, Nagaraju S, Powelson JA, Lutz A, Fridell JA. Pancreas transplantation following total pancreatectomy for chronic pancreatitis. *Clin Transplant* (2019) 33(12):e13731. doi: 10.1111/ctr.13731
6. Hering BJ, Clarke WR, Bridges ND, Eggerman TL, Alejandro R, Bellin MD, et al. Phase 3 Trial of Transplantation of Human Islets in Type 1 Diabetes Complicated by Severe Hypoglycemia. *Diabetes Care* (2016) 39(7):1230. doi: 10.2337/dc15-1988
7. Lablanche S, Vantyghem MC, Kessler L, Wojtusciszyn A, Borot S, Thivolet C, et al. Islet transplantation versus insulin therapy in patients with type 1 diabetes with severe hypoglycaemia or poorly controlled glycaemia after kidney transplantation (TRIMECO): a multicentre, randomised controlled trial. *Lancet Diabetes Endocrinol* (2018) 6(7):527–37. doi: 10.1016/S2213-8587(18)30078-0
8. Foster ED, Bridges ND, Feurer ID, Eggerman TL, Hunsicker LG, Alejandro R. Improved Health-Related Quality of Life in a Phase 3 Islet Transplantation Trial in Type 1 Diabetes Complicated by Severe Hypoglycemia. *Diabetes Care* (2018) 41(5):1001–8. doi: 10.2337/dc17-1779
9. Martín-Timón I, Del Cañizo-Gómez FJ. Mechanisms of hypoglycemia unawareness and implications in diabetic patients. *World J Diabetes* (2015) 6(7):912–26. doi: 10.4239/wjd.v6.i7.912
10. Cryer PE. Severe Hypoglycemia Predicts Mortality in Diabetes. *Diabetes Care* (2012) 35(9):1814–6. doi: 10.2337/dc12-0749
11. Emamaullee JA, Shapiro AM. Factors influencing the loss of beta-cell mass in islet transplantation. *Cell Transplant* (2007) 16(1):1–8. doi: 10.3727/000000007783464461
12. Pepper AR, Gala-Lopez B, Ziff O, Shapiro AMJ. Revascularization of transplanted pancreatic islets and role of the transplantation site. *Clin Dev Immunol* (2013) 2013:352315–5. doi: 10.1155/2013/352315
13. Sever CE, Demetris AJ, Zeng Y, Tzakis A, Fung JJ, Starzl TE, et al. Islet cell allotransplantation in diabetic patients. Histologic findings in four adults simultaneously receiving kidney or liver transplants. *Am J Pathol* (1992) 140 (5):1255–60.
14. Delaune V, Berney T, Lacotte S, Toso C, et al. Intraportal islet transplantation: the impact of the liver microenvironment. *Transplant Int* (2017) 30(3):227–38. doi: 10.1111/tri.12919
15. Maffi P, Balzano G, Ponzoni M, Nano R, Sordi V, Melzi R, et al. Autologous pancreatic islet transplantation in human bone marrow. *Diabetes* (2013) 62 (10):3523–31. doi: 10.2337/db13-0465
16. Rafael E, Tibell A, Rydén M, Lundgren T, Sävendahl L, Borgström B, et al. Intramuscular autotransplantation of pancreatic islets in a 7-year-old child: a 2-year follow-up. *Am J Transplant* (2008) 8(2):458–62. doi: 10.1111/j.1600-6143.2007.02060.x
17. Bertuzzi F, Colussi G, Lauterio A, De Carlis L. Intramuscular islet allotransplantation in type 1 diabetes mellitus. *Eur Rev Med Pharmacol Sci* (2018) 22(6):1731–6. doi: 10.26355/eurrev_201803_14588
18. Bellin MD, et al. Defective glucagon secretion during hypoglycemia after intrahepatic but not nonhepatic islet autotransplantation. *Am J Transplant* (2014) 14(8):1880–6. doi: 10.1111/ajt.12776
19. Baidal D, Berman DM, Ricordi C, Alvarez Gil AM, Padilla N, Ciancio G, et al. 259-LB: Stable Graft Function and Glycemic Control after Clinical Islet Transplantation on the Omentum. *Diabetes* (2019) 68(Supplement 1):259-LB. doi: 10.2337/db19-259-LB
20. Baidal D, Ricordi C, Berman DM, Pileggi A, Alvarez Gil AM, Padilla N, et al. Long-Term Function of Islet Allografts Transplanted on the Omentum Using a Biological Scaffold. *Diabetes* (2018) 67(Supplement 1):140–OR. doi: 10.2337/db18-140-OR
21. Maffi P, Nano R, Monti P, Melzi R, Sordi V, Mercalli A, et al. Islet Allotransplantation in the Bone Marrow of Patients With Type 1 Diabetes: A Pilot Randomized Trial. *Transplantation* (2019) 103(4):839–51. doi: 10.1097/TP.0000000000002416
22. Huang Q, Ma X, Wang Y, Niu Z, Wang R, Yang F, et al. IL-10 producing type 2 innate lymphoid cells prolong islet allograft survival. *EMBO Mol Med* (2020) p:e12305. doi: 10.15252/emmm.202012305
23. Ng NH, Tan W, Koh YX, Teo A. Human Islet Isolation and Distribution Efforts for Clinical and Basic Research. *OBM Transplantation* (2019) 3(2):31. doi: 10.21926/obm.transplant.1902068
24. Welsch CA, Rust WL, Csete M. Concise Review: Lessons Learned from Islet Transplant Clinical Trials in Developing Stem Cell Therapies for Type 1 Diabetes. *Stem Cells Transl Med* (2019) 8(3):209–14. doi: 10.1002/sctm.18-0156
25. Shapiro AMJ. Islet Transplantation - The Canadian Perspective. *CellR4–repair replacement regeneration reprogramming* (2019) 7:e2799. doi: 10.32113/cellr4_201911_2799
26. Israni AK, Zaun D, Hadley N, Rosendale JD, Schaffhausen C, McKinney W, et al. OPTN/SRTR 2018 Annual Data Report: Deceased Organ Donation. *Am J Transplant* (2020) 20(s1):509–41. doi: 10.1111/ajt.15678
27. You W-P, Henneberg M. Type 1 diabetes prevalence increasing globally and regionally: the role of natural selection and life expectancy at birth. *BMJ Open Diabetes Res & Care* (2016) 4(1):e000161. doi: 10.1136/bmjdrc-2015-000161
28. Rowley WR, Bezold C, Arikan Y, Byrne E, Krohe S. Diabetes 2030: Insights from Yesterday, Today, and Future Trends. *Population Health Manage* (2017) 20(1):6–12. doi: 10.1089/pop.2015.0181
29. Archibald PR, Williams DJ. Using the cost-effectiveness of allogeneic islet transplantation to inform induced pluripotent stem cell-derived β-cell therapy reimbursement. *Regener Med* (2015) 10(8):959–73. doi: 10.2217/rme.15.59
30. Wallner K, Shapiro AM, Senior PA, McCabe C. Cost effectiveness and value of information analyses of islet cell transplantation in the management of 'unstable' type 1 diabetes mellitus. *BMC Endocr Disord* (2016) 16:17. doi: 10.1186/s12902-016-0097-7
31. Bandeiras C, Hwa AJ, Cabral JMS, Ferreira FC, Finkelstein SN, Gabbay RA. Economics of Beta-Cell Replacement Therapy. *Curr Diabetes Rep* (2019) 19 (9):75. doi: 10.1007/s11892-019-1203-9
32. Nair GG, Tzanakakis ES, Hebrok M. Emerging routes to the generation of functional β-cells for diabetes mellitus cell therapy. *Nat Rev Endocrinol* (2020) 16(9):506–18. doi: 10.1038/s41574-020-0375-3
33. Wallner K, Pedroza RG, Awotwe I, Piret JM, Senior PA, Shapiro AMJ, et al. Stem cells and beta cell replacement therapy: a prospective health technology assessment study. *BMC Endocr Disord* (2018) 18(1):6. doi: 10.1186/s12902-018-0233-7
34. Bandeiras C, Cabral JMS, Gabbay RA, Finkelstein SN, Ferreira FC, et al. Bringing Stem Cell-Based Therapies for Type 1 Diabetes to the Clinic: Early Insights from Bioprocess Economics and Cost-Effectiveness Analysis. *Biotechnol J* (2019) 14(8):e1800563. doi: 10.1002/biot.201800563
35. Brown SA, Kovatchev BP, Raghinaru D, Lum JW, Buckingham BA, Kudva YC, et al. Six-Month Randomized, Multicenter Trial of Closed-Loop Control in Type 1 Diabetes. *N Engl J Med* (2019) 381(18):1707–17. doi: 10.1056/NEJMoa1907863
36. Tauschmann M, Thabit H, Bally L, Allen JM, Hartnell S, Wilinska ME, et al. Closed-loop insulin delivery in suboptimally controlled type 1 diabetes: a multicentre, 12-week randomised trial. *Lancet* (2018) 392(10155):1321–9. doi: 10.1016/S0140-6736(18)31947-0
37. Russell SJ, Hillard MA, Balliro C, Magyar KL, Selagamsetty R, Sinha M, et al. Day and night glycaemic control with a bionic pancreas versus conventional insulin pump therapy in preadolescent children with type 1 diabetes: a randomised crossover trial. *Lancet Diabetes Endocrinol* (2016) 4(3):233–43. doi: 10.1016/S2213-8587(15)00489-1
38. Bekiari E, Kitsios K, Thabit H, Tauschmann M, Athanasiadou E, Karagiannis T, et al. Artificial pancreas treatment for outpatients with type 1 diabetes: systematic review and meta-analysis. *BMJ* (2018) 361:k1310. doi: 10.1136/bmj.k1310
39. Boughton CK, Hovorka R. Is an artificial pancreas (closed-loop system) for Type 1 diabetes effective? *Diabetes Med* (2019) 36(3):279–86. doi: 10.1111/dme.13816

40. Tagougui S, Taleb N, Rabasa-Lhoret R. The Benefits and Limits of Technological Advances in Glucose Management Around Physical Activity in Patients Type 1 Diabetes. *Front Endocrinol (Lausanne)* (2019) 9(818). doi: 10.3389/fendo.2018.00818
41. McEwen LN, Casagrande SS, Kuo S, Herman WH. Why Are Diabetes Medications So Expensive and What Can Be Done to Control Their Cost? *Curr Diabetes Rep* (2017) 17(9):71. doi: 10.1007/s11892-017-0893-0
42. Rezania A, Bruin JE, Arora P, Rubin A, Batushansky I, Asadi A, et al. Reversal of diabetes with insulin-producing cells derived in vitro from human pluripotent stem cells. *Nat Biotechnol* (2014) 32(11):1121–33. doi: 10.1038/nbt.3033
43. Pagliuca FW, Millman JR, Gürtler M, Segel M, Van Dervort A, Ryu JH, et al. Generation of functional human pancreatic β cells in vitro. *Cell* (2014) 159 (2):428–39. doi: 10.1016/j.cell.2014.09.040
44. Russ HA, Parent AV, Ringler JJ, Hennings TG, Nair GG, Shveygert M, et al. Controlled induction of human pancreatic progenitors produces functional beta-like cells in vitro. *EMBO J* (2015) 34(13):1759–72. doi: 10.15252/embj.201591058
45. Rezania A, Bruin JE, Riedel MJ, Mojibian M, Asadi A, Xu J, et al. Maturation of human embryonic stem cell-derived pancreatic progenitors into functional islets capable of treating pre-existing diabetes in mice. *Diabetes* (2012) 61(8):2016–29. doi: 10.2337/db11-1711
46. Velazco-Cruz L, Song J, Maxwell KG, Goedegebuure MM, Augsornworawat P, Hogrebe NJ, et al. Acquisition of Dynamic Function in Human Stem Cell-Derived β Cells. *Stem Cell Rep* (2019) 12(2):351–65. doi: 10.1016/j.stemcr.2018.12.012
47. Millman JR, Xie C, Van Dervort A, Gürtler M, Pagliuca FW, Melton DA. Generation of stem cell-derived β-cells from patients with type 1 diabetes. *Nat Commun* (2016) 7:11463. doi: 10.1038/ncomms11463
48. Crisan M, Yap S, Casteilla L, Chen CW, Corselli M, Park TS, et al. A perivascular origin for mesenchymal stem cells in multiple human organs. *Cell Stem Cell* (2008) 3(3):301–13. doi: 10.1016/j.stem.2008.07.003
49. Kern S, Eichler H, Stoeve J, Klüter H, Bieback K. Comparative analysis of mesenchymal stem cells from bone marrow, umbilical cord blood, or adipose tissue. *Stem Cells* (2006) 24(5):1294–301. doi: 10.1634/stemcells.2005-0342
50. Morikawa S, Mabuchi Y, Niibe K, Suzuki S, Nagoshi N, Sunabori T, et al. Development of mesenchymal stem cells partially originate from the neural crest. *Biochem Biophys Res Commun* (2009) 379(4):1114–9. doi: 10.1016/j.bbrc.2009.01.031
51. Achilleos A, Trainor PA. Neural crest stem cells: discovery, properties and potential for therapy. *Cell Res* (2012) 22(2):288–304. doi: 10.1038/cr.2012.11
52. Slukvin II, Kumar A. The mesenchymoangioblast, mesodermal precursor for mesenchymal and endothelial cells. *Cell Mol Life Sci CMLS* (2018) 75 (19):3507–20. doi: 10.1007/s00018-018-2871-3
53. Song N, Scholtemeijer M, Shah K. Mesenchymal Stem Cell Immunomodulation: Mechanisms and Therapeutic Potential. *Trends Pharmacol Sci* (2020) 41(9):653–64. doi: 10.1016/j.tips.2020.06.009
54. Timper K, Seboek D, Eberhardt M, Linscheid P, Christ-Crain M, Keller U, et al. Human adipose tissue-derived mesenchymal stem cells differentiate into insulin, somatostatin, and glucagon expressing cells. *Biochem Biophys Res Commun* (2006) 341(4):1135–40. doi: 10.1016/j.bbrc.2006.01.072
55. Phadnis SM, Joglekar MV, Dalvi MP, Muthyala S, Nair PD, Ghaskadbi SM, et al. Human bone marrow-derived mesenchymal cells differentiate and mature into endocrine pancreatic lineage in vivo. *Cytotherapy* (2011) 13 (3):279–93. doi: 10.3109/14653249.2010.523108
56. Govindasamy V, Ronald VS, Abdullah AN, Nathan KR, Ab Aziz ZA, Abdullah M, et al. Differentiation of dental pulp stem cells into islet-like aggregates. *J Dent Res* (2011) 90(5):646–52. doi: 10.1177/0022034510396879
57. Tang DQ, Wang Q, Burkhardt BR, Litherland SA, Atkinson MA, Yang LJ. In vitro generation of functional insulin-producing cells from human bone marrow-derived stem cells, but long-term culture running risk of malignant transformation. *Am J Stem Cells* (2012) 1(2):114–27.
58. Prabakar KR, Domínguez-Bendala J, Molano RD, Pileggi A, Villate S, Ricordi C, et al. Generation of glucose-responsive, insulin-producing cells from human umbilical cord blood-derived mesenchymal stem cells. *Cell Transplant* (2012) 21(6):1321–39. doi: 10.3727/096368911X612530
59. Kanafi MM, Rajeshwari YB, Gupta S, Dadheech N, Nair PD, Gupta PK, et al. Transplantation of islet-like cell clusters derived from human dental pulp stem cells restores normoglycemia in diabetic mice. *Cytotherapy* (2013) 15 (10):1228–36. doi: 10.1016/j.jcyt.2013.05.008
60. Guo QS, Zhu MY, Wang L, Fan XJ, Lu YH, Wang ZW, et al. Combined transfection of the three transcriptional factors, PDX-1, NeuroD1, and MafA, causes differentiation of bone marrow mesenchymal stem cells into insulin-producing cells. *Exp Diabetes Res* (2012) 2012:672013. doi: 10.1155/2012/672013
61. Shivakumar SB, Lee HJ, Son YB, Bharti D, Ock SA, Lee SL, et al. In vitro differentiation of single donor derived human dental mesenchymal stem cells into pancreatic β cell-like cells. *Biosci Rep* (2019) 39(5). doi: 10.1042/BSR20182051
62. Perillo M, Paganos P, Mattiello T, Cocurullo M, Oliveri P, Arnone MI. New Neuronal Subtypes With a "Pre-Pancreatic" Signature in the Sea Urchin Stongylocentrotus purpuratus. *Front Endocrinol* (2018) 9:650–0. doi: 10.3389/fendo.2018.00650
63. Hansson M, Tonning A, Frandsen U, Petri A, Rajagopal J, Englund MC, et al. Artifactual insulin release from differentiated embryonic stem cells. *Diabetes* (2004) 53(10):2603–9. doi: 10.2337/diabetes.53.10.2603
64. Wu LF, Wang NN, Liu YS, Wei X. Differentiation of Wharton's jelly primitive stromal cells into insulin-producing cells in comparison with bone marrow mesenchymal stem cells. *Tissue Eng Part A* (2009) 15(10):2865–73. doi: 10.1089/ten.tea.2008.0579
65. Lechner A, Yang YG, Blacken RA, Wang L, Nolan AL, Habener JF. No evidence for significant transdifferentiation of bone marrow into pancreatic beta-cells in vivo. *Diabetes* (2004) 53(3):616–23. doi: 10.2337/diabetes.53.3.616
66. Choi JB, Uchino H, Azuma K, Iwashita N, Tanaka Y, Mochizuki H, et al. Little evidence of transdifferentiation of bone marrow-derived cells into pancreatic beta cells. *Diabetologia* (2003) 46(10):1366–74. doi: 10.1007/s00125-003-1182-9
67. Ezquer F, Ezquer M, Contador D, Ricca M, Simon V, Conget P, et al. The antidiabetic effect of mesenchymal stem cells is unrelated to their transdifferentiation potential but to their capability to restore Th1/Th2 balance and to modify the pancreatic microenvironment. *Stem Cells* (2012) 30(8):1664–74. doi: 10.1002/stem.1132
68. Dang LT, Bui AN, Le-Thanh Nguyen C, Truong NC, Bui AT, Kim NP, et al. Intravenous Infusion of Human Adipose Tissue-Derived Mesenchymal Stem Cells to Treat Type 1 Diabetic Mellitus in Mice: An Evaluation of Grafted Cell Doses. *Adv Exp Med Biol* (2018) 1083:145–56. doi: 10.1007/5584_2017_127
69. Li L, Hui H, Jia X, Zhang J, Liu Y, Xu Q, et al. Infusion with Human Bone Marrow-derived Mesenchymal Stem Cells Improves β-cell Function in Patients and Non-obese Mice with Severe Diabetes. *Sci Rep* (2016) 6:37894–4. doi: 10.1038/srep37894
70. Yaochite JN, Caliari-Oliveira C, de Souza LE, Neto LS, Palma PV, Covas DT, et al. Therapeutic efficacy and biodistribution of allogeneic mesenchymal stem cells delivered by intrasplenic and intrapancreatic routes in streptozotocin-induced diabetic mice. *Stem Cell Res Ther* (2015) 6(1):31. doi: 10.1186/s13287-015-0017-1
71. Kerby A, Jones ES, Jones PM, King AJ. Co-transplantation of islets with mesenchymal stem cells in microcapsules demonstrates graft outcome can be improved in an isolated-graft model of islet transplantation in mice. *Cytotherapy* (2013) 15(2):192–200. doi: 10.1016/j.jcyt.2012.10.018
72. Madec AM, Mallone R, Afonso G, Abou Mrad E, Mesnier A, Eljaafari A, et al. Mesenchymal stem cells protect NOD mice from diabetes by inducing regulatory T cells. *Diabetologia* (2009) 52(7):1391–9. doi: 10.1007/s00125-009-1374-z
73. Lee RH, Seo MJ, Reger RL, Spees JL, Pulin AA, Olson SD, et al. Multipotent stromal cells from human marrow home to and promote repair of pancreatic islets and renal glomeruli in diabetic NOD/scid mice. *Proc Natl Acad Sci* (2006) 103(46):17438–43. doi: 10.1073/pnas.0608249103
74. Khatri R, Mazurek S, Petry SF, Linn T. Mesenchymal stem cells promote pancreatic β-cell regeneration through downregulation of FoxO1 pathway. *Stem Cell Res Ther* (2020) 11(1):497–7. doi: 10.1186/s13287-020-02007-9
75. Dave SD, Vanikar AV, Trivedi HL, Thakkar UG, Gopal SC, Chandra T. Novel therapy for insulin-dependent diabetes mellitus: infusion of in vitro-

generated insulin-secreting cells. *Clin Exp Med* (2015) 15(1):41–5. doi: 10.1007/s10238-013-0266-1

76. Thakkar UG, Trivedi HL, Vanikar AV, Dave SD. Insulin-secreting adipose-derived mesenchymal stromal cells with bone marrow-derived hematopoietic stem cells from autologous and allogenic sources for type 1 diabetes mellitus. *Cytotherapy* (2015) 17(7):940–7. doi: 10.1016/j.jcyt.2015.03.608
77. Salgado AJ, Reis RL, Sousa NJ, Gimble JM. Adipose tissue derived stem cells secretome: soluble factors and their roles in regenerative medicine. *Curr Stem Cell Res Ther* (2010) 5(2):103–10. doi: 10.2174/157488810791268564
78. Chen J, Chen J, Cheng Y, Fu Y, Zhao H, Tang M, et al. Mesenchymal stem cell-derived exosomes protect beta cells against hypoxia-induced apoptosis via miR-21 by alleviating ER stress and inhibiting p38 MAPK phosphorylation. *Stem Cell Res Ther* (2020) 11(1):97. doi: 10.1186/s13287-020-01610-0
79. Najar M, Raicevic G, Crompot E, Fayyad-Kazan H, Bron D, Toungouz M, et al. The Immunomodulatory Potential of Mesenchymal Stromal Cells: A Story of a Regulatory Network. *J Immunother* (2016) 39(2):45–59. doi: 10.1097/CJI.0000000000000108
80. Mesples A, Majeed N, Zhang Y, Hu X. Early immunotherapy using autologous adult stem cells reversed the effect of anti-pancreatic islets in recently diagnosed type 1 diabetes mellitus: preliminary results. *Med Sci Monit* (2013) 19:852–7. doi: 10.12659/MSM.889525
81. Carlsson PO, Schwarcz E, Korsgren O, Le Blanc K. Preserved β-cell function in type 1 diabetes by mesenchymal stromal cells. *Diabetes* (2015) 64(2):587–92. doi: 10.2337/db14-0656
82. Gan J, Wang Y, Zhou X. Stem cell transplantation for the treatment of patients with type 1 diabetes mellitus: A meta-analysis. *Exp Ther Med* (2018) 16(6):4479–92. doi: 10.3892/etm.2018.6769
83. Hwang G, Jeong H, Yang HK, Kim HS, Hong H, Kim NJ, et al. Efficacies of Stem Cell Therapies for Functional Improvement of the β Cell in Patients with Diabetes: A Systematic Review of Controlled Clinical Trials. *Int J Stem Cells* (2019) 12(2):195–205. doi: 10.15283/ijsc18076
84. Bhansali A, Upreti V, Khandelwal N, Marwaha N, Gupta V, Sachdeva N, et al. Efficacy of autologous bone marrow-derived stem cell transplantation in patients with type 2 diabetes mellitus. *Stem Cells Dev* (2009) 18(10):1407–16. doi: 10.1089/scd.2009.0164
85. Jiang R, Han Z, Zhuo G, Qu X, Li X, Wang X, et al. Transplantation of placenta-derived mesenchymal stem cells in type 2 diabetes: a pilot study. *Front Med* (2011) 5(1):94–100. doi: 10.1007/s11684-011-0116-z
86. Bhansali A, Asokumar P, Walia R, Bhansali S, Gupta V, Jain A, et al. Efficacy and safety of autologous bone marrow-derived stem cell transplantation in patients with type 2 diabetes mellitus: a randomized placebo-controlled study. *Cell Transplant* (2014) 23(9):1075–85. doi: 10.3727/096368913X665576
87. Bhansali S, Dutta P, Kumar V, Yadav MK, Jain A, Mudaliar S, et al. Efficacy of Autologous Bone Marrow-Derived Mesenchymal Stem Cell and Mononuclear Cell Transplantation in Type 2 Diabetes Mellitus: A Randomized, Placebo-Controlled Comparative Study. *Stem Cells Dev* (2017) 26(7):471–81. doi: 10.1089/scd.2016.0275
88. Thomson JA, Itskovitz-Eldor J, Shapiro SS, Waknitz MA, Swiergiel JJ, Marshall VS, et al. Embryonic stem cell lines derived from human blastocysts. *Science* (1998) 282(5391):1145–7. doi: 10.1126/science.282.5391.1145
89. Takahashi K, Yamanaka S. Induction of pluripotent stem cells from mouse embryonic and adult fibroblast cultures by defined factors. *Cell* (2006) 126 (4):663–76. doi: 10.1016/j.cell.2006.07.024
90. Gerace D, Martiniello-Wilks R, Habib R, Ren B, Nassif NT, O'Brien BA, et al. Ex Vivo Expansion of Murine MSC Impairs Transcription Factor-Induced Differentiation into Pancreatic β-Cells. *Stem Cells Int* (2019) 2019:1395301. doi: 10.1155/2019/1395301
91. Kondo Y, Toyoda T, Inagaki N, Osafune K. iPSC technology-based regenerative therapy for diabetes. *J Diabetes Invest* (2018) 9(2):234–43. doi: 10.1111/jdi.12702
92. Kroon E, Martinson LA, Kadoya K, Bang AG, Kelly OG, Eliazer S, et al. Pancreatic endoderm derived from human embryonic stem cells generates glucose-responsive insulin-secreting cells in vivo. *Nat Biotechnol* (2008) 26 (4):443–52. doi: 10.1038/nbt1393
93. Schulz TC. Concise Review: Manufacturing of Pancreatic Endoderm Cells for Clinical Trials in Type 1 Diabetes. *Stem Cells Transl Med* (2015) 4 (8):927–31. doi: 10.5966/sctm.2015-0058
94. Henry RR, Pettus J, Wilensky JON, Shapiro AMJ, Senior PA, Roep B, et al. Initial Clinical Evaluation of VC-01TM Combination Product—A Stem Cell–Derived Islet Replacement for Type 1 Diabetes (T1D). *Diabetes* (2018) 67(Supplement 1):138–OR. doi: 10.2337/db18-138-OR
95. Shapiro AJ. Insulin Expression and Glucose-Responsive Circulating C-Peptide in Type 1 Diabetes Patients Implanted Subcutaneously with Pluripotent Stem Cell-Derived Pancreatic Endoderm Cells in a Macro-Device. In: *The Lancet* (2019). doi: 10.2139/ssrn.3501034
96. Kirk K, Hao E, Lahmy R, Itkin-Ansari P. Human embryonic stem cell derived islet progenitors mature inside an encapsulation device without evidence of increased biomass or cell escape. *Stem Cell Res* (2014) 12(3):807–14. doi: 10.1016/j.scr.2014.03.003
97. Stock AA, Manzoli V, De Toni T, Abreu MM, Poh YC, Ye L, et al. Conformal Coating of Stem Cell-Derived Islets for β Cell Replacement in Type 1 Diabetes. *Stem Cell Rep* (2020) 14(1):91–104. doi: 10.1016/j.stemcr.2019.11.004
98. Bochenek MA, Veiseh O, Vegas AJ, McGarrigle JJ, Qi M, Marchese E, et al. Alginate encapsulation as long-term immune protection of allogeneic pancreatic islet cells transplanted into the omental bursa of macaques. *Nat Biomed Eng* (2018) 2(11):810–21. doi: 10.1038/s41551-018-0275-1
99. Vegas AJ, Veiseh O, Gürtler M, Millman JR, Pagliuca FW, Bader AR, et al. Long-term glycemic control using polymer-encapsulated human stem cell–derived beta cells in immune-competent mice. *Nat Med* (2016) 22(3):306–11. doi: 10.1038/nm.4030
100. Yu M, Agarwal D, Korutla L, May CL, Wang W, Griffith NN, et al. Islet transplantation in the subcutaneous space achieves long-term euglycaemia in preclinical models of type 1 diabetes. *Nat Metab* (2020) 2(10):1013–20. doi: 10.1038/s42255-020-0269-7
101. Zhu K, Dong L, Wang J, Li D, Chen M, Jiang C, et al. Enhancing the functional output of transplanted islets in diabetic mice using a drug-eluting scaffold. *J Biol Eng* (2018) 12:5–5. doi: 10.1186/s13036-018-0098-3
102. Pathak S, Acharya S, Regmi S, Shrestha P, You Z, Bae YK, et al. Particulate-Based Single-Dose Local Immunosuppressive Regimen for Inducing Tolerogenic Dendritic Cells in Xenogeneic Islet Transplantation. *Adv Healthc Mater* (2020) 10(2):e2001157 e2001157. doi: 10.1002/adhm.202001157
103. Memon B, Abdelalim EM. Stem Cell Therapy for Diabetes: Beta Cells versus Pancreatic Progenitors. *Cells* (2020) 9(2):283. doi: 10.3390/cells9020283
104. Johnston NR, Mitchell RK, Haythorne E, Pessoa MP, Semplici F, Ferrer J, et al. Beta Cell Hubs Dictate Pancreatic Islet Responses to Glucose. *Cell Metab* (2016) 24(3):389–401. doi: 10.1016/j.cmet.2016.06.020
105. Nair GG, Liu JS, Russ HA, Tran S, Saxton MS, Chen R, et al. Recapitulating endocrine cell clustering in culture promotes maturation of human stem-cell-derived β cells. *Nat Cell Biol* (2019) 21(2):263–74. doi: 10.1038/s41556-018-0271-4
106. Peterson QP, Veres A, Chen L, Slama MQ, Kenty JHR, Hassoun S, et al. A method for the generation of human stem cell-derived alpha cells. *Nat Commun* (2020) 11(1):2241. doi: 10.1038/s41467-020-16049-3
107. El Khatib MM, Sakuma T, Tonne JM, Mohamed MS, Holditch SJ, Lu B, et al. β-Cell-targeted blockage of PD1 and CTLA4 pathways prevents development of autoimmune diabetes and acute allogeneic islets rejection. *Gene Ther* (2015) 22(5):430–8. doi: 10.1038/gt.2015.18
108. Figueiredo C, Blasczyk R. A future with less HLA: potential clinical applications of HLA-universal cells. *Tissue Antigens* (2015) 85(6):443–9. doi: 10.1111/tan.12564
109. Koga K, Wang B, Kaneko S. Current status and future perspectives of HLA-edited induced pluripotent stem cells. *Inflammation Regener* (2020) 40:23. doi: 10.1186/s41232-020-00132-9
110. Lee J, Sheen JH, Lim O, Lee Y, Ryu J, Shin D, et al. Abrogation of HLA surface expression using CRISPR/Cas9 genome editing: a step toward universal T cell therapy. *Sci Rep* (2020) 10(1):17753. doi: 10.1038/s41598-020-74772-9
111. Yagyu S, Hoyos V, Del Bufalo F, Brenner MK. An Inducible Caspase-9 Suicide Gene to Improve the Safety of Therapy Using Human Induced Pluripotent Stem Cells. *Mol Ther J Am Soc Gene Ther* (2015) 23(9):1475–85. doi: 10.1038/mt.2015.100

5

Use of Induced Pluripotent Stem Cells to Build Isogenic Systems and Investigate Type 1 Diabetes

Lucas H. Armitage[1,2], *Scott E. Stimpson*[1,2], *Katherine E. Santostefano*[1,3,4], *Lina Sui*[5], *Similoluwa Ogundare*[1,2], *Brittney N. Newby*[1,2], *Roberto Castro-Gutierrez*[6], *Mollie K. Huber*[1,2], *Jared P. Taylor*[1], *Prerana Sharma*[7], *Ilian A. Radichev*[7], *Daniel J. Perry*[1,2], *Natalie C. Fredette*[1,3], *Alexei Y. Savinov*[7], *Mark A. Wallet*[1,2,8], *Naohiro Terada*[1,3], *Todd M. Brusko*[1,2], *Holger A. Russ*[6], *Jing Chen*[1,2], *Dieter Egli*[5] *and Clayton E. Mathews*[1,2,3*]

[1] *Department of Pathology, Immunology, and Laboratory Medicine, College of Medicine, University of Florida, Gainesville, FL, United States,* [2] *University of Florida Diabetes Institute, University of Florida, Gainesville, FL, United States,* [3] *Center for Cellular Reprogramming, College of Medicine, University of Florida, Gainesville, FL, United States,* [4] *Century Therapeutics, iPSC Biology, Philadelphia, PA, United States,* [5] *Department of Pediatrics, Naomi Berrie Diabetes Center, Columbia Stem Cell Initiative, Columbia University, New York, NY, United States,* [6] *Barbara Davis Center for Diabetes, University of Colorado School of Medicine, Aurora, CO, United States,* [7] *Children's Health Research Center, Sanford Research, Sioux Falls, SD, United States,* [8] *Century Therapeutics, Immunology, Philadelphia, PA, United States*

***Correspondence:**
Clayton E. Mathews
cxm@ufl.edu

Type 1 diabetes (T1D) is a disease that arises due to complex immunogenetic mechanisms. Key cell-cell interactions involved in the pathogenesis of T1D are activation of autoreactive T cells by dendritic cells (DC), migration of T cells across endothelial cells (EC) lining capillary walls into the islets of Langerhans, interaction of T cells with macrophages in the islets, and killing of β-cells by autoreactive $CD8^+$ T cells. Overall, pathogenic cell-cell interactions are likely regulated by the individual's collection of genetic T1D-risk variants. To accurately model the role of genetics, it is essential to build systems to interrogate single candidate genes in isolation during the interactions of cells that are essential for disease development. However, obtaining single-donor matched cells relevant to T1D is a challenge. Sourcing these genetic variants from human induced pluripotent stem cells (iPSC) avoids this limitation. Herein, we have differentiated iPSC from one donor into DC, macrophages, EC, and β-cells. Additionally, we also engineered T cell avatars from the same donor to provide an *in vitro* platform to study genetic influences on these critical cellular interactions. This proof of concept demonstrates the ability to derive an isogenic system from a single donor to study these relevant cell-cell interactions. Our system constitutes an interdisciplinary approach with a controlled environment that provides a proof-of-concept for future studies to determine the role of disease alleles (e.g. *IFIH1, PTPN22, SH2B3, TYK2)* in regulating cell-cell interactions and cell-specific contributions to the pathogenesis of T1D.

Keywords: type 1 diabetes mellitus, isogenic, human, beta cells, immunity, autoimmunity, induced pluripotent stem cells

INTRODUCTION

Type 1 diabetes (T1D) is a debilitating autoimmune disease that is caused by T cell-mediated destruction of β-cells in the islets of Langerhans in the pancreas. This results in lifelong dependence on exogenous insulin and can lead to many complications that degrade quality of life for patients living with T1D. With the advent of genome-wide association studies (GWAS) a big picture view of the genetic contributions of T1D has come into focus. While human leukocyte antigen (HLA) loci, encoding the major histocompatibility (MHC) proteins, make up the bulk of genetic risk for T1D, over 100 non-HLA loci have been identified with polymorphisms that increase or decrease risk for T1D (1–12). However, studying the phenotypic outcome of these polymorphisms has proven difficult due to the heterogeneous nature of the human population and the inaccessibility of pancreatic tissue from individuals with signs of islet autoimmunity (i.e. islet autoantibodies) and individuals with clinical T1D.

Over the past half-century, immune and non-immune cells have been implicated in T1D (13–15). Key interactions involved in the pathogenesis of T1D in humans (**Figure 1**) include; A) the initial interactions between dendritic cells (DC) and β-cells where DC uptake β-cell antigens from dead, dying, or stressed β-cells, B) the activation of autoreactive CD4$^+$ and CD8$^+$ T cells by the β-cell antigen-loaded DC, C) the interaction between autoreactive CD8$^+$ T cells and the endothelial cells (EC) lining the capillaries that supply the islets of Langerhans as autoreactive CD8$^+$ T cells undergo transendothelial migration to access the islets, D) the targeting of β-cells for cell-mediated lysis by autoreactive CD8$^+$ T cells, and E) the *in situ* amplification of T cell responses by DC and macrophages in the islets. Genetics influence T1D pathogenesis and deleterious cellular function; however, genes implicated in T1D do not function in isolation. Numerous implicated genes cluster into pathways likely to create situations where different combinations of candidate genes can contribute equally to increased T1D risk (16). The development of a simple and flexible human model is essential to deconvolute genetic risk and make progress in understanding human T1D. We propose the use of models constructed from single-donor iPSC-derived cells. The utility of iPSC to investigate the relationship between rare and common gene variants and gene expression in pluripotent cells was recently highlighted (17).

Human iPSC exhibit similar gene expression patterns to human embryonic stem cells (ESC) and have the capacity to differentiate to all cell types of the human body. Generating iPSC and differentiating them to the cell types of interest (**Figure 1**) from a single donor preserves the genetic background and integrity of disease alleles, potentially making for an excellent isogenic model system to study T1D pathogenesis. In support of these isogenic systems, advances in stem cell technologies have

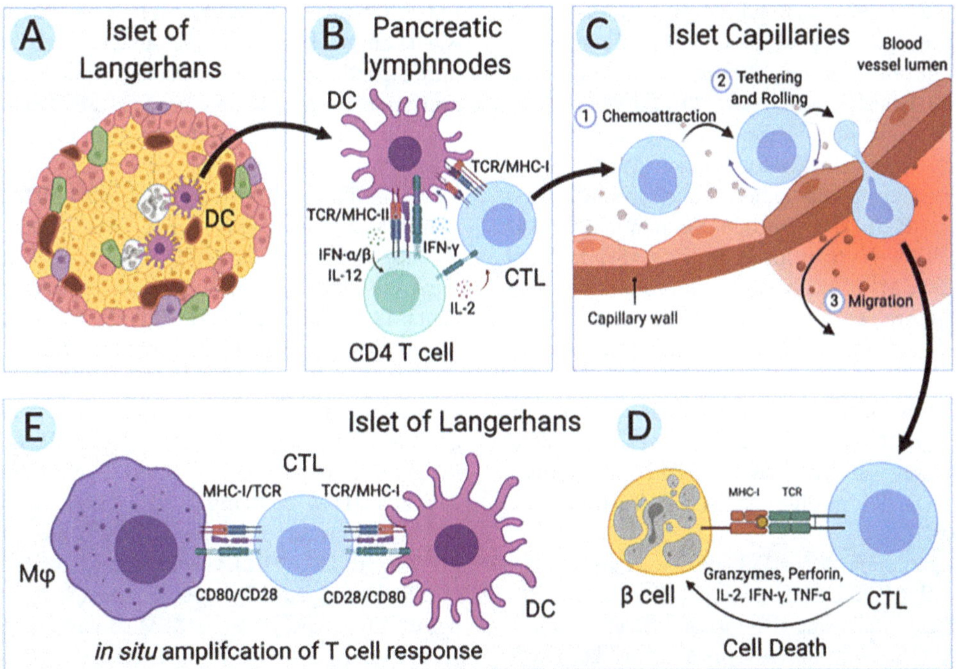

FIGURE 1 | Key cellular interactions involved in pathogenesis of T1D in humans. **(A)** First dendritic cells (DC) uptake antigen from dead, dying, and stressed β-cells in the islets of Langerhans in individuals at risk for type 1 diabetes (T1D). **(B)** Second, these β-cell-antigen-loaded DC migrate to the pancreatic lymph nodes where they present β-cell antigen to autoreactive CD4$^+$ and CD8$^+$ T cells, initiating an adaptive immune response. **(C)** Third, these autoreactive CD8$^+$ T cells, also called cytotoxic T lymphocytes (CTLs), migrate back to the pancreas and undergo transendothelial migration across the endothelium lining capillaries that supply the islets of Langerhans. **(D)** Fourth, these CTLs directly target and lyse β-cells while **(E)** resident and infiltrating macrophages and DC amplify the CTL response *in situ*. Created with BioRender.com.

resulted in protocols allowing for differentiation of iPSC to functional replicas of cells essential for T1D, including the hard to access β-cells. Additionally, methods for engineering primary human T cells have advanced our understanding of immunity and autoimmunity (18–20). Here we selected three candidate iPSC lines to build the isogenic system, 1-018, 1-023, and 2395, that had previously been differentiated to stem cell derived beta-like cells (sBC), as this is the most time intensive and thus the rate limiting step for construction of these isogenic models (21–23). Based on HLA requirements and necessity of successful differentiation into pancreatic β-cells, endothelial cells, macrophages, and dendritic cells, line 2395 was selected. In addition, T cells were purified from this same donor. Employing the most recent and advanced methods for culturing and differentiating iPSC that have similar functionality to cells *in vivo*, we show that these iPSC-derived cells can be used for modeling and interrogating cell-cell interactions that are relevant to the pathogenesis of T1D (**Figure 1**). Utilizing these cellular approaches will allow investigators to examine relationships of genotype with cell function and cell-cell interactions.

MATERIALS AND METHODS

Reprogramming Somatic Cells to iPSC

The generation and characterization of the iPSC lines 1-018 and 1-023 was previously described (21–23). Derivation of iPSC from donor 2395 was as previously described (24). Briefly, peripheral blood was collected from donor 2395 under informed consent by the University of Florida Diabetes Institute Study Bank with Institutional Review Board approval (IRB201400703). $CD34^+$ peripheral stem cells were isolated from peripheral blood and expanded using the Complete Kit for Human Whole Blood $CD34^+$ Cells (Stem Cell Technologies, Vancouver, Canada) following the manufacturer's protocol. Expanded $CD34^+$ peripheral stem cells were transduced with the Sendai viral vector, SeVdp(KOSM)302L, which encodes the four reprogramming factors, OCT4, SOX2, KLF4, and c-MYC (20, 25, 26). Following transduction, cells were cultured on hESC qualified Matrigel- (Corning, Corning, NY) coated dishes in ReproTeSR medium (Stem Cell Technologies) and medium was changed daily until iPSC colonies were observed. iPSC colonies were manually isolated then expanded on Matrigel-coated dishes in mTeSR1 medium (StemCell Technologies, Vancouver, Canada).

Following reprogramming, donor 2395 iPSC were assessed for pluripotency using the StemDiff Trilineage Differentiation Kit (StemCell Technologies) following the manufacturer's protocol. iPSC were subjected to karyotyping to verify a normal karyotype by Cell Line Genetics (Madison, WI, USA). Undifferentiated donor 2395 iPSC were stained with anti-NANOG-AF647, anti-SSEA-4-PE, and anti-Oct-4-AF488 (BioLegend, San Diego, CA) using the Fixation/Permeabilization Solution kit from BD Biosciences (San Jose, CA) and analyzed on an Accuri C6 flow cytometer (BD Biosciences). Information for all antibodies used during the included studies are contained in **Supplementary Table 1**.

Differentiation of iPSC to Stem Cell-Derived Beta-Like Cells

The iPSC lines used in this study, 1-018, 1-023, and 2395, were initially selected because they had previously been differentiated to stem cell derived beta-like cells (sBC). The iPSC lines, 1-018 and 1-023, were differentiated to sBC as previously described (22). Donor 2395 iPSC were maintained on Matrigel-coated 6 well plates in mTeSR+ medium (StemCell Technologies) and differentiated to sBC as previously described (27, 28). Differentiation to sBC was carried out in suspension-based, low attachment suspension culture plates as described (29) or in an ABLE bioreactor magnetic stirring system (Reprocell, Beltsville, MD, USA) as follows. For differentiations, 70 - 80% confluent iPSC cultures were washed with PBS and incubated in TrypLE Express Enzyme (Gibco, Watham, MD, USA) for 8 min at 37°C followed by quenching with mTeSR+. Bioreactors were seeded at 0.5×10^6 cells/mL in mTeSR+ medium supplemented with 10 µM ROCK inhibitor. Bioreactors were placed on a magnetic stirring system set at 60 RPM in a cell culture incubator at 5% CO_2 to induce sphere formation for 48-72 hours. To begin the differentiation process, spheres were collected in a 50 mL Falcon tube, allowed to settle by gravity, washed once with RPMI (Gibco) + 0.2% FBS, and re-suspended in day 0 medium (RPMI containing 0.2% FBS, 1:5,000 ITS (Gibco), 100 ng/mL Activin-A (R&D Systems), and 3 µM CHIR (StemCell Technologies)). The differentiation medium was changed daily by letting spheres settle by gravity for 3-10 min. Approximately 80% of spent supernatant was removed by aspiration, fresh medium was added, and bioreactors were placed back on stirrer system. Differentiation media are as follows: day 1-2, RPMI containing 0.2% FBS, 1:2,000 ITS, and 100 ng/mL Activin A; day 3-4, RPMI containing 2% FBS, 1:1,000 ITS, and 50 ng/mL KGF (Peprotech); day 5, DMEM with 4.5 g/L D-glucose (Gibco) containing 1:100 SM1 (StemCell Technologies), 1:100 NEAA (Gibco), 1 mM Sodium Pyruvate (Gibco), 1:100 GlutaMAX (Gibco), 3 nM TTNPB, (R&D Systems), 250 nM Sant-1 (R&D Systems), 250 nM LDN (StemCell Technologies), 30 nM PMA (Sigma Aldrich), 50 µg/mL 2-phospho-L-ascorbic acid trisodium salt (VitC; Sigma); day 6, DMEM with 4.5 g/L D-glucose containing 1:100 SM1, 1:100 NEAA, 1 mM Sodium Pyruvate, 1:100 GlutaMAX, 3 nM TTNPB and 50 µg/mL VitC; day 7, DMEM containing 1:100 SM1, 1:100 NEAA, 1 mM Sodium Pyruvate, 1:100 GlutaMAX, 3 nM TTNPB and 50 µg/mL VitC; day 8-9, DMEM containing 1:100 SM1, 1:100 NEAA, 1 mM Sodium Pyruvate, 1:100 GlutaMAX, 200 ng/ml EGF and 50 ng/mL KGF; days 10-16, DMEM containing 2% fraction V BSA, 1:100 NEAA, 1 mM Sodium Pyruvate, 1:100 GlutaMAX, 1:100 ITS, 10 µg/ml Heparin (Sigma), 2 mM N-Acetyl-L-cysteine (Cysteine; Sigma), 10 µM Zinc sulfate heptahydrate (Zinc; Sigma), 1x BME, 10 µM Alk5i II RepSox (R&D Systems), 2 µM 3,3',5-Triiodo-L-thyronine sodium salt (T3; Sigma), 0.5 µM LDN, 1 µM Gamma Secretase Inhibitor XX (XXi) (AsisChem, Waltham, MA, USA) and 1:250 1 M NaOH to adjust pH to ~7.4;

day 17+, CMRL (Gibco) containing 1% BSA, 1:100 NEAA, 1 mM Sodium Pyruvate, 1:100 GlutaMAX, 10 µg/mL Heparin, 2 mM Cysteine, 10 µM Zinc, 1x BME, 10 µM Alk5i II RepSox, 1 µM T3, 50 µg/mL VitC, and 1:250 NaOH to adjust pH to ~7.4 (also referred to as maturation medium). All media contained 1x PenStrep (Gibco). Medium was changed every other day starting day 11. Validation and function of sBC, including staining for glucagon and insulin as well as glucose-stimulated insulin secretion were performed as previously described (22, 28, 29).

HLA Typing of iPSC

DNA was isolated from iPSC lines, 1-018, 1-023, and 2395, and subjected to genotyping at 974,650 unique loci using a custom SNP array. The Axiom Precision Medicine Research Array (ThermoFisher Scientific) was modified to include content from the ImmunoChip v2.0 (30) as well as previously reported credible T1D risk variants (10), and additional content to create the UFDIchip. UFDIchips were processed on an Affymetrix Gene Titan instrument with external sample handling on a BioMek FX dual arm robotic workstation. Data processing and quality control pipelines included standard quality control procedures at the SNP, sample, and plate levels using Axiom™ Analysis Suite 3.0 (ThermoFisher Scientific) set to the default stringency thresholds as recommended, as well as for genetic versus reported sex. The analysis pipeline also includes race imputation using EthSeq (31), relatedness using KING (32), and imputation to 40M SNPs using the Human Reference Consortium (version 1.1) with the Michigan Imputation Server (33). The array also includes dense coverage of the highly polymorphic HLA region allowing accurate imputation of HLA haplotypes to 4-digit resolution (34). SNP data are available upon request. HLA alleles arranged into putative extended haplotypes using the Allele Frequency Net Database (http://allelefrequencies.net/).

Differentiation of iPSC to Monocytes, Monocyte-Derived Macrophages, and Monocyte-Derived Dendritic Cells

Donor 2395 iPSC were differentiated to monocytes, monocyte-derived dendritic cells (moDC), and monocyte-derived macrophages (MDM), as described (35). Briefly, iPSC were differentiated to hematopoietic progenitors using the STEMdiff Hematopoietic Kit (StemCell Technologies) over 12 days. On day 12, hematopoietic progenitors were harvested and the hemogenic endothelium left behind in the 12-well plate was cultured in X-VIVO15 (Lonza, Basel, Switzerland) supplemented with β-mercaptoethanol (Gibco), Glutamax (Gibco), 25 ng/mL recombinant human IL-3 (rhIL-3, PeproTech, Rocky Hill, NJ), and 100 ng/mL recombinant human M-CSF (rhM-CSF, PeproTech). The hemogenic endothelium transitioned into monocyte factories and produced non-adherent CD14$^+$ monocytes that can be harvested repeatedly starting around day 21. Thereafter, CD14$^+$ cells are harvested every 2-7 days. Monocyte factories were replaced on Day 94. iPSC-monocytes were differentiated to iPSC-MDM as previously described (24, 35). Briefly, iPSC-monocytes were plated in high-glucose DMEM supplemented with 10% FBS (Genesee Scientific, San Diego, CA) or 10% human serum (Atlanta Biologicals, Flowery Branch, GA), 1000 U/mL Penicillin/streptomycin (Genessee Scientific), Glutamax, and 10 ng/ml of rhM-CSF for 7 days before use in experiments. iPSC-monocytes from Donor 2395 were differentiated to iPSC-moDC as previously described (36). Briefly, the CD14$^+$ iPSC-monocytes were plated in GMP DC medium (CellGenix, Freiburg im Breisgau, Germany) supplemented with Glutamax, 50 ng/mL recombinant human IL-4 (PeproTech), and 50 ng/mL recombinant human GM-CSF (PeproTech) and cultured for 6 days. On day 7, iPSC-moDC were plated for experiments.

Differentiation of iPSC to Endothelial Cells

Donor 2395 iPSC were differentiated to iPSC-derived endothelial cells (iPSC-EC) as previously described (37). This newly published protocol was adapted from a previously published protocol (38). Briefly, donor 2395 iPSC colonies were dissociated to single cells with Accutase (StemCell Technologies) and plated on Matrigel-coated plates at a density of 5 x 10^5 cells per cm^2 in mTESR1 medium (StemCell Technologies) supplemented with 10 ng/mL of the RHO/ROCK pathway inhibitor, Y-27632 (StemCell Technologies) on day 0. On day 1, the medium was changed to STEMdiff Mesoderm Induction medium (StemCell Technologies) and the medium was changed daily. On day 3, medium was replaced with STEMdiff APEL medium (StemCell Technologies) supplemented with 25 ng/mL BMP-4 (PeproTech) and 50 ng/mL VEGF (Peprotech). On day 5, the medium was replaced with fresh STEMdiff APEL medium supplemented with 25 ng/mL BMP-4 and 50 ng/mL VEGF. On day 7, CD34$^+$ cells were enriched with the EasySep Human Cord Blood CD34 Positive Selection Kit II (StemCell Technologies) without using the RosetteSep™ Human Cord Blood CD34 Pre-Enrichment Cocktail. The enriched iPSC-derived CD34$^+$ cells were plated at a density of 3.5 x 10^5 cells per cm^2 onto fibronectin (Millipore) coated dishes (5 ng/cm^2) and cultured in Endothelial Growth Medium 2 (EGM2, Lonza) supplemented with 50 ng/mL VEGF with medium changes every other day until they reached confluency. After the 1st passage, cells were cultured in regular EGM2 without extra VEGF and split at a 1:3 ratio onto fibronectin-coated plates (5 ng/cm^2) as they approached confluency. iPSC-EC were cryopreserved in 0.5mL of CryoStor CS10 Freeze Media (BioLife Solutions, Bothell, WA) at 1 x 10^6 cells/vial according to the manufacturer's protocol. Phase contrast photomicrographs of 2395 iPSC-EC were collected using an EVOS FL Cell Imaging System (Fisher Scientific).

Generation of CD8$^+$ T Cell Avatars From Donor 2395

Antigen-specific CD8$^+$ T Cell avatars were produced and validated as previously published (20). Briefly, CD8$^+$ T cells were isolated from peripheral blood of donor 2395 *via* negative selection using the RosetteSep Human CD8$^+$ T Cell Enrichment Cocktail (StemCell Technologies) according to the manufacturer's protocol. Naïve CD8$^+$ T cells (CD45RA$^+$CD45RO$^-$) were isolated by FACS with a

FACSAria III Cell Sorter (BD Biosciences). Naïve $CD8^+$ T cells were activated for two days with Human T-Activator CD3/CD28 Dynabeads (ThermoFisher Scientific, Waltham, MA). After two days, cells were transduced with lentiviral vectors, either pCCL.IGRPopt.eGFP, encoding a TCR recognizing an HLA-A*02-01-restricted epitope derived from the T1D-relevant autoantigen, glucose-6-phosphatase 2 (G6PC2) commonly referred to as islet-specific G6CP-related protein (IGRP) (39, 40) or LV.Mart1.TCR.RK, encoding a TCR recognizing an HLA-A*02-01-restricted epitope derived from the non-T1D-relevant auto-antigen, MART-1 (18, 19). Following transduction, the IGRP and the MART-1 CD8+ T cell avatars (IGRP avatars and MART-1 avatars) were expanded for 7 more days then cryopreserved.

Co-Culturing Donor 2395 iPSC-moDC and iPSC-MDM With Donor 2395 $CD8^+$ T Cells

To verify the ability of iPSC-moDC and iPSC-MDM to induce antigen-specific T cell expansion, iPSC-moDC and MDM were treated with 10 ng/mL of LPS (Sigma-Aldrich, St. Louis, MO) and 10 ng/mL of IFNγ (Peprotech) for 48 hours. After iPSC-moDC and MDM were activated for 48 hours, they were loaded for 2 hours with both CEFX Ultra SuperStim Pool MHC-I and MHC-II Subsets (JPT innovative peptide solutions, Berlin, Germany), for a total of 80 MHC Class I restricted epitopes and 68 MHC Class II restricted epitopes derived from a broad range of common human viruses, as well as *Clostridium tetani* and *Toxoplasma gondii*. After 2 hours, iPSC-moDC and MDM were washed with T cell expansion medium (RPMI supplemented with 10% HyClone FBS (GE Healthcare, Chicago, IL, USA), 1000 U/mL Penicillin/streptomycin, 1mM Sodium Pyruvate (Corning), 10mM HEPES (Corning), nonessential amino acids (Corning), Glutamax, and β-mercaptoethanol) and the medium was replaced with T cell expansion medium. Freshly isolated total T cells from donor 2395 were stained with CellTrace CFSE (ThermoFisher), added to the CEFX peptide-loaded iPSC-moDC or MDM and co-cultured for 5 days. After 5 days, expansion of total T cells was analyzed *via* flow cytometry on an Accuri C6 Flow Cytometer.

In Vitro Hydrodynamic Flow Chamber Adhesion Assay

The *in vitro* hydrodynamic flow chamber adhesion assay was performed as previously described (15, 41). Briefly, iPSC-EC were grown in monolayers to 95-100% confluence on a 0.2 mm Luer μ-slide (channel slide) pre-coated with fibronectin. The iPSC-EC monolayers were then loaded with either vehicle solution, $IGRP_{265-273}$ (the antigenic peptide recognized by the IGRP avatars), or Melan-A_{26-35} (the antigenic peptide recognized by the MART-1 avatars) at a final concentration of 10 μg/ml and incubated overnight. The next day, the iPSC-EC-containing channel slide(s) were loaded on the stage of a confocal microscope heated to 37°C. Ten thousand MART-1 or IGRP avatars labeled with 200 nM CellTracker Green (Molecular Probes, Eugene, OR) were added to the channel slide and allowed to settle for 30 minutes. HBSS buffer with $MgCl_2$ and $CaCl_2$ was then flowed across the channel slide by a syringe-driven pump at a series of increasing flow rates ranging from 1 to 80 dyne/cm^2. After passing the buffer through the channel slide for 1 minute for each shear flow rate, bright field and fluorescent images of the slide were acquired by confocal microscopy. T cells that remained attached after each flow rate were then counted using ImageJ software (42).

Chromium Release Assay

The chromium release assay to assess antigen-specific targeting of sBC or the immortalized human β-cell line, BetaLox5 cells (βL5), by donor 2395 avatars was performed as previously described (20, 43). Briefly, sBC clusters from donor 2395, 1-018, or 1-023 were dispersed in enzyme-free Cell Dissociation Buffer (Gibco) for 10 minutes at 37°C with gentle pipetting. Dispersed sBC were plated at 40,000 cells/well or βL5 cells were plated at 10,000 cells/well in 96-well flat-bottom plates. Target cells were loaded with Melan-A_{27-35}. Target cells were then labeled with ^{51}Cr at 1 μCi/well for 3 hours and then washed 3 times with culture medium. Donor 2395 MART-1 avatars were added at effector to target (E:T) ratios of 0:1, 5:1, and 10:1 and co-cultured for 16 hours. Afterwards, the supernatants were collected, cells were lysed in 2% SDS, and cell lysate was collected. ^{51}Cr release was measured with a gamma counter and specific lysis was calculated as follows:

$$\%Specific\ Lysis = Experimental \frac{(\#Release)}{(\#Release) + (\#Lysate)} - Spontaneous \frac{(\#Release)}{(\#Release) + (\#Lysate)}$$

where spontaneous release is the 0:1 E:T ratio and experimental release is measured in the 5:1 and 10:1 E:T ratios.

Statistical Analysis

All statistical analyses were performed in GraphPad Prism version 8.4.3 (San Diego, CA) using unpaired Student's t test, one-way ANOVA, or two-way ANOVA, as indicated in the figure legends.

RESULTS

Reprogramming Donor 2395 $CD34^+$ Peripheral Stem Cells to iPSC

Our proposed isogenic system for T1D (**Figure 1**) requires constant supplies of β-cells, DCs, T cells, endothelial cells, and macrophages from the same donor. The major goal of this effort was to determine if iPSC lines had potential to differentiate into all of these cell types allowing construction of a fully isogenic system. For the work described here all of these cell types, except for T cells, were to be differentiated from donor iPSC lines. Before working with these iPSC lines, they were all verified to express pluripotency markers and have a normal karyotype. iPSC from lines 1-023 and 1-018 were validated and published previously (21–23). iPSC generated from donor 2395 showed similar morphology to normal hESCs (**Figure 2A**) and expressed the pluripotency markers, Oct4, SSEA-4, and NANOG (**Figure 2B**). 2395 iPSC were pluripotent, they differentiated into $Nestin^+$ ectoderm, $FoxA2^+Sox17^+$ endoderm, and $NCAM^+Brachyury^+$ mesoderm (**Figure 2C**).

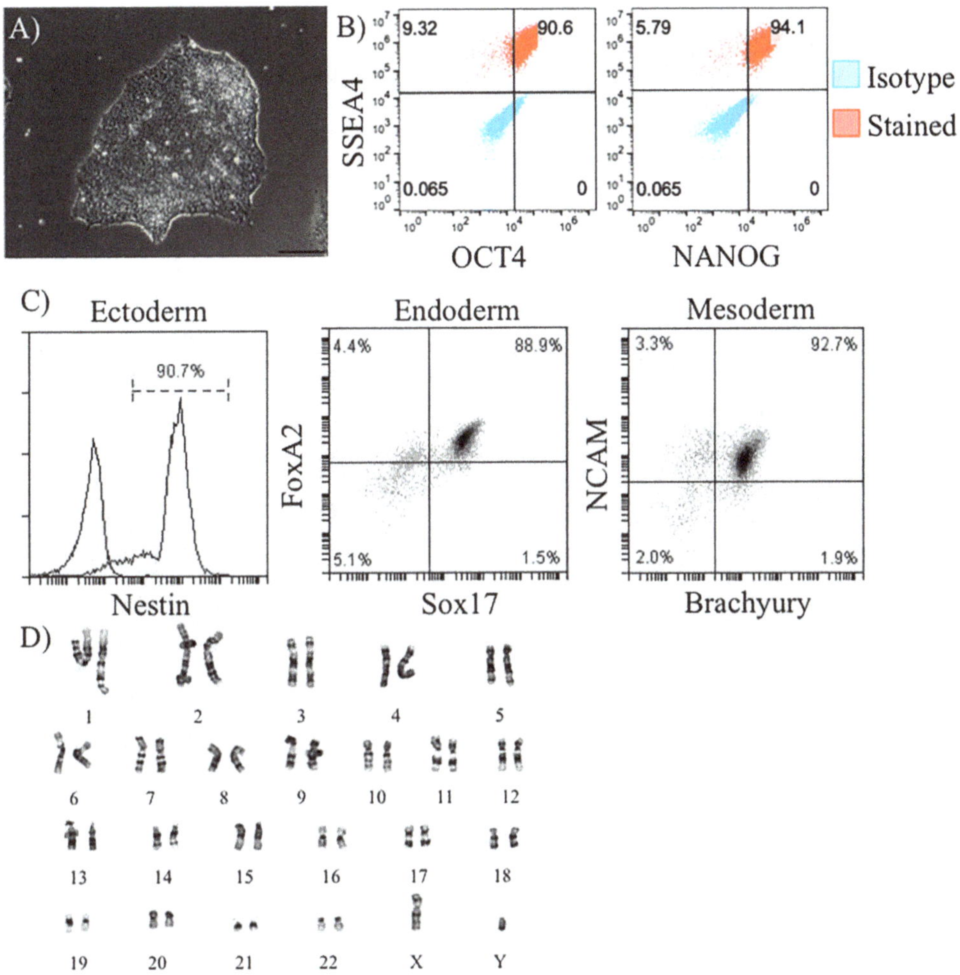

FIGURE 2 | Verification of pluripotency of donor 2395 iPSC. **(A)** A phase contrast image of an iPSC colony from donor 2395 showing densely packed cells with a defined border characteristic of iPSC. **(B)** Donor 2395 iPSC are positive for Oct4, SSEA4, and NANOG expression. **(C)** Donor 2395 iPSC express Nestin when differentiated to ectoderm, express FoxA2 and Sox17 when differentiated to endoderm, and express NCAM and Brachyury when differentiated to mesoderm. **(D)** Karyotypic analysis of donor 2395 iPSC reveals a normal karyotype.

Karyotyping analysis of 2395 iPSC showed a normal chromosome count of 46,XY (**Figure 2D**).

Donor 2395 iPSC Differentiate Into Insulin Producing Beta-Like Cells That Respond to Glucose

β-cells participate in the pathogenesis of T1D by shedding antigen (**Figure 1A**) to initiate the autoimmune response and are also targeted for destruction by the immune system in T1D (**Figure 1D**). Any fully isogenic system to study these steps in T1D (**Figure 1**) will require a renewable source of islets or β-cells. Stem cell-derived β-cells (sBC) can fulfill this need. With published protocols for differentiation of β-cells available, we proceeded to differentiate iPSC from the three lines (1-018, 1-023, and 2395) into sBC. Lines 1-018 and 1-023 have previously been differentiated into validated sBC in islet like clusters with expression of c-peptide and secretion of insulin (22, 44). iPSC from line 2395 were differentiated to sBC by first differentiating to definitive endoderm (DE). Cultures were only moved forward if they reached >80% SOX17$^+$ at the DE stage (**Figure 3A**). At the β-like stage, cells reached 28.1 ± 3.0% insulin$^+$glucagon$^-$ (**Figures 3B, C**; n = 3 separate differentiations). Fluorescent micrographs (**Figure 3D**) demonstrate an islet-like cluster of sBC from 2395 with cells expressing NXK6.1, PDX-1, and c-peptide. To verify that the sBC were functional, those differentiated from 1-018, 1-023, and 2395 were subjected to static glucose-stimulated insulin secretory assays and compared to primary human islets. Basal insulin release at 3mM glucose in human islets [2.3 ± 1.8 μg insulin/10 ng DNA (n=6)] was higher than that observed from all three sBC clusters [1-023 (0.051 ± 0.043 μg insulin/10 ng DNA (n=3)], 2395 [0.04 ± 0.01 μg insulin/10 ng DNA (n=6)], and 1-018 [0.007 ± 0.005 μg insulin/10 ng DNA (n=3)]. Clusters from all three lines secreted insulin, albeit less than the average stimulation index of primary islets (**Figure 3E**).

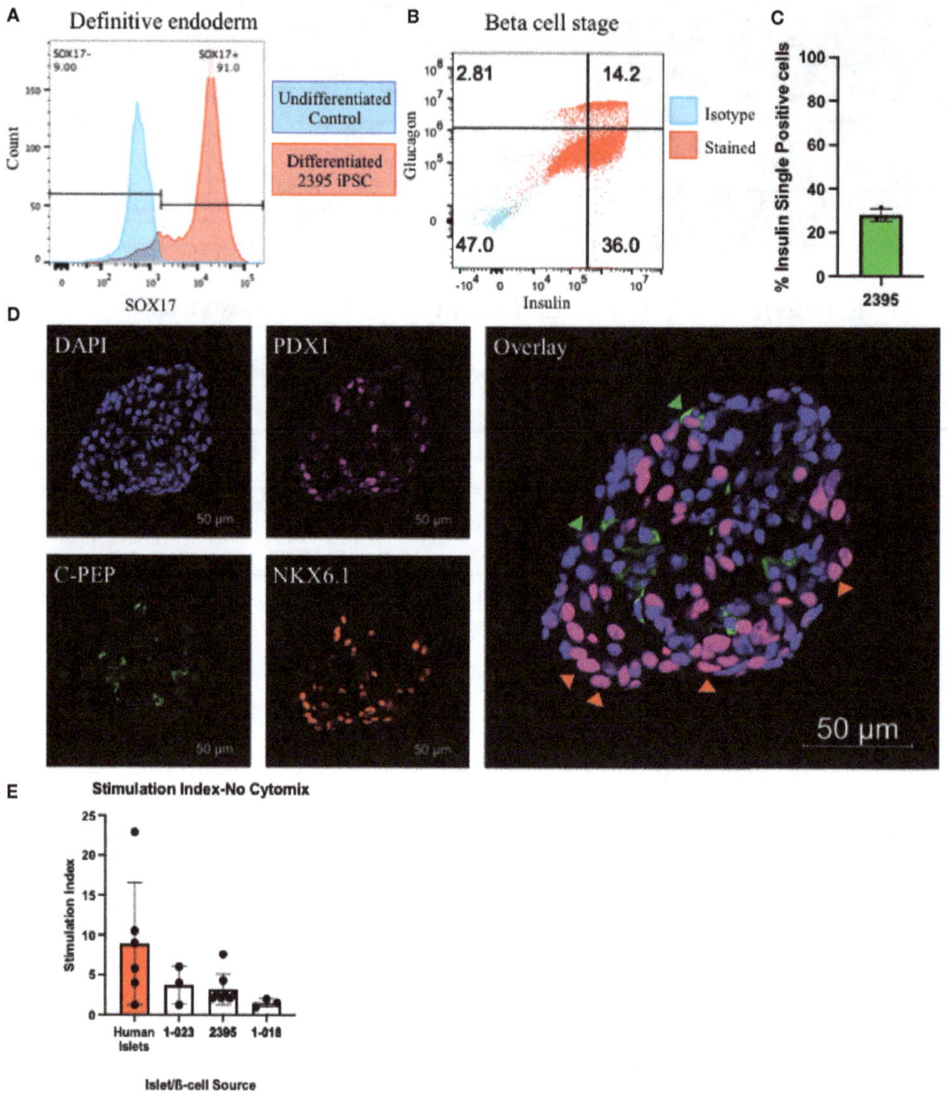

FIGURE 3 | Donor 2395 iPSC differentiate to stem cell derived beta-like cells (sBC). **(A)** Donor 2395 iPSC upregulate SOX17 when differentiated to definitive endoderm. **(B)** When donor 2395 definitive endoderm is differentiated to the β-cell stage, many polyhormonal (insulin$^+$glucagon$^+$) and monohormonal (insulin$^+$) cells arise. **(C)** The percent of insulin$^+$glucagon$^-$ cells from 3 repeats of 2395 sBC differentiations. **(D)** An sBC from donor 2395 was fixed, labeled with DAPI, stained with antibodies against PDX1, C-peptide, and NKX6.1 revealing some C-peptide/NKX6.1 double positive β-cells (green arrow heads) as well as many NKX6.1/PDX1 double positive cells (red arrow heads). **(E)** Donor 2395 sBC as well as iPS-β-cells from lines 1-018 and 1-023 exhibit glucose-stimulated insulin secretion. Stimulation index = insulin secretion in 16.7mM glucose/insulin secretion in 3mM glucose.

While there were no significant differences, 1-023 and 2395 outperformed clusters from 1-018 on average. In summary these lines differentiated into glucose-responsive sBC allowing a first step towards building *in vitro* systems for studying the interactions of isogenic immune cells with β-cells.

HLA Typing of iPSC 1-018, 1-023, and 2395

Class II HLA alleles confer the strongest genetic risk for pathogenesis of T1D, highlighting the importance of CD4$^+$ T cells. CD8$^+$ T cells are the predominant cell in human insulitis emphasizing that HLA Class I alleles play a major role in T1D development. HLA Class I regulates many of the key steps thought to be vital for T1D initiation such as T cell activation and targeted deletion of pancreatic β-cells (**Figure 1**). For these reasons it is essential that iPSC harbor useful HLA alleles for the study of T1D. HLA alleles were determined for all three iPSC lines (**Table 1**). The studies here will focus on the interactions of HLA-A*0201 restricted CD8$^+$ T cell avatars (20) with iPSC-derived cells expressing HLA-A*0201 with cognate peptide. Lines 2395 and 1-023 carry HLA-A*0201 while 1-018 does not (**Table 1**). Therefore, we prioritized the former two iPSC lines for further studies.

TABLE 1 | iPSC Donor Demographics and HLA Alleles for lines 1-018, 1-023, and 2395.

Donor	Age at Study	Sex	T1D Status	Age at Onset	HLA-A	HLA-B	HLA-DQA1	HLA-DQB1	HLA-DRB1
1-018	32	Female	T1D	10	0101	1801	0301	0302	0402
					3101	7301	0301	0201	0405
1-023	23	Male	Control	Not applicable	0201	3501	0102	0502	1101
					6801	5101	0501	0301	1506
2395	52	Male	Control	Not applicable	2901	4403	0201	0202	0701
					0201	4002	0505	0301	0701

Alleles are arranged into putative extended haplotypes using the Allele Frequency Net Database).

Donor 2395 iPSC Efficiently Differentiate to Monocytes, MDM, and MoDC

While the initiating events in T1D are not decisively characterized, initiating a productive T cell response requires at least two signals during the interaction of APC with T cells (**Figure 1B**). The first signal is binding of the TCR on the T cell to its cognate antigen/MHC complex on an APC and the second signal is the costimulatory signal in the form of co-receptors on the T cell binding their respective ligands on an APC (e.g. CD28 binding CD80/CD86, CD27 binding CD70, CD226 binding CD155, OX-40 binding OX-40L, etc.) (45, 46). APC such as DC and macrophages serve to initiate T cell responses in the lymph nodes and enhance T cell responses at sites of inflammation. In order to model interactions between macrophages and T cells or DC and T cells, we differentiated donor 2395 iPSC to monocytes, then MDM and moDC (**Figure 4**). Similar to primary classical peripheral blood monocytes, these iPSC-monocytes are $CD14^{+}CD64^{+}CD16^{-}$ (**Figure 4A**). Also similar to their primary counterparts, iPSC-MDM are $CD11b^{+}CD68^{+}$ (**Figure 4B**) and iPSC-moDC are $MHC\text{-}II^{+}CDllc^{+}$ (**Figure 4C**). Attempts were made to differentiate the iPSC line, 1-023, to monocytes however they never produced $CD34^{+}$ hematopoietic progenitors following hematopoietic differentiation and they also never produced $CD14^{+}$ monocytes following monocyte induction. Due to the inability of 1-023 iPSC to differentiate into Monocytes, MDM, and MoDC, we prioritized iPSC line 2395 for the remaining studies.

Donor 2395 iPSC-moDC and iPSC-MDM Elicit Donor 2395 CD4^{+} and CD8^{+} T Cell Expansion in an Antigen Specific Manner

To determine if iPSC-moDC and MDM are able to elicit an antigen specific $CD4^{+}$ and $CD8^{+}$ T cell response, iPSC-moDC or MDM from donor 2395 were activated with LPS/IFNγ, loaded with CEFX Ultra SuperStim Pool MHC-I and MHC-II subsets, and co-cultured with CFSE (ThermoFisher) stained total T cells from donor 2395. When 2395 total T cells were cultured with iPSC-MDM alone (**Figure 4D**; closed black circles), or LPS/IFNγ-activated iPSC-MDM (**Figure 4D**; open black circles) there was little T cell expansion observed. However, when 2395 total T cells were cultured with CEFX peptide-loaded MDM (**Figure 4D**; closed red squares), or LPS/IFNγ-treated, CEFX peptide-loaded iPSC-MDM (**Figure 4D**; open red squares), there was significantly more expansion than when they were cultured with iPSC-MDM without peptide loading. iPSC-moDC also induced T cell expansion in an antigen specific manner. When total T cells were cultured with peptide-loaded iPSC-moDC (**Figure 4E**; closed red diamonds), expansion was increased compared to T cells cultured with iPSC-moDC alone (**Figure 4E**; closed black triangles). However, when iPSC-moDC were activated with LPS/IFNγ first, there was no significant difference in T cell expansion when iPSC-moDC were peptide loaded (**Figure 4E**; open red diamond) compared to iPSC-moDC without peptide loading (**Figure 4E**; open black triangles).

Donor 2395 CD8^{+} T Cell Avatars Bind to Donor 2395 EC in an Antigen Specific Manner

After T cells are activated, they must traffic to the islets of Langerhans. To exit the islet capillaries and enter the islets, T cells must migrate across the endothelial cell (EC) layer lining the insides of the capillaries in a process called extravasation [(47) and **Figure 1C**]. In this process, T cells first migrate to the site of inflammation *via* chemotaxis, they then must tether and roll across activated ECs before adhering, and finally transmigrating across the epithelium (47). To model the interactions of $CD8^{+}$ T cells with vascular endothelium, we assessed $CD8^{+}$ T cell avatars adhesion to an antigen-loaded iPSC-EC monolayer using cells from the same donor. As described recently (37), 2395 iPSC differentiate into iPSC-EC with morphology (**Supplementary Figure 1A**), cell surface marker profile (**Supplementary Figure 1B**), and functional capabilities similar to human aortic endothelial cells (HAEC). In this study 2395 iPSC efficiently differentiated into iPSC-EC with high a percentage of the cells being double positive for both CD31 and CD144 (84% ± 8, n=4). Here, an *in vitro* hydrodynamic flow chamber adhesion assay was performed with monolayers of iPSC-EC from donor 2395 and either IGRP avatars or MART-1 avatars from 2395. When the 2395 iPSC-EC monolayer was not presenting antigenic peptide, the 2395 IGRP avatars failed to adhere strongly and were displaced at higher laminar flow rates (**Figure 5A**; black). Similarly, when the 2395 iPSC-EC monolayer was pulsed with the peptide antigen recognized by the MART-1 TCR, Melan-$A_{26\text{-}35}$ (EAAGIGILTV), 2395 IGRP avatars detached at higher flow rates (**Figure 5A**; red). In contrast, 2395 IGRP avatars formed strong adhesions with the $IGRP_{265\text{-}273}$ peptide-pulsed iPSC-EC monolayer and remained attached at higher flow rates (**Figure 5A**; gray). We confirmed the system's utility with a

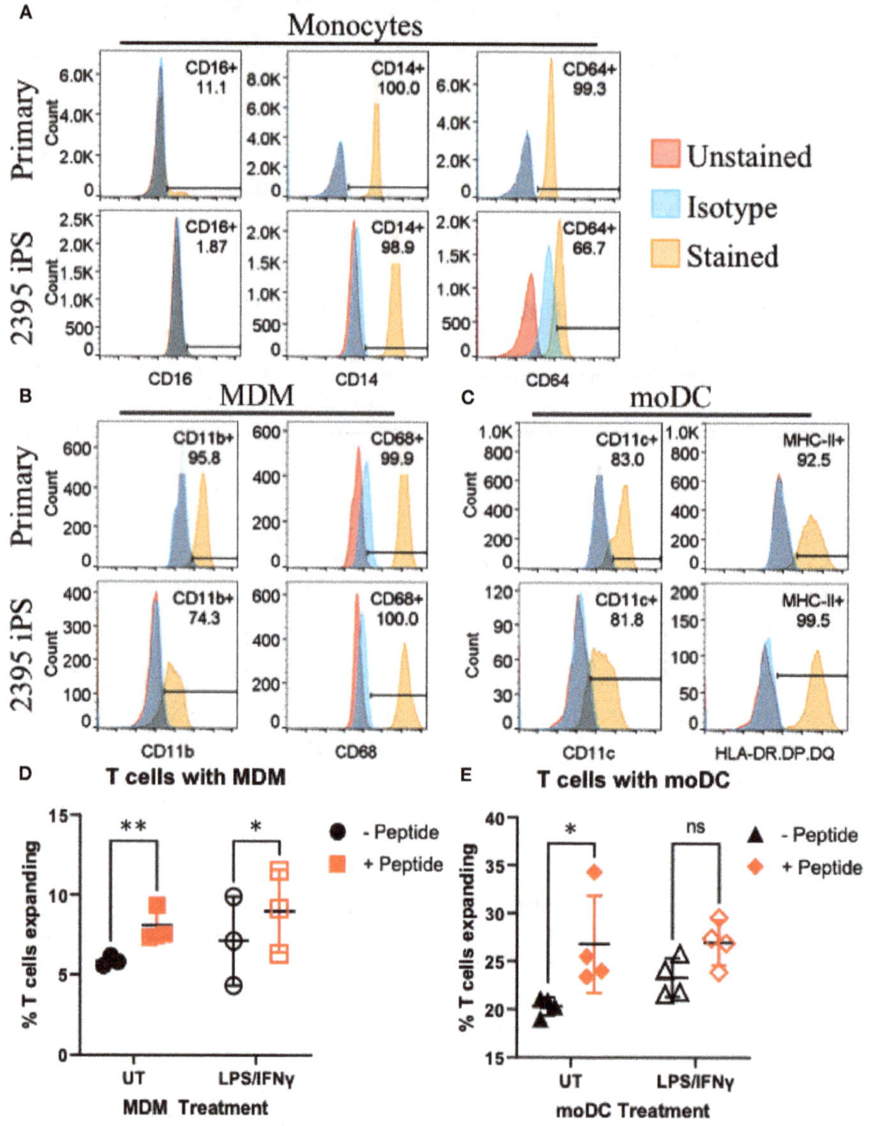

FIGURE 4 | Donor 2395 iPSC differentiate efficiently to monocytes, monocyte-derived macrophages (MDM) and monocyte-derived dendritic cells (moDC) and induce antigen specific $CD4^+$ and $CD8^+$ T cell expansion. **(A)** Donor 2395 iPSC-derived monocytes are $CD16^-$, $CD14^+$, and $CD64^+$ similar to primary classical monocytes isolated from blood. **(B)** Donor 2395 iPSC-derived MDM express CD11b and CD68 similar to primary MDM. **(C)** Donor 2395 iPSC-derived moDC express CD11c and MHC-II similar to primary moDC. **(D)** iPSC-MDM expand Total T cells in an antigen specific manner. When antigen (CEFX peptide pools) is present, there is a significantly increased expansion of T cells compared to when Total T cells are cultured with APC alone (2-way ANOVA $^{**}p = 0.0011$ with 3 experimental repeats). **(E)** iPSC-moDC expand Total T cells in an antigen specific manner. When antigen is present, there is a significantly increased expansion of T cells compared to when Total T cells are cultured with APC alone (2-way ANOVA $^{*}p = 0.0104$ with 4 experimental repeats). Significance shown on the graphs were from *post-hoc* analysis utilizing the Šidák method to correct for multiple comparisons. * indicates that $p < 0.05$, ** indicates that $p < 0.01$, and ns indicates not significant.

second TCR and antigen, 2395 MART-1 avatars and the peptide it recognizes, Melan-$A_{26\text{-}35}$. Using the laminar flow system without peptide the MART-1 avatars failed to form firm adhesions with the iPSC-EC monolayer (**Figure 5B**; black). Addition of $IGRP_{265\text{-}273}$ did not facilitate adhesion as this peptide is not recognized by the MART-1 avatars (**Figure 5A**; gray). In contrast, 2395 MART-1 avatars made firm adhesions with the Melan-$A_{26\text{-}35}$ peptide-pulsed iPSC-EC monolayer as observed by binding at higher flow rates (**Figure 5A**; red). As vascular adhesion interactions are critical for T cell trafficking into pancreatic islets for targeted destruction of β-cells in T1D (15, 48–50), this proof-of-concept study demonstrates the utility of an antigen-dependent system for assessing interactions of T cells with endothelial cells. The ability to model these interactions with human cells allows for intensive interrogation of genetics regulating vascular inflammation or therapeutics meant to disrupt lymphocyte adhesion with vascular endothelial cells (**Figure 1C**).

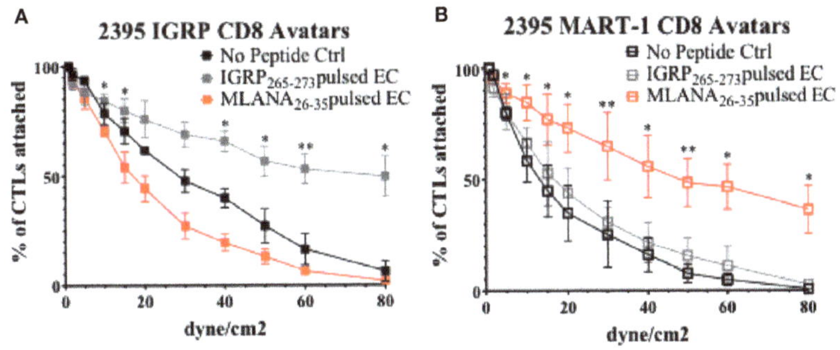

FIGURE 5 | Donor 2395 CD8+ T cell avatars exhibit antigen specific binding to donor 2395 iPS-EC. **(A)** When donor 2395 IGRP CD8+ T cell avatars are flowed across a donor 2395 iPSC-derived endothelial cell (iPS-EC) monolayer loaded with IGRP peptide, the IGRP CD8+ T cell avatars adhere more than MART-1 CD8+ T cell avatars or when the iPS-EC is not loaded with peptide. **(B)** When donor 2395 MART-1 CD8+ T cell avatars are flowed across a donor 2395 iPSC-derived endothelial cell (iPS-EC) monolayer loaded with Melan-A_{26-35}, the MART-1 CD8+ T cell avatars adhere more than IGRP CD8+ T cell avatars or when the iPS-EC is not loaded with peptide. All *in vitro* hydrodynamic flow assay experiments were repeated 3 times. Significance shown on the graphs were from *post-hoc* analysis utilizing the Fisher's LSD test. * indicates that $p < 0.05$ and ** indicates that $p < 0.01$.

Donor 2395 CD8+ T Cell Avatars Efficiently Target Donor 2395 sBC in an Antigen Specific Manner

To determine if sBC are able to be targeted by CD8+ T cells in an antigen specific manner, a cell mediated lymphocytocixity (CML) assay was run with dispersed clusters of sBC differentiated from 1-023, 2395, or 1-018 as targets. The HLA-A*0201+ human βL5 cell line and dispersed HLA-A*0201+ human islet cells were used as positive controls (**Figure 6**). These groups were co-cultured with donor 2395 T cell avatars expressing a T cell receptor that recognizes a peptide from IGRP in the context of HLA-A*0201 (39, 40). The 2395 IGRP-avatars were able to target and kill the HLA-A*0201 expressing βL5 and primary human islets (**Figure 6A**). In addition, cells differentiated from 1-023 and 2395 were lysed by the avatars (**Figure 6A**) as they carry HLA-A*0201 (**Table 1**). Cells from 1-018 were not killed by the avatars (**Figure 6A**) as these cells do not have the correct HLA restriction (**Table 1**). We also assessed the ability of 2395 MART-1 T cell avatars that recognize a non-β-cell peptide derived from Melan-A_{26-35} (20, 51) to target and destroy cells that were pulsed with this Melan-A peptide or left untreated. MART-1-avatars were able to target and lyse both βL5 and 2395 sBC when the peptide was present in a fashion that was regulated by the number of MART-1 avatars added (**Figure 6B**). However, when the βL5 or 2395 sBC were not pulsed with Melan-A_{26-35}, neither population of cells was lysed. These results indicate that sBC can be utilized to study the interactions of β-cells with antigen specific T cells. In addition, as 2395 T cell avatars kill 2395 sBC in an antigen-specific fashion (**Figure 6B**) this isogenic system is a viable method to study interactions of autoreactive T cells with β-cells.

DISCUSSION AND CONCLUSION

While T1D develops in genetically at-risk individuals, our current understanding of the pathogenesis of T1D in humans continues to evolve due to the complexities of genetic risk in combination with environmental factors (52). Recent GWAS efforts have dramatically expanded the number of loci that are associated with both increased or decreased risk for T1D (1–12). The mechanisms linking the majority of these polymorphisms, alleles, and allotypes to the development of T1D are not clear. Disease heterogeneity in T1D may result from the dissimilar combinations of genes inherited by at risk individuals that instigate variations in onset age and disease endotypes. Indeed, specific combinations of risk HLA haplotypes have been linked to earlier age of onset as well as changes in the immune cell components of the insulitic lesion (53). Due to the impact of total combined genetic risk, it is necessary to sufficiently power mechanistic studies with large sample sizes to determine the responsibility of a single polymorphism to T1D and the cell types influenced by the greater than 100 polymorphisms that are linked to T1D (1–12). With the development of isogenic systems, such as that described here, methodologies have advanced that allow for interrogation of the effect of a single polymorphism in the absence of other genetic differences. Models derived from primary cells can lack reproducibility because certain cell types (i.e. β-cells) are not renewable; utilizing iPSC circumvents this issue. Herein we have shown that key cell types involved in the pathogenesis of T1D (**Figure 1**) can be derived from a single donor and used to construct an isogenic system to interrogate relevant cell-cell interactions. This renewable system provides a powerful platform to interrogate how specific genetic polymorphisms influence both cell function and cell-cell interactions in isolation of other genetic differences.

Thus far, we have observed that 2395 iPSC are pluripotentusing *via* trilineage differentiation, demonstrating that 2395 iPSC are capable of forming all 3 germ layers *in vitro* (**Figure 2C**). Further, 2395 iPSC express the pluripotency markers, OCT4 and NANOG (**Figure 2B**), and differentiate into functional cells from the endoderm (**Figures 3, 5, 6**) and mesoderm (**Figure 4**). However, prior to use of 2395 cells an *in vivo* model it will be essential to determine teratoma formation and tumorigenic potential. In this

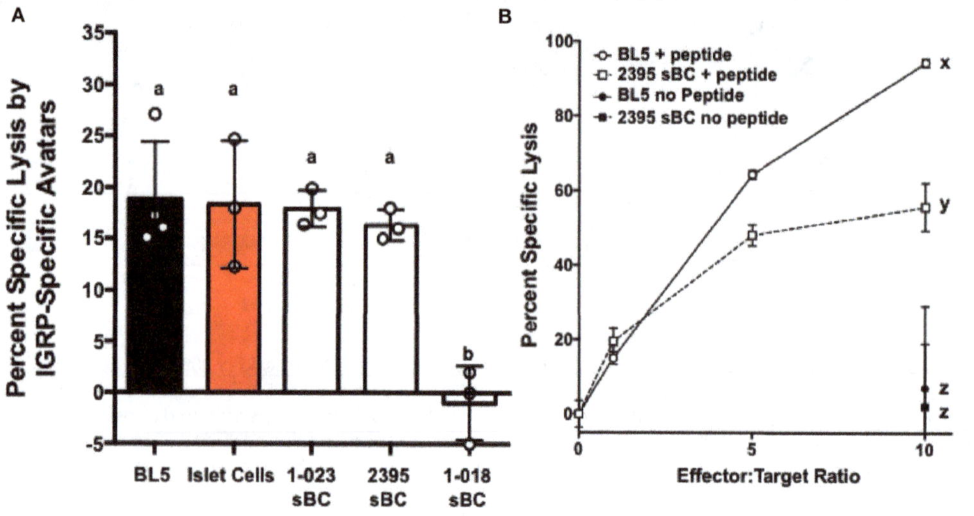

FIGURE 6 | sBC are efficiently targeted by donor 2395 CD8⁺ T cell avatars in an antigen specific and HLA restricted manner. **(A)** Donor 2395 IGPR T cell avatars were co-cultured with βL5 cells, primary islet cells from a HLA-A*0201 positive donor, 1-023 sBC, or 2395 sBC the cells were effectively lysed at an effector to target (E:T) ratio of 10:1. As 1-018 sBC do not harbor HLA-A*0201 these cells were not lysed by these T cell avatars as they require the IGRP peptide to be presented by HLA-A*0201. **(B)** Donor 2395 MART-1-avatars were co-cultured with either βL5 cells or donor 2395 sBC at increasing E:T ratios (0:1, 1:1, 5:1, and 10:1). When pulsed with Melan-A_{26-35}, lysis of both βL5 and 2395 sBC increased with escalating E:T ratios. In the absence of Melan-A_{26-35}, no lysis was measured. *Letters denote significance: groups/lines with different letters were statistically significant ($P<0.05$), whereas those with the same letter were not statistically different. These data are from at least 3 independent studies performed in triplicate.

isogenic system derived from donor 2395, we have shown that both macrophages and DC can expand total T cells (**Figure 4**), that CD8⁺ T cell avatars bind to EC in an antigen specific manner (**Figure 5**), and that CD8⁺ T cell avatars can target and lyse sBC that is dependent on the trimolecular complex (**Figure 6**). The current system affords power for interrogating these simple cell-cell interactions as well as more complex cell-cell interactions such as macrophage, T cell, and β-cell interactions. While the system has great utility, it does however have limitations.

One of the current requirements of the current system is the need to recall specific donors to obtain T cells. This necessitates proper IRB approval and flexibility of research participants and investigators to schedule blood draws. Currently there are protocols for obtaining T cells from iPSC however these produce predominantly CD8⁺ T cells and very few CD4⁺ T cells (54, 55). The addition of iPSC-derived naïve autoreactive CD8⁺ T cells to this system can be implemented to provide yet another way for interrogation of cell-cell interactions and eliminate the need to recall donors to obtain CD8⁺ T cells. iPSC can be transduced with lentiviral vectors (56, 57), therefore a system to develop antigen specific iPSC-T cells could be employed wherein the iPSC are engineered to express a specific TCR to recognize self or nominal antigens.

Another shortcoming of this system and iPSC in general is that not all iPSC lines efficiently differentiate into all necessary cell types. For example, 1-018 sBC were not as functional as sBC from lines 1-023 or 2395 (**Figure 3E**). In addition, while 1-023 differentiated well into sBC, this iPSC line did not produce monocytes. Lack of monocytes prevented downstream differentiation to macrophages and dendritic cells. Other groups have observed that lineage differentiation capacity is dependent on the iPSC line necessitating protocol optimization for lines that fail to differentiate effectively (58). As such, methodological modifications specific for 1-023 may resolve the observed dearth of monocytes from this line. However, the necessity of individualized differentiation protocols could impact the throughput of any system. Another important consideration and potential difficulty results from the complexity of T1D genetics. This requires careful consideration of the cells utilized for differentiation and study. Due to the strong genetic influence of the HLA super locus on T1D and the need for specific HLA alleles to be present for antigen specific studies, genotyping and selecting iPSC with HLA alleles (**Table 1**) that match TCRs to be utilized is essential. Here we eliminated 1-018 from the system as this line did not have the necessary HLA-A*0201 allele to interact with the T cell avatars (**Table 1**). Indeed, 1-018 sBC were not targeted and lysed by the IGRP-Avatars (**Figure 6A**). The HLA Class I and Class II loci are the most polymorphic coding genes in the human genome (59), therefore matching HLA for the current study was prioritized. However, advances in gene editing technology may, in future studies, allow for extensive modification of HLA loci and obviate the requirement for excluding cells based on HLA allele types. Regardless of the more difficult requirements of the current system, there are many developing technologies that will improve future iterations.

One of the currently developing technologies is gene editing in iPSC. Gene editing can increase the flexibility of this system. There are numerous protocols for editing SNPs or knocking out genes in iPSC. While knockout is straightforward and there are high-efficiency protocols available for iPSC, easy to implement and high-efficiency protocols for scarless editing of single bases, such as SNPs, are still being developed (60–66). Once these high-efficiency protocols are developed, editing SNPs in iPSC will allow us to interrogate how credible risk or resistance SNPs influence these cell-cell interactions. The ability to edit iPSC donor lines to harbor both the T1D-risk and T1D-resistance alleles at a single locus will potentially suppress variation by allowing edited donor lines to be compared to each other. These methods would have exceptional value to study the genetic influences in cells, organoids, or organs that are difficult to access or genetically modify in a homogenous fashion. This would greatly facilitate the study of rare genetic variants (e.g. *rs34536443* in *TYK2*, *rs2476601* in *PTPN22*, *rs35744605* in *IFIH1*, etc.) as editing iPSC lines to the rare minor alleles at T1D risk loci would allow for generation of numerous lines from different donors that could be studied in a pairwise fashion versus identifying and specifically recruiting donors with minor allele frequency (MAF) <0.05. For example, the frequency of individuals homozygous for the rare allele of *rs34536443* in *TYK2* was recently reported to be 213 per 100,000 with MAF = 0.04 (67, 68). The ability to edit a credible risk or resistance SNP will facilitate determination of the impact of these rare variants and likely bolster efforts seeking to resolve the actual SNP(s) in linkage disequilibrium (LD) that influence disease risk (e.g. the 233kb LD block on chromosome 16p13 that contains *CLEC16A*) (69–71). In addition, other technologies in development, such as ESC and iPSC-derived thymii that can output functional naïve T cells, are novel tools for studying human disease etiology. As thymic epithelial cells (iPSC-TEC) (72, 73) have the potential to regulate T cell production from progenitors, use of these evolving technologies in combination (gene editing and iPSC-TEC) would allow for studies to dissect how T1D-linked loci, or those linked to other diseases, regulate T cell development. These and other currently advancing technologies will greatly enhance isogenic systems, improving their capacity to model human diseases.

Isogenic systems derived from iPSC will greatly augment our ability to interrogate key cellular interactions involved in genetic regulation of T1D pathogenesis. While the system built here is focused on T1D, it can be utilized to study cell-cell interactions and genetic regulation in the context of other human diseases. In summary, we have presented an isogenic model that can currently be used to interrogate many T1D-relevant cell-cell interactions. As more technologies for gene editing and differentiating iPSC continue to be developed these tools can be implemented to build a more relevant model capable of answering more complex questions.

AUTHOR CONTRIBUTIONS

LHA, SES, CEM, and MAW conceived the idea, designed the research plan, and participated in all aspects of writing the manuscript. KS, LS, SO, BN, MH, JT, NF, PS, IR, RC-G, DP, HR, AS, NT, TB, JC, and DE performed experiments, analyzed data, interpreted results of experiments, and edited the manuscript. All authors contributed to the article and approved the submitted version.

FUNDING

This work was supported by research grants from the National Institutes of Health UC4 DK104194 (CM), R01 DK127497 (CM), UG3 DK122638 (TB, CM), P01 AI042288 (TB, CM), R01 DK103585 (DE), UC4 DK104207 (DE), R24 GM119977 (NT), F30 DK105788 (BN), T32 DK108736 (LA), UL1TR001427 (CTSI Pilot award to NF), R01 DK120444 (HR), R21 AI140044 (HR), a HIRN new investigator award (HR), a Culshaw Junior Investigator Award in Diabetes (HR), the Children's Diabetes Foundation (HR), the Sebastian Family Endowment for Diabetes Research (CM), the American Heart Association (16GRNT3098002 to NT), the JDRF 2-SRA-2019-781-S-B (HR), and the Leona and Harry Helmsley Charitable Trust (DE).

ACKNOWLEDGMENTS

The University of Florida Center for Immunology and Transplantation, The University of Florida Center for Cellular Reprogramming, The University of Florida Diabetes Institute, The University of Florida Clinical and Translational Science Institute, and the NIH funded Integrated Islet Distribution Program were all essential to the completion of these studies. This intra-institutional collaboration was facilitated by the Human Islet Research Network.

REFERENCES

1. Barrett JC, Clayton DG, Concannon P, Akolkar B, Cooper JD, Erlich HA, et al. Genome-Wide Association Study and Meta-Analysis Find That Over 40 Loci Affect Risk of Type 1 Diabetes. *Nat Genet* (2009) 41:703–7.. doi: 10.1038/ng.381
2. Bradfield JP, Qu H-Q, Wang K, Zhang H, Sleiman PM, Kim CE, et al. A Genome-Wide Meta-Analysis of Six Type 1 Diabetes Cohorts Identifies Multiple Associated Loci. *PloS Genet* (2011) 7:e1002293. doi: 10.1371/journal.pgen.1002293
3. Chiou J, Geusz RJ, Okino M-L, Yun Han J, Miller M, Huang S, et al. Large-Scale Genetic Association and Single Cell Accessible Chromatin Mapping Defines Cell Type-Specific Mechanisms of Type 1 Diabetes Risk 2. (2021) 3:426472. doi: 10.1101/2021.01.13.426472
4. Cooper JD, Smyth DJ, Smiles AM, Plagnol V, Walker NM, Allen JE, et al. Meta-Analysis of Genome-Wide Association Study Data Identifies

Additional Type 1 Diabetes Risk Loci. *Nat Genet* (2008) 40:1399–401. doi: 10.1038/ng.249

5. Grant SFA, Qu H-Q, Bradfield JP, Marchand L, Kim CE, Glessner JT, et al. Follow-Up Analysis of Genome-Wide Association Data Identifies Novel Loci for Type 1 Diabetes. *Diabetes* (2009) 58:290–5. doi: 10.2337/db08-1022
6. Hakonarson H, Grant SFA, Bradfield JP, Marchand L, Kim CE, Glessner JT, et al. A Genome-Wide Association Study Identifies KIAA0350 as a Type 1 Diabetes Gene. *Nature* (2007) 448:591–4. doi: 10.1038/nature06010
7. Hakonarson H, Qu H-Q, Bradfield JP, Marchand L, Kim CE, Glessner JT, et al. A Novel Susceptibility Locus for Type 1 Diabetes on Chr12q13 Identified by a Genome-Wide Association Study. *Diabetes* (2008) 57:1143–6. doi: 10.2337/db07-1305
8. Huang J, Ellinghaus D, Franke A, Howie B, Li Y. 1000 Genomes-Based Imputation Identifies Novel and Refined Associations for the Wellcome Trust Case Control Consortium Phase 1 Data. *Eur J Hum Genet* (2012) 20:801–5. doi: 10.1038/ejhg.2012.3
9. Noble JA, Valdes AM. Genetics of the HLA Region in the Prediction of Type 1 Diabetes. *Curr Diab Rep* (2011) 11:533–42. doi: 10.1007/s11892-011-0223-x
10. Onengut-Gumuscu S, Chen W-M, Burren O, Cooper NJ, Quinlan AR, Mychaleckyj JC, et al. Fine Mapping of Type 1 Diabetes Susceptibility Loci and Evidence for Colocalization of Causal Variants With Lymphoid Gene Enhancers. *Nat Genet* (2015) 47:381–6. doi: 10.1038/ng.3245
11. Robertson CC, Inshaw JRJ, Onengut-Gumuscu S, Wei-Min, Chen W-M, Santa Cruz DF, Yang H, et al. Fine-Mapping, Trans-Ancestral and Genomic Analyses Identify Causal Variants, Cells, Genes and Drug Targets for Type 1 Diabetes. *Nat Genet* (2021) 53:962–71. doi: 10.1038/s41588-021-00880-5
12. Todd JA, Walker NM, Cooper JD, Smyth DJ, Downes K, Plagnol V, et al. Robust Associations of Four New Chromosome Regions From Genome-Wide Analyses of Type 1 Diabetes. *Nat Genet* (2007) 39:857–64. doi: 10.1038/ng2068
13. Drexhage HA, Dik WA, Leenen PJM, Versnel MA. The Immune Pathogenesis of Type 1 Diabetes: Not Only Thinking Outside the Cell But Also Outside the Islet and Out of the Box. *Diabetes* (2016) 65:2130–3. doi: 10.2337/dbi16-0030
14. Mathews CE, Graser RT, Savinov A, Serreze DV, Leiter EH. Unusual Resistance of ALR/Lt Mouse Beta Cells to Autoimmune Destruction: Role for Beta Cell-Expressed Resistance Determinants. *Proc Natl Acad Sci USA* (2001) 98:235–40. doi: 10.1073/pnas.98.1.235
15. Whitener RL, Gallo Knight L, Li J, Knapp S, Zhang S, Annamalai M, et al. The Type 1 Diabetes-Resistance Locus Idd22 Controls Trafficking of Autoreactive CTLs Into the Pancreatic Islets of NOD Mice. *J Immunol* (2017) 199:3991–4000. doi: 10.4049/jimmunol.1602037
16. Driver JP, Chen Y-GG, Mathews CE. Comparative Genetics: Synergizing Human and NOD Mouse Studies for Identifying Genetic Causation of Type 1 Diabetes. *Rev Diabet Stud* (2012) 9:169–87. doi: 10.1900/RDS.2012.9.169
17. Bonder MJ, Smail C, Gloudemans MJ, Frésard L, Jakubosky D, D'Antonio M, et al. Identification of Rare and Common Regulatory Variants in Pluripotent Cells Using Population-Scale Transcriptomics. *Nat Genet* (2021) 53:313–21. doi: 10.1038/s41588-021-00800-7
18. Johnson LA, Morgan RA, Dudley ME, Cassard L, Yang JC, Hughes MS, et al. Gene Therapy With Human and Mouse T-Cell Receptors Mediates Cancer Regression and Targets Normal Tissues Expressing Cognate Antigen. *Blood* (2009) 114:535–46. doi: 10.1182/blood-2009-03-211714
19. Morgan RA, Dudley ME, Wunderlich JR, Hughes MS, Yang JC, Sherry RM, et al. Cancer Regression in Patients After Transfer of Genetically Engineered Lymphocytes. *Science* (2006) 314:126–9. doi: 10.1126/science.1129003
20. Newby BN, Brusko TM, Zou B, Atkinson MA, Clare-Salzler M, Mathews CE. Type 1 Interferons Potentiate Human CD8+ T-Cell Cytotoxicity Through a STAT4- and Granzyme B-Dependent Pathway. *Diabetes* (2017) 66:3061–71. doi: 10.2337/db17-0106
21. Johannesson B, Sagi I, Gore A, Paull D, Yamada M, Golan-Lev T, et al. Comparable Frequencies of Coding Mutations and Loss of Imprinting in Human Pluripotent Cells Derived by Nuclear Transfer and Defined Factors. *Cell Stem Cell* (2014) 15:634–42. doi: 10.1016/j.stem.2014.10.002
22. Sui L, Danzl N, Campbell SR, Viola R, Williams D, Xing Y, et al. β-Cell Replacement in Mice Using Human Type 1 Diabetes Nuclear Transfer Embryonic Stem Cells. *Diabetes* (2018) 67:26–35. doi: 10.2337/db17-0120
23. Yamada M, Johannesson B, Sagi I, Burnett LC, Kort DH, Prosser RW, et al. Human Oocytes Reprogram Adult Somatic Nuclei of a Type 1 Diabetic to Diploid Pluripotent Stem Cells. *Nature* (2014) 510:533–6. doi: 10.1038/nature13287
24. Taylor JP, Cash MN, Santostefano KE, Nakanishi M, Terada N, Wallet MA. CRISPR/Cas9 Knockout of USP18 Enhances Type I IFN Responsiveness and Restricts HIV-1 Infection in Macrophages. *J Leukoc Biol* (2018) 103:1225–40. doi: 10.1002/JLB.3MIA0917-352R
25. Nishimura K, Sano M, Ohtaka M, Furuta B, Umemura Y, Nakajima Y, et al. Development of Defective and Persistent Sendai Virus Vector: A Unique Gene Delivery/Expression System Ideal for Cell Reprogramming. *J Biol Chem* (2011) 286:4760–71. doi: 10.1074/jbc.M110.183780
26. Nishimura K, Ohtaka M, Takada H, Kurisaki A, Tran NVK, Tran YTH, et al. Simple and Effective Generation of Transgene-Free Induced Pluripotent Stem Cells Using an Auto-Erasable Sendai Virus Vector Responding to microRNA-302. *Stem Cell Res* (2017) 23:13–9. doi: 10.1016/j.scr.2017.06.011
27. Docherty FM, Riemondy KA, Castro-Gutierrez R, Dwulet JM, Shilleh AH, Hansen MS, et al. ENTPD3 Marks Mature Stem Cell Derived Beta Cells Formed by Self-Aggregation *in Vitro*. *Diabetes* (2021) db200873. doi: 10.2337/db20-0873
28. Russ HA, Parent AV, Ringler JJ, Hennings TG, Nair GG, Shveygert M, et al. Controlled Induction of Human Pancreatic Progenitors Produces Functional Beta-Like Cells *In Vitro*. *EMBO J* (2015) 34:1759–72. doi: 10.15252/embj.201591058
29. Faleo G, Russ HA, Wisel S, Parent AV, Nguyen V, Nair GG, et al. Mitigating Ischemic Injury of Stem Cell-Derived Insulin-Producing Cells After Transplant. *Stem Cell Rep* (2017) 9:807–19. doi: 10.1016/j.stemcr.2017.07.012
30. Cortes A, Brown MA. Promise and Pitfalls of the Immunochip. *Arthritis Res Ther* (2011) 13:101. doi: 10.1186/ar3204
31. Romanel A, Zhang T, Elemento O, Demichelis F. EthSEQ: Ethnicity Annotation From Whole Exome Sequencing Data. *Bioinformatics* (2017) 33:2402–4. doi: 10.1093/bioinformatics/btx165
32. Manichaikul A, Mychaleckyj JC, Rich SS, Daly K, Sale M, Chen W-M. Robust Relationship Inference in Genome-Wide Association Studies. *Bioinformatics* (2010) 26:2867–73. doi: 10.1093/bioinformatics/btq559
33. Das S, Forer L, Schönherr S, Sidore C, Locke AE, Kwong A, et al. Next-Generation Genotype Imputation Service and Methods. *Nat Genet* (2016) 48:1284–7. doi: 10.1038/ng.3656
34. Jie Huang MMP, Yuan Lin MP. Genotyping Technologies and Applications in the Era of Precision Medicine. *North Am J Med Sci* (2017) 10:176–80.
35. Armitage LH, Khosravi-Maharlooei M, Meacham A, Butfiloski EJ, Viola R, Egli D, et al. High-Yield Monocyte, Macrophage, and Dendritic Cell Differentiation From Induced Pluripotent Stem Cells Running Title: High-Yield Monocyte Differentiation From hiPSC. *BioRxiv* (2021). 2021.04.29.441947.
36. Vopenkova K, Mollova K, Buresova I, Michalek J. Complex Evaluation of Human Monocyte-Derived Dendritic Cells for Cancer Immunotherapy. *J Cell Mol Med* (2012) 16:2827–37. doi: 10.1111/j.1582-4934.2012.01614.x
37. Fredette NC, Malik E, Mukhtar ML, Prossnitz ER, Terada N. A Hypertension Patient-Derived iPSC Model Demonstrates a Role for G Protein-Coupled Estrogen Receptor in Hypertension Risk and Development. *Am J Physiol Cell Physiol* (2020) 319:C825–38. doi: 10.1152/ajpcell.00350.2019
38. Tan JY, Sriram G, Rufaihah AJ, Neoh KG, Cao T. Efficient Derivation of Lateral Plate and Paraxial Mesoderm Subtypes From Human Embryonic Stem Cells Through GSKi-Mediated Differentiation. *Stem Cells Dev* (2013) 22:1893–906. doi: 10.1089/scd.2012.0590
39. Brusko TM, Koya RC, Zhu S, Lee MR, Putnam AL, McClymont SA, et al. Human Antigen-Specific Regulatory T Cells Generated by T Cell Receptor Gene Transfer. *PloS One* (2010) 5:e11726. doi: 10.1371/journal.pone.0011726
40. Driver JP, Racine JJ, Ye C, Lamont DJ, Newby BN, Leeth CM, et al. Interferon-γ Limits Diabetogenic CD8+ T-Cell Effector Responses in Type 1 Diabetes. *Diabetes* (2017) 66:710–21. doi: 10.2337/db16-0846
41. Sharp CD, Huang M, Glawe J, Patrick DR, Pardue S, Barlow SC, et al. Stromal Cell-Derived Factor-1/CXCL12 Stimulates Chemorepulsion of NOD/LtJ T-Cell Adhesion to Islet Microvascular Endothelium. *Diabetes* (2008) 57:102–12. doi: 10.2337/db07-0494
42. Schneider CA, Rasband WS, Eliceiri KW. Image to ImageJ: 25 Years of Image Analysis. *Nat Methods* (2012) 9:671–5. doi: 10.1038/nmeth.2089
43. Chen J, Grieshaber S, Mathews CE. Methods to Assess Beta Cell Death Mediated by Cytotoxic T Lymphocytes. *J Vis Exp* (2011) (52):e2724. doi: 10.3791/2724

44. Sui L, Leibel RL, Egli D. Pancreatic Beta Cell Differentiation From Human Pluripotent Stem Cells. *Curr Protoc Hum Genet* (2018) 99:e68. doi: 10.1002/cphg.68
45. Shapiro MR, Yeh W-I, Longfield JR, Gallagher J, Infante CM, Wellford S, et al. CD226 Deletion Reduces Type 1 Diabetes in the NOD Mouse by Impairing Thymocyte Development and Peripheral T Cell Activation. *Front Immunol* (2020) 11:2180. doi: 10.3389/fimmu.2020.02180
46. Sharpe AH. Mechanisms of Costimulation. *Immunol Rev* (2009) 229:5–11. doi: 10.1111/j.1600-065X.2009.00784.x
47. Muller WA. Getting Leukocytes to the Site of Inflammation. *Vet Pathol* (2013) 50:7–22. doi: 10.1177/0300985812469883
48. Martin S, van den Engel NK, Vinke A, Heidenthal E, Schulte B, Kolb H. Dominant Role of Intercellular Adhesion Molecule-1 in the Pathogenesis of Autoimmune Diabetes in Non-Obese Diabetic Mice. *J Autoimmun* (2001) 17:109–17. doi: 10.1006/jaut.2001.0526
49. Phillips JM, Haskins K, Cooke A. MAdCAM-1 Is Needed for Diabetes Development Mediated by the T Cell Clone, BDC-2.5. *Immunology* (2005) 116:525–31. doi: 10.1111/j.1365-2567.2005.02254.x
50. Savinov AY, Strongin AY. Targeting the T-Cell Membrane Type-1 Matrix Metalloproteinase-CD44 Axis in a Transferred Type 1 Diabetes Model in NOD Mice. *Exp Ther Med* (2013) 5:438–42. doi: 10.3892/etm.2012.821
51. Blanchet JS, Valmori D, Dufau I, Ayyoub M, Nguyen C, Guillaume P, et al. A New Generation of Melan-A/MART-1 Peptides That Fulfill Both Increased Immunogenicity and High Resistance to Biodegradation: Implication for Molecular Anti-Melanoma Immunotherapy. *J Immunol* (2001) 167:5852–61. doi: 10.4049/jimmunol.167.10.5852
52. Wasserfall C, Nead K, Mathews C, Atkinson MA. The Threshold Hypothesis: Solving the Equation of Nurture *vs* Nature in Type 1 Diabetes. *Diabetologia* (2011) 54:2232–6. doi: 10.1007/s00125-011-2244-z
53. Inshaw JRJ, Cutler AJ, Crouch DJM, Wicker LS, Todd JA. Genetic Variants Predisposing Most Strongly to Type 1 Diabetes Diagnosed Under Age 7 Years Lie Near Candidate Genes That Function in the Immune System and in Pancreatic β-Cells. *Diabetes Care* (2020) 43:169–77. doi: 10.2337/dc19-0803
54. Kaneko S. *In Vitro* Generation of Antigen-Specific T Cells From Induced Pluripotent Stem Cells of Antigen-Specific T Cell Origin. *Methods Mol Biol* (2016) 1393:67–73. doi: 10.1007/978-1-4939-3338-9_6
55. Nishimura T, Kaneko S, Kawana-Tachikawa A, Tajima Y, Goto H, Zhu D, et al. Generation of Rejuvenated Antigen-Specific T Cells by Reprogramming to Pluripotency and Redifferentiation. *Cell Stem Cell* (2013) 12:114–26. doi: 10.1016/j.stem.2012.11.002
56. Naujok O, Diekmann U, Elsner M. Gene Transfer Into Pluripotent Stem Cells *via* Lentiviral Transduction. *Methods Mol Biol* (2016) 1341:67–85. doi: 10.1007/7651_2015_221
57. Zare M, Soleimani M, Mohammadian M, Akbarzadeh A, Havasi P, Zarghami N. Efficient Biotechnological Approach for Lentiviral Transduction of Induced Pluripotent Stem Cells. *Artif Cells Nanomed Biotechnol* (2016) 44:743–8. doi: 10.3109/21691401.2014.982804
58. Koyanagi-Aoi M, Ohnuki M, Takahashi K, Okita K, Noma H, Sawamura Y, et al. Differentiation-Defective Phenotypes Revealed by Large-Scale Analyses of Human Pluripotent Stem Cells. *Proc Natl Acad Sci USA* (2013) 110:20569–74. doi: 10.1073/pnas.1319061110
59. Beck S, Geraghty D, Inoko H, Rowen L, Aguado B, Bahram S, et al. Complete Sequence and Gene Map of a Human Major Histocompatibility Complex. The MHC Sequencing Consortium. *Nature* (1999) 401:921–3. doi: 10.1038/44853
60. Hou Z, Zhang Y, Propson NE, Howden SE, Chu L-F, Sontheimer EJ, et al. Efficient Genome Engineering in Human Pluripotent Stem Cells Using Cas9 From Neisseria Meningitidis. *Proc Natl Acad Sci USA* (2013) 110:15644–9. doi: 10.1073/pnas.1313587110
61. Ben Jehuda R, Shemer Y, Binah O. Genome Editing in Induced Pluripotent Stem Cells Using CRISPR/Cas9. *Stem Cell Rev Rep* (2018) 14:323–36. doi: 10.1007/s12015-018-9811-3
62. Li X-L, Li G-H, Fu J, Fu Y-W, Zhang L, Chen W, et al. Highly Efficient Genome Editing *via* CRISPR-Cas9 in Human Pluripotent Stem Cells Is Achieved by Transient BCL-XL Overexpression. *Nucleic Acids Res* (2018) 46:10195–215. doi: 10.1093/nar/gky804
63. Maguire JA, Cardenas-Diaz FL, Gadue P, French DL. Highly Efficient CRISPR-Cas9-Mediated Genome Editing in Human Pluripotent Stem Cells. *Curr Protoc Stem Cell Biol* (2019) 48:e64. doi: 10.1002/cpsc.64
64. Paquet D, Kwart D, Chen A, Sproul A, Jacob S, Teo S, et al. Efficient Introduction of Specific Homozygous and Heterozygous Mutations Using CRISPR/Cas9. *Nature* (2016) 533:125–9. doi: 10.1038/nature17664
65. Xu X, Gao D, Wang P, Chen J, Ruan J, Xu J, et al. Efficient Homology-Directed Gene Editing by CRISPR/Cas9 in Human Stem and Primary Cells Using Tube Electroporation. *Sci Rep* (2018) 8:11649. doi: 10.1038/s41598-018-30227-w
66. Yumlu S, Bashir S, Stumm J, Kühn R. Efficient Gene Editing of Human Induced Pluripotent Stem Cells Using CRISPR/Cas9. *Methods Mol Biol* (2019) 1961:137–51. doi: 10.1007/978-1-4939-9170-9_10
67. Dendrou CA, Cortes A, Shipman L, Evans HG, Attfield KE, Jostins L, et al. Resolving TYK2 Locus Genotype-to-Phenotype Differences in Autoimmunity. *Sci Transl Med* (2016) 8:363ra149. doi: 10.1126/scitranslmed.aag1974
68. Mero I-L, Lorentzen AR, Ban M, Smestad C, Celius EG, Aarseth JH, et al. A Rare Variant of the TYK2 Gene Is Confirmed to be Associated With Multiple Sclerosis. *Eur J Hum Genet* (2010) 18:502–4. doi: 10.1038/ejhg.2009.195
69. Bakay M, Pandey R, Grant SFA, Hakonarson H. The Genetic Contribution to Type 1 Diabetes. *Curr Diab Rep* (2019) 19:116. doi: 10.1007/s11892-019-1235-1
70. Gingerich MA, Sidarala V, Soleimanpour SA. Clarifying the Function of Genes at the Chromosome 16p13 Locus in Type 1 Diabetes: CLEC16A and DEXI. *Genes Immun* (2020) 21:79–82. doi: 10.1038/s41435-019-0087-7
71. Zuvich RL, Bush WS, McCauley JL, Beecham AH, De Jager PLInternational Multiple Sclerosis Genetics Consortium, et al. Interrogating the Complex Role of Chromosome 16p13.13 in Multiple Sclerosis Susceptibility: Independent Genetic Signals in the CIITA-CLEC16A-SOCS1 Gene Complex. *Hum Mol Genet* (2011) 20:3517–24. doi: 10.1093/hmg/ddr250
72. Chhatta AR, Cordes M, Hanegraaf MAJ, Vloemans S, Cupedo T, Cornelissen JJ, et al. *De Novo* Generation of a Functional Human Thymus From Induced Pluripotent Stem Cells. *J Allergy Clin Immunol* (2019) 144:1416–1419.e7. doi: 10.1016/j.jaci.2019.05.042
73. Parent AV, Russ HA, Khan IS, LaFlam TN, Metzger TC, Anderson MS, et al. Generation of Functional Thymic Epithelium From Human Embryonic Stem Cells That Supports Host T Cell Development. *Cell Stem Cell* (2013) 13:219–29. doi: 10.1016/j.stem.2013.04.004

6

Selection for CD26⁻ and CD49A⁺ Cells from Pluripotent Stem Cells-Derived Islet-Like Clusters Improves Therapeutic Activity in Diabetic Mice

Kfir Molakandov[1], Denise A. Berti[1], Avital Beck[1], Ofer Elhanani[2], Michael D. Walker[2], Yoav Soen[2], Karina Yavriyants[1], Michal Zimerman[1], Ella Volman[1], Itzik Toledo[1], Anna Erukhimovich[1], Alon M. Levy[1], Arik Hasson[1], Joseph Itskovitz-Eldor[1], Judith Chebath[1] and Michel Revel[1,3]*

[1] Kadimastem Ltd., Weizmann Science Park, Ness Ziona, Israel, [2] Department of Biomolecular Sciences, Weizmann Institute of Science, Rehovot, Israel, [3] Department of Molecular Genetics (emeritus), Weizmann Institute of Science, Rehovot, Israel

Correspondence:
Kfir Molakandov
k.molakandov@kadimastem.com

Background: Cell therapy of diabetes aims at restoring the physiological control of blood glucose by transplantation of functional pancreatic islet cells. A potentially unlimited source of cells for such transplantations would be islet cells derived from an *in vitro* differentiation of human pluripotent stem cells (hESC/hiPSC). The islet-like clusters (ILC) produced by the known differentiation protocols contain various cell populations. Among these, the β-cells that express both insulin and the transcription factor Nkx6.1 seem to be the most efficient to restore normoglycemia in diabetes animal models. Our aim was to find markers allowing selection of these efficient cells.

Methods: Functional Cell-Capture Screening (FCCS) was used to identify markers that preferentially capture the cells expressing both insulin and Nkx6.1, from hESC-derived ILC cells. In order to test whether selection for such markers could improve cell therapy in diabetic mouse models, we used ILC produced from a clinical-grade line of hESC by a refined differentiation protocol adapted to up-scalable bioreactors. Re-aggregated MACS sorted cells were encapsulated in microspheres made of alginate modified to reduce foreign body reaction. Implantation was done intraperitoneally in STZ-treated C57BL/6 immuno-competent mice.

Results: CD49A (integrin alpha1) was identified by FCCS as a marker for cells that express insulin (or C-peptide) as well as Nkx6.1 in ILC derived by hESC differentiation. The ILC fraction enriched in CD49A⁺ cells rapidly reduced glycemia when implanted in diabetic mice, whereas mice receiving the CD49A depleted population remained highly diabetic. CD49A-enriched ILC cells also produced higher levels of human C-peptide in the blood of transplanted mice. However, the difference between CD49A-enriched and total ILC cells remained small. Another marker, CD26 (DPP4), was identified by FCCS as binding insulin-expressing cells which are Nkx6.1 negative. Depletion of CD26⁺ cells followed by enrichment for CD49A⁺ cells increased insulin⁺/Nkx6.1⁺ cells fraction to ~70%. The

$CD26^-/CD49A^+$ enriched ILC exhibited improved function over non-sorted ILC or $CD49A^+$ cells in diabetic mice and maintain prolonged blood C-peptide levels.

Conclusions: Refining the composition of ILC differentiated from hPSC by negative selection to remove cells expressing CD26 and positive selection for CD49A expressing cells could enable more effective cell therapy of diabetes.

Keywords: human ESC-derived insulin producing cells, islet-like clusters (ILC), functional cell capture screening, integrin alpha1 (CD49A), DPP4 (CD26), alginate encapsulation, STZ-treated C57BL/6 mice diabetes models

INTRODUCTION

Diabetic conditions due to destruction (in type I) or dysfunction (in type II) of pancreatic islets of Langerhans, have detrimental impacts on the quality of life and lifespan. For the hundred million patients depending on frequent insulin injections for controlling blood glucose levels, islet cells transplantation would have the advantage to restore a physiological regulation of glycemia. The Edmonton protocol (1–3) based on transplantation through the portal vein of human islets obtained from brain dead donors has allowed patients to become insulin-free for significant periods of time. Yet, the availability of such islet donations is too limited to meet the transplantation demand. An alternative, more abundant supply may be generated by differentiation of islet-like clusters (ILC) from large-scale cultures of human pluripotent cells (hPSC). Recently developed multistage protocols of differentiation produce pancreatic islet cell populations including mature β-cells that function in reducing glycemia in animal models of diabetes (4–6). New methods for microencapsulation in modified alginate support long term function after implantation in diabetic animals (7–9).

A drawback is that *in vitro* differentiated ILC contain additional populations of cells, which may not be necessary for islet function, and even may impair the efficacy or the safety of the transplanted cells for future clinical treatment of diabetes. To alleviate this problem, we investigated which cell surface marker could help identify and isolate the ILC cells that have higher capacity to normalize glycemia in diabetes models. We screened a large array of antibodies to cell-surface proteins by the Functional Cell-Capture Screening (FCCS) previously developed to identify cell-surface markers selective for endoderm and non-endoderm populations of differentiating hESC (10). The same procedure was later used to characterize surface markers of cells in post-mortem islets from adult humans (11). The present study was aimed to generate more potent ILC by using antibodies that were identified by the FCCS as selecting for insulin-producing cells in ILC differentiated from human ESC. Starting from a clinical-grade hESC line, we refined a 30-day protocol, in spinner suspension cultures or controlled bioreactors, which converts 3D clusters of these highly pluripotent stem cells into pancreatic islet-like clusters. The clusters were dissociated into single cells and fractionated according to their capacity to bind, or not bind, to two antibodies found by FCCS to capture insulin-producing cells. One antibody against integrin-alpha1 (CD49A) was found to bind the insulin-producing cells that express Nkx6.1, a transcription factor essential for the formation and function of mature β-cell s (12, 13), but to bind also insulin-producing cells lacking Nkx6.1. Another antibody, against Dipeptidyl peptidase-4 (CD26), was found to bind only insulin-producing cells that do not express Nkx6.1. A novel sorting strategy is suggested by first removing cells expressing CD26 and then enriching for CD49A positive cells. Thus, it is possible to increase the *in vivo* therapeutic activity of ILC to normalize glycemia in diabetic mice by two consecutive selections and supports the feasibility of functional enrichment strategies to improve the activity of hPSC-derived ILC for the treatment of diabetes.

Abbreviations: AFP, alpha fetoprotein; BSA, bovine serum albumin; CMRL, CMRL serum free medium; DPP4, dipeptidyl peptidase-4; ES, embryonic stem; FACS, fluorescent activated cell sorting; FBS, fetal bovine serum; FCCS, functional cell capture screening; GCG, glucagon; GLP1R, glucagon-like peptide-1 receptor; GSIS, glucose stimulated insulin secretion; HPRT, hypoxanthine phosphoribosyl transferase; hPSC, human pluripotent stem cell; hESC, human embryonic stem cell; IAPP, islet amyloid polypeptide; ILC, islet-like cluster; ip, intraperitoneal; IPGTT, intraperitoneal glucose tolerance test; ITGA1, integrin alpha1; MACS, magnetic activated cell sorting; MafA, Pancreatic B-Cell-Specific Transcriptional Activator; Nkx6.1, Nk6 homeobox-1;NMR, nuclear magnetic resonance; NPTX2, neuronal pentraxin-2; $PBS^{-/-}$, phosphate buffered saline, no calcium, no magnesium; PCSK1, prohormone convertase; PFA, paraformaldehyde; PPY, pancreatic polypeptide Y; qPCR, quantitative Polymerase Chain Reaction; RT, room temperature; SST, somatostatin; STZ, streptozotocin; TMTD, triazole-thiomorpholine-dioxide; TPH1, tryptophan hydroxylase-1; UCN3, urocortin-3.

METHODS

Human ES Cell Expansion and Differentiation to Islet-Like Clusters

Highly pluripotent, clinical grade, human ES cells HADC-100 (14) (provided by Professor Benjamin Reubinoff, Hadassah Medical School, Jerusalem, Israel) were grown to confluent monolayers in essential E8 medium (Gibco, Cat#A1517001), with addition of penicillin and streptomycin (PS, Gibco, Cat#15140-122) on vitronectin-coated flasks (Gibco, Cat# A14700). Differentiation was performed on cell aggregates formed in spinner flasks during 2 days in dynamic suspension cultures. In brief, 48 h before starting the differentiation protocol (day-2), non-differentiated cells were dissociated with Versene (Gibco, Cat#15040033). Single cells washed with $PBS^{-/-}$ (Gibco, Cat#14190-094), were seeded in 500 ml disposable spinner flasks (Corning, Cat#CZ-3153), filled with 250 ml E8 medium containing 10 μM Rock Inhibitor Y27632 (Cayman Chemical, #10005583-10), at concentration of 0.8–1 × 10^6 cells/ml. The spinner flasks were placed on magnetic stirrer

(DURA-MAG, 9 position stirrer, Chemglass) at speed of 70 rpm in a humidified incubator set at 5% CO_2 and 37°C. This resulted in the formation of ES cell clusters in suspension, as well as in cell proliferation. On day -1, 80% of the E8 medium was replaced. On Day 0, the E8 medium was washed away (15) by letting the aggregates settle for 5 min and removing the supernatant with a pipette. Cells were washed with 250 ml $PBS^{-/-}$; after 3 min stirring in the incubator, $PBS^{-/-}$ was replaced by 250 ml of stage 1 differentiation medium. The media for the seven-stage differentiation protocol, refined on the basis of several published protocols (5, 16–19), are detailed in additional files 1 and 2 (**Tables S1** and **S2**).

For Bioreactors, the hESC aggregation and differentiation was similarly done in the DASbox® mini system (Eppendorf) with online monitoring of culture parameters. Up to four parallel bottles containing 150 ml medium were seeded with 0.8–1 × 10^6 cells/ml 48 h before and washed with PBS just before differentiation as above. Medium changes were done batch-wise in the semi-closed system using peristaltic pumps.

Real-Time Quantitative PCR

RNA was isolated from cells using the RNAeasy micro kit (Qiagen #74004) and purified from genomic DNA with RNase-free DNase kit (Qiagen #79254). The cDNA synthesis was done with the high-capacity cDNA Reverse Transcription kit (Applied Biosystems 4368814). Transcript levels were measured by real-time qPCR using Taqman Fast advanced master mix (Applied Biosystems #4444557). The level of each gene was normalized to endogenous HPRT gene, using the $2^{-\Delta\Delta}CT$ method. The probes used for qPCR are listed in **Table S3**. The MARIS procedure (20) is described in Additional file 4.

Flow Cytometry

Samples from settled aggregates during or at the end of the differentiation process were dissociated with Accumax (Sigma, Cat# A7089) at 0.35 ml per 300,000 cells for 8–10 min, after which the enzyme was blocked in 1 ml $PBS^{-/-}$ with 10% FBS, and the cells were centrifuged (350 × g, 3 min). For external cell membrane labeling, cells were washed in $PBS^{-/-}$, and antibodies (e.g. anti-CD49A, as listed in **Table S4**) were directly added to the cell suspension in FACS buffer (0.5% BSA in PBS) followed by incubation at 4°C for 30 min. For internal antigen labeling (e.g. anti-human C-peptide and Nkx6.1, **Table S4**), cells after centrifugation were washed once in 1 ml PBS and fixed in 0.4 ml of 4% paraformaldehyde (PFA, EMS Cat# 15710), for 20 min at 4°C. After two washes with PBS cells were incubated for 1 h at 4°C in blocking solution [PBS with 5% Bovine serum albumin (BSA) and 3% horse serum] containing 0.3% Triton X-100 (Sigma, T6878), and washed once with PBS. The cell pellet suspended in 0.1 ml blocking solution (with 0.1% Triton x-100) containing antibodies, was incubated overnight at 4°C, or for 1 h at room temperature (RT), and washed with PBS. When fluorescent tag-conjugated primary antibodies were used, fluorescence was read after this step in Flow Cytometer-BD FACS Canto II. Otherwise, cells were incubated in blocking solution containing 1:100 dilutions of the fluorescent tag-conjugated secondary antibody and washed with PBS before reading fluorescence.

Functional Cell Capture Screening (FCCS) on Antibody Arrays

Antibody arrays were printed in a Microgrid printer with solid pins (Total array Systems, BioRobotics, Cambridge, UK) on hydrogel-coated slides (Full Moon Biosystems, Sunnyvale, CA, USA) using a panel of 235 monoclonal mouse anti-human antibodies (BD biosciences), each antibody being spotted at five different places in the array, as described before (10, 11) (US patent 2018/0369290 A1, Item 0081). The cell clusters were dissociated using TrypLE Express (Invitrogen Cat#12604) for 4 min, followed by quenching with 10% FBS in PBS, centrifugation, and resuspension in CMRL. The printed area of the array was blocked for 3 min with 1% BSA in PBS solution, before cell seeding, at about 0.5 × 10^6 cells/ml in 0.25–0.5 ml of CMRL medium, supplemented with 2 μl of DNase I (Ambion 2 U/μl) and incubation was for 1 h at 37°C. Excess cells were removed in a large volume of PBS and the arrays were fixed in 4% PFA for 10 min. Cells on the array were permeabilized in PBS, 0.2% Triton X-100 for 20 min, washed twice with PBS, and blocked for 45 min in blocking buffer (2% FBS, 2% BSA, 50 mM glycine in PBS). After blocking, arrays were washed twice with PBS and incubated for 2 h at RT in blocking buffer with 0.1% of Triton X-100 containing the primary antibody guinea-pig anti-insulin (DAKO, A0564). Primary antibodies were removed, and arrays were washed three times with working buffer. Then, secondary antibodies were added in working buffer for 45 min at room temp: cy5 donkey anti-guinea–pig (Jackson ImmunoResearch 706-175 -148). Arrays were washed three times in working buffer and imaged using automated, high content fluorescence microscopy (IXmicro, MDC). Total cells in each spot were counted by phase microscopy and the percent of insulin positive cells was calculated. Three repeats were performed with different batches of ES-derived cells at stage 7 of the differentiation protocol. The significance of the amount of cell binding to surface antibody was evaluated by two sample paired T-test (*P* value less than 0.05). In another set of experiments, cells were reacted with antibodies against insulin as above but also with antibodies against Pdx1 and against Nkx6.1 (see **Table S4**). The number of cells stained for PDX1 and Insulin, and for insulin and Nkx6.1 was counted.

Magnetic Activated Cell Sorting (MACS)

Cell clusters after stage 7 of differentiation were washed in $PBS^{-/-}$, dissociated with Accumax (10 ml for 25 × 10^6 cells) for 10 min at 37°C and washed with CMRL 2% BSA. Dissociated cell suspension was passed through a 30 μm MACS filter previously washed with $PBS^{-/-}$ and counted using Nucleocounter® NC-200. Part of the clusters was set aside for control (non-dissociated, non-selected cells), seeded at 10^6 cells per ml in ultra-low binding six-well-plates (Corning Cat#CLS3471) and left in the incubator on orbital shaker (NovaShake-B32X) set at 95 rpm for 3 days in medium $CMRL^+$ (**Table S2**), before implantation. Cells were suspended in MACS buffer ($PBS^{-/-}$, 2% BSA, 2 mM EDTA, sterile, degassed), 100 μl per 10^7 cells, for reaction with anti-CD49A-PE (Myltenyi, cat#

130-101-397) using 10 µl per 10^7 cells for 10 min at 4°C, followed by washing with 5 ml MACS buffer. Cells suspended in MACS buffer (80 µl per 10^7 cells) were reacted with 20 µl per 10^7 cells of anti-PE magnetic microbeads (Miltenyi; cat# 5181214192) for 15 min at 4°C. After washing with 5 ml cold MACS buffer and centrifugation, cells were suspended in MACS buffer and applied to pre-separation filters and LS MACS column(s) as recommended by the manufacturer. Prior to implantation, all single cells fractions were re-aggregated in suspension in non-TC treated six-well plates (Corning Cat#CLS3471-24EA) in CMRL$^+$, 10 µM RI Y27632 and 2 µg/ml Laminin (Bio Lamina Cat#MX521CTG), on orbital shaker.

For removal of CD26 positive cells prior to CD49A enrichment, the cells were dissociated as above, incubated for 10 min at 4°C with anti-CD26-PE (cat#302706), (10 µl/10^7 cells, in 100 µl MACS buffer for 10^7 cells), and after washing, reacted with anti-PE microbeads as described above. The mixture, after washing and resuspension in MACS buffer, was applied to LS column(s) and the flow through fraction (CD26 depleted) kept for further fractionation by MACS with anti-CD49A antibody.

Diabetes Induction in Mice and ILC Implantation

Six-to-8-week-old immune-competent mice C57BL/6JOlaHsd (Harlan, Israel) were rendered diabetic by intraperitoneal injection of streptozotocin (STZ) (Sigma, Cat#S0130), using 4 × 50 mg STZ/kg after 6 h daily fasting. Implantation was performed in diabetic mice, defined by blood glucose higher than 250 mg/dl for three consecutive tests. Blood glucose was measured by a glucometer, on tail vein blood. Intraperitoneal Glucose Tolerance Test (IPGTT) was done after fasting the mice overnight by i.p. injection of glucose (2 g/kg). Blood glucose was then monitored during a 2 h period.

For implantation of micro-encapsulated ILC cells (see below), mice were anesthetized by an IP injection of ketamine/xylazine (Sigma, K4138) at 87.5 mg/kg ketamine/12.5 mg/kg xylazine and then mounted on a surgical pad. The skin was prepared by shaving with electric clippers, application of Polydin, and then 70% ethanol solution. An abdominal incision (1 cm), and peritoneal incision (0.5 cm) allowed to insert microencapsulated ILC into the peritoneal cavity of the mouse using a 1 ml sterile plastic tip (about 0.5 ml total volume). The peritoneum and the skin were closed with sutures and cleaned with Polydin. The mice were kept warm by a heating pad till they woke up. The cell doses implanted were between 1.0 and 2 × 10^6 cells, as indicated.

Human C-Peptide ELISA Assays

The levels of human C-peptide, which reflect the levels of insulin secreted by the human ILC, were measured in blood samples collected after anesthesia from mouse retroorbital sinus. To test the insulin/C-peptide response to glucose, mice were withdrawn from food for 12 h, and injected intra-peritoneally with a glucose solution (25%, 41-302-500, Biological Industries). Blood was collected before and 30 min after the glucose injection, centrifuged, and stored at −20°C. ELISA assays were performed using ultra-sensitive ELISA kit for human C-peptide from Mercodia (#10-1141-01) according to instructions.

Preparation of TMTD-Modified Alginate and Microcapsules

Triazole thiomorpholine dioxide (TMTD Y1-Z15) preparation and coupling to alginate PRONOVA UP-MVG alginate (NovaMatrix) were done for Kadimastem, by Recipharm-Israel as described [8, 9]. After verification of the product structure by NMR, purification by filtration, dialysis, and desiccation, elemental analysis revealed that more than 50% alginate guluronic or mannuronic residues were coupled to TMTD. Solutions of 4.6% of TMTD-coupled alginate were used [80% in volume of 5% (w/v) TMTD-coupled UP MVG and 20% of 3% (w/v) UP MVG]. Stage 7 ILC (see **Table S1**), washed with KREBS buffer without Ca^{++} pH 7.4, were mixed with alginate solutions at the concentration of about 10^7 cells/ml in a 5 ml Eppendorf tube. The micro-encapsulator Buchi B395, located in a tissue culture laminar flow hood for sterility, was set up to obtain microcapsules of 1.5 mm diameter with 4.6% alginate polymerized in $CaCl_2$ (100 mM in HEPES pH 7.4).

ILC Immunostaining and Imaging

Stage 7 cell clusters were fixed in 4% PFA and washed in PBS$^{-/-}$. A minimum volume of warm 1% agarose was added to the pellet of clusters. After agarose became solid, the block was embedded in paraffin, and 10 µm thick sections were produced and bound to glass slides. After removal of paraffin by alternative baths of xylene and ethanol, antigen retrieval was performed by heating slides in 10 mM citrate buffer pH 6.0 (ZYTOMED systems), for 15 min in pressure cooker (Bio TintoRetriever). Blocking and permeabilization was done by incubation with PBS containing 5% BSA, 3% horse serum (blocking solution) supplemented with 0.3% Triton X-100 for 1 h at RT. Antibodies against PDX1, C-peptide, Nkx6.1 (**Table S4**) were diluted blocking solution supplemented with 0.1% Triton X-100 and incubated overnight at 4°C in humidified chambers. After two washes in PBS, incubation with secondary antibodies was done for 1 h at room temperature and washes were done similarly. Nuclei were stained with DAPI (1 µg/ml). The slides were mounted with aqueous mounting medium and covered with coverslip. Images were obtained using Nikon Eclipse 80i fluorescence microscope.

RESULTS

Islet-Like Clusters Differentiated From hESC Contain Mature β-Cells

Highly pluripotent hESCs were differentiated according to a seven-stage stepwise protocol carried out in suspension culture conditions (3D), in spinner flasks as well as in controlled bioreactors (as detailed in *Methods*). After stage 7 (days 30–34), 100–200 µm Islet-like clusters (**Figure 1A**) contain hormone-positive cells, especially cells producing insulin (as well as the C-peptide fragment processed from human proinsulin) (**Figure 1B**). These insulin-producing cells typically amount to 60% of the total cells, with about 10% of cells producing glucagon and 2% producing somatostatin (**Figure 1B**). In addition to hormone-producing cells, the hESC-derived ILC still contain precursor cells, since over 90%

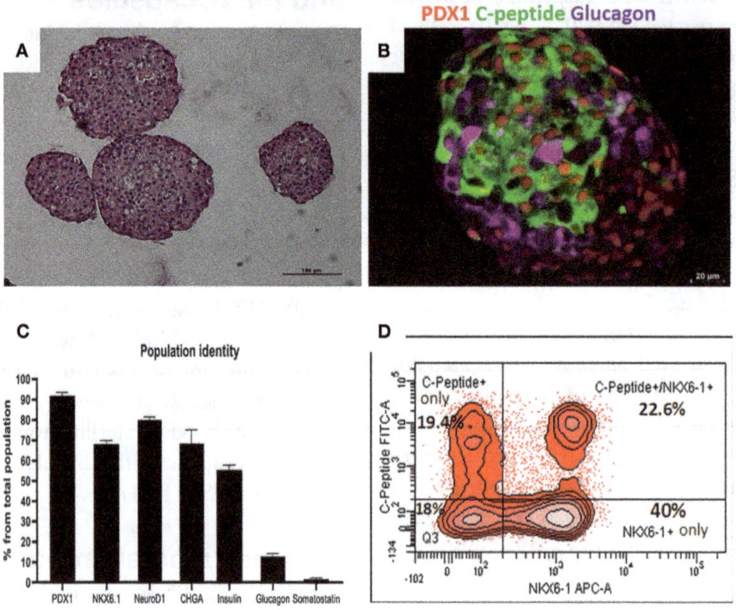

FIGURE 1 | Cell populations identity in Islet-like clusters at the end of the differentiation process. **(A)** H&E of microencapsulated ILC, demonstrating clusters ranging from 100 to 200 µm. **(B)** Immunostaining of an ILC for human C-peptide (green), Glucagon (purple), PDX1 (red, nuclear stain). **(C)** Percentage of ILC cells stained by antibodies to indicated proteins, calculated from flow cytometry analysis of a representative ILC preparations (n = 10, different batches, antibodies listed in **Table S4**). **(D)** Flow-cytometry analysis of dissociated total ILC cells, fixed and stained for human C-peptide and Nkx6.1. The percentage of cells with C-peptide only (no Nkx6.1), with C-peptide and Nkx6.1, with Nkx6.1 only (no C-peptide) is shown. Q3 is the double negative fraction; C-peptide$^-$/Nkx6.1$^-$.

of cells express the key transcription factor for pancreatic development PDX1 and 70–80% express transcription factors important for β-cell function such as Nkx6.1 and NeuroD1 (**Figure 1C**). Nkx6.1 is of particular importance, being essential for development and function of mature β-cells (12, 13), and serving as a marker of mono-hormonal insulin-producing β-cells (21). In the hESC-derived ILC, these β-cells can be identified by flow cytometry (FACS) as double positive for Nkx6.1 and human C-peptide (**Figure 1D**). This C-peptide$^+$/Nkx6.1$^+$ double positive fraction usually represents around 20–30% of the total population, the rest being C-peptide$^+$/Nkx6.1$^-$ (C-pep$^+$ only), C-peptide$^-$/Nkx6.1$^+$ cells (Nkx6.1$^+$ only), and C-peptide$^-$/Nkx6.1$^-$ (Negative, Q3) cells (**Figure 1D**). The four subpopulations were characterized by gene expression (**Figure 2**). After separation by preparative FACS, RNA from each of the fixed and stained cell fractions was extracted and analyzed using the MARIS method (20). The qPCR data relative to the unsorted cells confirmed that the C-peptide$^+$/Nkx6.1$^+$ double positive compartment is enriched for cells that primarily express insulin. On the other hand, the C-peptide-only fraction is enriched for cells expressing Insulin but also Glucagon (GCG), Somatostatin (SST), and Pancreatic Polypeptide (PPY), identifying these cells with the reported polyhormonal precursors (22–24). Among the four sorted cell fractions, the C-peptide$^+$/Nkx6.1$^+$ double positive cells had, in addition to insulin, the highest expression level of the transcription factors MafA and Nkx6.1, of the prohormone convertase PCSK1 and the GLP1 receptor (GLP1R) (**Figure 2**).

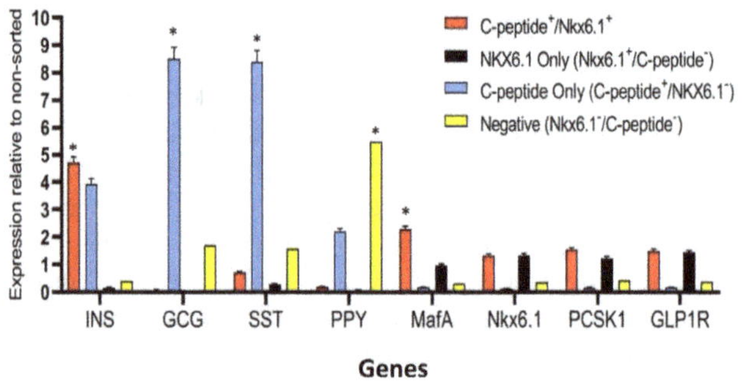

FIGURE 2 | Gene expression analysis of Islet-like clusters stained for C-peptide and NKX6.1. A method for analyzing RNA after intracellular staining (MARIS) was applied to ILC cells. Shortly, dissociated cells were fixed and stained for intracellular C-peptide and NKX6.1 antigens. Following labeling with secondary antibodies, the designated subpopulations were sorted by preparative FACS and RNA from each subpopulation was extracted. The relative expression levels of indicated genes was analyzed by qPCR relative to the non-sorted total ILC cells taken as 1. Results (expressed as mean ± SEM) demonstrated signature genes for specific sorted populations (*significance $p < 0.05$, Student t test): Insulin and MAFA is highly expressed in C-peptide$^+$/Nkx6.1$^+$ population; Glucagon and Somatostatin are highly expressed in C-peptide$^+$/Nkx6.1$^-$ population (poly-hormonal cells fraction), PPY in C-peptide$^-$/Nkx6.1$^-$ population. The expression of GLP1R, NKX6.1, and PCSK1 genes is not significantly enriched in one specific population but notably expressed in similar levels in both C-peptide$^+$/Nkx6.1$^+$ and C-peptide$^-$/Nkx6.1$^+$ populations.

These are characteristics of more mature β-cells. The Nkx6.1-only fraction also showed enrichment for PCSK1, GLP1R, and Nkx6.1 expression, but had lower MafA and very low insulin, suggesting that these are pre-hormonal progenitors. The C-peptide$^-$/Nkx6.1$^-$ population contained cells expressing GCG, SST, and PPY (**Figure 2**), suggesting the presence of maturing α, δ, and PP islet cells, respectively.

Identification of CD49A as a Marker for Mature β-Cell

We used the FCCS platform (10) for identifying antibodies to cell surface proteins that preferentially capture insulin-producing cells (**Figure 3A**). Single cells from dissociated ILC were incubated on the array and antibody-bound cells were then stained for insulin. Out of the 235 antibodies in the array, 61 of them captured some ILC cells. Of these antibodies, anti-CD49A consistently captured insulin$^+$ cells in amounts exceeding 33% of the total cells captured (the median value for all antibodies being 13%). As illustrated in **Figure 3A**, there were other antibodies binding insulin$^+$ cells (e.g. CD99) and several that captured almost only insulin-negative cells (e.g. CD66C, CD73). These latter antibodies may serve for negative selection to remove cells that do not express insulin.

As a marker for positive selection of β-cells, we chose CD49A (Integrin alpha-1) since further experiments showed that this marker is predominantly present in the C-peptide$^+$/Nkx6.1$^+$ double positive subpopulation (**Figure 3B**, gray in upper panel). Magnetic activated cell sorting (MACS) with antibodies to CD49A, was used to fractionate live ILC cells into CD49A enriched and CD49A depleted populations. A marked increase in the percentage of C-peptide$^+$/Nkx6.1$^+$ double positive cells was observed by FACS in the CD49A enriched fraction compared to CD49A depleted fraction (**Figure 3B**, lower panel) or to non-sorted ILC cells (**Table 1**, line 1). On the other hand, there was

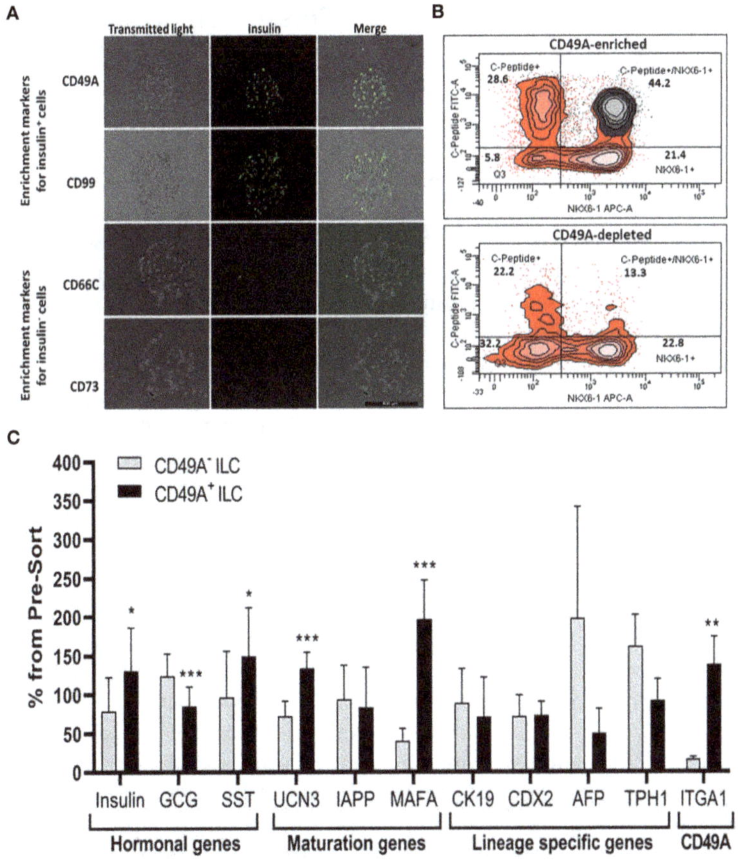

FIGURE 3 | Functional Cell-Capture Screening on antibody array. **(A)** In the array, the spots containing CD49A and CD99 antibodies attach ILC cells that comprise a high proportion of Insulin-positive cells, while spots with CD66c and CD73 antibodies attached cells that do not express insulin. The spots shown were part of arrays of 235 antibodies against cell-surface proteins, reacted with dissociated ILC cells and then stained with anti-insulin antibodies and Cy-5 labeled secondary antibodies. The number of insulin-positive cells was compared to the total cells captured (phase contrast). **(B)** Upper panel: FACS plot of ILC cells selected by MACS for binding to the CD49A antibody (CD49A enriched) and stained for human C-peptide, Nkx6.1, and CD49A, shows that CD49A positive cells (gray) are mainly in the C-peptide$^+$/Nkx6.1$^+$ double positive fraction. Lower panel: same for fraction that did not bind to the MACS CD49A antibody column (CD49A depleted). For comparative percentage of cells in each FACS subpopulation see **Table 1**. **(C)** qPCR analysis of RNA extracted from ILC cells fractionated by MACS shows that the cells binding to CD49A antibodies (CD49A enriched) have higher Insulin, UCN3, and mainly MAFA expression level. Results calculated in comparison to non-sorted cells, taken as basal 100% reference. The results are expressed as mean ± SEM. Student t test. $^*p < 0.05$; $^{**}p < 0.01$; $^{***}p < 0.001$.

TABLE 1 | Proportions of C-peptide and Nkx6.1 positive subpopulations in ILC cells selected for binding to CD49A antibodies.

Cell Fraction:	Expt 1			Expt 2		
	Total Non-sorted	CD49A enriched	CD49A depleted	Total Non-sorted	CD49A enriched	CD49A depleted
Cells positive for markers:			Percent of cells			
1. C-peptide/Nkx6.1	17.8	46.2	13.3	22.6	44.2	16.2
2. C-peptide only	24.8	19.7	22.2	25.4	28.6	26.8
3. NKX6.1 only	29.2	10.7	22.8	27.9	15.4	50.4
4. CD49A (overall)	23.8	61.4	9.5	26.7	52.8	10
5. C-peptide/Nkx6.1/CD49A	ND	40.1	5.0	ND	35.9	2.9

Table 1*. The C-peptide/Nkx6.1 double positive population is increased in the cells binding to CD49A antibodies (CD49A-enriched) compared to cells that did not bind (CD49A depleted). Quantitation of FACS plots (as in* ***Figure 2B****) of ILC cells either non-sorted or sorted by MACS with anti-CD49a antibody and further reacted with anti-human C-peptide, anti-Nx6.1, and anti-CD49 antibodies. Two different experiments are shown. ND, not determined.*

almost no change in the C-peptide-only cells (**Table 1**, line 2). Thus, within the CD49A-enriched fraction, the C-peptide$^+$/Nkx6.1$^+$ double positive cells became relatively more abundant than the C-peptide-only (i.e. Nkx6.1 negative) and Nkx6.1-only (i.e C-peptide negative) cells. While the proportion of CD49A$^+$ cells increased to about 50–60% in the enriched fraction, there were still around 10% of CD49A$^+$ cells in the depleted fraction (**Table 1**, line 4). This indicates that even after two passages on a MACS column with anti-CD49A, the separation was not complete. Yet, there may also be heterogeneity in the distribution of CD49A. For example, when the CD49A-enriched fraction was triple stained for CD49A, C-peptide and Nkx6.1, it was found that 81–86% of the C-peptide$^+$/Nkx6.1$^+$ cells were positive for CD49A (compare lines 5 to line 1 in **Table 1**). However, the same comparison shows that most of the double positive cells remaining in the CD49A-depleted fraction did not express CD49A (only 18–38% of them scoring positive for CD49A). This suggests that the C-peptide$^+$/Nkx6.1$^+$ double positive population may be heterogeneous, some with and some without CD49A.

The CD49A-enriched MACS fraction had significant differences in gene expression as compared to the CD49A depleted fraction (**Figure 3C**). After enrichment for CD49A$^+$ cells (confirmed by the increase in ITGA1 transcripts encoding CD49A), there was an increase in mRNA for insulin, urocortin-3 (UCN3), and most significantly for MafA. The expression level of glucagon mRNA was somewhat decreased but, unexpectedly, somatostatin mRNA was slightly increased, suggesting enrichment of the relatively small population of δ-cells. Interestingly, transcripts of a hepatic lineage gene (AFP), of which low amounts still remain in the hESC-derived ILC, are further decreased in the CD49A enriched MACS fraction (**Figure 3C**). In addition, TPH1, a gene of the serotonin synthesis pathway, was also reduced (**Figure 3C**), suggesting that CD49A enrichment removes non β-cells producing serotonin inhibiting insulin secretion (25). Overall, the gene expression data confirm that selection for the CD49A cell surface antigen helps to enrich for functional mature β-cells.

In order to be able to transplant cells in large enough amounts, the selection method needs to perform in large-scale preparations. Cells dissociated from ILC at day 35 of differentiation (~200*10^6 cells), were fractionated by two consecutive passage on MACS columns with anti-CD49A antibodies. The twice retained fraction contained 75 million cells. The quality of this CD49A enriched preparation, evaluated by qPCR, was similar to that of small-scale preparations, with increased Insulin, MafA, and UCN3 mRNA, decreased GCG and AFP mRNAs, relative to non-sorted and CD49A depleted cells. This made it possible to study the *in vivo* activity of the sorted ILC cells to reduce glycemia in diabetes model.

CD49A Selection Separates ILC Cells That Normalize Glycemia in Diabetic Mice From Inactive Cells

Current methods for implantation of human ILC in immunocompetent mice are based on micro-encapsulation in alginate spheres, so as to reduce direct contact of the cells with host immune cells. While the introduction of alginate spheres into the peritoneal cavity of C57BL/6 mice has been shown to elicit foreign body reaction (FBR) and fibrosis even without ILC (8, 9), this reaction can be inhibited by using chemically modified alginate. In particular, long-term functionality was demonstrated for ILC transplants that were encapsulated in low-viscosity SLG20 alginate carrying triazole-thiomorpholine dioxide (TMTD-alginate) and implanted intraperitoneally (i.p.) in mice with streptozotocin (STZ)-induced diabetes (9). In the present study, we used an ultra-pure alginate of medium viscosity (UP MVG alginate) coupled with TMTD for ILC microencapsulation. Microencapsulated ILC, sorted or total, were implanted i.p. into STZ-induced diabetes C57BL/6 mice. The function of non-sorted (Total ILC), CD49A-enriched (CD49A$^+$ ILC), and CD49A-depleted (CD49A$^-$ ILC) cells was compared by measuring of blood human C-peptide levels and effects on glycemia. Mice implanted with CD49A$^-$ ILC fraction cells did not exhibit reduction in BGL levels (**Figure 4A**, purple line), in direct correlation to the low human C-peptide secretion level (**Figure 4B**). Nevertheless, even the low amounts of human insulin secreted improved the viability of CD49A$^-$ ILC implanted mice compared to non-implanted diabetic control, which remained highly hyperglycemic and died around day 40 of the follow-up period. The mice implanted with either non-sorted ILC (Blue line, **Figure 4A**) or CD49A$^+$ enriched ILC (Red line, **Figure 4A**) exhibited a rapid decrease of blood sugar following implantation, reaching within ~6 days to the normoglycemic range (**Figure 4A**), with the CD49A$^+$ enriched ILC exhibited significantly lower BGL values than non-sorted ILC only at the first week. The average non-fasting blood glucose remained

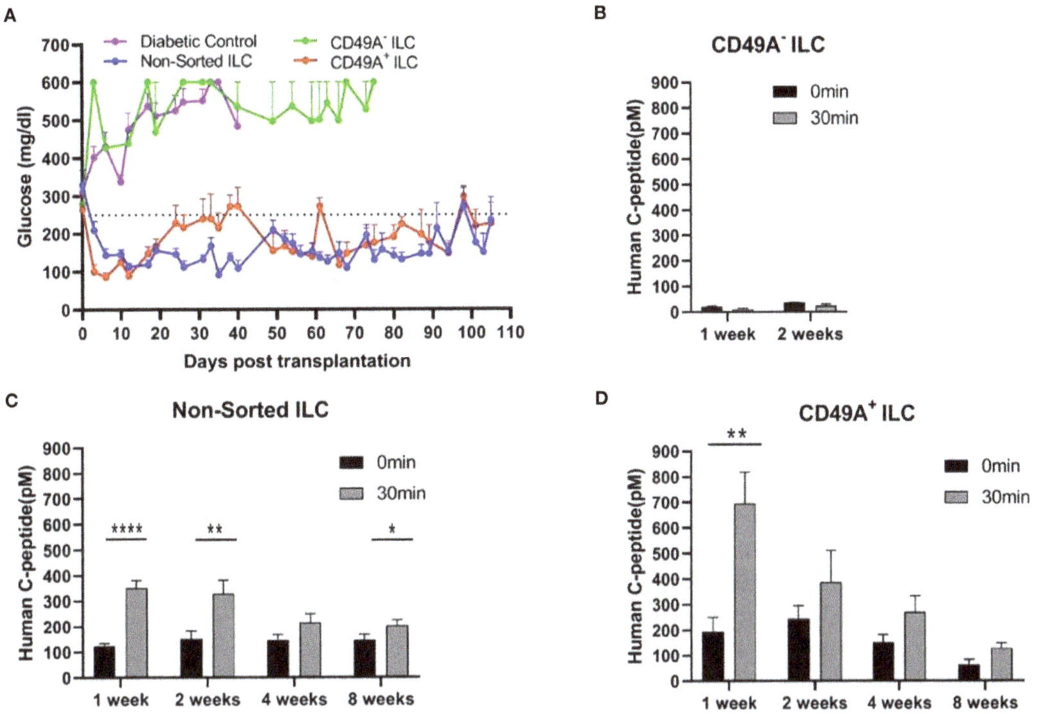

FIGURE 4 | Glycemia of diabetic mice is reduced by CD49A$^+$ enriched ILC cells but not by CD49A depleted cells. **(A)** ILC cells at the end of S7 differentiation protocol were fractionated by MACS column with CD49A antibodies. Cells of the bound fraction (CD49A enriched) and of the unbound (CD49A depleted) were reaggregated for 2 days, encapsulated in TMTD-MVG alginate, and implanted by i.p. route (~2*10^6 cells) into STZ-diabetic mice (n ≥ 5). Non-sorted ILC were similarly implanted. Shortly following transplantation, a reduction of hyperglycemia was observed only in the non-sorted and CD49A$^+$ enriched ILC groups, and blood glucose levels of diabetic mice were normalized for more than 100 days. **(B–D)** Glycemia reduction is correlated to human C-peptide glucose dependent secretion. At designated timepoints, mice were tested for human C-peptide concentration levels in blood. Human C-peptide was measured starting from 7 days post-implantation in the three groups of mice after overnight food deprivation (t = 0, black bars) and 30 min after i.p injection of 2 g/kg glucose (t = 30, gray bars). Average blood C-peptide level in pM ± SEM are shown for each group. The glucose-dependent C-peptide levels are significantly higher at day 7 in CD49A$^+$ ILC treated mice (p = 0.002), but at later time points, there is no significant functional difference **(C, D)**. The CD49A$^-$ ILC treated mice differ significantly from CD49A$^+$ enriched and non-sorted treated mice, exhibiting significantly lower levels of human C-peptide **(B)**, suggesting that this ILC population can be depleted (p = 0.034) (Student *t* test). *p < 0.05; **p < 0.01; ***p < 0.001.

mostly in the normal range (below 250 mg/dl) over a follow-up period of 100 days with some fluctuations. These temporal fluctuations can be attributed to changes in ILC composition *in vivo*. The effect on glycemia was confirmed by detection of serum human c-peptide. Albeit CD49A-enriched ILC produced a ~2 fold higher levels of C-peptide than implants of non-sorted ILC at the first week post implantation (**Figures 4C, D**), eventually at later stages, the *in vivo* function results were similar between the groups. Both, non-sorted and CD-49A-enriched, exhibited stimulated human c-peptide secretion 30 min after glucose injection already 1-week post transplantation, demonstrating typical β-cells response to glucose. Overall, these data establish that selection for the CD49A surface marker identifies and separates the ILC cells that control glycemia from inactive ones. On the other hand, the data did not show a clear functional difference between the CD49A-enriched ILC to the non-sorted ILC (**Figure 4A**) during the analysis period. The conclusion from these experiments was that MACS sorting for CD49A by itself was not sufficient to markedly improve the therapeutic activity in the diabetes model, in line with the equivalent levels in the amount of human C-peptide found in the blood of the mice.

Combining Negative Selection for CD26 With Positive Selection for CD49A

Selection for the CD49A marker increases the percentage of C-peptide$^+$/Nkx6.1$^+$ double positive cells but does not remove the C-peptide-only fraction (**Figure 3B** upper panel and **Table 1**). To improve the selection, we attempted to identify surface markers of ILC cells expressing both Insulin and Nkx6.1 simultaneously. In a new FCCS, each spot of the antibody array was evaluated for capture of insulin$^+$ cells and then for the percentage of Nkx6.1$^+$ cells among them. We were unable to find an antibody that captured exclusively C-peptide$^+$/Nkx6.1$^+$ double positive cells, but observed that in the *in vitro* differentiated ILC, an antibody to CD26 (Dipeptidyl peptidase-4, DPP4), captured efficiently insulin$^+$ cells (70% of all the captured cells), which only 3.5% of it had Nkx6.1 (**Table S6**). The presence of CD26 on these cells was unexpected since CD26 has been described as a marker of ductal cells and glucagon producing α-cells, but not of insulin-producing β-cells in natural islets isolated from pancreas (26–28). Our finding that in hESC-derived ILC cells, CD26 binds the cells expressing insulin but not Nkx6.1 (**Figure 5**, gray in the uppermost panel) offered the possibility to remove these cells and enrich for cells positive for Nkx6.1 and insulin (or C-peptide). Indeed, MACS selection with anti-CD26 showed that in the unbound cell fraction (CD26 depleted) the C-peptide$^+$/Nkx6.1$^-$ cells were markedly reduced whereas the C-peptide$^+$/Nkx6.1$^+$ double positive cells were increased (**Figure 5** and **Table 2**). By a subsequent selection for the CD49A marker (**Figure 5**, lower left panel), the percentage of double positive cells raised and could reach 70% in the CD26 depleted/CD49A enriched fraction (**Table 2** line 1). Gene expression analysis of the CD26$^-$/CD49A$^+$ ILC cells fraction demonstrated significant differences in gene expression as compared to the CD49A+ enriched population (**Figure 6**). The prior removal of CD26$^+$ population resulted in an increase mRNA for insulin, and most significantly for MafA. The expression level of glucagon mRNA was decreased by 70%, and somatostatin mRNA continued to increase, suggesting additional enrichment of the relatively small population of δ-cells. Moreover, depletion of CD26$^+$ cells further decreased AFP mRNA levels by ~90% from non-sorted cells (*vs.* 50% decrease by CD49A$^+$ enrichment). Overall, the gene expression data confirm that the preparative depletion of CD26$^+$ cells from ILC and subsequent selection for the CD49A$^+$ cells augment the functional mature β-cell phenotype characteristics.

To evaluate their activity for diabetes therapy, implants of CD26-depleted/CD49A-enriched ILC were compared to total non-sorted ILC. We observed that even when the dose of cells implanted was reduced to 1 million cells, the CD26 depleted/CD49A enriched ILC still showed efficiency to rapidly reduce glycemia and maintain it in the normal level (**Figure 7A**, green squares). Under these conditions, the non-sorted ILC did not bring down glycemia to normal levels, but only reduced it slightly and maintained it at an intermediate diabetic state (**Figure 7A**, red circles). The CD49A depleted fraction obtained from the CD26 depleted cells, did not maintain the glycemia, which rose to highly diabetic levels (**Figure 7A**, blue triangles). The levels of human C-peptide in the blood of mice implanted with CD26 depleted/CD49A enriched ILC were higher than with the other types of ILC, throughout the follow-up period of 8 weeks (**Figure 7B**). This was observed in fasting mice and after stimulation of C-peptide secretion by glucose. The increased therapeutic activity of the CD26 depleted/CD49A enriched ILC cells over non-sorted total cells was also demonstrated by an intraperitoneal glucose tolerance test (IPGTT) on day 46 after implantation (**Figure 7C**), as the area under the curve being reduced by 50% (p = 0.046).

DISCUSSION

Our aim was to fractionate the cell populations produced by differentiation of human pluripotent stem cells into pancreatic islet-like clusters, in order to select the insulin-producing cells which are most active to normalize glycemia in diabetes models. To find markers for selection, we used a functional cell capture screening (FCCS) assay on a microarray of large set of antibodies to cell surface proteins (10). In a first series of screens, we identified an antibody to integrin-alpha1 (CD49A, ITGA1) as efficiently binding insulin-producing cells. We show that by binding to anti-CD49A, one can separate cells that reduce glycemia in diabetic mice from cells completely lacking this capacity, although they express insulin. However, MACS selection with CD49A antibodies was not sufficient to clearly improve the therapeutic activity over that of non-sorted ILC. We, therefore, made a new series of screens in which we compared the

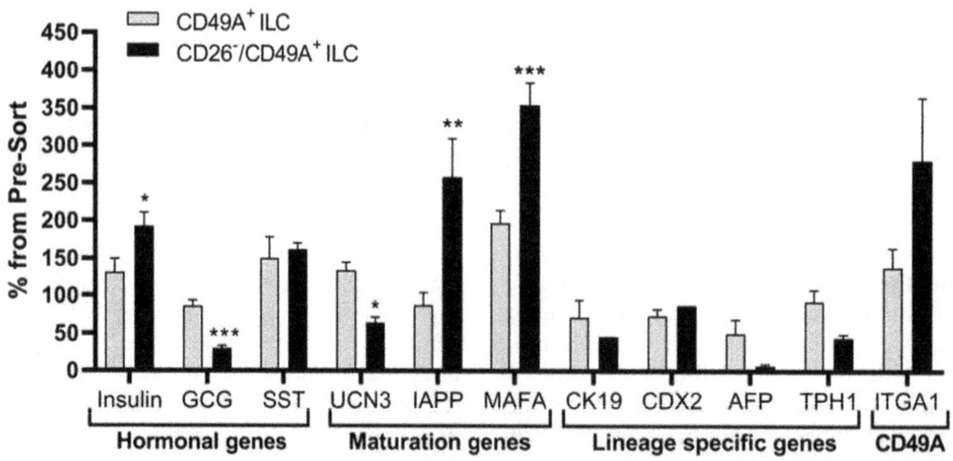

FIGURE 5 | CD26 depletion followed by CD49A selection improves mature β-cells phenotype *in vitro*. qPCR analysis of RNA extracted from ILC cells fractionated by CD26$^-$/CD49A$^+$ MACS (n ≥ 6 different batches) shows that the depletion/selection strategy yields an improved gene expression phenotype with increased expression of mature islet signature genes as MAFA, Insulin, and IAPP compared to CD49A enrichment alone. In addition, the removal of CD26 results in significantly lower expression levels of Glucagon and AFP mRNAs. Results calculated in comparison to non-sorted cells, taken as basal 100% reference. The results are expressed as mean ± SEM. Student t test. $^*p < 0.05$; $^{**}p < 0.01$; $^{***}p < 0.001$.

TABLE 2 | Proportions of C-peptide and Nkx6.1 positive subpopulations in ILC cells double selected with antibodies to CD26 and to CD49A.

Expt 1 **Cell fraction**	**Total Non-sorted**	**CD26 enriched**	**CD26 depleted**	**CD26 depleted CD49A enriched**	**CD26 depleted CD49A depleted**
Cells positive for:	Percent of cells				
1. C-peptide and NKX6.1	24.2	19.7	33.8	71.5	16
2. C-peptide only	22.9	43.0	4.8	10.7	4.0
3. NKX6.1 only	35.8	18.0	52.8	12.1	69.5
4. Negative (Q3)	17.1	19.3	8.5	5.7	10.5

Expt 2 Cell fraction	Total non-sorted	Total reaggregated	CD26 enriched	CD26 depleted	CD26 depleted CD49A enriched	CD26 depleted CD49A depleted
Cell positive for:	Percent of cells					
1. C-peptide and NKX6.1	22.5	32.6	6.1	33.7	66.6	21.2
2. C-peptide only	20.2	21.7	43.8	8.4	11.4	7.0
3. NKX6.1 only	13.3	21.6	6	24.3	13.3	42.7
4. Negative (Q3)	44.1	24.2	44.1	33.6	8.7	29.1

The C-peptide/Nkx6.1 double positive cells predominate in CD26 depleted/CD49 enriched fraction. After double MACS sorting and staining for human C-peptide and Nkx6.1 cells were analyzed in FACS plots as illustrated in ***Figure 5****. Quantitation from fractionation of two independent batches are shown in two experiments. In experiment 2, non-sorted ILC cells after dissociation and reaggregation without fractionation are also shown.*

antibodies for capture of insulin-producing cells and of Nkx6.1$^+$ cells together. We found that an antibody to dipeptidyl peptidase-4 (CD26, DPP4) bound ILC cells expressing insulin (and human C-peptide) but lacking Nkx6.1, and did no bind C-peptide$^+$/Nkx6.1$^+$ cells (**Figure 5**). By depletion of CD26$^+$ cells followed by enrichment for CD49A$^+$ cells, ILC populations containing 60–70% C-peptide$^+$/Nkx6.1$^+$ cells could be obtained. Such purified ILC fractions had an improved efficacy to reduce glycemia and secrete human C-peptide, relative to non-sorted ILC, or cells selected only for CD49A.

CD49A, the Integrin alpha1 chain, had been reported to have a function in the development of β-cells in human fetal and adult pancreas since, in combination with Integrin-β1, it forms the primary collagen-binding receptor (in particular for Collagen IV) and thereby contributes to β-cell adhesion and motility, as well as to insulin secretion (29). In recent studies using single-cell RNA Seq analysis of gene expression (30, 31) in the various cell populations formed by differentiating hPSC toward islet-like cells, Veres et al. (31) found that CD49A (ITGA1) is mostly (but not exclusively) a marker of β-cells expressing the INS, Nkx6.1, NPTX2, PCSK1 genes. These authors reported that MACS with CD49A antibodies can isolate these β-cells at 60–80% purity and that, *in vitro*, glucose-stimulated insulin secretion (GSIS) was improved in these cells. Our approach was different and based on screening of a large array of antibodies against membrane proteins to identify antibodies that bound insulin-producing hESC-derived β-like cells. CD49A emerged from this screen and we proceeded to study the *in vivo* therapeutic activity of cells selected for this marker in models of STZ-induced diabetes in C57BL/6 mice.

CD26 had been reported to be a marker of ductal cells and of alpha cells, but not of β-cells, in islets isolated from human pancreas (26, 27). However, we unexpectedly found that antibodies to CD26 efficiently bound insulin-producing ILC cells lacking Nkx6.1. This finding allowed us to further purify the hESC-derived ILC cells by depletion for CD26 and enrichment for CD49A and thereby obtain ILC with improved anti-diabetic activity in STZ-treated diabetic mice. The correlation between the cell types seen in the selected fractions and the anti-diabetic activity is interesting to investigate. Overall, our data support the assumption that it is the abundance of C-peptide$^+$/Nkx6.1$^+$ double positive cells which contributes most to the therapeutic activity. In the total non-sorted ILC used in the different experiments, the percentage of double positive cells was in the range of 18–24%, which increased to 34–35% after removal of cells binding to anti-CD26 antibodies and to 60–71% after further purification of the cells binding to CD49A. If the CD49A selection was done alone, only 44–46% double positive were obtained. Hence, the combined CD26 and CD49A selection is better and this is what was observed in term of therapeutic activity in diabetic mice. Similarly, compared to non-sorted cells, the combined CD26 and CD49A selection gave a larger increase in the levels of insulin and MafA mRNA, supporting the effect of enrichment on functional maturation phenotype. Besides enriching for more active β-cells, the purification also removes unnecessary cells. Thus, depleting CD26$^+$ cells removed the insulin-producing C-peptide$^+$ cells lacking Nkx6.1, which correspond to polyhormonal progenitors that do not seem to contribute to β-cell functions *in vivo* an even to become alpha cells (22) delta cells (13). The beneficial or detrimental role of the other cell types in the ILC remains to be clarified. For example, the CD49A-enrichment also removes cells expressing TPH1 (**Figure 2C**), which probably correspond to cells producing serotonin (entero-chromaffin cells) that inhibit insulin secretion (25). Removing CD26$^+$ cells may reduce the amount of DPP4 enzyme which degrades GLP-1, an important stimulator of β-cell function (32). So, removing these populations could eliminate cells that do not participate in the control of glycemia or disturb this control. In addition, and not less important, the purification helps reducing the number of *in vitro* differentiated ILC cells that need to be implanted for cell therapy of diabetes. Our double selection, depleting CD26$^+$ cells and enriching for CD49A$^+$ cells, removes

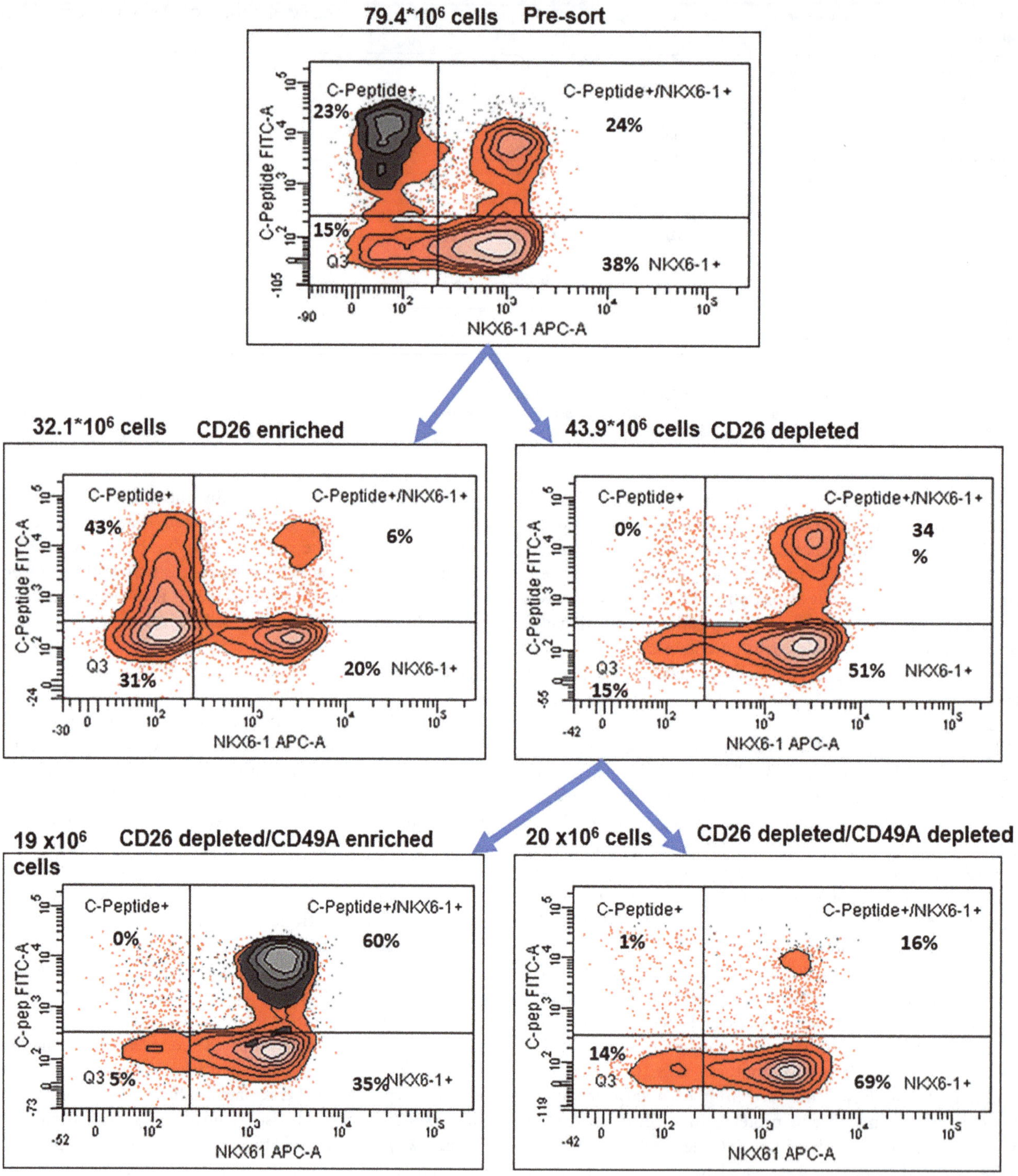

FIGURE 6 | Preparative MACS selection using novel sorting strategy. After selection by MACS with CD26 antibodies (middle two panels), the CD26 depleted cells have lost the C-peptide only fraction (right panel). After subsequent selection of the CD26 depleted fraction by MACS with CD49A antibodies (lower two panels), the C-peptide/Nkx6.1 double positive fraction is increased (lower left panel with CD49A stain in gray). For comparative percentage of cells in each FACS subpopulation see **Table 2**. The total number of cells at each step is shown next to the panels.

many cells (**Figure 5**) and allowed to reduce the number of cells needed for normalizing glycemia in our experiments with diabetic mice. This reduction could be of practical importance to decrease the volume of hPSC-derived ILC that will be required to treat human diabetic patients. Performing MACS purification requires dissociation of the differentiated islet-like clusters produced in our

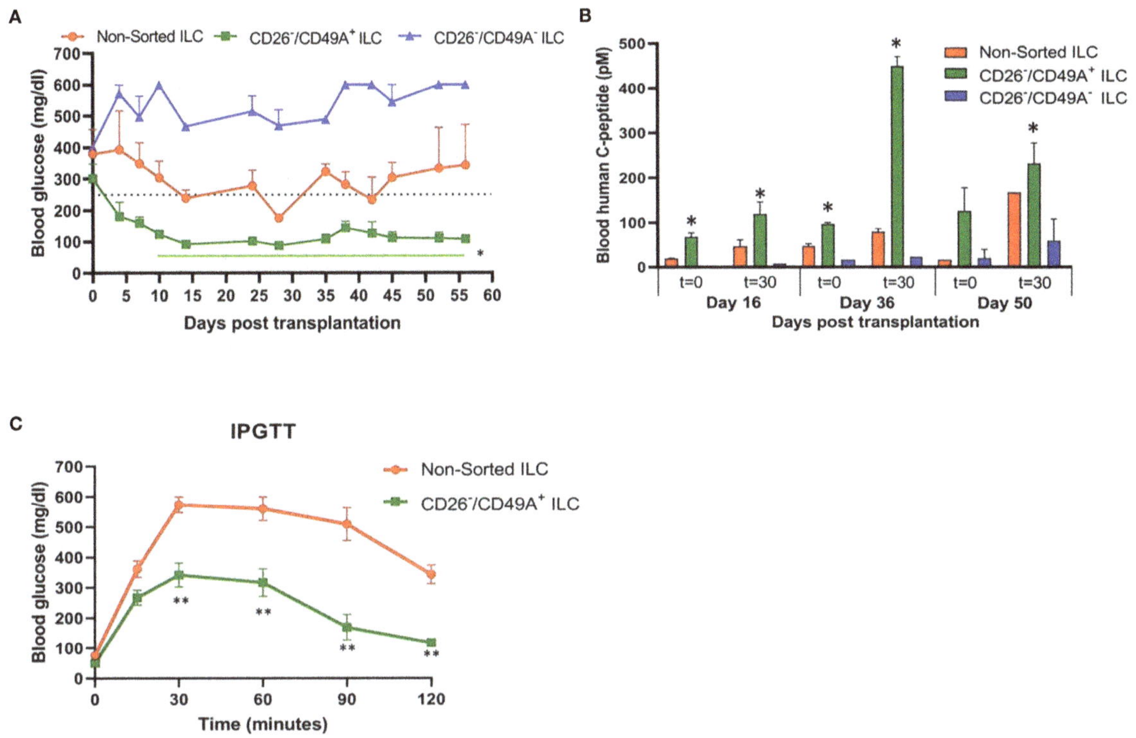

FIGURE 7 | Improved activity of CD26 depleted/CD49A enriched ILC to normalize glycemia in diabetic mice. **(A)** CD26 depleted/CD49A enriched ILC cells (squares with green line) implanted in diabetic C57BL/6 mice, reduce blood glucose better than non-sorted cells (red circles) or than CD26 depleted/CD49A depleted cells (dark blue triangles). After a first MACS with CD26 antibodies, the $CD26^-$ flow-through cells were separated by MACS with CD49A antibodies. The non-sorted and the sorted cell fractions were reaggregated into ILC prior to micro-encapsulation in TMTD-MVG alginate and, from each, a dose of 1×10^6 cells was implanted i.p. in C57BL/6 mice ($n \geq 5$). Average blood glucose ± SEM is shown. Starting from day 10, the average blood glucose results of $CD26^-/CD49A^+$ group is significantly lower than non-sorted cells (Student t test. $^*p < 0.05$, horizontal green line). **(B)** In the same mice, blood C-peptide was measured at indicated days, after overnight food deprivation (t = 0) and 30 min after i.p. injection of 2 g/kg glucose (t = 30). Average blood levels of human C-peptide in pM ± SEM are shown for mice with implants of ILC either non-sorted (red), CD26 depleted/CD49A enriched (green), or CD26 depleted/CD49A depleted (blue). The significantly higher values for each timepoint are marked (Student t test. $^*p < 0.05$). **(C)** Intraperitoneal Glucose Tolerance Test (IPGTT) shows that CD26 depleted/CD49A enriched ILC cells (squares with green line) reduce more the diabetic state than non-sorted ILC cells (red circles). On day 45 after implantation the mice were injected i.p with 2 g/kg glucose and blood glucose was measured at different times for a 2-h period. The results are expressed as BGL mean ± SEM. Student t test. $^{**}p < 0.01$ from t = 30 onwards. The AUC with the CD26 depleted/CD49A enriched cells was 50% lower than that with the non-sorted re-aggregated cells (p = 0.046).

3D-suspension cultures. The dissociated cells, MACS-fractionated or not, were reaggregated by a few days of culture before implantation. This procedure does not alter the composition of the ILC, nor their activity in diabetes models (33). For implantation, we encapsulated the ILC in medium-viscosity MVG alginate coupled with TMTD, and not the very low-viscosity SLG20 alginate as used by Vegas et al. (9). When encapsulated in TMTD-MVG microspheres (of 1.5 mm diameter) and implanted in the peritoneal cavity of the diabetic mice, the total and the purified ILC caused a rapid decrease in glycemia starting soon after implantation. There was no 3–6 month delay as seen when non-mature progenitors are implanted (12). This indicates that the ILC fully differentiated *in vitro*, provide the functionally mature β and other cell types needed to restore and maintain glycemia in the normal range, shortly after implantation. A reservation is that since the MACS fractionation is not complete, the purified fractions are likely to still contain other cell types that influence the therapeutic effect. Indeed, the presence of glucagon-producing alpha-like cells is probably important to prevent hypoglycemic events. The aim is to implant cell clusters that mimic the activities of natural islets, and not pure β-cells. The FCCS assay had previously shown that CD56 and CD9 are markers of β-cell s in islets from human post-mortem donations (11). With our hESC-derived ILC, CD56 was also high on the list of antibodies binding insulin-producing cells, while CD9 was not found (not shown). The CD200 marker reported to isolate endocrine cells at the pancreatic progenitor stage of hESC differentiation (22) was also in our FCCS (not shown). The FCCS assay identified additional antibodies that may be candidates to select cells active in restoring and maintaining normoglycemia in diabetics, and/or remove inactive cells by negative sorting (data not shown). Development of better means of large-scale cell sorting with combinations of antibodies, such as shown here, appears as an important endeavor to produce the amounts of ILC needed for treating the many millions of insulin-dependent diabetic patients by regenerative cell therapy.

CONCLUSIONS

In vitro differentiation of islet-like clusters from human pluripotent stem cells represents a potentially unlimited source of cells that could restore physiological control of blood glucose in diabetic patients requiring insulin. The CD49A surface protein (integrin-alpha1) was identified as a selective marker of the ILC cells that are active to normalize glycemia in a diabetic mice model. Removal of cells expressing CD26 (DPP4) prior to enrichment of $CD49A^+$ cells further improved their therapeutic activity and reduced the number of ILC cells needed to normalize glycemia. The ILC purification described appears as a promising strategy to improve cell therapy of diabetes with hPSC-derived ILC.

AUTHOR CONTRIBUTIONS

KM, JC and MR directed the research work and wrote the manuscript. OE, MW and YS conceived and contributed the Functional Cell Capture Screening. DB, AB, KY, MZ, IT, AE and EV performed the experiments. AL, KM supervised the animal experimentation. AH and JI were senior advisors. All authors contributed to the article and approved the submitted version.

ACKNOWLEDGMENTS

We acknowledge with thanks the valuable contribution of Dr. Raanan Margalit, CEO, Science In Action Ltd., where all animal experimentations were performed.

REFERENCES

1. Shapiro AM, Pokrywczynska M, Ricordi C. Clinical Pancreatic Islet Transplantation. *Nat Rev Endocrinol* (2017) 13(5):268–77. doi: 10.1038/nrendo.2016.178
2. Gamble A, Pepper AR, Bruni A, Shapiro AMJ. The Journey of Islet Cell Transplantation and Future Development. *Islets* (2018) 10(2):80–94. doi: 10.1080/19382014.2018.1428511
3. Shapiro AMJ. Islet Transplantation - The Canadian Perspective. *CellR4 Repair Replace Regener Reprogram* (2019) 7. doi: 10.32113/cellr4_201911_2799
4. Pagliuca FW, Millman JR, Gurtler M, Segel M, Van Dervort A, Ryu JH, et al. Generation of Functional Human Pancreatic β-Cell s In Vitro. *Cell* (2014) 159 (2):428–39. doi: 10.1016/j.cell.2014.09.040
5. Rezania A, Bruin JE, Arora P, Rubin A, Batushansky I, Asadi A, et al. Reversal of Diabetes With Insulin-Producing Cells Derived In Vitro From Human Pluripotent Stem Cells. *Nat Biotechnol* (2014) 32(11):1121–33. doi: 10.1038/nbt.3033
6. Russ HA, Parent AV, Ringler JJ, Hennings TG, Nair GG, Shveygert M, et al. Controlled Induction of Human Pancreatic Progenitors Produces Functional β-Like Cells In Vitro. *EMBO J* (2015) 34(13):1759–72. doi: 10.15252/embj.201591058
7. Farah S, Doloff JC, Muller P, Sadraei A, Han HJ, Olafson K, et al. Long-Term Implant Fibrosis Prevention in Rodents and non-Human Primates Using Crystallized Drug Formulations. *Nat Mater* (2019) 18(8):892–904. doi: 10.1038/s41563-019-0377-5
8. Vegas AJ, Veiseh O, Doloff JC, Ma M, Tam HH, Bratlie K, et al. Combinatorial Hydrogel Library Enables Identification of Materials That Mitigate the Foreign Body Response in Primates. *Nat Biotechnol* (2016) 34 (3):345–52. doi: 10.1038/nbt.3462
9. Vegas AJ, Veiseh O, Gurtler M, Millman JR, Pagliuca FW, Bader AR, et al. Long-Term Glycemic Control Using Polymer-Encapsulated Human Stem Cell-Derived β-Cell s in Immune-Competent Mice. *Nat Med* (2016) 22 (3):306–11. doi: 10.1038/nm.4030
10. Sharivkin R, Walker MD, Soen Y. Proteomics-Based Dissection of Human Endoderm Progenitors by Differential Cell Capture on Antibody Array. *Mol Cell Proteomics* (2012) 11(9):586–95. doi: 10.1074/mcp.M111.016840
11. Sharivkin R, Walker MD, Soen Y. Functional Proteomics Screen Enables Enrichment of Distinct Cell Types From Human Pancreatic Islets. *PloS One* (2015) 10(2):e0115100. doi: 10.1371/journal.pone.0115100
12. Rezania A, Bruin JE, Xu J, Narayan K, Fox JK, O'Neil JJ, et al. Enrichment of Human Embryonic Stem Cell-Derived NKX6.1-expressing Pancreatic Progenitor Cells Accelerates the Maturation of Insulin-Secreting Cells In Vivo. *Stem Cells* (2013) 31(11):2432–42. doi: 10.1002/stem.1489
13. Taylor BL, Liu FF, Sander M. Nkx6.1 is Essential for Maintaining the Functional State of Pancreatic β-Cell s. *Cell Rep* (2013) 4(6):1262–75. doi: 10.1016/j.celrep.2013.08.010
14. Tannenbaum SE, Turetsky TT, Singer O, Aizenman E, Kirshberg S, Ilouz N, et al. Derivation of Xeno-Free and GMP-grade Human Embryonic Stem Cells–Platforms for Future Clinical Applications. *PloS One* (2012) 7(6): e35325. doi: 10.1371/journal.pone.0035325
15. D'Amour KA, Agulnick AD, Eliazer S, Kelly OG, Kroon E, Baetge EE. Efficient Differentiation of Human Embryonic Stem Cells to Definitive Endoderm. *Nat Biotechnol* (2005) 23(12):1534–41. doi: 10.1038/nbt1163
16. Agulnick AD, Ambruzs DM, Moorman MA, Bhoumik A, Cesario RM, Payne JK, et al. Insulin-Producing Endocrine Cells Differentiated in Vitro From Human Embryonic Stem Cells Function in Macroencapsulation Devices In Vivo. *Stem Cells Transl Med* (2015) 4(10):1214–22. doi: 10.5966/sctm.2015-0079
17. D'Amour KA, Bang AG, Eliazer S, Kelly OG, Agulnick AD, Smart NG, et al. Production of Pancreatic Hormone-Expressing Endocrine Cells From Human Embryonic Stem Cells. *Nat Biotechnol* (2006) 24(11):1392–401. doi: 10.1038/nbt1259
18. Kroon E, Martinson LA, Kadoya K, Bang AG, Kelly OG, Eliazer S, et al. Pancreatic Endoderm Derived From Human Embryonic Stem Cells Generates Glucose-Responsive Insulin-Secreting Cells In Vivo. *Nat Biotechnol* (2008) 26 (4):443–52. doi: 10.1038/nbt1393
19. Schulz TC, Young HY, Agulnick AD, Babin MJ, Baetge EE, Bang AG, et al. A Scalable System for Production of Functional Pancreatic Progenitors From Human Embryonic Stem Cells. *PloS One* (2012) 7(5):e37004. doi: 10.1371/journal.pone.0037004
20. Hrvatin S, Deng F, O'Donnell CW, Gifford DK, Melton DA. MARIS: Method for Analyzing RNA Following Intracellular Sorting. *PloS One* (2014) 9(3): e89459. doi: 10.1371/journal.pone.0089459
21. Nostro MC, Sarangi F, Ogawa S, Holtzinger A, Corneo B, Li X, et al. Stage-Specific Signaling Through Tgfβ Family Members and WNT Regulates Patterning and Pancreatic Specification of Human Pluripotent Stem Cells. *Development* (2011) 138(5):861–71. doi: 10.1242/dev.055236
22. Kelly OG, Chan MY, Martinson LA, Kadoya K, Ostertag TM, Ross KG, et al. Cell-Surface Markers for the Isolation of Pancreatic Cell Types Derived From Human Embryonic Stem Cells. *Nat Biotechnol* (2011) 29(8):750–6. doi: 10.1038/nbt.1931
23. Katsuta H, Akashi T, Katsuta R, Nagaya M, Kim D, Arinobu Y, et al. Single Pancreatic β-Cell s Co-Express Multiple Islet Hormone Genes in Mice. *Diabetologia* (2010) 53(1):128–38. doi: 10.1007/s00125-009-1570-x
24. Bruin JE, Erener S, Vela J, Hu X, Johnson JD, Kurata HT, et al. Characterization of Polyhormonal Insulin-Producing Cells Derived In Vitro From Human Embryonic Stem Cells. *Stem Cell Res* (2014) 12(1):194–208. doi: 10.1016/j.scr.2013.10.003
25. Cataldo LR, Mizgier ML, Bravo Sagua R, Jana F, Cardenas C, Llanos P, et al. Prolonged Activation of the Htr2b Serotonin Receptor Impairs Glucose Stimulated Insulin Secretion and Mitochondrial Function in MIN6 Cells. *PloS One* (2017) 12(1):e0170213. doi: 10.1371/journal.pone.0170213
26. Arda HE, Li L, Tsai J, Torre EA, Rosli Y, Peiris H, et al. Age-Dependent Pancreatic Gene Regulation Reveals Mechanisms Governing Human β-Cell Function. *Cell Metab* (2016) 23(5):909–20. doi: 10.1016/j.cmet.2016.04.002
27. Augstein P, Naselli G, Loudovaris T, Hawthorne WJ, Campbell P, Bandala-Sanchez E, et al. Localization of Dipeptidyl Peptidase-4 (CD26) to Human Pancreatic Ducts and Islet Alpha Cells. *Diabetes Res Clin Pract* (2015) 110 (3):291–300. doi: 10.1016/j.diabres.2015.10.010
28. Dorrell C, Schug J, Lin CF, Canaday PS, Fox AJ, Smirnova O, et al. Transcriptomes of the Major Human Pancreatic Cell Types. *Diabetologia* (2011) 54(11):2832–44. doi: 10.1007/s00125-011-2283-5

29. Kaido T, Yebra M, Cirulli V, Montgomery AM. Regulation of Human β-Cell Adhesion, Motility, and Insulin Secretion by Collagen IV and its Receptor Alpha1β1. *J Biol Chem* (2004) 279(51):53762–9. doi: 10.1074/jbc.M411202200
30. Baron M, Veres A, Wolock SL, Faust AL, Gaujoux R, Vetere A, et al. A Single-Cell Transcriptomic Map of the Human and Mouse Pancreas Reveals Inter- and Intra-cell Population Structure. *Cell Syst* (2016) 3(4):346–60.e4. doi: 10.1016/j.cels.2016.08.011
31. Veres A, Faust AL, Bushnell HL, Engquist EN, Kenty JH, Harb G, et al. Charting Cellular Identity During Human In Vitro β-Cell Differentiation. *Nature* (2019) 569(7756):368–73. doi: 10.1038/s41586-019-1168-5
32. Klemann C, Wagner L, Stephan M, von Horsten S. Cut to the Chase: A Review of CD26/dipeptidyl Peptidase-4's (DPP4) Entanglement in the Immune System. *Clin Exp Immunol* (2016) 185(1):1–21. doi: 10.1111/cei.12781
33. Song J, Millman JR. Economic 3D-Printing Approach for Transplantation of Human Stem Cell-Derived β-Like Cells. *Biofabrication* (2016) 9(1):015002. doi: 10.1088/1758-5090/9/1/015002

7

Human Pluripotent Stem Cells: A Unique Tool for Toxicity Testing in Pancreatic Progenitor and Endocrine Cells

*Erin M. MacFarlane and Jennifer E. Bruin**

Department of Biology & Institute of Biochemistry, Carleton University, Ottawa, ON, Canada

Correspondence:
Jennifer E. Bruin
jenny.bruin@carleton.ca

Diabetes prevalence is increasing worldwide, and epidemiological studies report an association between diabetes incidence and environmental pollutant exposure. There are >84,000 chemicals in commerce, many of which are released into the environment without a clear understanding of potential adverse health consequences. While *in vivo* rodent studies remain an important tool for testing chemical toxicity systemically, we urgently need high-throughput screening platforms in biologically relevant models to efficiently prioritize chemicals for in depth toxicity analysis. Given the increasing global burden of obesity and diabetes, identifying chemicals that disrupt metabolism should be a high priority. Pancreatic endocrine cells are key regulators of systemic metabolism, yet often overlooked as a target tissue in toxicology studies. Immortalized β-cell lines and primary human, porcine, and rodent islets are widely used for studying the endocrine pancreas *in vitro*, but each have important limitations in terms of scalability, lifespan, and/or biological relevance. Human pluripotent stem cell (hPSC) culture is a powerful tool for *in vitro* toxicity testing that addresses many of the limitations with other β-cell models. Current *in vitro* differentiation protocols can efficiently generate glucose-responsive insulin-secreting β-like cells that are not fully mature, but still valuable for high-throughput toxicity screening *in vitro*. Furthermore, hPSCs can be applied as a model of developing pancreatic endocrine cells to screen for chemicals that influence endocrine cell formation during critical windows of differentiation. Given their versatility, we recommend using hPSCs to identify potential β-cell toxins, which can then be prioritized as chemicals of concern for metabolic disruption.

Keywords: stem cells, diabetes, pollution, beta cells, pancreas development, toxicology

INTRODUCTION

Diabetes Pathogenesis

Diabetes is a chronic disease characterized by high blood sugar levels and devastating secondary health complications (1). In 2019, there were >460 million people with diabetes worldwide, which translates to roughly 1 in 11 adults ages 20-79 years. For those over the age of 65, diabetes rates further increase to 1 in 5 (1). Moreover, the International Diabetes Federation projects that diabetes incidence will increase by 51% over the next 25 years to exceed 700 million adults worldwide.

Glucose homeostasis is maintained by the exquisite balance of hormones secreted from pancreatic islets. The predominant islet cell type is the β-cell, which secretes insulin in a tightly regulated manner in response to glucose and other stimuli (2, 3). Type 1 diabetes (T1D), accounting for ~10% of patients with diabetes, is caused by autoimmune destruction of β-cells leading to insufficient insulin production. Type 2 diabetes (T2D) accounts for ~90% of cases and was classically thought of as a disease of insulin resistance. However, we now appreciate that β-cell dysfunction and loss of β-cell mass are also central to T2D pathogenesis (4–6). The critical role of β-cells in driving diabetes risk is further confirmed by genome-wide association studies, which find that most loci influencing T2D risk are involved in regulating insulin secretion (7–9). These studies also emphasize that rising diabetes rates cannot simply be explained by genetics, but rather must be influenced by environmental factors (9, 10). For example, there is strong epidemiological evidence linking exposure to persistent organic pollutants (POPs) with increased T2D incidence (11–29) and β-cell dysfunction (28–30) in humans. However, basic research in clinically relevant models is needed to understand the potential causal role for environmental contaminants in diabetes pathogenesis and to explore underlying tissue- and cell-specific mechanisms of toxicity.

Environmental Contaminants

Environmental pollutants are a major global concern due to their wide-ranging acute and chronic adverse effects on human health (31). With over 84,000 chemicals in commerce, there is an urgent need to develop tools for extensive chemical screening and toxicity testing (32). Environmental contaminants fall within a wide range of classes, including but not limited to POPs (e.g., pesticides, polychlorinated biphenyls (PCBs), dioxin-like compounds), estrogen analogues (e.g., bisphenol A (BPA), used in polycarbonate plastics), phthalates (used in cosmetics, paints, textiles), heavy metals, perfluorinated chemicals (e.g., perfluorooctane sulfonate (PFOS) used in food packaging and fire-fighting foams), and flame retardants (e.g., polybrominated diphenyl ethers, organohalogen compounds, organophosphates esters) (33–35). Contaminants can be further classified by their mechanism of action. For example, chemicals that impair proper hormone function are referred to as endocrine-disrupting chemicals (EDCs) (33, 36–40) and those that disrupt metabolism are classified as metabolism-disrupting chemicals (MDCs) (41–43). Despite restrictions on many environmental pollutants, these chemicals continue to persist in the environment, contaminating food and water sources, and remain detectable in human tissues (44, 45).

Biomonitoring is essential for tracking human contaminant exposure and predicting adverse health outcomes (46, 47). However, this is a reactive approach to evaluating the impact of toxins on human health. Ideally, we need to efficiently screen chemicals for toxicity in relevant model systems prior to their release into the environment. Since pollutants often accumulate in tissues, effectively creating a chemical mixture cocktail (48), we also need to consider the combined effects of complex chemical mixtures. Dose and duration of exposure add additional layers of complexity. For example, POPs have long half-lives of years to decades (49), but the shorter lifespan of other chemicals such as BPA and phthalates is also not trivial. Much like hormones, EDCs can exert their effects on the human body at low concentrations over an extended period of time (50). Despite being excreted within days, frequent consumption of these pollutants results in chronic, low dose exposure over time (51, 52). Furthermore, nonlinear dose-responses are frequently seen with EDCs, so acute high dose studies may not accurately predict adverse health outcomes of chronic or subacute low dose exposures (42, 53, 54). The need to consider chemical exposures ranging from acute high doses to chronic low doses, as well as individual chemicals and complex mixtures, further emphasizes the importance of scaling up toxicity testing capacity.

Developmental Origins of Disease

Another important consideration for toxicology studies is the timing of exposure to environmental contaminants. Gestational or early life stressors, such as undernutrition or overnutrition, are linked to a variety of adult-onset diseases - termed developmental origins of health and disease (DOHaD) (55, 56). For instance, low birth weight and early life "catch up" growth are well-established risk factors for developing metabolic disease later in life (57, 58). Maternal-fetal exposure to POPs has been linked to adverse outcomes such as reduced birth weight, disruption of hormone levels in cord blood, and changes in epigenetic markers of development (59–61). There is also mounting epidemiological evidence suggesting a possible link between early-life environmental contaminant exposure and long-term metabolic dysfunction (62–66). More epidemiology is needed and important cohort studies like the Maternal-Infant Research on Environmental Chemicals (MIREC) continue to track long-term metabolic outcomes in offspring (67–70). However, it takes decades to truly establish a link between early-life exposure and long-term adverse health outcomes. *In vitro* model systems that allow for toxicity screening in developing human cells will be a powerful starting point for studying DOHaD.

Perspective Overview

There is an urgent need to identify environmental contaminants, specifically EDCs or MDCs, that contribute to diabetes pathogenesis. To do so, we must consider non-classical toxicological endpoints in a wide variety of tissues involved in regulating metabolic homeostasis. This means thinking beyond typical hepatoxicity endpoints and considering diverse metabolic targets such as neuroendocrine cells, enteroendocrine cells, white or brown adipocytes, skeletal muscle, thyroid gland, and pancreatic endocrine cells (38, 40, 71). While injury to any of these tissues would potentially disrupt energy homeostasis, we propose that pancreatic endocrine cells should be a high priority for toxicity testing to identify MDCs of concern for diabetes pathogenesis. In this Perspective Article, we discuss a range of endpoints that could be considered in the context of β-cell toxicity. We also discuss various model systems available for toxicity testing, including the numerous advantages of human pluripotent stem cells (hPSCs). In particular, we propose hPSCs as a unique model system for evaluating toxicity both during critical windows of β-cell development and in glucose-responsive adult β-like cells (**Figure 1**).

TOXICITY TESTING IN PANCREATIC β-CELLS

Despite mounting evidence implicating pollutants as metabolic disruptors, the pancreas has not been extensively studied in the toxicology field (40, 42). Interestingly, the occasional biodistribution studies that include pancreas tissue report a slower elimination of lipophilic pollutants in the pancreas compared to liver or adipose (72, 73). Xenobiotic metabolism enzymes, such as cytochrome P450 (Cyp) enzymes, are useful biomarkers for direct cellular exposure to pollutants. We have reported induction of *Cyp1a1* in mouse and human islets following direct exposure to TCDD/dioxin or dioxin-like pollutants *in vitro* and in mouse islets following systemic administration of TCDD *in vivo* (73). Moreover, in pregnant TCDD-exposed mice, *Cyp1a1* was induced 17-fold in pancreas compared to only 3-fold and 7-fold in liver and adipose, respectively (74). Therefore, pancreatic cells are not only directly exposed to pollutants *in vivo*, but may even act as a "sink" for long-term storage of lipophilic chemicals, similar to adipose depots.

There is mounting evidence that a wide range of environmental contaminants can directly impact β-cell function. For example, BPA, a non-persistent additive commonly used in plastic products, acutely increases insulin secretion in mouse and human islets *via* inhibition of K_{ATP} channels and increased Ca^{2+} signaling (75), whereas longer-term BPA exposure inhibits Ca^{2+} entry and reduces insulin secretion (76). Newer BPA-replacement chemicals, BPS and BPF, also disrupt mouse β-cell function (77). Exposure to POPs, including organochlorine pesticides and a PCB mixture, directly inhibited insulin secretion in a rat β-cell line (INS-1E cells) (30). A "northern contaminant mixture", containing 20 different POPs at environmentally relevant concentrations, also suppressed insulin secretion in rats *in vivo* and in a rodent β-cell line (MIN6 cells) *in vitro* (78). Additionally, we and others have shown that dioxin suppresses insulin secretion in rodent islets (73, 79, 80) and human islets (73). Interestingly, acute high-dose dioxin exposure caused β-cell apoptosis in male mice but not female mice (81), whereas prolonged low-dose dioxin exposure led to impaired β-cell adaptation to high fat diet feeding in female but not male mice (74, 82).

Given the critical role for β-cells in diabetes pathogenesis, environmental toxins that adversely impact β-cells are likely to disrupt overall glucose homeostasis or at minimum, increase diabetes risk. Importantly, there are numerous plausible avenues for toxin-induced β-cell injury that could lead to adverse metabolic health outcomes. Below, we propose key toxicity endpoints for β-cells. This is not intended to be a comprehensive list of potential adverse outcomes, but rather examples that should be considered as a starting point for identifying MDCs that act as β-cell toxins.

Insulin Secretory Defects

Pancreatic β-cells are highly specialized to synthesize, process, store, and secrete insulin rapidly and sustainably in response to numerous physiological stimuli, including glucose (2). There is a rapid first phase of insulin secretion within minutes of a glucose stimulus, followed by a sustained second phase that can last for several hours. Glucose-stimulated insulin secretion is amplified by other nutrients, such as fatty acids and amino acids, and binding of gut-derived hormones (GLP-1, GIP) to incretin receptors. Dysregulated glucose-induced insulin secretion is a well-established feature of T2D (83–87) and also reported in T1D patients prior to disease onset (88–92), suggesting a possible link between insulin secretory defects and activation of inappropriate autoimmune responses. Therefore, environmental contaminants that interfere with any aspect of the complex insulin secretory pathway (e.g., glucose sensing, mitochondrial metabolism, ion flux, exocytotic machinery, sensitivity to amplification signals) could adversely affect the fine-tuned ability of β-cells to couple insulin secretion with a nutrient secretagogue. Furthermore, defects in the timing of insulin release, either the rapid first-phase or the sustained second-phase response, could impact overall glycemic control. Importantly, β-cell dysfunction is not just insufficient or decreased insulin secretion, but also refers to overproduction of insulin. Hyperinsulinemia is not only an adaptation to insulin resistance but can also be the primary defect that drives obesity and insulin resistance (93–96). Therefore, toxins that increase insulin secretion - either inappropriate insulin release under basal glucose conditions or excessive insulin secretion following a nutrient stimulus - should also be considered potential MDCs.

Loss of β-Cell Mass

Patients with T1D display near complete loss of β-cell mass (> 80% reduction) at the time of disease onset (6). Although less extreme, individuals with T2D also have reduced β-cell mass (5, 6, 97, 98), which may be caused by a combination of increased β-cell death, insufficient β-cell proliferation, or impaired β-cell neogenesis. β-cell mass is generally increased in overweight or obese non-diabetic subjects compared to lean controls but reduced by 24%–65% in patients with T2D (5, 6, 99). Loss of β-cell mass can be detected by measuring the α-cell to β-cell ratio, which is consistently higher in patients with T2D (97). Therefore, environmental toxins that disrupt β-cell mass, for

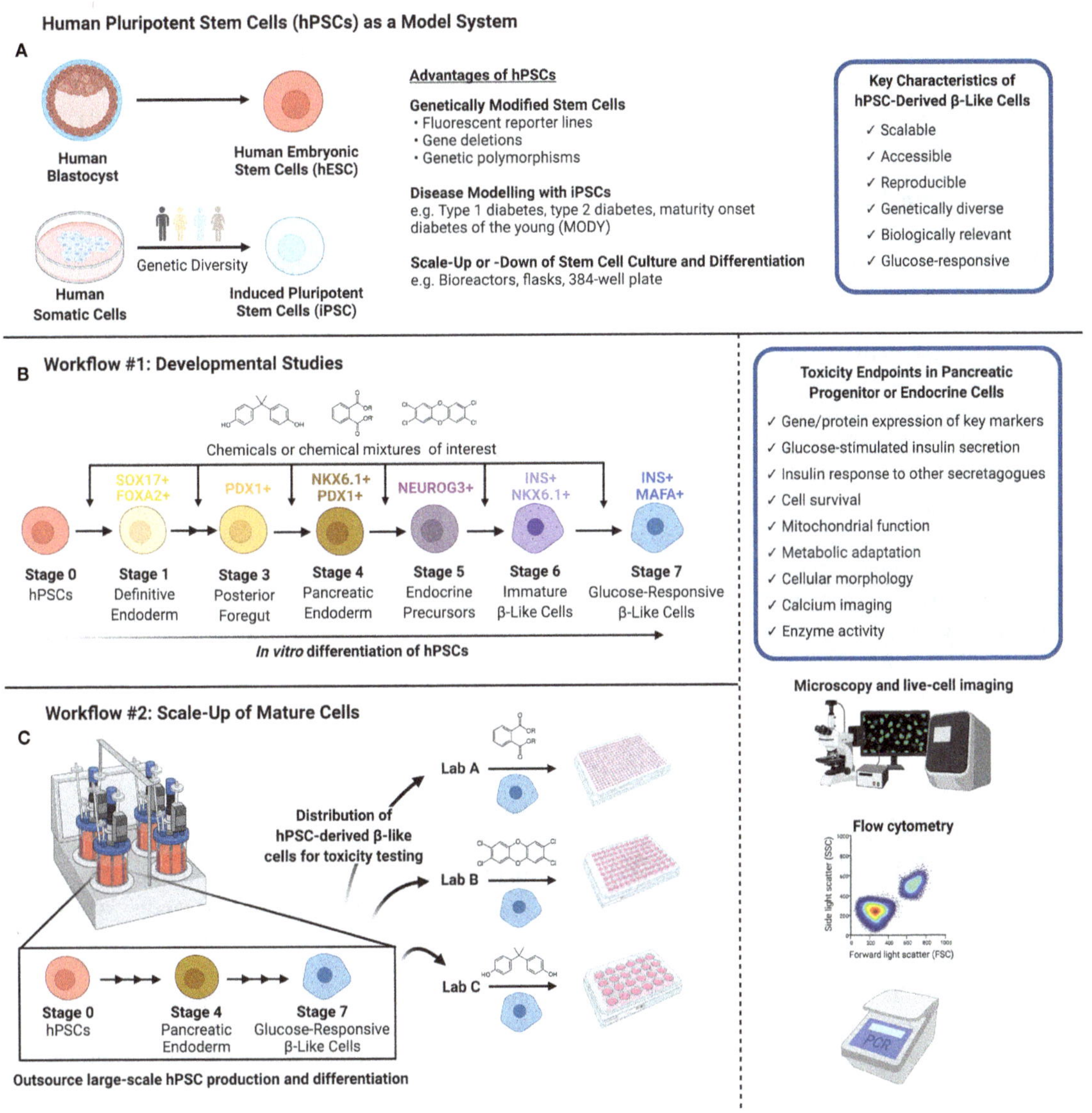

FIGURE 1 | **(A)** Human pluripotent stem cells (hPSCs) can be isolated from the inner cell mass of a human blastocyst (human embryonic stem cells, hESCs) or obtained *via* reprogramming of human somatic cells obtained from genetically diverse donors (induced pluripotent stem cells, iPSCs). hPSCs are versatile in their capacity for genetic modifications and disease modeling and may be scaled up or down to suit a variety of experimental conditions. **(B)** Workflow #1 illustrates how hPSCs may be used to screen chemicals or chemical mixtures of interest throughout pancreas development. hPSCs can be differentiated into pancreatic endoderm using published protocols or commercially available differentiation kits, and further into maturing, glucose-responsive β-like cells. Chemicals can be introduced at different days or stages of differentiation to mimic environmental exposures at different windows of pancreas development. **(C)** Workflow #2 demonstrates the capacity to outsource hPSC expansion and large-scale differentiation, allowing individual labs to conduct toxicity screening of specific chemicals/chemical mixtures using glucose-responsive β-like cells generated in a central location. **(B, C)** We suggest a number of potential toxicity endpoints, such as cell survival, insulin secretion, and mitochondrial function. Common analytical methods include but are not limited to microscopy and live-cell imaging, flow cytometry to quantify cell populations throughout differentiation, and PCR to assess gene expression (Created with BioRender.com).

example by inducing β-cell apoptosis or preventing β-cell expansion, should be flagged as potential MDCs.

Impaired β-Cell Adaptation

Healthy β-cells have fine-tuned mechanisms for adapting to fluctuations in energy supply and insulin demand to maintain appropriate glucose homeostasis (100). Examples of complex compensatory mechanisms include a) regulation of key β-cell transcription factors that control the β-cell transcriptome, b) altered activity or expression of metabolic enzymes to allow for flexibility in nutrient metabolism, and c) increased β-cell proliferation to expand functional mass of β-cells. Failed compensatory insulin secretion and expansion of β-cell mass during insulin resistance are important predictors of diabetes susceptibility (99, 101, 102). Therefore, toxicology studies should consider the ability of β-cells to appropriately compensate under conditions of fasting or insulin resistance. If only direct cellular toxicity is tested without considering interactions with other metabolic challenges, potential MDCs will be overlooked.

Impaired β-Cell Development

The number of pancreatic progenitors present throughout embryonic development is a critical determinant of β-cell mass and pancreas size in adulthood, unlike the closely related endoderm-derived liver which can fully compensate following partial progenitor cell ablation (103). Therefore, an infant born with reduced β-cell mass may have a compromised ability to adapt to metabolic stressors later in life. Additionally, overproduction of insulin at birth caused by inappropriate β-cell expansion or accelerated maturation could lead to the development of insulin resistance and obesity (as described above) (94–96, 104). Therefore, a starting point for predicting long-term adverse metabolic health outcomes following intrauterine chemical exposure would be screening for chemicals that alter β-cell development. We propose that "developmental MDCs" could be prioritized, in part, based on whether they influence the formation of β-cells - either decreasing or increasing numbers - during critical windows of development.

TRADITIONAL MODELS FOR TOXICITY TESTING IN PANCREATIC ENDOCRINE CELLS

In Vivo Rodent Models

In vivo rodent models are an important tool for toxicity testing, but pose a significant technical barrier to high throughput screening (105) and are limited in their ability to predict human outcome. In a largescale study of pharmaceutical toxicity testing, rodents were predictive of human toxicity for only 43% of tested compounds, and demonstrated poor concordance for liver and endocrine toxicity (106). Further, human populations are genetically diverse and exist amongst variable exogenous factors, whereas laboratory animals are genetically uniform and housed within controlled environments to support reproducibility. While *in vivo* testing is necessary for assessing the impact of chemicals on a whole organism rather than just a single cell or tissue type in isolation, there has been a shift towards first prioritizing chemicals using *in vitro* model systems with higher throughput capacity (107, 108).

Immortalized β-Cells

Rodent Cell Lines

Immortalized rodent β-cell lines are robust in culture and highly proliferative, making them a useful tool for large-scale studies. Unfortunately, their replication capacity limits their applicability as a model for human β-cells, which have minimal ability to proliferate (99, 109). In addition, commonly used insulin-secreting rodent β-cell lines, including β-TC-6 (mouse), MIN6 (mouse), and INS-1 (rat) cells, have varying degrees of glucose-responsiveness (110, 111). Immortalized cell lines also tend to be genetically unstable in culture for extended periods of time. Clonal INS-1E cells have higher stability than the INS-1 parental line and maintain their insulin content in passages >40 (112), but both INS-1/INS-1E cells are cultured with the toxic reducing agent, 2-mercaptoethanol, which further limits their biological relevance for toxicity testing.

Despite these limitations, INS-1 823/13 cells were comprehensively evaluated as a potential pollutant-screening system. This cell line was found to be adequately glucose-responsive, but the insulin secretory responses to key control compounds and pollutants deviated substantially from previous reports (113). The authors concluded that INS-1 823/13 cells were lacking key β-cell characteristics and deemed inadequate as a diabetogenic pollutant screening system (113). We reached a similar conclusion after testing immortalized pancreatic endocrine cells for their response to TCDD/dioxin, a POP that acts *via* the aryl hydrocarbon receptor (AhR). In primary mouse and human islets, TCDD significantly upregulated CYP1A1 gene expression and enzyme activity, whereas immortalized β-cell (INS-1, MIN6, β-TC6) and α-cell (α-TC1 and α-TC3) lines showed no evidence of AhR activation by TCDD (73). Our findings confirm that important discrepancies exist in the cellular machinery between primary and immortalized cell lines.

Human Cell Lines

Novel engineered human β-cell lines, EndoC-βH1 and EndoC-βH2, are a useful tool for studying β-cell physiology and drug responses (114, 115). EndoC-βH1 cells were engineered from human fetal pancreatic buds transduced with an SV40LT-expressing lentiviral vector under the control of an insulin promoter. These cell lines are glucose-responsive *in vitro* and have minimal expression of other pancreatic genes (114–116). Their main drawback is the limited capacity for expansion *in vitro*, which is more biologically appropriate than rodent β-cell lines, but less practical. With a doubling time of approximately 7 days, their potential for high-throughput toxicity studies is limited. While proliferation can be stimulated with SerpinA6, STC1, and APOH (114), using compounds that alter cellular physiology is not ideal for toxicity testing. Additionally, as EndoC β-cells are a product of oncogenesis, there is undoubtedly alterations to normal pathways (117).

Primary Isolated Islets

Non-Human Islets

Primary rodent islets are more biologically relevant than immortalized rodent β-cell lines, but there are important distinctions between human and rodent islets that must be considered. For example, the distribution and composition of endocrine cells, vasculature, innervation, proliferation, and insulin secretion all differ between human and rodent islets (118, 119). Pig islets are more similar to human islets (120). Interestingly, islets isolated from juvenile porcine pancreata have greater expression of β-cell-related genes compared to those isolated from adult pigs (121), but are functionally immature and require *in vitro* maturation following isolation (122–124). Further, it is possible to isolate up to 5,000 islets/g juvenile porcine pancreas (122), compared to a typical yield of ~200–400 islets total per mouse pancreas and 600–800 islets per rat pancreas depending on the strain (125–127). Thus, pig islets may be a useful tool for toxicity screening, although species differences will always remain a concern for translation.

Human Islets

Human islets are currently the gold standard for a physiologically relevant model to study the endocrine pancreas *in vitro* due to their cellular composition, human origin, and genetic diversity. Human islets are harvested from deceased organ donors and great strides have been made to ensure that high quality donor islets are broadly available for research (128, 129). However, even with the highest quality isolation procedures, the pancreas begins to autodigest after death, resulting in decreased cell viability and sample quality (119). Human islets have a limited functional *in vitro* lifespan with current tissue culture protocols, although advances in the field are ongoing. For example, islets cultured on specific matrices maintain glucose-stimulated insulin secretion for at least 7 days in culture (130). The number of purified islets per donor also varies; while an average healthy adult pancreas houses over 3 million islets (118), between 200,000 and 500,000 islets typically remain post-purification (131). This, coupled with the limited proliferation of human β-cells, presents a critical barrier to scalability and longevity for toxicology testing. Therefore, human islets are an excellent resource for *in vitro* chemical testing at a smaller scale, wherein endpoints such as glucose-stimulated insulin secretion, islet morphology, mitochondrial function, and gene expression can be assessed in biologically diverse organ donors. Numerous factors, such as donor sex, age, and body mass index, will influence islet function *ex vivo* and thus impact biological reproducibility. However, the genetic and environmental diversity of human organ donors (132, 133) also offer a unique opportunity for toxicity testing.

STEM CELLS ARE A UNIQUE TOOL FOR PANCREAS TOXICOLOGY STUDIES

HPSC culture offers a unique *in vitro* solution to address the need for high-throughput screening of environmental toxins in a variety of biologically relevant mature cell types, as well as in differentiating or "developing" immature progenitor cells. HPSCs can be obtained from either the inner cell mass of a human blastocyst, termed human embryonic stem cells (hESCs), or from human somatic cells that have been reprogrammed to a pluripotent state, termed induced pluripotent stem cells (iPSCs) (**Figure 1A**).

Remarkable progress has been made over the past decade unraveling the developmental cues involved in islet cell formation. We now have robust step-wise differentiation protocols that mimic the key fate decisions for directing hPSCs into pancreatic endocrine cells using small molecules and growth factors *in vitro* (134–139). These differentiation protocols efficiently guide hPSCs towards pancreatic endoderm cells ($PDX1^+/NKX6.1^+$) in four "stages", followed by commitment to the pancreatic endocrine lineage ($NEUROG3^+$), then insulin-secreting endocrine cells ($INS^+/NKX6.1^+$), and finally to β-like cells capable of glucose-induced insulin secretion ($INS^+/MAFA^+/UCN3^+$) (3) (**Figure 1B**). The challenge in recent years has been understanding the final stages of human β-cell maturation so we can generate fully mature β-cells with a rapid and robust insulin secretory response to various secretagogues. Despite these limitations, we believe that hPSCs are an excellent tool for studying adverse effects of environmental contaminants both during pancreas development and in adult pancreatic endocrine cells. Indeed, a recent study by Zhou et al. in *Nature Communications* beautifully demonstrated the diverse and powerful applicability of hPSCs for high-content screening of potential β-cell toxins, exploring gene-environment interactions, and comparing toxicity in diverse cell types (140). The authors differentiated hESCs into INS^+ cells in a 384-well plate format and screened a U.S. Environmental Protection Agency (EPA) ToxCast library of ~ 2,000 compounds for "hits" that impaired survival of INS^+ cells (140). Using this study as an example, we highlight the numerous benefits of using hPSCs, whether hESCs or iPSCs, for exploring MDC toxicity.

Flexibility to Model Developing or Adult Cells

HPSCs offer a flexible model to test for MDCs that impact either the early formation of β-cells during fetal development or the function and survival of adult insulin-secreting β-like cells. Depending on the research question, we propose two different workflow approaches. For developmental studies (**Figure 1B**, Workflow #1), environmental toxin(s) can be introduced to differentiating hPSCs at critical days or "stages" of differentiation. Thus, a critical aspect of this workflow is establishing hPSC differentiation protocols within the toxicology lab conducting chemical testing. The impact of toxins can be assessed by measuring key pancreatic cell markers by flow cytometry, image-based analysis, qPCR, or other techniques that are amenable to high throughput analysis. For example, the proportion of cells expressing markers of pancreatic commitment (% $PDX1^+$), pancreatic endoderm (% $PDX1^+/NKX6.1^+$), and induction of the endocrine program (% $NEUROG3^+$) are excellent benchmarks for early stages of differentiation. At later stages, the proportion of cells that acquire insulin (% $INS^+/NKX6.1^+$) is as an indicator of commitment to the β-cell lineage, and subsequently the proportion of INS^+ cells co-expressing critical β-cell markers such as MAFA is an important indicator of β-cell maturity. Fluorescent reporter hPSC lines generated by genome editing – for example, NEUROG3-EGFP or INS-GFP hESCs – will be particularly useful for efficient image-based screening or high-content flow cytometry applications to identify MDCs that disrupt the formation of key pancreatic cell populations (141–144).

The workflow and endpoints for toxicity studies in adult cells could differ considerably from developmental studies (**Figure 1C**, Workflow #2). First, it is feasible for hPSC-derived β-like cells to be mass-produced in large quantities at a central location to generate a reproducible starting point for toxicology screening studies. Once hPSC-derived β-like cells are validated, they can be distributed to toxicology laboratories for testing of individual chemicals or complex chemical mixtures. This is important because it separates the need for toxicity testing capacity and stem cell differentiation expertise to be housed within the same lab. As with the progenitor cell model, there are numerous potential outcomes that could be assessed in a high-throughput screening platform, such as the expression of key β-cell markers using live-cell imaging or flow cytometry, β-cell survival as in Zhou et al. (140), and basal or glucose-induced insulin secretion. Any of the outcome measures described in the section on "Toxicity Testing in Pancreatic β-cells" could be assessed in hPSC-derived β-like cells, although not necessarily in a high-content format.

It is important to recognize that current differentiation protocols generate human β-like cells with a blunted insulin secretory response to glucose compared to primary human islets (134–139). For the purpose of identifying MDCs that cause β-cell dysfunction or apoptosis, we propose that generating fully mature human β-cells *in vitro* may not be a necessary milestone. Instead, the benefits of a large-scale source of expandable stem cells that can generate large quantities of moderately glucose-responsive insulin-secreting cells outweighs the downside of working with a slightly immature β-like cell. This has certainly proven true for toxicity studies in other cell types, such as cardiomyocytes, where differentiation protocols currently generate immature cardiomyocytes, but recapitulate sufficient features of adult cells to study adverse drug reactions in specific aspects of cardiotoxicity (145–148).

Scalability and Reproducibility

The scalability of hPSCs is a significant advantage for high-content screening. Importantly, hPSCs share the proliferative advantage of immortalized β-cells, but subsequently lose this capacity as they differentiate into pancreatic lineage cells (139). The highly proliferative nature of hPSCs allows them to be substantially expanded before differentiation, 50–100 fold per week, particularly when grown in suspension format (149). Large batches of hPSCs can then be differentiated into a mass-produced cell product, which can be carefully validated with well-defined QA/QC protocols before being frozen down and distributed for toxicity testing (**Figure 1C**, Workflow #2). This is similar to the model proposed by the diabetes cell therapy field for mass-production of a GMP-grade cell product for transplantation (3, 149). Alternatively, more modest scale-up approaches can be established within the same lab that will perform toxicity endpoint assessments (**Figure 1B**, Workflow #1). For example, Zhou and colleagues expanded hESCs in a more traditional adherent format with Matrigel-coated plates before seeding dissociated cells into 384-well plates for pancreatic differentiation and chemical screening (140). The incredible flexibility to both scale-up hPSC production and differentiation or to miniaturize pancreatic differentiation is an important benefit of using hPSCs for toxicity studies.

Unique Capacity for Disease Modeling

Stem cells offer remarkable capacity for disease modeling through both the natural genetic diversity of iPSCs (150) and the ability to create isogenic hPSC lines using genome editing (151). The use of hPSCs for disease modeling in diabetes has been reviewed elsewhere (152), but here, we briefly discuss the benefits of toxicity testing in human β-cells with diverse genetic backgrounds. There is much to be learned from comparing the impact of environmental contaminants on β-like cells generated using iPSCs from a spectrum of patients with different types of diabetes (T1D, T2D, maturity onset diabetes of the young (MODY), or neonatal diabetes) or known genetic risk factors for diabetes (153–160), relative to iPSCs derived from control subjects. One particularly exciting avenue to explore in the context of T1D is how environmental toxins influence immune interactions between iPSC-derived β-cells and autologous immune cells from the same donor (159). Zhou and colleagues also demonstrated the potential for using iPSCs to explore mechanisms of toxicity (140). They used 10 different iPSC lines with heterogeneous expression of a phase 2 xenobiotic metabolism enzyme, *GSTT1*, and found that pesticide-induced INS^+ cell death was significantly higher in lines lacking *GSTT1* compared to those with at least one copy of *GSTT1* (140). Their results were also validated in isogenic hESC lines with *GSTT1* deletion by CRISPR-based genome editing; INS^+ cells generated from $GSTT1^{-/-}$ hESCs were more susceptible to pesticide-induced cell death than INS^+ cells from wildtype hESCs. Importantly, with the advent of CRISPR-Cas9 technology, modifying the genome of hPSCs has become broadly accessible and the number of gene-edited hPSC lines that effectively recapitulate different aspects of diabetes-related phenotypes is increasing rapidly (140, 151, 155, 161–164).

One final consideration for disease modeling is that despite being reprogrammed back to their embryonic/pluripotent state, iPSCs retain DNA methylation marks, lineage bias, and other memory of previous environmental exposures (165). For this reason, there is a strong argument for developmental models of pancreatic toxicity being limited to hESCs rather than iPSCs. On the other hand, the genetic variability of iPSCs, combined with the ability to create targeted genome-edited hPSC lines with isogenic wildtype controls, should be harnessed to explore the biological diversity of gene-environment interactions in adult β-like cells.

Diversity of Human Cell Types

Another unique advantage of hPSCs is their ability to be directed into diverse cell types. For example, the toxicology field is already using hPSCs to test for adverse drug reactions in iPSC-derived hepatocyte-like cells to model hepatoxicity (166–168) and iPSC-derived cardiomyocytes to model cardiotoxicity (145–147, 168–172). While our Perspective focused on the application of hPSCs for toxicity testing in pancreatic lineage cells specifically, there is immense value in a more integrated approach to screen for MDCs that adversely impact different metabolic target tissues, all derived from the same hPSC source. For example, Zhou and colleagues differentiated hESCs into $CD29^+/CD73^+$ mesenchymal stem cells, $CTNT^+$ cardiomyocytes, $A1AT^+$ hepatocytes, and HuC/D^+ neurons (140). They found that much like hESC-derived INS^+ cells, HuC/D^+ neurons were also highly susceptible to pesticide-induced cell death, suggesting that the pesticide flagged in their high-content screening could be involved in the pathogenesis of both diabetes and Parkinson's disease. An even more complex, but intriguing application of hPSCs is the potential to develop multi-organ systems in a microfluidic device (173) or other platform containing numerous hPSC-derived metabolic tissues such as liver, adipose, and β-cells to determine how environmental contaminants influence metabolic tissue cross-talk.

CONCLUSION

Despite the critical importance of pancreatic endocrine cells for maintaining metabolic homeostasis, the pancreas has not traditionally been studied as a key target tissue of chemical toxicity. Given the metabolic-disrupting nature of many environmental pollutants, we propose that islet toxicity should be considered a key toxicological endpoint. With the staggering number of poorly studied chemicals in commerce, physiologically relevant models that can be scaled up for efficient chemical screening are urgently needed. Human stem cells offer a unique solution to many of the limitations posed by other *in vitro* model systems of pancreatic endocrine cells. Most importantly, hPSCs are scalable and amenable to high-throughput screening for assessing the impact of environmental contaminants on either adult β-like cells or critical windows of pancreas development.

AUTHOR CONTRIBUTIONS

EM and JB conceived the review topic and wrote the manuscript. All authors contributed to the article and approved the submitted version.

ACKNOWLEDGMENTS

We thank Myriam Hoyeck for her valuable insight during the review of this manuscript.

REFERENCES

1. International Diabetes Federation. *IDF Diabetes Atlas, 9th Edition.* (2019).
2. Rorsman P, Braun M. Regulation of insulin secretion in human pancreatic islets. *Annu Rev Physiol* (2013) 75:155–79. doi: 10.1146/annurev-physiol-030212-183754
3. Bruin JE, Rezania A, Kieffer TJ. Replacing and safeguarding pancreatic beta cells for diabetes. *Sci Transl Med* (2015) 7:316ps23. doi: 10.1126/scitranslmed.aaa9359
4. Kahn SE, Zraika S, Utzschneider KM, Hull RL. The beta cell lesion in type 2 diabetes: there has to be a primary functional abnormality. *Diabetologia* (2009) 52:1003–12. doi: 10.1007/s00125-009-1321-z
5. Butler AE, Janson J, Bonner-Weir S, Ritzel R, Rizza RA, Butler PC. Beta-cell deficit and increased beta-cell apoptosis in humans with type 2 diabetes. *Diabetes* (2003) 52:102–10. doi: 10.2337/diabetes.52.1.102
6. Chen C, Cohrs CM, Stertmann J, Bozsak R, Speier S. Human beta cell mass and function in diabetes: Recent advances in knowledge and technologies to understand disease pathogenesis. *Mol Metab* (2017) 6:943–57. doi: 10.1016/j.molmet.2017.06.019
7. Gaulton KJ, Ferreira T, Lee Y, Raimondo A, Magi R, Reschen ME, et al. Genetic fine mapping and genomic annotation defines causal mechanisms at type 2 diabetes susceptibility loci. *Nat Genet* (2015) 47:1415–25. doi: 10.1038/ng.3437
8. DIAbetes Genetics Replication Meta-analysis (DIAGRAM) Consortium. Asian Genetic Epidemiology Network Type 2 Diabetes (AGEN-T2D) Consortium, South Asian Type 2 Diabetes (SAT2D) Consortium, Mexican American Type 2 Diabetes (MAT2D) Consortium, Type 2 Diabetes Genetic Exploration by Next-generation sequencing in multi-Ethnic Samples (T2D-GENES) Consortium. Genome-wide trans-ancestry meta-analysis provides insight into the genetic architecture of type 2 diabetes susceptibility. *Nat Genet* (2014) 46:234–44. doi: 10.1038/ng.2897
9. Franks PW, McCarthy MI. Exposing the exposures responsible for type 2 diabetes and obesity. *Science* (2016) 354:69–73. doi: 10.1126/science.aaf5094
10. Maher B. Personal genomes: The case of the missing heritability. *Nature* (2008) 456:18–21. doi: 10.1038/456018a
11. Bertazzi PA, Consonni D, Bachetti S, Rubagotti M, Baccarelli A, Zocchetti C, et al. Health effects of dioxin exposure: a 20-year mortality study. *Am J Epidemiol* (2001) 153:1031–44. doi: 10.1093/aje/153.11.1031
12. Wang SL, Tsai PC, Yang CY, Guo YL. Increased risk of diabetes and polychlorinated biphenyls and dioxins: a 24-year follow-up study of the Yucheng cohort. *Diabetes Care* (2008) 31:1574–9. doi: 10.2337/dc07-2449
13. Vena J, Boffetta P, Becher H, Benn T, Bueno-de-Mesquita HB, Coggon D, et al. Exposure to dioxin and nonneoplastic mortality in the expanded IARC international cohort study of phenoxy herbicide and chlorophenol production workers and sprayers. *Environ Health Perspect* (1998) 106 (Suppl 2):645–53. doi: 10.1289/ehp.98106645
14. Calvert GM, Sweeney MH, Deddens J, Wall DK. Evaluation of diabetes mellitus, serum glucose, and thyroid function among United States workers exposed to 2,3,7,8-tetrachlorodibenzo-p-dioxin. *Occup Environ Med* (1999) 56:270–6. doi: 10.1136/oem.56.4.270
15. Yi SW, Hong JS, Ohrr H, Yi JJ. Agent Orange exposure and disease prevalence in Korean Vietnam veterans: the Korean veterans health study. *Environ Res* (2014) 133:56–65. doi: 10.1016/j.envres.2014.04.027
16. Kim JS, Lim HS, Cho SI, Cheong HK, Lim MK. Impact of Agent Orange exposure among Korean Vietnam veterans. *Ind Health* (2003) 41:149–57. doi: 10.2486/indhealth.41.149
17. Henriksen GL, Ketchum NS, Michalek JE, Swaby JA. Serum dioxin and diabetes mellitus in veterans of Operation Ranch Hand. *Epidemiology* (1997) 8:252–8. doi: 10.1097/00001648-199705000-00005
18. Longnecker MP, Michalek JE. Serum dioxin level in relation to diabetes mellitus among Air Force veterans with background levels of exposure. *Epidemiology* (2000) 11:44–8. doi: 10.1097/00001648-200001000-00010
19. Kuo CC, Moon K, Thayer KA, Navas-Acien A. Environmental chemicals and type 2 diabetes: an updated systematic review of the epidemiologic evidence. *Curr Diabetes Rep* (2013) 13:831–49. doi: 10.1007/s11892-013-0432-6
20. Porta M. Persistent organic pollutants and the burden of diabetes. *Lancet* (2006) 368:558–9. doi: 10.1016/S0140-6736(06)69174-5

21. Thayer KA, Heindel JJ, Bucher JR, Gallo MA. Role of environmental chemicals in diabetes and obesity: a National Toxicology Program workshop review. *Environ Health Perspect* (2012) 120:779–89. doi: 10.1289/ehp.1104597
22. Lee DH, Lee IK, Song K, Steffes M, Toscano W, Baker BA, et al. A strong dose-response relation between serum concentrations of persistent organic pollutants and diabetes: results from the National Health and Examination Survey 1999-2002. *Diabetes Care* (2006) 29:1638–44. doi: 10.2337/dc06-0543
23. Pal S, Blais JM, Robidoux MA, Haman F, Krummel E, Seabert TA, et al. The association of type 2 diabetes and insulin resistance/secretion with persistent organic pollutants in two First Nations communities in northern Ontario. *Diabetes Metab* (2013) 39:497–504. doi: 10.1016/j.diabet.2013.01.006
24. Everett CJ, Frithsen IL, Diaz VA, Koopman RJ, Simpson WMJr., Mainous AG. 3rd. Association of a polychlorinated dibenzo-p-dioxin, a polychlorinated biphenyl, and DDT with diabetes in the 1999-2002 National Health and Nutrition Examination Survey. *Environ Res* (2007) 103:413–8. doi: 10.1016/j.envres.2006.11.002
25. Lee DH, Porta M, Jacobs DRJr., Vandenberg LN. Chlorinated persistent organic pollutants, obesity, and type 2 diabetes. *Endocr Rev* (2014) 35:557–601. doi: 10.1210/er.2013-1084
26. Taylor KW, Novak RF, Anderson HA, Birnbaum LS, Blystone C, Devito M, et al. Evaluation of the association between persistent organic pollutants (POPs) and diabetes in epidemiological studies: a national toxicology program workshop review. *Environ Health Perspect* (2013) 121:774–83. doi: 10.1289/ehp.1205502
27. Roh E, Kwak SH, Jung HS, Cho YM, Pak YK, Park KS, et al. Serum aryl hydrocarbon receptor ligand activity is associated with insulin resistance and resulting type 2 diabetes. *Acta Diabetol* (2015) 52:489–95. doi: 10.1007/s00592-014-0674-z
28. Dirinck EL, Dirtu AC, Govindan M, Covaci A, Van Gaal LF, Jorens PG. Exposure to persistent organic pollutants: relationship with abnormal glucose metabolism and visceral adiposity. *Diabetes Care* (2014) 37:1951–8. doi: 10.2337/dc13-2329
29. Jorgensen ME, Borch-Johnsen K, Bjerregaard P. A cross-sectional study of the association between persistent organic pollutants and glucose intolerance among Greenland Inuit. *Diabetologia* (2008) 51:1416–22. doi: 10.1007/s00125-008-1066-0
30. Lee YM, Ha CM, Kim SA, Thoudam T, Yoon YR, Kim DJ, et al. Low-Dose Persistent Organic Pollutants Impair Insulin Secretory Function of Pancreatic beta-Cells: Human and In Vitro Evidence. *Diabetes* (2017) 66:2669–80. doi: 10.2337/db17-0188
31. Government of Canada. *The Canadian Environmental Protection Act, 1999.* (1999). Available at: https://www.canada.ca/en/environment-climate-change/services/canadian-environmental-protection-act-registry/publications/canadian-environmental-protection-act-1999.html.
32. United States Government Accountability Office. Toxic substances: EPA has increased efforts to assess and control chemicals but could strengthen its approach. GAO-13-249. (2013). Available at: https://www.gao.gov/products/GAO-13-249GAO-13-249.
33. Schug TT, Janesick A, Blumberg B, Heindel JJ. Endocrine disrupting chemicals and disease susceptibility. *J Steroid Biochem Mol Biol* (2011) 127:204–15. doi: 10.1016/j.jsbmb.2011.08.007
34. Manzetti S, van der Spoel ER, van der Spoel D. Chemical properties, environmental fate, and degradation of seven classes of pollutants. *Chem Res Toxicol* (2014) 27:713–37. doi: 10.1021/tx500014w
35. Howdeshell KL, Hotchkiss AK, Thayer KA, Vandenbergh JG, vom Saal FS. Exposure to bisphenol A advances puberty. *Nature* (1999) 401:763–4. doi: 10.1038/44517
36. Lecomte S, Habauzit D, Charlier TD, Pakdel F. Emerging Estrogenic Pollutants in the Aquatic Environment and Breast Cancer. *Genes (Basel)* (2017) 8:1–21. doi: 10.3390/genes8090229
37. Gao H, Yang BJ, Li N, Feng LM, Shi XY, Zhao WH, et al. and hormone-associated cancers: current progress and perspectives. *Med (Baltimore)* (2015) 94:e211. doi: 10.1097/MD.0000000000000211
38. Nadal A, Quesada I, Tuduri E, Nogueiras R, Alonso-Magdalena P. Endocrine-disrupting chemicals and the regulation of energy balance. *Nat Rev Endocrinol* (2017) 13:536–46. doi: 10.1038/nrendo.2017.51
39. Meeker JD. Exposure to environmental endocrine disruptors and child development. *Arch Pediatr Adolesc Med* (2012) 166:E1–7. doi: 10.1001/archpediatrics.2012.241
40. Diamanti-Kandarakis E, Bourguignon JP, Giudice LC, Hauser R, Prins GS, Soto AM, et al. Endocrine-disrupting chemicals: an Endocrine Society scientific statement. *Endocr Rev* (2009) 30:293–342. doi: 10.1210/er.2009-0002
41. Mimoto MS, Nadal A, Sargis RM. Polluted Pathways: Mechanisms of Metabolic Disruption by Endocrine Disrupting Chemicals. *Curr Environ Health Rep* (2017) 4:208–22. doi: 10.1007/s40572-017-0137-0
42. Heindel JJ, Blumberg B, Cave M, Machtinger R, Mantovani A, Mendez MA, et al. Metabolism disrupting chemicals and metabolic disorders. *Reprod Toxicol* (2017) 68:3–33. doi: 10.1016/j.reprotox.2016.10.001
43. Sargis RM, Heindel JJ, Padmanabhan V. Interventions to Address Environmental Metabolism-Disrupting Chemicals: Changing the Narrative to Empower Action to Restore Metabolic Health. *Front Endocrinol (Lausanne)* (2019) 10:33. doi: 10.3389/fendo.2019.00033
44. Jing R, Fusi S, Kjellerup BV. Remediation of Polychlorinated Biphenyls (PCBs) in Contaminated Soils and Sediment: State of Knowledge and Perspectives. *Front Environ Sci* (2018) 6:79. doi: 10.3389/fenvs.2018.00079
45. U.S. Department of Health and Human Services, Centers for Disease Control and Prevention. Fourth National Report on Human Exposure to Environmental Chemicals, Updated Tables Volume 1 and 2. (2019). Available at: https://www.cdc.gov/exposurereport/index.html.
46. Government of Canada. Uses of human biomonitoring in data risk assessment. (2018). Available at: https://www.canada.ca/en/health-canada/services/chemical-substances/fact-sheets/human-biomonitoring-data-risk-assessment.html.
47. Centers for Disease Control and Prevention. National biomonitoring program. (2017). Available at: https://www.cdc.gov/biomonitoring/about.html.
48. Dewailly E, Ayotte P, Pereg D, Dery S, Dallaire R, et al. Exposure to heavy metals and persistant organic pollutants in Nunavik: the Nunavik Health Study (Human Health). S. Smith and J. Stow (Eds) Synopsis of Research conducted under the 2005–2006 Northern Contaminants Program Indian and Northern Affairs Canada (2006) 44–64. Available at: http://pubs.aina.ucalgary.ca/ncp/60582.pdf.
49. Ashraf MA. Persistent organic pollutants (POPs): a global issue, a global challenge. *Environ Sci Pollut Res Int* (2017) 24:4223–7. doi: 10.1007/s11356-015-5225-9
50. Vandenberg LN, Chahoud I, Heindel JJ, Padmanabhan V, Paumgartten FJ, Schoenfelder G. Urinary, circulating, and tissue biomonitoring studies indicate widespread exposure to bisphenol A. *Environ Health Perspect* (2010) 118:1055–70. doi: 10.1289/ehp.0901716
51. Meeker JD, Sathyanarayana S, Swan SH. Phthalates and other additives in plastics: human exposure and associated health outcomes. *Philos Trans R Soc Lond B Biol Sci* (2009) 364:2097–113. doi: 10.1098/rstb.2008.0268
52. Main KM, Mortensen GK, Kaleva MM, Boisen KA, Damgaard IN, Chellakooty M, et al. Human breast milk contamination with phthalates and alterations of endogenous reproductive hormones in infants three months of age. *Environ Health Perspect* (2006) 114:270–6. doi: 10.1289/ehp.8075
53. Zoeller RT, Vandenberg LN. Assessing dose-response relationships for endocrine disrupting chemicals (EDCs): a focus on non-monotonicity. *Environ Health* (2015) 14:42. doi: 10.1186/s12940-015-0029-4
54. Vandenberg LN, Colborn T, Hayes TB, Heindel JJ, Jacobs DRJr., Lee DH, et al. Hormones and endocrine-disrupting chemicals: low-dose effects and nonmonotonic dose responses. *Endocr Rev* (2012) 33:378–455. doi: 10.1210/er.2011-1050
55. Barker DJ. The origins of the developmental origins theory. *J Intern Med* (2007) 261:412–7. doi: 10.1111/j.1365-2796.2007.01809.x
56. Vickers MH. Developmental programming and transgenerational transmission of obesity. *Ann Nutr Metab* (2014) 64(Suppl 1):26–34. doi: 10.1159/000360506
57. Barker DJ, Hales CN, Fall CH, Osmond C, Phipps K, Clark PM. Type 2 (non-insulin-dependent) diabetes mellitus, hypertension and hyperlipidaemia (syndrome X): relation to reduced fetal growth. *Diabetologia* (1993) 36:62–7. doi: 10.1007/BF00399095

58. Eriksson JG, Osmond C, Kajantie E, Forsen TJ, Barker DJ. Patterns of growth among children who later develop type 2 diabetes or its risk factors. *Diabetologia* (2006) 49:2853–8. doi: 10.1007/s00125-006-0459-1
59. Ouidir M, Buck Louis GM, Kanner J, Grantz KL, Zhang C, Sundaram R, et al. Association of Maternal Exposure to Persistent Organic Pollutants in Early Pregnancy With Fetal Growth. *JAMA Pediatr* (2019) 174(2):149–61. doi: 10.1001/jamapediatrics.2019.5104
60. Papalou O, Kandaraki EA, Papadakis G, Diamanti-Kandarakis E. Endocrine Disrupting Chemicals: An Occult Mediator of Metabolic Disease. *Front Endocrinol (Lausanne)* (2019) 10:112. doi: 10.3389/fendo.2019.00112
61. Kappil MA, Li Q, Li A, Dassanayake PS, Xia Y, Nanes JA, et al. In utero exposures to environmental organic pollutants disrupt epigenetic marks linked to fetoplacental development. *Environ Epigenet* (2016) 2:1–7. doi: 10.1093/eep/dvv013
62. Inadera H. Developmental origins of obesity and type 2 diabetes: molecular aspects and role of chemicals. *Environ Health Prev Med* (2013) 18:185–97. doi: 10.1007/s12199-013-0328-8
63. Iszatt N, Stigum H, Govarts E, Murinova LP, Schoeters G, Trnovec T, et al. Perinatal exposure to dioxins and dioxin-like compounds and infant growth and body mass index at seven years: A pooled analysis of three European birth cohorts. *Environ Int* (2016) 94:399–407. doi: 10.1016/j.envint.2016.04.040
64. Smink A, Ribas-Fito N, Garcia R, Torrent M, Mendez MA, Grimalt JO, et al. Exposure to hexachlorobenzene during pregnancy increases the risk of overweight in children aged 6 years. *Acta Paediatr* (2008) 97:1465–9. doi: 10.1111/j.1651-2227.2008.00937.x
65. Karmaus W, Osuch JR, Eneli I, Mudd LM, Zhang J, Mikucki D, et al. Maternal levels of dichlorodiphenyl-dichloroethylene (DDE) may increase weight and body mass index in adult female offspring. *Occup Environ Med* (2009) 66:143–9. doi: 10.1136/oem.2008.041921
66. Verhulst SL, Nelen V, Hond ED, Koppen G, Beunckens C, Vael C, et al. Intrauterine exposure to environmental pollutants and body mass index during the first 3 years of life. *Environ Health Perspect* (2009) 117:122–6. doi: 10.1289/ehp.0800003
67. Lee WC, Fisher M, Davis K, Arbuckle TE, Sinha SK. Identification of chemical mixtures to which Canadian pregnant women are exposed: The MIREC Study. *Environ Int* (2017) 99:321–30. doi: 10.1016/j.envint.2016.12.015
68. Rawn DFK, Sadler AR, Casey VA, Breton F, Sun WF, Arbuckle TE, et al. Dioxins/furans and PCBs in Canadian human milk: 2008-2011. *Sci Total Environ* (2017) 595:269–78. doi: 10.1016/j.scitotenv.2017.03.157
69. Fisher M, Arbuckle TE, Liang CL, LeBlanc A, Gaudreau E, Foster WG, et al. Concentrations of persistent organic pollutants in maternal and cord blood from the maternal-infant research on environmental chemicals (MIREC) cohort study. *Environ Health* (2016) 15:59. doi: 10.1186/s12940-016-0143-y
70. Arbuckle TE, Fraser WD, Fisher M, Davis K, Liang CL, Lupien N, et al. Cohort profile: the maternal-infant research on environmental chemicals research platform. *Paediatr Perinat Epidemiol* (2013) 27:415–25. doi: 10.1111/ppe.12061
71. Marraudino M, Bonaldo B, Farinetti A, Panzica G, Ponti G, Gotti S. Metabolism Disrupting Chemicals and Alteration of Neuroendocrine Circuits Controlling Food Intake and Energy Metabolism. *Front Endocrinol (Lausanne)* (2018) 9:766. doi: 10.3389/fendo.2018.00766
72. Diliberto JJ, Akubue PI, Luebke RW, Birnbaum LS. Dose-response relationships of tissue distribution and induction of CYP1A1 and CYP1A2 enzymatic activities following acute exposure to 2,3,7,8-tetrachlorodibenzo-p-dioxin (TCDD) in mice. *Toxicol Appl Pharmacol* (1995) 130:197–208. doi: 10.1006/taap.1995.1025
73. Ibrahim M, MacFarlane EM, Matteo G, Hoyeck MP, Rick KRC, Farokhi SB, et al. Functional cytochrome P450 1a enzymes are induced in mouse and human islets following pollutant exposure. *Diabetologia* (2020) 63:162–78. doi: 10.1007/s00125-019-05035-0
74. Hoyeck MP, Merhi RC, Blair HL, Spencer CD, Payant MA, Martin Alfonso DI, et al. Female mice exposed to low doses of dioxin during pregnancy and lactation have increased susceptibility to diet-induced obesity and diabetes. *Mol Metab* (2020) 101104:1–4. doi: 10.1016/j.molmet.2020.101104
75. Soriano S, Alonso-Magdalena P, Garcia-Arevalo M, Novials A, Muhammed SJ, Salehi A, et al. Rapid insulinotropic action of low doses of bisphenol-A on mouse and human islets of Langerhans: role of estrogen receptor beta. *PLoS One* (2012) 7:e31109. doi: 10.1371/journal.pone.0031109
76. Villar-Pazos S, Martinez-Pinna J, Castellano-Munoz M, Alonso-Magdalena P, Marroqui L, Quesada I, et al. Molecular mechanisms involved in the non-monotonic effect of bisphenol-a on ca2+ entry in mouse pancreatic beta-cells. *Sci Rep* (2017) 7:11770. doi: 10.1038/s41598-017-11995-3
77. Martinez-Pinna J, Marroqui L, Hmadcha A, Lopez-Beas J, Soriano S, Villar-Pazos S, et al. Bisphenol-S and Bisphenol-F alter mouse pancreatic β-cell ion channel expression and activity and insulin release through an estrogen receptor ERβ mediated pathway. *Chemosphere* (2020) 129051. doi: 10.1016/j.chemosphere.2020.129051
78. Mailloux R, Fu A, Florian M, Petrov I, Chen Q, Coughlan MC, et al. A Northern contaminant mixture impairs pancreas function in obese and lean JCR rats and inhibits insulin secretion in MIN6 cells. *Toxicology* (2015) 334:81–93. doi: 10.1016/j.tox.2015.06.001
79. Novelli M, Piaggi S, De Tata V. 2,3,7,8-Tetrachlorodibenzo-p-dioxin-induced impairment of glucose-stimulated insulin secretion in isolated rat pancreatic islets. *Toxicol Lett* (2005) 156:307–14. doi: 10.1016/j.toxlet.2004.12.004
80. Kurita H, Yoshioka W, Nishimura N, Kubota N, Kadowaki T, Tohyama C. Aryl hydrocarbon receptor-mediated effects of TCDD on glucose-stimulated insulin secretion in mice. *J Appl Toxicol* (2009) 29:689–94. doi: 10.1002/jat.1459
81. Hoyeck M, Blair H, Ibrahim M, Solanky S, Elsawy M, Matteo G, et al. Long-term metabolic consequences of acute dioxin exposure differ between male and female mice. *Sci Reports* (2020) 10:1448. doi: 10.1038/s41598-020-57973-0
82. Matteo G, Hoyeck M, Blair H, Rick KRC, Williams A, Buick JK, et al. Chronic low-dose dioxin exposure accelerates high fat diet-induced hyperglycemia in female mice. *BioRxiv* (2020) 2020.09.12.294587. doi: 10.1101/2020.09.12.294587
83. Fu J, Githaka JM, Dai X, Plummer G, Suzuki K, Spigelman AF, et al. A glucose-dependent spatial patterning of exocytosis in human beta-cells is disrupted in type 2 diabetes. *JCI Insight* (2019) 5:e127896. doi: 10.1101/534669
84. Alejandro EU, Gregg B, Blandino-Rosano M, Cras-Meneur C, Bernal-Mizrachi E. Natural history of beta-cell adaptation and failure in type 2 diabetes. *Mol Aspects Med* (2015) 42:19–41. doi: 10.1016/j.mam.2014.12.002
85. Camunas-Soler J, Dai XQ, Hang Y, Bautista A, Lyon J, Suzuki K, et al. Patch-Seq Links Single-Cell Transcriptomes to Human Islet Dysfunction in Diabetes. *Cell Metab* (2020) 31:1017–31.e4. doi: 10.1016/j.cmet.2020.04.005
86. Del Guerra S, Lupi R, Marselli L, Masini M, Bugliani M, Sbrana S, et al. Functional and molecular defects of pancreatic islets in human type 2 diabetes. *Diabetes* (2005) 54:727–35. doi: 10.2337/diabetes.54.3.727
87. Marchetti P, Del Guerra S, Marselli L, Lupi R, Masini M, Pollera M, et al. Pancreatic islets from type 2 diabetic patients have functional defects and increased apoptosis that are ameliorated by metformin. *J Clin Endocrinol Metab* (2004) 89:5535–41. doi: 10.1210/jc.2004-0150
88. Scott FW, Pound LD, Patrick C, Eberhard CE, Crookshank JA. Where genes meet environment-integrating the role of gut luminal contents, immunity and pancreas in type 1 diabetes. *Transl Res* (2017) 179:183–98. doi: 10.1016/j.trsl.2016.09.001
89. Krogvold L, Skog O, Sundstrom G, Edwin B, Buanes T, Hanssen KF, et al. Function of Isolated Pancreatic Islets From Patients at Onset of Type 1 Diabetes: Insulin Secretion Can Be Restored After Some Days in a Nondiabetogenic Environment In Vitro: Results From the DiViD Study. *Diabetes* (2015) 64:2506–12. doi: 10.2337/db14-1911
90. Koskinen MK, Helminen O, Matomaki J, Aspholm S, Mykkanen J, Makinen M, et al. Reduced beta-cell function in early preclinical type 1 diabetes. *Eur J Endocrinol* (2016) 174:251–9. doi: 10.1530/EJE-15-0674
91. Keskinen P, Korhonen S, Kupila A, Veijola R, Erkkila S, Savolainen H, et al. First-phase insulin response in young healthy children at genetic and immunological risk for Type I diabetes. *Diabetologia* (2002) 45:1639–48. doi: 10.1007/s00125-002-0981-8
92. Chase HP, Cuthbertson DD, Dolan LM, Kaufman F, Krischer JP, Schatz DA, et al. First-phase insulin release during the intravenous glucose tolerance test as a risk factor for type 1 diabetes. *J Pediatr* (2001) 138:244–9. doi: 10.1067/mpd.2001.111274

93. Templeman NM, Clee SM, Johnson JD. Suppression of hyperinsulinaemia in growing female mice provides long-term protection against obesity. *Diabetologia* (2015) 58:2392–402. doi: 10.1007/s00125-015-3676-7
94. Templeman NM, Skovso S, Page MM, Lim GE, Johnson JD. A causal role for hyperinsulinemia in obesity. *J Endocrinol* (2017) 232:R173–R83. doi: 10.1530/JOE-16-0449
95. Thomas DD, Corkey BE, Istfan NW, Apovian CM. Hyperinsulinemia: An Early Indicator of Metabolic Dysfunction. *J Endocr Soc* (2019) 3:1727–47. doi: 10.1210/js.2019-00065
96. Mehran AE, Templeman NM, Brigidi GS, Lim GE, Chu KY, Hu X, et al. Hyperinsulinemia drives diet-induced obesity independently of brain insulin production. *Cell Metab* (2012) 16:723–37. doi: 10.1016/j.cmet.2012.10.019
97. Yoon KH, Ko SH, Cho JH, Lee JM, Ahn YB, Song KH, et al. Selective beta-cell loss and alpha-cell expansion in patients with type 2 diabetes mellitus in Korea. *J Clin Endocrinol Metab* (2003) 88:2300–8. doi: 10.1210/jc.2002-020735
98. Hanley SC, Austin E, Assouline-Thomas B, Kapeluto J, Blaichman J, Moosavi M, et al. {beta}-Cell mass dynamics and islet cell plasticity in human type 2 diabetes. *Endocrinology* (2010) 151:1462–72. doi: 10.1210/en.2009-1277
99. Linnemann AK, Baan M, Davis DB. Pancreatic beta-cell proliferation in obesity. *Adv Nutr* (2014) 5:278–88. doi: 10.3945/an.113.005488
100. Wortham M, Sander M. Mechanisms of beta-cell functional adaptation to changes in workload. *Diabetes Obes Metab* (2016) 18(Suppl 1):78–86. doi: 10.1111/dom.12729
101. Cnop M, Vidal J, Hull RL, Utzschneider KM, Carr DB, Schraw T, et al. Progressive loss of beta-cell function leads to worsening glucose tolerance in first-degree relatives of subjects with type 2 diabetes. *Diabetes Care* (2007) 30:677–82. doi: 10.2337/dc06-1834
102. Weyer C, Bogardus C, Mott DM, Pratley RE. The natural history of insulin secretory dysfunction and insulin resistance in the pathogenesis of type 2 diabetes mellitus. *J Clin Invest* (1999) 104:787–94. doi: 10.1172/JCI7231
103. Stanger BZ, Tanaka AJ, Melton DA. Organ size is limited by the number of embryonic progenitor cells in the pancreas but not the liver. *Nature* (2007) 445:886–91. doi: 10.1038/nature05537
104. Elsakr JM, Gannon M. Developmental programming of the pancreatic islet by in utero overnutrition. *Trends Dev Biol* (2017) 10:79–95.
105. Soldatow VY, Lecluyse EL, Griffith LG, Rusyn I. In vitro models for liver toxicity testing. *Toxicol Res (Camb)* (2013) 2:23–39. doi: 10.1039/C2TX20051A
106. Olson H, Betton G, Robinson D, Thomas K, Monro A, Kolaja G, et al. Concordance of the toxicity of pharmaceuticals in humans and in animals. *Regul Toxicol Pharmacol* (2000) 32:56–67. doi: 10.1006/rtph.2000.1399
107. Sass N. Humane endpoints and acute toxicity testing. *ILAR J* (2000) 41:114–23. doi: 10.1093/ilar.41.2.114
108. Krewski D, Acosta D Jr., Andersen M, Anderson H, Bailar JC,3, Boekelheide K, et al. Toxicity testing in the 21st century: a vision and a strategy. *J Toxicol Environ Health B Crit Rev* (2010) 13:51–138. doi: 10.1080/10937404.2010.483176
109. Gregg BE, Moore PC, Demozay D, Hall BA, Li M, Husain A, et al. Formation of a human beta-cell population within pancreatic islets is set early in life. *J Clin Endocrinol Metab* (2012) 97:3197–206. doi: 10.1210/jc.2012-1206
110. Ishihara H, Asano T, Tsukuda K, Katagiri H, Inukai K, Anai M, et al. Pancreatic beta cell line MIN6 exhibits characteristics of glucose metabolism and glucose-stimulated insulin secretion similar to those of normal islets. *Diabetologia* (1993) 36:1139–45. doi: 10.1007/BF00401058
111. Poitout V, Olson LK, Robertson RP. Insulin-secreting cell lines: classification, characteristics and potential applications. *Diabetes Metab* (1996) 22:7–14.
112. Merglen A, Theander S, Rubi B, Chaffard G, Wollheim CB, Maechler P. Glucose sensitivity and metabolism-secretion coupling studied during two-year continuous culture in INS-1E insulinoma cells. *Endocrinology* (2004) 145:667–78. doi: 10.1210/en.2003-1099
113. Hectors TL, Vanparys C, Pereira-Fernandes A, Martens GA, Blust R. Evaluation of the INS-1 832/13 cell line as a beta-cell based screening system to assess pollutant effects on beta-cell function. *PLoS One* (2013) 8: e60030. doi: 10.1371/journal.pone.0060030
114. Tsonkova VG, Sand FW, Wolf XA, Grunnet LG, Kirstine Ringgaard A, Ingvorsen C, et al. The EndoC-betaH1 cell line is a valid model of human beta cells and applicable for screenings to identify novel drug target candidates. *Mol Metab* (2018) 8:144–57. doi: 10.1016/j.molmet.2017.12.007
115. Andersson LE, Valtat B, Bagge A, Sharoyko VV, Nicholls DG, Ravassard P, et al. Characterization of stimulus-secretion coupling in the human pancreatic EndoC-betaH1 beta cell line. *PLoS One* (2015) 10:e0120879. doi: 10.1371/journal.pone.0120879
116. Ravassard P, Hazhouz Y, Pechberty S, Bricout-Neveu E, Armanet M, Czernichow P, et al. A genetically engineered human pancreatic beta cell line exhibiting glucose-inducible insulin secretion. *J Clin Invest* (2011) 121:3589–97. doi: 10.1172/JCI58447
117. Oleson BJ, McGraw JA, Broniowska KA, Annamalai M, Chen J, Bushkofsky JR, et al. Distinct differences in the responses of the human pancreatic beta-cell line EndoC-betaH1 and human islets to proinflammatory cytokines. *Am J Physiol Regul Integr Comp Physiol* (2015) 309:R525–34. doi: 10.1152/ajpregu.00544.2014
118. Da Silva Xavier G. The Cells of the Islets of Langerhans. *J Clin Med* (2018) 7:1–17. doi: 10.3390/jcm7030054
119. Hart NJ, Powers AC. Use of human islets to understand islet biology and diabetes: progress, challenges and suggestions. *Diabetologia* (2019) 62:212–22. doi: 10.1007/s00125-018-4772-2
120. Kim S, Whitener RL, Peiris H, Gu X, Chang CA, Lam JY, et al. Molecular and genetic regulation of pig pancreatic islet cell development. *Development* (2020) 147:1–14. doi: 10.1242/dev.186213
121. Nagaya M, Hayashi A, Nakano K, Honda M, Hasegawa K, Okamoto K, et al. Distributions of endocrine cell clusters during porcine pancreatic development. *PLoS One* (2019) 14:e0216254. doi: 10.1371/journal.pone.0216254
122. Zhu HT, Wang WL, Yu L, Wang B. Pig-islet xenotransplantation: recent progress and current perspectives. *Front Surg* (2014) 1:7. doi: 10.3389/fsurg.2014.00007
123. Lamb M, Laugenour K, Liang O, Alexander M, Foster CE, Lakey JR. In vitro maturation of viable islets from partially digested young pig pancreas. *Cell Transpl* (2014) 23:263–72. doi: 10.3727/096368912X662372
124. Hardikar AA, Wang XY, Williams LJ, Kwok J, Wong R, Yao M, et al. Functional maturation of fetal porcine beta-cells by glucagon-like peptide 1 and cholecystokinin. *Endocrinology* (2002) 143:3505–14. doi: 10.1210/en.2001-211344
125. Saliba Y, Bakhos JJ, Itani T, Fares N. An optimized protocol for purification of functional islets of Langerhans. *Lab Invest* (2017) 97:70–83. doi: 10.1038/labinvest.2016.123
126. Yin N, Chen T, Yu Y, Han Y, Yan F, Zheng Z, et al. Facile mechanical shaking method is an improved isolation approach for islet preparation and transplantation. *Exp Ther Med* (2016) 12:3658–64. doi: 10.3892/etm.2016.3844
127. Carter JD, Dula SB, Corbin KL, Wu R, Nunemaker CS. A practical guide to rodent islet isolation and assessment. *Biol Proced Online* (2009) 11:3–31. doi: 10.1007/s12575-009-9021-0
128. Lyon J, Manning Fox JE, Spigelman AF, Kim R, Smith N, O'Gorman D, et al. Research-Focused Isolation of Human Islets From Donors With and Without Diabetes at the Alberta Diabetes Institute IsletCore. *Endocrinology* (2016) 157:560–9. doi: 10.1210/en.2015-1562
129. Brissova M, Niland JC, Cravens J, Olack B, Sowinski J, Evans-Molina C. The Integrated Islet Distribution Program answers the call for improved human islet phenotyping and reporting of human islet characteristics in research articles. *Diabetologia* (2019) 62:1312–4. doi: 10.1007/s00125-019-4876-3
130. Sigmundsson K, Ojala JRM, Ohman MK, Osterholm AM, Moreno-Moral A, Domogatskaya A, et al. Culturing functional pancreatic islets on alpha5-laminins and curative transplantation to diabetic mice. *Matrix Biol* (2018) 70:5–19. doi: 10.1016/j.matbio.2018.03.018
131. Yeh CC, Wang LJ, McGarrigle JJ, Wang Y, Liao CC, Omami M, et al. Effect of Manufacturing Procedures on Human Islet Isolation From Donor Pancreata Standardized by the North American Islet Donor Score. *Cell Transpl* (2017) 26:33–44. doi: 10.3727/096368916X692834
132. Ihm SH, Matsumoto I, Sawada T, Nakano M, Zhang HJ, Ansite JD, et al. Effect of donor age on function of isolated human islets. *Diabetes* (2006) 55:1361–8. doi: 10.2337/db05-1333
133. Wang Y, Danielson KK, Ropski A, Harvat T, Barbaro B, Paushter D, et al. Systematic analysis of donor and isolation factor's impact on human islet

yield and size distribution. *Cell Transpl* (2013) 22:2323–33. doi: 10.3727/096368912X662417

134. Velazco-Cruz L, Song J, Maxwell KG, Goedegebuure MM, Augsornworawat P, Hogrebe NJ, et al. Acquisition of Dynamic Function in Human Stem Cell-Derived beta Cells. *Stem Cell Rep* (2019) 12:351–65. doi: 10.1016/j.stemcr.2018.12.012
135. Hogrebe NJ, Augsornworawat P, Maxwell KG, Velazco-Cruz L, Millman JR. Targeting the cytoskeleton to direct pancreatic differentiation of human pluripotent stem cells. *Nat Biotechnol* (2020) 38:460–70. doi: 10.1038/s41587-020-0430-6
136. Pagliuca FW, Millman JR, Gurtler M, Segel M, Van Dervort A, Ryu JH, et al. Generation of functional human pancreatic beta cells in vitro. *Cell* (2014) 159:428–39. doi: 10.1016/j.cell.2014.09.040
137. Russ HA, Parent AV, Ringler JJ, Hennings TG, Nair GG, Shveygert M, et al. Controlled induction of human pancreatic progenitors produces functional beta-like cells in vitro. *EMBO J* (2015) 34:1759–72. doi: 10.15252/embj.201591058
138. Veres A, Faust AL, Bushnell HL, Engquist EN, Kenty JH, Harb G, et al. Charting cellular identity during human in vitro beta-cell differentiation. *Nature* (2019) 569:368–73. doi: 10.1038/s41586-019-1168-5
139. Rezania A, Bruin JE, Arora P, Rubin A, Batushansky I, Asadi A, et al. Reversal of diabetes with insulin-producing cells derived in vitro from human pluripotent stem cells. *Nat Biotechnol* (2014) 32:1121–33. doi: 10.1038/nbt.3033
140. Zhou T, Kim TW, Chong CN, Tan L, Amin S, Sadat Badieyan Z, et al. Chen S. A hPSC-based platform to discover gene-environment interactions that impact human beta-cell and dopamine neuron survival. *Nat Commun* (2018) 9:4815. doi: 10.1038/s41467-018-07201-1
141. Krentz NA, Nian C, Lynn FC. TALEN/CRISPR-mediated eGFP knock-in add-on at the OCT4 locus does not impact differentiation of human embryonic stem cells towards endoderm. *PLoS One* (2014) 9:e114275. doi: 10.1371/journal.pone.0114275
142. Krentz NAJ, van Hoof D, Li Z, Watanabe A, Tang M, Nian C, et al. Phosphorylation of NEUROG3 Links Endocrine Differentiation to the Cell Cycle in Pancreatic Progenitors. *Dev Cell* (2017) 41:129–42.e6. doi: 10.1016/j.devcel.2017.02.006
143. Micallef SJ, Li X, Schiesser JV, Hirst CE, Yu QC, Lim SM, et al. INS(GFP/w) human embryonic stem cells facilitate isolation of in vitro derived insulin-producing cells. *Diabetologia* (2012) 55:694–706. doi: 10.1007/s00125-011-2379-y
144. Krentz NAJ, Lee MYY, Xu EE, Sproul SLJ, Maslova A, Sasaki S, et al. Single-Cell Transcriptome Profiling of Mouse and hESC-Derived Pancreatic Progenitors. *Stem Cell Rep* (2018) 11:1551–64. doi: 10.1016/j.stemcr.2018.11.008
145. Breckwoldt K, Letuffe-Breniere D, Mannhardt I, Schulze T, Ulmer B, Werner T, et al. Differentiation of cardiomyocytes and generation of human engineered heart tissue. *Nat Protoc* (2017) 12:1177–97. doi: 10.1038/nprot.2017.033
146. Ye L, Zhang S, Greder L, Dutton J, Keirstead SA, Lepley M, et al. Effective cardiac myocyte differentiation of human induced pluripotent stem cells requires VEGF. *PLoS One* (2013) 8:e53764. doi: 10.1371/journal.pone.0053764
147. Zhang J, Klos M, Wilson GF, Herman AM, Lian X, Raval KK, et al. Extracellular matrix promotes highly efficient cardiac differentiation of human pluripotent stem cells: the matrix sandwich method. *Circ Res* (2012) 111:1125–36. doi: 10.1161/CIRCRESAHA.112.273144
148. Denning C, Borgdorff V, Crutchley J, Firth KS, George V, Kalra S, et al. Cardiomyocytes from human pluripotent stem cells: From laboratory curiosity to industrial biomedical platform. *Biochim Biophys Acta* (2016) 1863:1728–48. doi: 10.1016/j.bbamcr.2015.10.014
149. Schulz TC, Young HY, Agulnick AD, Babin MJ, Baetge EE, Bang AG, et al. A scalable system for production of functional pancreatic progenitors from human embryonic stem cells. *PLoS One* (2012) 7:e37004. doi: 10.1371/journal.pone.0037004
150. Warren CR, Cowan CA. Humanity in a Dish: Population Genetics with iPSCs. *Trends Cell Biol* (2018) 28:46–57. doi: 10.1016/j.tcb.2017.09.006
151. Balboa D, Prasad RB, Groop L, Otonkoski T. Genome editing of human pancreatic beta cell models: problems, possibilities and outlook. *Diabetologia* (2019) 62:1329–36. doi: 10.1007/s00125-019-4908-z
152. Balboa D, Saarimaki-Vire J, Otonkoski T. Concise Review: Human Pluripotent Stem Cells for the Modeling of Pancreatic beta-Cell Pathology. *Stem Cells* (2019) 37:33–41. doi: 10.1002/stem.2913
153. Teo AK, Windmueller R, Johansson BB, Dirice E, Njolstad PR, Tjora E, et al. Derivation of human induced pluripotent stem cells from patients with maturity onset diabetes of the young. *J Biol Chem* (2013) 288:5353–6. doi: 10.1074/jbc.C112.428979
154. Braverman-Gross C, Nudel N, Ronen D, Beer NL, McCarthy MI, Benvenisty N. Derivation and molecular characterization of pancreatic differentiated MODY1-iPSCs. *Stem Cell Res* (2018) 31:16–26. doi: 10.1016/j.scr.2018.06.013
155. Balboa D, Saarimaki-Vire J, Borshagovski D, Survila M, Lindholm P, Galli E, et al. Insulin mutations impair beta-cell development in a patient-derived iPSC model of neonatal diabetes. *Elife* (2018) 7:e38519. doi: 10.7554/eLife.38519
156. Shang L, Hua H, Foo K, Martinez H, Watanabe K, Zimmer M, et al. beta-cell dysfunction due to increased ER stress in a stem cell model of Wolfram syndrome. *Diabetes* (2014) 63:923–33. doi: 10.2337/db13-0717
157. Teo AK, Lau HH, Valdez IA, Dirice E, Tjora E, Raeder H, et al. Early Developmental Perturbations in a Human Stem Cell Model of MODY5/HNF1B Pancreatic Hypoplasia. *Stem Cell Rep* (2016) 6:357–67. doi: 10.1016/j.stemcr.2016.01.007
158. Hosokawa Y, Toyoda T, Fukui K, Baden MY, Funato M, Kondo Y, et al. Insulin-producing cells derived from 'induced pluripotent stem cells' of patients with fulminant type 1 diabetes: Vulnerability to cytokine insults and increased expression of apoptosis-related genes. *J Diabetes Investig* (2017) 9 (3):481–93. doi: 10.1111/jdi.12727
159. Leite NC, Sintov E, Meissner TB, Brehm MA, Greiner DL, Harlan DM, et al. Modeling Type 1 Diabetes In Vitro Using Human Pluripotent Stem Cells. *Cell Rep* (2020) 32:107894. doi: 10.1016/j.celrep.2020.107894
160. Millman JR, Xie C, Van Dervort A, Gurtler M, Pagliuca FW, Melton DA. Generation of stem cell-derived beta-cells from patients with type 1 diabetes. *Nat Commun* (2016) 7:11463. doi: 10.1038/ncomms11463
161. Beer NL, Gloyn AL. Genome-edited human stem cell-derived beta cells: a powerful tool for drilling down on type 2 diabetes GWAS biology. *F1000Res* (2016) 5:1–12. doi: 10.12688/f1000research.8682.1
162. Zhu Z, Li QV, Lee K, Rosen BP, Gonzalez F, Soh CL, et al. Genome Editing of Lineage Determinants in Human Pluripotent Stem Cells Reveals Mechanisms of Pancreatic Development and Diabetes. *Cell Stem Cell* (2016) 18:755–68. doi: 10.1016/j.stem.2016.03.015
163. McGrath PS, Watson CL, Ingram C, Helmrath MA, Wells JM. The Basic Helix-Loop-Helix Transcription Factor NEUROG3 Is Required for Development of the Human Endocrine Pancreas. *Diabetes* (2015) 64:2497–505. doi: 10.2337/db14-1412
164. Guo D, Liu H, Ruzi A, Gao G, Nasir A, Liu Y, et al. Modeling Congenital Hyperinsulinism with ABCC8-Deficient Human Embryonic Stem Cells Generated by CRISPR/Cas9. *Sci Rep* (2017) 7:3156. doi: 10.1038/s41598-017-03349-w
165. Liang G, Zhang Y. Genetic and epigenetic variations in iPSCs: potential causes and implications for application. *Cell Stem Cell* (2013) 13:149–59. doi: 10.1016/j.stem.2013.07.001
166. Goldring C, Antoine DJ, Bonner F, Crozier J, Denning C, Fontana RJ, et al. Stem cell-derived models to improve mechanistic understanding and prediction of human drug-induced liver injury. *Hepatology* (2017) 65:710–21. doi: 10.1016/j.toxlet.2017.07.124
167. Lynch S, Pridgeon CS, Duckworth CA, Sharma P, Park BK, Goldring CEP. Stem cell models as an in vitro model for predictive toxicology. *Biochem J* (2019) 476:1149–58. doi: 10.1042/BCJ20170780
168. Grimm FA, Iwata Y, Sirenko O, Bittner M, Rusyn I. High-Content Assay Multiplexing for Toxicity Screening in Induced Pluripotent Stem Cell-Derived Cardiomyocytes and Hepatocytes. *Assay Drug Dev Technol* (2015) 13:529–46. doi: 10.1089/adt.2015.659
169. Lahti AL, Kujala VJ, Chapman H, Koivisto AP, Pekkanen-Mattila M, Kerkela E, et al. Model for long QT syndrome type 2 using human iPS cells demonstrates arrhythmogenic characteristics in cell culture. *Dis Model Mech* (2012) 5:220–30. doi: 10.1242/dmm.008409
170. Matsa E, Rajamohan D, Dick E, Young L, Mellor I, Staniforth A, et al. Drug evaluation in cardiomyocytes derived from human induced pluripotent stem cells carrying a long QT syndrome type 2 mutation. *Eur Heart J* (2011) 32:952–62. doi: 10.1093/eurheartj/ehr073

171. Itzhaki I, Maizels L, Huber I, Gepstein A, Arbel G, Caspi O, et al. Modeling of catecholaminergic polymorphic ventricular tachycardia with patient-specific human-induced pluripotent stem cells. *J Am Coll Cardiol* (2012) 60:990–1000. doi: 10.1016/j.jacc.2012.02.066
172. Sirenko O, Cromwell EF, Crittenden C, Wignall JA, Wright FA, Rusyn I. Assessment of beating parameters in human induced pluripotent stem cells enables quantitative in vitro screening for cardiotoxicity. *Toxicol Appl Pharmacol* (2013) 273:500–7. doi: 10.1016/j.taap.2013.09.017
173. Luni C, Giulitti S, Serena E, Ferrari L, Zambon A, Gagliano O, et al. High-efficiency cellular reprogramming with microfluidics. *Nat Methods* (2016) 13:446–52. doi: 10.1038/nmeth.3832

8

Monogenic Diabetes Modeling: *In Vitro* Pancreatic Differentiation from Human Pluripotent Stem Cells Gains Momentum

Juan Ignacio Burgos[1], *Ludovic Vallier*[2*] *and Santiago A. Rodríguez-Seguí*[1*]

[1] *Departamento de Fisiología, Biología Molecular y Celular, Facultad de Ciencias Exactas y Naturales, Universidad de Buenos Aires and Instituto de Fisiología, Biología Molecular y Neurociencias (IFIBYNE), CONICET-Universidad de Buenos Aires, Ciudad Universitaria, Buenos Aires, Argentina,* [2] *Wellcome-Medical Research Council Cambridge Stem Cell Institute and Department of Surgery, University of Cambridge, Cambridge, United Kingdom*

***Correspondence:**
Ludovic Vallier
lv225@cam.ac.uk
Santiago A. Rodríguez-Seguí
srodriguez@fbmc.fcen.uba.ar

The occurrence of diabetes mellitus is characterized by pancreatic β cell loss and chronic hyperglycemia. While Type 1 and Type 2 diabetes are the most common types, rarer forms involve mutations affecting a single gene. This characteristic has made monogenic diabetes an interesting disease group to model *in vitro* using human pluripotent stem cells (hPSCs). By altering the genotype of the original hPSCs or by deriving human induced pluripotent stem cells (hiPSCs) from patients with monogenic diabetes, changes in the outcome of the *in vitro* differentiation protocol can be analyzed in detail to infer the regulatory mechanisms affected by the disease-associated genes. This approach has been so far applied to a diversity of genes/diseases and uncovered new mechanisms. The focus of the present review is to discuss the latest findings obtained by modeling monogenic diabetes using hPSC-derived pancreatic cells generated *in vitro*. We will specifically focus on the interpretation of these studies, the advantages and limitations of the models used, and the future perspectives for improvement.

Keywords: pancreas, beta cell, human, pluripotent stem cell, monogenic, modeling, diabetes, *in vitro* differentiation

INTRODUCTION

Diabetes mellitus (DM) is characterized by pancreatic β cell loss and chronic hyperglycemia. Type 1 diabetes (T1D) is caused by the autoimmune reaction against β cells (1), and Type 2 diabetes originates from insulin resistance and β cell overload (2–4). In addition, rarer monogenic forms of diabetes account for approximately 1–5% of diabetes cases, depending on the population studied (5, 6). Over 30 subtypes of monogenic diabetes have been identified to date, each having a characteristic phenotype and a specific pattern of inheritance (6, 7). The identification of genes implicated in the pathogenesis of monogenic diabetes, including components of the insulin secretory pathway and transcription factors, has provided important insights into human pancreas and β cell development and function.

Monogenic diabetes is caused by either splice-site, non-sense, missense, or frame-shift mutations, and more rarely partial or full deletions, affecting a single gene (8–14). The disease phenotype and associated extra-pancreatic features vary depending on the affected gene (15). These characteristics have made monogenic diabetes an interesting disease subtype to model using human pluripotent stem cells (hPSCs). Indeed, hPSCs can be differentiated into pancreatic cells following key steps of differentiation induced by well-established combinations of growth factors and small molecules, thereby respecting a natural path of development [recently reviewed in (16, 17)]. Thus, differentiation of human induced pluripotent stem cells (hiPSCs) either derived from patients with monogenic diabetes or genetically edited to carry the mutation of interest can be used to study the potential regulatory mechanisms affected by each of the disease-associated genes (**Figure 1**). Applying this approach to a diversity of genes has led to the discovery of new mechanisms associated with specific regulators of pancreatic development. The focus of this review is to discuss the latest findings obtained by modeling monogenic diabetes using hPSC-derived pancreatic cells generated *in vitro*. We will focus on the interpretation of these studies, the advantages and limitations of the models used, and the future perspectives for improvement. Of note, the reader is referred to recent reviews concerning: 1) the state-of-the-art knowledge in pancreatic β cell development in mice and humans (18–21); 2) the tools for hPSC genome editing (22, 23); 3) a comparison of the *in vitro* pancreatic differentiation protocols including the latest advances to achieve functional β cells from hPSCs (16, 17, 24); 4) analyzing the intrinsic variation in the protocol outcomes from different sources of hPSCs (25, 26); and 5) the use of *in vitro* pancreatic differentiation from hPSCs to discover new mechanisms underlying human pancreas development (24, 27), or to model other types of diabetes (23, 28–31). The later also summarize findings on monogenic diabetes modeling. Our review adds up on top of these by exclusively focusing on the modeling of monogenic diabetes, discussing in more detail the different approaches taken and including extremely recent works which provide insightful information for the interpretation of the results published so far.

HUMAN PANCREAS DEVELOPMENT

Lessons Learned From Mice and Current Challenges

Pancreas development begins with the establishment of the pancreatic bud containing multipotent pancreatic progenitor cells (MPCs) at ~E8.5 in the mouse (19, 20) or ~29 days post conception in humans (20, 21, 32) and progresses until E18.5 in the mouse (19, 20) or 24 weeks post conception (wpc) estimated in humans (20, 21, 33–35). By this time, most of the pancreatic progenitor cells are terminally fate-committed. The MPCs are capable of differentiating along the three main lineages of the adult pancreas, namely the ductal, the exocrine (comprising acinar cells that secrete digestive enzymes), and the endocrine (including the β cells that produce insulin, but also the α, δ, γ, and PP cells) (19, 36, 37). As the pancreas develops, MPCs differentiate into acinar or endocrine-ductal bipotent progenitor

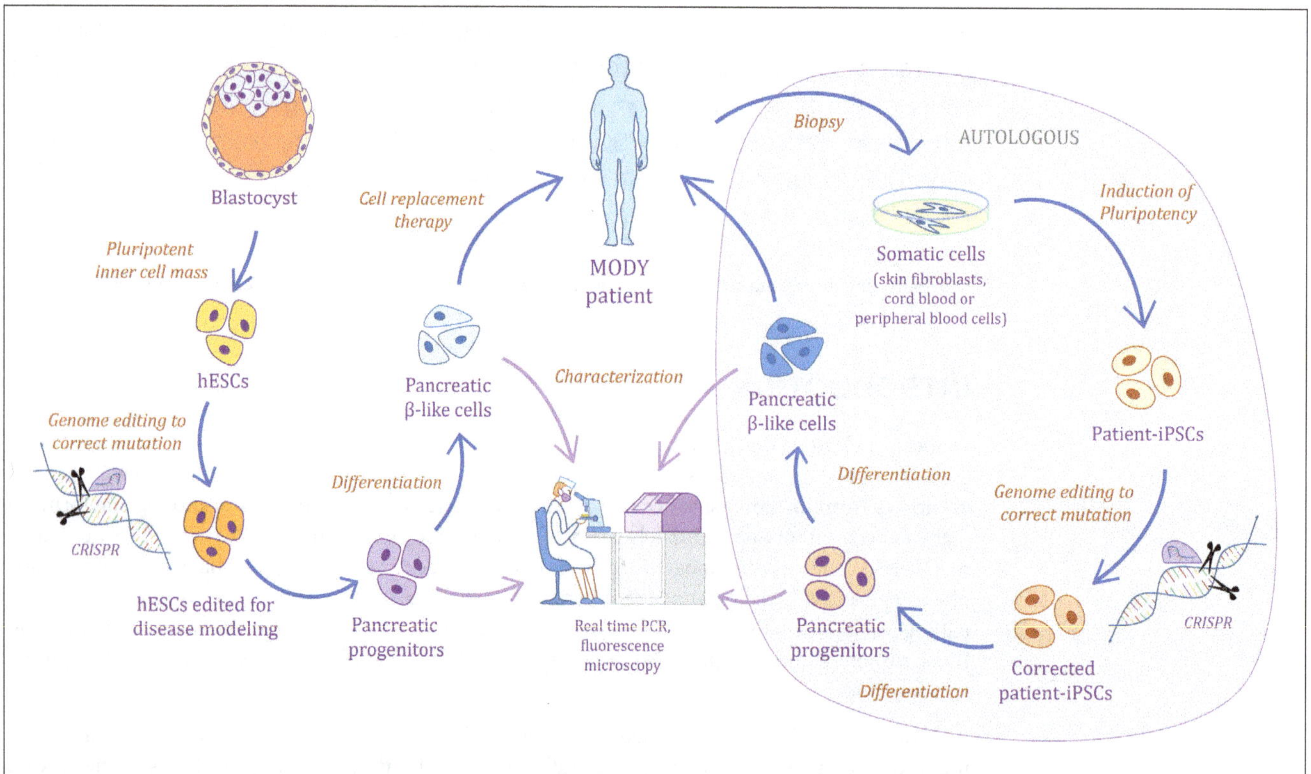

FIGURE 1 | Schematic illustrating the pipeline for monogenic diabetes disease modeling using human pluripotent stem cells.

(BP) cells, and eventually to endocrine-committed progenitors (EPs) that will give rise to β cells. Importantly, recent single cell RNA-seq (scRNA-seq) studies have described the transcriptional profiles that characterize these pancreatic progenitor cell stages in the mouse and identified additional progenitor cell sub stages (38–42). Thus, we can now rely on a precise transcriptomic fingerprint for several progenitors that arise during mouse pancreas development.

Cumulative knowledge has revealed the role and stage-specific functions of signaling pathways in the pancreatic developmental program, including Wnt, TGF-β, Notch, FGF and, more recently, the Hippo pathway (19, 36, 37, 43–46). The pancreatic mesenchyme has also been shown to play important roles during development, by fine-tuning the crosstalk with the pancreatic epithelium through these pathways (47–50). This knowledge has been exploited to develop protocols to differentiate human embryonic stem cells (hESCs) into the pancreatic lineage (51–55), opening the possibility to produce large quantities of β cells for cell-based therapies and also providing a new avenue for research in human pancreas development.

Differentiation protocols currently available to produce β cells from hPSCs take advantage of the cell signaling events that occur during fetal development (18–21). Despite this knowledge-based approach, the generation of fully functional β cells *in vitro* has remained elusive. We thus refer to the cells produced *in vitro* so far as β-like cells. This limitation could be in part explained by the fact that the function of these pathways during pancreas development has been mainly studied in rodents, and it has recently been reported that human pancreas development could differ in several aspects (20, 21, 32, 56, 57). We are still lacking systematic studies comparing islet development between human and mouse, although recent reports are moving forward to address this gap (57–60). It is expected that a deeper understanding of these inter-species differences in islet development will probably be critical for the production of fully functional and mature β cells from hPSCs. As well, this will allow a more precise dissection of the molecular mechanisms driving diabetes predisposition by genetic mutations and/or external stimuli. Noteworthy, recent protocols allow the derivation of monohormonal insulin-producing cells expressing key β cell transcription factors, including *PDX1*, *NKX6.1*, and *MAFA* (61–65). Fine tuning of such protocols could further improve the glucose-response of *in vitro* derived β-like cells (66–69) without the need to involve a step of cell transplantation in mice to ensure proper maturation.

Lessons Learned From *In Vitro* Human PSC Pancreatic Differentiation

The use of hPSCs as *in vitro* model system to study human pancreatic development has gained momentum and several important discoveries have been made using this approach [recently reviewed in (27)].

A Role for TEAD and YAP in Pancreas Development

By comparing the transcriptomes and key epigenomic features of MPCs derived *in vitro* from hESCs with human fetal primary pancreatic tissue of six wpc embryos, we were able to show that *in vitro* derived cells closely recapitulated the main expression profile and regulatory landscape of their *in vivo* counterparts (45). Furthermore, this epigenomic characterization was extended to include ChIP-seq profiling of several transcription factors by taking advantage of the *in vitro* system. A combined analysis of these data led to the discovery that TEAD1 was an integral component of the enhancer network in human embryonic pancreatic progenitors (45, 46). The relevance of TEAD protein binding for the activation of MPC enhancers was mechanistically validated using the platform provided by hPSC-derived pancreatic cells *in vitro*. We concluded that, while highly tissue specific enhancers were defined by co-binding of pancreas-specific transcription factors, TEAD proteins conferred these regions the ability to be regulated by YAP (an effector of the Hippo pathway) during human pancreas development. These results were consistent with reports for a role of the Hippo pathway in mouse pancreas development (43, 44). More recent reports support these findings by showing that YAP and other components of the Hippo pathway are active and highly enriched within the SOX9+/PTF1A+ progenitor cells of the human fetal pancreas (58).

The human *in vitro* differentiation system has provided additional mechanistic evidence for the relevance of the Hippo pathway in pancreas differentiation by showing that YAP links extracellular matrix-mediated mechano-signals to regulate gene expression. Integration of such signaling plays a key role in the fate choice of bipotent pancreatic progenitors, whereby YAP downregulation favors endocrine cell commitment (70). In agreement, sustained *in vitro* YAP activation impairs β cell differentiation while inhibition of YAP enhances differentiation of functional β cells derived from hPSCs (71). Combined together, these results illustrate how *in vitro* pancreatic differentiation can help in the discovery of new regulatory mechanisms that are relevant for human β cell development.

A Role for Polycomb Group-Mediated Repression in Pancreas Development

Mapping the dynamic changes in histone modifications and chromatin accessibility across different stages of the *in vitro* pancreatic differentiation protocol has provided important insights into the suitability and current limitations of this model (72–75). It has to be noted that, despite successful applications, the *in vitro* pancreatic differentiation system does not exactly replicate all the epigenomic features of their *in vivo* counterparts. Sander and colleagues have profiled selected chromatin modifications and the transcriptome of these cells, at different stages of the pancreatic endocrine differentiation protocol using hESCs (72). They showed that removal of Polycomb group (PcG)-mediated repression on stage-specific genes was a key mechanism for the induction of developmental regulators in the *in vitro* system, consistent with the *in vivo* relevance of this mechanism in mouse endocrine pancreas development (76–78). However, they also reported that elimination of PcG-mediated repression on endocrine-specific genes was not fully recapitulated by the *in vitro* derived endocrine cells. This was particularly evident at genes involved in organ morphogenesis, underscoring a current limitation (*i.e.* the lack of tissue-specific contextual cell signaling) of the *in vitro* protocols for studying some aspects of *in vivo* development. Noteworthy, these experiments were performed using 2D *in vitro* differentiation protocols which have been shown to be less efficient to produce functional β-like cells than three-dimensional suspension culture systems (61–63, 66–68). This "second generation" protocols could be more efficient in allowing the proper deposition of epigenetic marks.

Distinct Progenitor Cell Populations Could Differentiate Into Monohormonal β Cells

Another example of the complexity of the *in vitro* endocrine differentiation process was provided by Petersen et al. who profiled single-cell transcripts by qRT-PCR at selected stages of the protocol used to produce β-like cells from hPSCs (79). This analysis identified two distinct progenitor cell populations with the potential to differentiate into monohormonal β-like cells. *NKX6.1* expression prior or after the onset of *NEUROG3* (the gene coding for the EP transcription factor marker NGN3) was the main difference between these progenitors. Building up on these results, Ramond et al. performed a combined analysis of single-cell qRT-PCR datasets obtained from pancreatic progenitor and endocrine cells from *in vitro* and *in vivo* samples (59, 60). Their observations suggest that these distinct progenitor cell populations identified *in vitro* could indeed exist during *in vivo* development. This finding contrasts with knowledge gained from mouse studies where Nkx6 factors systematically specify endocrine cell fate upstream of Ngn3 in the MPC stage (80, 81). Still, the function and characteristics of these two populations of progenitors in the developing human pancreas remain to be fully elucidated.

Taking advantage of novel single-cell technologies can help to match some of the transcriptional signatures from *in vivo* pancreatic progenitor cell stages identified in the mouse (38–41), with pancreatic progenitors derived from hPSCs (38, 66, 82). However, such exercise remains challenging due to inter-species differences and also the impact of the *in vitro* culture. New tools are being quickly developed to address this limitation, making the bioinformatic analyses of these data an exciting area of research (83, 84). More recently, the first scRNA-seq experiments using human embryonic pancreas from 15.2 and 17.1 wpc have been reported (58, 85). Integration of these scRNA-seq datasets with those derived from the mouse embryonic pancreas will help to identify differences and similarities in the transcriptional fingerprints of the distinct pancreatic progenitor cell types. This will ultimately contribute to validate the identity of pancreatic progenitors produced from hPSCs.

MODELING MONOGENIC DIABETES WITH HUMAN PLURIPOTENT STEM CELLS

The most frequently affected maturity-onset diabetes of the young (MODY) genes include the enzyme glucokinase (*GCK*, MODY2) (86, 87) and the transcription factor genes hepatic nuclear factor 1 alpha (*HNF1A*, MODY3) (88), hepatic nuclear factor 4 alpha (*HNF4A*, MODY1) (89), and hepatic nuclear factor 1 beta (*HNF1B*, MODY5) (90). Other MODY genes include *PDX1* (MODY4), *NEUROD1* (MODY6), *KLF11* (MODY7), *CEL* (MODY8), *PAX4* (MODY9), *INS* (MODY10), *BLK* (MODY11), *ABCC8* (MODY12), *KCNJ11* (MODY13), *APPL1* (MODY14) (91). On the other hand, homozygous mutations at several lineage determining transcription factors, such as *PTF1A*, *PDX1*, *NEUROG3*, *RFX6*, *NEUROD1*, *MNX1*, *NKX2.2* and *GLIS3* result in permanent neonatal diabetes mellitus (PNDM) in humans (13, 92–100). Interestingly, heterozygous mutations in these genes rarely result in diabetes in mice, thereby suggesting an important divergence in the activity or function for these factors between human and mouse (19, 21, 101). The importance of haploinsufficiency and the mechanisms by which a decrease in transcription factor activity causes a disease in humans is poorly understood, mostly due to the lack of an appropriate model system. As an example, MODY5 diabetes (*HNF1B*-associated) can be induced by a diversity of mutations including several splice-site, non-sense, missense, and frame-shift mutations or whole gene deletions, all of which result in a diabetes (102). The heterozygous mutation in mouse has no effect on pancreatic development, while homozygous mutation blocks foregut specification thereby masking its downstream function in the differentiation of MPCs. Ultimately, haploinsufficiency may reflect the functional effects of different gene anomalies, stochastic variation in temporal gene expression during early development or additional genetic and/or environmental modifiers that may influence the disease phenotype (102–104).

As mentioned above, mouse models often do not recapitulate the disease phenotype associated with heterozygous mutations of *HNF1A*, *HNF4A*, or *HNF1B* in humans. The genetic discrepancy between the mouse and monogenic diabetes gene haploinsufficient patients and the difficulty in accessing patient samples have reinforced the interest in using hPSCs (**Figure 1**). Genome-editing tools combined with directed differentiation of hPSCs offer a unique platform for generating patient-specific disease models to elucidate novel genes and molecular pathways that underlie monogenic diseases with complex traits, such as diabetes, and ultimately lead to the development of novel therapeutic strategies [recently reviewed in (22, 25)]. Several studies in the last decade have used genetically engineered hPSC culture systems for differentiation into pancreatic cells to further expand our understanding of the roles of various genes associated with monogenic diabetes. Their findings are summarized in **Tables 1**, **2**, and these will be discussed in more detail next.

WFS1

Egli and colleagues provided the first example for the use of hiPSCs to create insulin-producing cells from patients with Wolfram Syndrome (WS) (119). hiPSCs were generated from individuals with diabetes caused by mutations in the *WFS1* gene and healthy-donor controls. Differentiation of these cells towards β-like cells revealed increased levels of ER stress molecules and decreased insulin content in WFS1-deficient β-like cells. Overall, insulin processing and secretion in response to various

TABLE 1 | Summary of reports modeling maturity-onset diabetes of the young (MODY) mutation effects.

Gene studied	Pancreatic defects reported in humans	Effects recapitulated in mice	Genome editing approach	Differentiation protocol	Type of human pluripotent stem cell	*In vitro* phenotypes	Ref.
***HNF4A* (MODY1)**	*HNF4A* heterozygous mutations affect both liver and pancreas development. MODY1 patients present neonatal hyperinsulinemia and impairment in β cell function. They present normal insulin sensitivity but decreased insulin secretion.	Rodent models do not accurately recapitulate the MODY1 phenotype in humans. The available *Hnf4a* general knockout murine model is embryonic lethal, while heterozygous mice present normal glucose tolerance and do not show any diabetic features.	NA	(62)	hiPSCs were derived from MODY1 mutation carriers. Their family members, without the mutation, were used as controls.	The *HNF4A* mutation studied did not prevent the formation of insulin+ cells *in vitro*. Also, no defects in β-like cells differentiated from *HNF4A* mutant hiPSCs were found.	(105)
			NA	Adapted from (62).	Control hiPSC lines (CSES7 and IPSO lines) and MODY1 patient-derived hiPSCs.	Researchers report that cells from the MPC stage show increased expression of endocrine progenitor transcription factors, including *PAX6*, *NEUROD1* and *NEUROG3*.	(106)
			Site-directed mutagenesis.	(61)	hiPSCs were derived from non-diabetic and MODY1 patients.	Key developmental genes such as *HNF1B*, *PDX1*, *GATA4*, and *RFX6* are downregulated at the foregut progenitor stage, prior to MPC specification. Still, terminally differentiated β-like cells can be produced and express selective β cell markers and C-peptide. The functional capacity of these cells could not be appropriately elucidated due to limitations of the *in vitro* protocol used.	(107)
***GCK* (MODY2)**	Patients with *GCK* heterozygous mutations present progressive β-cell dysfunction, fasting hyperglycemia and reduced insulin secretion. These result in a mild diabetes phenotype that generally does not require anti-diabetes medication.	Homozygous mutant mice exhibit growth retardation and die soon after birth as consequence of severe hyperglycemia. Heterozygous mutant mice only present slightly elevated blood glucose levels from birth, with disturbed glucose tolerance and glucose-induced insulin secretion.	NA	NA	Non-edited MODY2 and PNDM patient-derived hiPSCs.	This work reports the generation of iPSCs from MODY2 patients. The researchers did not analyze differentiation into the pancreatic lineage.	(108)
HNF1A (MODY3)	Patients with *HNF1A* heterozygous mutations show β cell dysfunction and hyperglycemia due to insufficient insulin release in response to increased blood glucose levels.	Mouse models do not fully mimic the human disease phenotype. Mice with heterozygous mutations in *Hnf1a* are healthy and mice with homozygous null mutations present a diabetic phenotype.	CRISPR-CAS9 system.	(62), with minor modifications.	Genome-edited hESCs (MEL1 and H1) and human β-cell lines (EndoC-BH).	Differentiation from $HNF1A^{+/-}$ hESC show reduced number of INS+ cells. β-like cells present defects in mitochondrial function and the glycolysis process. Decreased expression of β cell transcription factors and genes associated with insulin synthesis. Reduced β cell proliferation and increased apoptosis.	(109)
			NA	(61), with some modifications.	hiPSCs were derived from MODY3 patients. hiPSCs derived from a healthy donor were used as control.	*HNF1A* MODY3 mutations caused decreased *GLUT2* expression, which was associated with reduced glucose uptake and ATP production. The mutant *HNF1A* β-like cells present decreased insulin secretion in response to high glucose.	(110)

(Continued)

TABLE 1 | Continued

Gene studied	Pancreatic defects reported in humans	Effects recapitulated in mice	Genome editing approach	Differentiation protocol	Type of human pluripotent stem cell	*In vitro* phenotypes	Ref.
***PDX1* (MODY4)**	*PDX1* heterozygous mutations are associated with insulin secretion deficiency. Common point heterozygous mutations in the *PDX1* transactivation domain impair human pancreatic β cell formation and function, and contribute to increased risk for diabetes. Pancreatic developmental anomalies related to *PDX1* mutations are reported only in neonatal diabetes cases.	Homozygous *Pdx1*-deficient mice fail to generate a pancreas, while heterozygous animals develop a pancreas but become diabetic in adulthood due to β cell apoptosis.	TALEN and CRISPR/Cas9.	Adapted from (52, 54).	Genome-edited hESCs (HUES8).	Monoallelic *PDX1* mutations are associated with decreased PDX1 protein expression. These compromise endocrine differentiation and lead to reduction in the number of INS+ cells derived *in vitro*.	(111)
			CRISPR/Cas9.	Based on (62).	Genome-edited hiPSCs and patient-derived hiPSCs.	Heterozygous mutations impair *in vitro* β cell differentiation and function. Homozygous point mutations in the *PDX1* transactivation domain do not only impact pancreatic endocrine lineage development, but also impair glucose-responsive function of β cells through misregulation of several PDX1 target genes.	(112)
***HNF1B* (MODY5)**	Patients with *HNF1B* heterozygous mutations commonly exhibit pancreatic hypoplasia, β-cell dysfunction and insulin resistance.	$Hnf1b^{-/-}$ mice present pancreatic agenesis, exhibiting loss of expression of several pancreatic genes, including Pax6, which regulate β-cell function. In contrast with MODY5 patients, $Hnf1b^{+/-}$ mice do not develop diabetes.	NA	Adapted from (52).	MODY5 patient-derived hiPSCs.	Upregulation of multiple key pancreatic transcription factors at the DE and MPC stage, including *FOXA2*, *PDX1*, *GATA4* and *GATA6*. Interestingly, expression of HNF1B itself was induced in mutant hiPSC-derived MPCs. Reduction of PAX6 expression.	(113)

NA, not applicable.

secretagogues was comparable to healthy controls, but the former displayed increased activity of unfolded protein response (UPR) pathways.

More recently, Maxwell et al. used CRISPR/Cas9 to correct a diabetes-causing pathogenic variant in *WFS1* hiPSCs (120). Noteworthy, *β*-like cells differentiated from *WFS1*-corrected hiPSCs showed robust and dynamic insulin secretion in response to glucose, and reversed streptozocin-induced diabetes when transplanted into mice. Single-cell RNA-seq transcriptome profiling showed that indeed these cells displayed increased insulin levels and decreased expression of genes associated with endoplasmic reticulum stress. Taken together, these studies illustrate the potential of *in vitro* pancreatic differentiation from hPSCs to study how mechanisms related to cellular stress can affect diabetes onset.

PDX1

Homozygous null mutations in *PDX1* result in pancreatic agenesis both in mice and humans (13, 121–123). Human patients with *PDX1* heterozygous inactivating mutations exhibit MODY4 diabetes caused by defects in *β* cell function and/or the maintenance of *β* cell mass in adults (36). In rodents, it has been reported that $Pdx1^{+/-}$ mice can develop a functional pancreas (121, 122) but become diabetic in adulthood due to *β* cell apoptosis (124).

Another pioneer study to model monogenic diabetes was reported by Huangfu and colleagues, who used TALEN and CRISPR-Cas-mediated gene editing combined with hPSC-directed differentiation. These researchers provide a systematic analysis of the role for PDX1 and seven additional pancreatic transcription factors (RFX6, PTF1A, GLIS3, MNX1, NGN3, HES1 and ARX) in pancreatic cell commitment (111). Noteworthy, they created mono- or biallelic frameshift mutations in all these genes and used untargeted isogenic cell lines as controls. This analysis not only defined the specific developmental steps affected by these mutations in a model of human pancreas differentiation, but also revealed new mechanisms. **Tables 1, 2** show a summary of their results for the genes previously associated with MODY and/or PNDM. An interesting finding of this work was that monoallelic frameshift translation mutations disrupting the PDX1 protein sequence cause a reduction (up to 65%) in the number of insulin+ cells derived *in vitro*. These findings suggest a haploinsufficient requirement for *PDX1* in pancreatic endocrine development. Importantly, this phenotype correlates with the observation that patients with heterozygous mutations in *PDX1* present with diabetes from an early age (125). These results further validate that decreased amounts of PDX1 could lead to *β* cell dysfunction, a decrease in *β* cell mass during fetal development and/or the maintenance of *β* cell mass in adults (124, 126, 127).

TABLE 2 | Summary of reports modeling monogenic mutations associated with permanent neonatal diabetes mellitus (PNDM) or pancreatic agenesis.

Gene studied	Pancreatic defects reported in humans	Effects recapitulated in mice	Genome editing approach	Differentiation protocol	Type of human pluripotent stem cell	*In vitro* phenotypes	Ref.
GATA6	*GATA6* heterozygous inactivating mutations result in pancreatic agenesis.	*Gata6* heterozygous mice are fertile and phenotypically normal. *Gata6* null mice are embryonic lethal. Biallelic loss of *Gata6* and its paralog *Gata4* result in a phenotype similar to human PNDM *GATA6*-mutated patients.	CRISPR/Cas9-mediated genome editing.	Adapted from (52).	Patient-derived hiPSCs and genome-edited hESCs. Isogenic, mutation-corrected, hiPSCs were used as controls.	*GATA6* homozygous mutations lead to impaired DE differentiation. Rescue of DE defects in these cells by re-expression of other GATA family members allows β-like cell production with a lower efficiency. hPSCs with *GATA6* heterozygous mutations show defects in DE differentiation. β-like cells produced in both cases are defective in the GSIS and in insulin processing.	(114)
			CRISPR/Cas9-mediated genome editing.	(61, 62, 111), with some modifications.	Genome-edited hESCs (H1 and HUES8).	Differentiation of $GATA6^{-/-}$ hPSCs revealed impaired DE commitment and pancreatic endocrine differentiation. No defects in DE differentiation from $GATA6^{+/-}$ hPSCs, but a lower number of *PDX1*+ *NKX6.1*+ pancreatic progenitors and β-like cells was produced.	(115)
			TALENs	(55), adapted from (52).	hiPSCs derived from pancreatic agenesis patients with *GATA6* heterozygous mutations. Genome-edited hESCs (H9) and hiPSCs. Non-mutated hESCs and hiPSCs were used as isogenic controls.	*GATA6* heterozygous hPSCs present a modest decrease in the generation of DE, which differentiate less efficiently into MPCs and EPs. *GATA6*-null hPSCs fail to enter the DE lineage.	(116)
			CRISPR-CAS9-mediated genome editing.	Adapted from (61, 62, 54).	hiPSCs derived from a patient with pancreatic agenesis. Isogenic, mutation-corrected hiPSCs were used as control.	hiPSCs with *GATA6* heterozygous mutations present reduced efficiency for generation of pancreatic progenitor cells *in vitro*. Correction of these mutations allowed identifying a non-coding SNP that additionally contributes to the phenotype observed.	(117)
PDX1	Homozygous mutations in *PDX1* result in pancreatic agenesis. PDX1 heterozygous patients exhibit diabetes caused by defects in β cell function and/or the maintenance of β cell mass in adults.	Homozygous mutations in *Pdx1* cause pancreatic agenesis, while heterozygous animals develop a pancreas but become diabetic in adulthood due to β cell apoptosis.	TALEN ad CRISPR/ Cas9.	Adapted from (52, 54).	Genome-edited hESCs (HUES8).	Differentiation of $PDX1^{+/-}$ mutant hESCs present a 65% reduction of INS+ cells at the β-like cell stage, which are mainly polyhormonal cells using the protocol described in this study.	(111)

(Continued)

TABLE 2 | Continued

Gene studied	Pancreatic defects reported in humans	Effects recapitulated in mice	Genome editing approach	Differentiation protocol	Type of human pluripotent stem cell	*In vitro* phenotypes	Ref.
RFX6	Patients carrying biallelic *RFX6* inactivating mutations present a reduction in the pancreas size and obstruction of the small intestine. These patients present defects in the formation of pancreatic progenitors and their further differentiation into functional endocrine cells.	Similar to humans, *Rfx6*-null mice show variable degrees of pancreatic hypoplasia and premature death.	TALEN ad CRISPR/Cas9.	Adapted from (52, 54).	Genome-edited hESCs (HUES8).	Differentiation of *RFX6*$^{-/-}$ mutant hESCs show a reduction in the number of *PDX1*+ pancreatic progenitor cells. Severe reduction in β-like cells and complete absence of α cells.	(111)
			CRISPR/Cas9-mediated genome editing.	Adapted from (62)	hiPSCs were derived from patients with MRS and from their healthy, heterozygous father. hESCs (H9) was used as control.	hiPSCs with *RFX6* homozygous mutations show normal DE and PFG differentiation, but fail to robustly activate *PDX1*. MPCs and endocrine-competent progenitors differentiate less efficiently from these cells.	(118)
PTF1A	Homozygous inactivating mutations in *PTF1A* cause pancreatic and cerebellar agenesis.	*Ptf1a*-null mice present a complete absence of exocrine pancreatic tissue, but all islet endocrine cell types are present until the late stages of embryogenesis.	TALEN ad CRISPR/Cas9.	Adapted from (52, 54).	Genome-edited hESCs (HUES8).	Differentiation of *PTF1A*$^{-/-}$ mutant hESCs do not present defects in pancreatic endocrine differentiation using the protocol described.	(111)
GLIS3	Biallelic mutations of *GLIS3* underlie a rare clinical syndrome, characterized by neonatal diabetes and congenital hypothyroidism.	Global *Glis3*$^{-/-}$ mice die of severe neonatal diabetes shortly after birth. Minor differences in gene dosage of Glis3 produce substantive changes in the expression levels of Ngn3 and Ins1, leading to a variable phenotype among the multiple *Glis3*-KO mouse lines.	TALEN ad CRISPR/Cas9.	Adapted from (52, 54).	Genome-edited hESCs (HUES8).	Differentiation of *GLIS3*$^{-/-}$ mutant hESCs do not present defects in pancreatic endocrine differentiation using the protocol described.	(111)
			CRISPR/Cas9-mediated genome editing.	(64)	Genome-edited hESCs.	Differentiation of *GLIS3*$^{-/-}$ mutant hESCs show impaired expression of pancreatic endocrine-associated genes, including *PDX1*, *NEUROD1*, *NKX6.1*, and *MAFA*, and present increased β-like cell death. A chemical screen identified a drug candidate that rescues mutant GLIS3-associated β-cell death both *in vitro* and *in vivo*.	(64)

(Continued)

TABLE 2 | Continued

Gene studied	Pancreatic defects reported in humans	Effects recapitulated in mice	Genome editing approach	Differentiation protocol	Type of human pluripotent stem cell	*In vitro* phenotypes	Ref.
MNX1	Homozygous mutations in *MNX1* are associated with the occurrence of diabetes in infancy without evidence of exocrine pancreatic dysfunction. Reduced number of pancreatic endocrine cells, including β cells.	*Mnx1*-deficient mice show pancreatic dorsal-lobe agenesis and smaller pancreatic islets, while *Mnx1* gain-of-function in the pancreas leads to aberrant pancreatic development.	TALEN ad CRISPR/ Cas9.	Adapted from (52, 54).	Genome-edited hESCs (HUES8).	Differentiation of $MNX1^{-/-}$ mutant hESCs do not present defects in pancreatic endocrine differentiation using the protocol described.	(111)

DE, definitive endoderm; MPC, multipotent pancreatic progenitor cells; PFG, posterior foregut; GSIS, glucose-stimulated insulin secretion; MRS, Mitchell-Riley syndrome; KO, knock out; ER, endoplasmic reticulum.

More recently, Lickert and colleagues generated hiPSCs from two patients with heterozygous missense mutations in the *PDX1* coding region ($PDX1^{P33T/+}$ and $PDX1^{C18R/+}$) leading to single amino acid exchanges in its transactivation domain (112). By comparing with a control hiPSC line derived from a healthy donor, the authors showed that MPC differentiation was not affected in patient-derived hiPSCs. However, the *PDX1* heterozygous point mutations impaired the differentiation of *β*-like cells and affected their response to glucose. A more severe effect was observed when artificially introducing the same point mutations in homozygosis (*i.e.* $PDX1^{P33T/P33T}$ and $PDX1^{C18R/C18R}$) in isogenic cell lines derived from the original control cell. Interestingly, this resulted in impaired *NKX6.1* induction in MPCs just in one of the cell lines ($PDX1^{P33T/P33T}$). Nevertheless, when differentiated towards insulin producing cells, both homozygous cell lines yielded a decreased number of *β*-like cells with impaired glucose response. The authors also generated additional isogenic lines carrying different heterozygous mutations in the *PDX1* transactivation domain, to generate a frame-shift mutation ($PDX1^{+/-}$). This created a more severe phenotype to the one observed in the patient-derived hiPSCs, leading to similar outcomes as obtained from the homozygous isogenic $PDX1^{P33T/P33T}$ point mutated cells. Further transcriptomic analyses of MPCs differentiated from these cell lines ascribed the observed effects to downregulation of key PDX1-bound genes including *MEG3* and *NNA*, which are involved in pancreas development and insulin secretion.

Taken together, these results illustrate how predisposition to develop diabetes can be provoked at the stage of pancreatic endocrine lineage development by genetic mutations on a gene that plays a key role at this timepoint. These anomalies could impair the glucose-responsive function of *β*-like cells through misregulation of genes involved in *β* cell development, maturation, and function. These results also emphasize that the choice between patient-derived hiPSCs or healthy donor hiPSCs with mutations artificially introduced, as well as the choice of the control cell line used, can affect experimental outcomes and their interpretations. In this context, patient-derived hiPSCs could carry additional mutations in non-coding regulatory regions and/or other genes which might further impair the *in vitro* differentiation outcomes. This effects have been elegantly exposed in a recent work by Gadue and colleagues (117), which will be discussed in more detail below, in the *GATA6* section of this review. In contrast, the use of healthy donor hiPSCs with mutations artificially introduced has the advantage of enabling the use of isogenic cell lines (*i.e.* non-mutated hiPSCs) to exclude additional effects of the genetic background.

RFX6

Lack of Rfx6 in mice blocks differentiation of all islet cell types, with the exception of pancreatic-polypeptide-producing cells, while *RFX6* mutations in humans result in PNDM (93, 100, 128). Modeling of the RFX6 requirement for human endocrine pancreas development has been addressed by Zhu et al. Their findings, in agreement with current knowledge, show a reduction of endocrine cell commitment from pancreatic progenitor cells derived from $RFX6^{-/-}$ mutant hPSCs (111).

In a more recent study, Trott et al. used hiPSCs derived from individuals with Mitchell–Riley syndrome (MRS) to specifically associate the role of *RFX6* mutations and the lack of pancreatic endocrine cells in a human model of pancreas development (118). X-ray microtomography of one of these patients confirmed the spectrum of congenital defects typical of MRS (loss of the pancreas body and tail), and exome sequencing identified a homozygous non-sense mutation in *RFX6*. hiPSCs derived from this patient and differentiated along the pancreatic cell lineage revealed that these cells efficiently differentiate into posterior foregut cells but exhibited a reduction in the pancreatic endoderm differentiation, which was accompanied by expression of genes associated with mesoderm differentiation. These findings indicate that RFX6 is crucial for maintaining the transcriptional program that specifies early pancreatic endoderm in humans.

NEUROG3

While loss of Ngn3 function has been associated with complete lack of pancreatic endocrine cells in mice (129), the phenotype in humans is variable [recently reviewed in (21)]. In this sense, while some patients with homozygous or compound heterozygous *NEUROG3* mutations show glycemic control into adulthood, indicating a functional endocrine pancreas, others present neonatal diabetes (96, 130, 131). A recent study suggests that each mutation could have unique effects on the structure and function of NGN3 (132). To further understand this divergence, the requirement of NGN3 for the generation of insulin-producing cells during human development has been addressed using hPSC differentiation. Zhu et al. reported that *in vitro* endocrine pancreatic differentiation of hPSCs with biallelic mutations in *NEUROG3* formed some insulin-producing cells (111), whereas another study reported a total lack of endocrine cells differentiated from *NEUROG3*$^{-/-}$ hPSCs (133). The latter work described that as little as 10% *NEUROG3* expression is sufficient for the formation of pancreatic endocrine cells, supporting that NGN3 is essential for endocrine pancreas development in humans. The divergence between differentiation protocols used in each laboratory and the influence of genetic background could explain the varied phenotypes observed between these two studies. Interestingly, a new adult mouse islet resident pancreatic endocrine progenitor cell population has been recently reported (134). These cells express the surface marker *Procr*, are *Neurog3* negative and, when isolated and co-cultured with endothelial cells, are able to give rise to islet-like clusters containing all endocrine cell types. Apparently, differentiation of this adult progenitor cell population into endocrine cells does not involve *Neurog3* expression, raising the intriguing question of whether such a population exists in humans and, if so, whether *in vitro* pancreatic differentiation from hPSCs is able to follow this "alternative" path for endocrine cell production. Such possibility could explain the divergence between different reports concerning the requirement of NGN3 in endocrine cell production. Taken together, these studies illustrate the complexity, as well as the potential, associated with hPSC differentiation for modeling the impact of genetic mutations on human development.

GLIS3

It has been reported that global *Glis3*$^{-/-}$ mice die of severe neonatal diabetes shortly after birth (135). Minor differences in gene dosage of *Glis3* produce substantive changes in the expression levels of *Neurog3* and *Ins1*, leading to a variable phenotype among the multiple *Glis3*-KO mouse lines (136). In agreement with these phenotypes, human biallelic mutations in *GLIS3* underlie a rare clinical syndrome, characterized by neonatal diabetes and congenital hypothyroidism (92).

The first report of the *in vitro* modeling for the requirement of GLIS3 in human pancreas development was provided by Zhu et al. These researchers did not find defects in pancreatic endocrine differentiation using *GLIS3*$^{-/-}$ mutant hESCs, when using a first generation *in vitro* pancreatic differentiation protocol that allows producing poly-hormonal cells (111).

More recently, Amin et al. developed an improved differentiation protocol that allowed the production of monohormonal β-like cells with enhanced functionality (64). Noteworthy, this protocol allowed the generation of robust *GLIS3* expression at the PDX1+/NKX6.1+ pancreatic progenitor cell stage, in contrast with previously reported protocols (52, 111). Using this improved protocol, they were able to demonstrate that differentiation of *GLIS3*$^{-/-}$ mutant hESCs presented impaired expression of pancreatic endocrine-associated genes, including *PDX1*, *NEUROD1*, *NKX6.1*, and *MAFA*. These cells also showed increased β-like cell death. These findings contrast with those reported by Zhu et al. (111). The difference could be explained by the improvements in the differentiation protocol, which allow a closer recapitulation of the differentiation steps to produce β-like cells. Furthermore, providing an illustrative example of the utility of the *in vitro* β cell differentiation protocols, these researchers performed a chemical screen that allowed the identification of a novel drug candidate that rescued mutant *GLIS3*-associated β-cell death both *in vitro* and *in vivo* (64).

HNF1A

Hnf1a has been shown to regulate the expression pattern of islet-specific genes involved in key functions of this tissue (137). In the mouse, while homozygous knockout (*Hnf1a*$^{-/-}$) results in insulin secretory defects and higher blood glucose concentrations, heterozygous knockout (*Hnf1a*$^{+/-}$) do not display this phenotype (138). This is in sharp contrast with the MODY3 pathology in humans, in which heterozygous mutations result in diabetes (139). In an attempt to elucidate the mechanisms by which dysfunctional HNF1A affects pancreatic development and/or β cell function, Gadue and colleagues have modeled MODY3 using CRISPR-Cas9 genome-edited hESCs and EndoC-BH human cell lines (109). Loss of HNF1A function was accomplished by deletion and premature termination in one or both *HNF1A* alleles, resulting in heterozygous and homozygous KO mutations. Their results suggest that HNF1A plays an essential role in endocrine cell development, as its loss leads to abnormal expression of genes related to β cell function and diabetes. Noteworthy, complete loss of HNF1A did not impair the production of pancreatic progenitors, but this factor was necessary for proper endocrine cell development as revealed by decreased expression of *PAX4*, and impaired insulin expression and secretion. Interestingly, HNF1A loss of function (deletion in one or both alleles of *HNF1A*) led to increased expression of α cell markers, including glucagon. The authors suggest that the increase found in α cells derived from this model system appears to be human-specific, since *Hnf1a* knockout mice do not display this phenotype.

Another key finding of this work was the identification of a previously unannotated human-specific long intergenic non-coding RNA (*lncRNA*). The *LINC01139*, designated *LINKA*, was shown to act as a downstream target of HNF1A. *In vitro* endocrine pancreatic differentiation of *LINKA*-deficient hESCs showed no effect on the production of pancreatic progenitors, but revealed a limited bias towards the production of α cells. Furthermore, β-like cells produced from *LINKA*-deficient hESCs showed a decrease in maximal respiration capacity to a similar extent as seen in the *HNF1A* heterozygous cells. Taken together, their findings point to a role for *LINKA* in the regulation of a subset of HNF1A target genes with implications in cellular respiration. The *in vivo* relevance of *LINKA* for diabetes onset remains to be explored. Of note, a significant variability was observed in the expression changes among the hESC lines used in this study. These could be partially explained by the impact of the genetic background, which could lead to differences in the efficiency of differentiation protocol when applied to each cell line.

A more recent report was provided by Teo and colleagues (110). These researchers used MODY3 patient-derived hiPSCs to study the impact of a recently reported patient-specific heterozygous HNF1A$^{+/H126D}$ mutation (140). The authors used hiPSCs reprogrammed from a healthy donor and H9 hESCs as two independent wild type controls. Molecular dynamics simulations predicted that the H126D mutation could compromise DNA binding and gene target transcription. Indeed, RNA-seq and ChIP-seq analyses performed on MODY3 hiPSC-derived endocrine progenitors revealed that the expression of several HNF1A gene targets was affected by the mutation. An in-depth analysis of the effects on the β-like cells derived from HNF1A$^{+/H126D}$ hiPSCs demonstrated that the *HNF1A* mutation causes a GLUT2 deficiency, that is associated with reduced glucose uptake and ATP production. Their findings reveal the importance of HNF1A in regulating GLUT2 and several genes involved in the MODY3 pathology that may partly account for the lack of insulin secretion clinically observed in these patients. This report extends the findings reported by Cardenas-Diaz et al. (109) by revealing additional mechanisms triggered by the *HNF1A* mutations on the rest of the stimulus-secretion coupling pathway and on HNF1A transcriptional targets in human β-like cells. Noteworthy, Teo and colleagues performed RNA-seq and ChIP-seq at the endocrine progenitor cell stage. They did not found a differential regulation of the *LINC01139* (*LINKA*) at this stage, and unfortunately the expression of this lncRNA in β-like cells derived from HNF1A$^{+/H126D}$ hiPSCs is not reported. It remains to be elucidated whether *LINC01139* is also downregulated in the latter model. Potential discrepancies on the regulation of this lncRNA could be accounted by the different approaches followed in each work to evaluate the effects of HNF1A haploinsufficiency. On one hand Cardenas-Diaz et al. artificially introduced KO mutations by generating a genomic deletion leading to premature termination in one or both *HNF1A* alleles, and non-mutated isogenic cell lines were used as controls. This approach has the advantage of using an isogenic control cell line, which neutralizes contributions from the genomic background. However, the mutations introduced generate a strong HNF1A loss of function that might not appropriately recapitulate the mechanisms that take place in MODY3 patients. On the other hand, Teo and colleagues used MODY3 patient-derived hiPSCs carrying a mutation that causes an amino acid substitution (HNF1A$^{+/H126D}$) and used hiPSCs derived from a healthy donor and H9 hESCs as wild type controls. This approach has the advantage of using hiPSCs derived from patient cells, accounting for a closer model to the MODY3 disease. However, the use of non-isogenic hPSCs as controls does not allow accounting for potential effects derived from the different genomic backgrounds. As presented in more detail in the next section, these might introduce an additional bias in the differentiation outcome. In summary, further studies are required to elucidate whether deregulation of *LINC01139* plays a relevant role in MODY3 diabetes.

GATA 6

Mono allelic mutations in *GATA6* have been linked with pancreas agenesis in humans (141) while the knockout of the same gene has little effect on pancreatic development in the mouse. Indeed, only knockout of both *Gata4* and *Gata6* results in pancreatic agenesis (142, 143). Thus, GATA6 seems to have a different or at least a more extensive function in human development. To confirm this observation, Shi et al. used CRISPR/Cas9 to create hPSCs carrying frameshift mutations in *GATA6*, alone or in combination with *GATA4* mutations (115). Their results show that GATA6$^{+/-}$ haploinsufficiency alters pancreatic progenitor cell differentiation leading to a reduced number of glucose-responsive β-like cells. Given that heterozygous inactivating mutations in *GATA6* have been linked with pancreas agenesis, these findings suggest that the severity of the phenotype could vary according to additional genetic, epigenetic, and/or environmental factors that were not accounted by the differentiation process. Interestingly, the authors also describe dosage-sensitive requirements for GATA6 and GATA4 in the formation of both definitive endoderm and pancreatic progenitor cells, confirming the complex interplays between these factors observed in genetic studies in the mouse.

In another study, Tiyaboonchai et al. used hiPSCs derived from a patient with pancreatic agenesis associated with a heterozygous *GATA6* frameshift mutation, which leads to production of a truncated protein. These researchers also used CRISPR/Cas9 genome editing to introduce this mutation on both alleles of the same hiPSC line (114). Noteworthy, hiPSC lines with homozygous mutations failed to differentiate into endoderm. Re-expression of *GATA6* or other *GATA* family members restored this defect. The use of endodermal progenitor cell lines established from the hiPSC allelic series, which expressed *GATA6* at lower levels but *GATA4* and *GATA3* at higher levels, allowed bypassing the endoderm defect and focusing on pancreatic β cell differentiation. The authors found that all mutant lines were able to differentiate into pancreatic β-like cells, but the response to glucose in these cells was functionally defective. Also, they showed that the clear decrease in pancreas specification and β-like cell generation was associated with limited endogenous retinoic acid signaling during *in vitro* pancreas induction using the *GATA6* mutant cell lines.

Additional information was provided by Chia et al. who combined both gene-edited and patient-derived hPSCs to study the function of GATA6 (116). These authors found that *GATA6* heterozygous hPSCs show a limited reduction in endoderm formation, while *GATA6*-null hPSCs can only form mesoderm-like cells. Thus, GATA6 seems to be upstream of the endoderm program in humans. Consistent with this hypothesis, genome-wide studies showed that GATA6 binds and cooperates with EOMES/SMAD2/3 to regulate the expression of master endoderm genes. In addition, the early deficit of $GATA6^{+/-}$ in definitive endoderm was accompanied by a significant reduction in PDX1+ pancreatic progenitors and C-peptide+ β-like cells. These findings show that, in humans, the formation of definitive endoderm and acquisition of pancreatic fate are exquisitely sensitive to *GATA6* gene dosage.

Taken together, the above-mentioned reports revealed different levels of requirement of GATA6 for pancreatic differentiation between protocols, labs and cell lines (**Table 2**). In this context, a very recent report by Gadue and colleagues provides an illustrative example which might help to understand this apparent divergence. These researchers generated a hiPSC line derived from a pancreatic agenesis patient, harboring a heterozygous 4 bp duplication in exon 2 of *GATA6* leading to a premature STOP codon, a genetically matched control line, and an identically artificially-mutated ESC line. Using these cell lines the authors identified a minor allele frequency of a SNP located downstream of *GATA6* which was associated with the level of expression of this gene (117). In their *in vitro* model, the expression of the GATA6 protein remained depressed in pancreatic progenitor cells even after correction of the coding mutation. Screening the regulatory regions of the *GATA6* gene in the patient cells and an additional pancreas agenesis hiPSC line revealed the above-mentioned SNP. Noteworthy, introducing this non-coding disease modifier SNP by CRISPR/Cas9 in control hESCs confirmed that it depressed *GATA6* expression in pancreas precursors. Thus, the phenotypic diversity found in *GATA6* heterozygous patients and the outcome of *in vitro* studies could be explained in part by this genetic variant.

The findings reported by Gadue and colleagues suggest that caution has to be taken when interpreting the results of monogenic diabetes modeling using patient-derived hiPSCs. Additional genomic variants might contribute to the *in vitro* differentiation outcomes, making it difficult to compare the results obtained by different groups. Nevertheless, some of the studies mentioned above did use the same hPSC line, including the original H9 line derived by JA Thomson and colleagues (144). In such cases, it is worth to underline that each group used different protocols of differentiation. Specific additives could compensate for the decrease in *GATA6* expression. For example, retinoic acid seems to support GATA6 function in pancreatic specification. Addition/increase of this morphogen could modulate the effect of *GATA6* haploinsufficiency. Taken together, these results illustrate the challenges and, at the same time, highlight the unique interest of investigating the function of key transcription factors in pancreatic development using hPSCs.

HNF4A

In mouse, it has been shown that full inactivation Hnf4a is embryonically lethal, while heterozygote knockout mice are normoglycemic and do not present diabetes features (145–147). In contrast, MODY1 patients carrying heterozygous mutations in *HNF4A* present diabetes due to impaired β cell function (148). Patient-derived hiPSCs have been recently used to address the potential mechanisms involved in this phenotype. Ræder and colleagues reported the use of hiPSCs derived from patients carrying a non-sense *HNF4A* mutation associated with MODY1 to study its effect on pancreas and β cell differentiation (105). Noteworthy, the mutation studied in this work (p. Ile271fs) generates a truncated HNF4A product from one of the alleles. The authors show that insulin-positive cells could be generated *in vitro* from these cells, suggesting that this human *HNF4A* mutation neither blocked the expression of the insulin gene nor the production of insulin-producing cells *in vitro*. However, they acknowledge that the insulin-producing cells derived are immature as a result of the β cell differentiation protocol *per se*, leaving open the possibility that HNF4A could have more subtle effects on the functionality of fully mature β cells.

In another study, Braverman-Gross et al. generated hiPSCs from MODY1 patients harboring a different non-sense mutation in the *HNF4A* gene and evaluated its differentiation along the pancreatic lineage (106). In this case, the mutation studied affects all *HNF4A* transcripts and impairs the protein dimerization and transactivation domains. Pancreatic progenitors differentiated from these cells exhibited an upregulation of other key pancreatic transcription factors, including *PAX6*, *NEUROD1*, and *NEUROG3*. The authors suggest that such gene expression increase could be a compensatory mechanism utilized by MODY1 cells to overcome the reduction in *HNF4A* expression. Interestingly, they also note that the differential expression of HNF4A target genes in posterior foregut progenitors derived from mutant cells is affected by the number of HNF4A DNA binding sites, its transcription start site distance, and the number of other transcription factor binding sites. Unfortunately, the authors of this work did not extend the differentiation protocol to evaluate proportion and functionality of β-like cells derived from these hiPSC samples.

MODY1 disease modeling was also more recently accomplished by Teo and colleagues using hiPSCs derived from patients with frameshift mutations that introduce a premature stop codon in *HNF4A*, leading to an unstable mRNA and overall lowered HNF4A levels (107). This mutation is the same one (p. Ile271fs) studied by Ræder and colleagues. Control hiPSC lines were derived from a non-diabetic patient family member. The resulting cell lines were differentiated into liver and pancreatic endocrine cells. Phenotypic analyses showed that *HNF4A* haploinsufficiency affects foregut endoderm gene expression signatures, contributing to long-term consequences on hepatic and pancreatic cell fates. While key developmental genes were perturbed by *HNF4A* haploinsufficiency at the pancreatic progenitor stage (including *HNF1B*, *PDX1*, *GATA4*, and *RFX6*), these mutant hiPSCs were still able to procure β-like cells expressing specific markers, including insulin and C-peptide. However, the β-like cells derived with the assayed *in vitro* protocol were not fully mature. More critical effects of *HNF4A* mutations taking place during the β cell maturation process or on already mature β cells could not be properly evaluated with the protocol described in this work.

Taken together, the results reported so far from hiPSC models used to study the effects of different *HNF4A* mutations suggest that the effects of such mutations might be more relevant at the functional level of the β cells produced. The generation of fully functional β-like cells from *in vitro* differentiation protocols still remains a challenge. Thus, evaluating the functionality of the β-like cells produced from control or patient-derived hiPSC cannot be appropriately assessed with the current differentiation protocols. On the other hand, it should be noted that while two of these studies evaluated the effects of the same mutation, the results described by Braverman-Gross et al. analyzed a different *HNF4A* mutation. These mutations lead to HNF4A loss-of-function through different mechanisms, thus potentially explaining the different outcomes obtained in each of the reports. Last, but not least, it should be kept in mind when using patient-derived hiPSCs that additional mutations in other genes or in *HNF4A* regulatory regions could also modulate the outcome of the *in vitro* differentiation experiments, as illustrated above for *GATA6*. To conclude, additional studies are necessary to address how *HNF4A* mutations cause MODY in humans, especially using the next generation of pancreatic differentiation protocols that improve the production of fully mature β-like cells.

HNF1B

Teo and colleagues established a well-controlled patient-derived hiPSC pancreatic differentiation model to elucidate the molecular mechanisms underlying MODY5 pancreatic hypoplasia (113). Differentiation of MODY5-hiPSCs into pancreatic progenitors showed that the HNF1B$^{S148L/+}$ mutation causes the up-regulation of several key endocrine pancreas-enriched transcription factors including *PDX1*. Pancreatic differentiation using these cells did not block *PDX1*, *PTF1A*, *GATA4*, and *GATA6* expression, suggesting that MODY5-mediated pancreatic hypoplasia in this case is mechanistically independent from the effect associated with these transcription factors. On the other hand, the point mutation in *HNF1B* caused an indirect reduction in the expression of the insulin gene activator *PAX6*, suggesting that loss of one copy of *HNF1B* in humans impairs β cell development and function. Although these findings are consistent with the potential occurrence of maturity-onset diabetes, they fail to uncover the mechanism by which *HNF1B* haploinsufficiency results in pancreatic hypoplasia.

To further address this question, we recently used an alternative hiPSC pancreatic differentiation model to elucidate the molecular mechanisms underlying HNF1B-associated diabetes (Khairi et al, manuscript submitted). To evaluate the transcriptional differences in the *HNF1B* haploinsufficient cells, we used bulk RNA-seq at several stages of the pancreatic differentiation protocol (from DE to β-like cells), immunofluorescence staining, and scRNA-seq at the MPC stage. Our analyses show that absence of HNF1B blocks the specification of the pancreatic fate from the foregut progenitor stage. In contrast, *HNF1B* haploinsufficiency allows differentiation of MPCs and the generation of functional β-like cells although at a lower frequency than the control isogenic cell line. We further report that *HNF1B* haploinsufficiency impairs cell proliferation in foregut progenitors and MPCs. Our results show that HNF1B plays a key role in the production and expansion of pancreatic progenitors and suggest that this factor could regulate the expression of several Hippo pathway components in MPCs. Thus, the level of HNF1B, combined with environmental stimuli, could define the number of pancreatic progenitor cells generated during development and therefore contribute to the susceptibility to diabetes during childhood/adulthood.

PTF1A

It has been described that homozygous inactivating mutations in *PTF1A* cause pancreatic and cerebellar agenesis (98). In agreement, *Ptf1a* null mice present a complete absence of exocrine pancreatic tissue, but all islet endocrine cell types are present until the late stages of embryogenesis (149). Zhu et al. reported the *in vitro* modeling of the PTF1A requirement for human pancreas development. Using *PTF1A*$^{-/-}$ hESCs and a first generation *in vitro* pancreatic differentiation protocol, these researchers did not find defects in pancreatic endocrine differentiation (111). This finding is in agreement with previous reports showing that *Ptf1a* is not required for the specification of Ngn3+ endocrine progenitors or the differentiation of mature β cells in mice (150).

The study of *PTF1A* regulation provides another example of how human *in vitro* pancreatic differentiation can guide the discovery of a developmental regulatory mechanism, in this case consisting in the identification of recessive mutations in a distal non-coding region (151). Identification of genetic mutations resulting in pancreatic agenesis can be challenging as these can be located in regulatory regions far away from known regulators. Accordingly, genome sequencing of a cohort of patients presenting pancreatic agenesis revealed several mutations in a distal non-coding region located >1 Mb upstream the *PTF1A* gene. Enhancer profiling in MPCs, derived *in vitro* from hPSCs, confirmed the functional importance of this regulatory sequence in humans (151). The mutation sites coincided with a FOXA2 binding site profiled by ChIP-seq in *in vitro* MPCs. Further mechanistic experiments performed *in vitro* confirmed that the targeted region acts as an enhancer in human MPCs, and that patient mutations affect PDX1 and FOXA2 binding. These findings allowed us to propose that the mutated enhancer region is in charge of triggering the early *PTF1A* expression in the gut region where the pancreas is specified. This study illustrates how human genetic and *in vitro* differentiation of hPSCs can be combined to define mechanisms driving developmental diseases.

INS

Balboa et al. generated a model based on hiPSCs from patients carrying *INS* mutations and engineered isogenic CRISPR-Cas9 mutation-corrected lines. These cells were differentiated to β-like cells (152). Using this model, the authors show that the *INS* mutations lead to accumulation of proinsulin misfolding, increased signs of ER-stress, and reduced proliferation in *INS*-mutant β-like cells compared with corrected controls. Following transplantation into mice, *INS*-mutant grafts presented reduced insulin secretion and further increased ER-stress, associated with decreased *PDX1* expression and β cell size, as well as mitochondrial alterations. The authors conclude that neonatal diabetes-associated *INS*-mutations lead to defective β cell mass expansion, contributing to neonatal diabetes development.

In another recent study, Egli and colleagues generated hiPSCs from fibroblasts of a patient with PNDM and undetectable insulin at birth due to a homozygous mutation in the translation start site of the insulin gene (153). Their results show that the differentiation of *INS* mutant cells resulted in hormone-negative hiPSCs, and the correction of this mutation by CRISPR-Cas9 restored insulin production and secretion to levels comparable to those of wild type endocrine cells. The authors also demonstrate that the insulin-producing cells of corrected patient hiPSCs protect mice from diabetes, providing a proof-of-principle study for the use of replacement therapy as a treatment for monogenic diabetes.

STAT3

Saarimäki-Vire et al. used hiPSCs derived from a patient with PNDM and pancreatic hypoplasia to investigate the effects of an activating *STAT3* mutation on pancreatic development (154). Noteworthy, the mutation studied has been identified as the cause of PNDM in association with early onset autoimmunity. These authors demonstrate that the mutation in *STAT3* leads to premature endocrine differentiation through binding and direct induction of *NEUROG3* by the increased nuclear shuttling of the mutated protein. They also showed that correction of the *STAT3* mutation using CRISPR/Cas9 completely reversed the disease phenotype. These results demonstrate that, in addition to the early onset autoimmunity, the same mutation leads to a primary developmental defect in pancreatic organogenesis.

CONCLUSION AND FUTURE DIRECTIONS

The field of hPSCs has allowed important advances in our understanding of the molecular mechanisms underlying the different forms of monogenic diabetes. Indeed, the establishment of hPSC-based *in vitro* platforms offers a unique opportunity to study pancreas development and to investigate the pathophysiology underlying monogenic diabetes. This basic knowledge paves the way to the development of new treatments, not only for diabetes induced by genetic mutations, but also more broadly for personalized medicine therapies in the context of type I and type II diabetes. Nonetheless, several challenges require attention. Current *in vitro* β-like cell differentiation protocols have been markedly improved and may be sufficient to recapitulate several of the MODY phenotypes in the hPSC-based model. However, one of their greatest limitations remains the lack of metabolic maturation of the β-like cells derived. A solution to turn the differentiated cells into fully mature and functional β cells has been their transplantation in mouse to allow for the latest steps of cell differentiation to take place *in vivo*. Alternative methods involve culture in 3D and cell self-aggregation into islet-like clusters to produce β-like cells with improved functionality. The emergence of scRNA-seq is expected to lead to the identification of new markers involved in pancreatic β cell maturation, thus allowing improved benchmarking of the *in vitro* differentiation protocol outcomes. Also, scRNA-seq applied to human embryonic pancreatic tissue might provide additional insights into the developmental cues that differ among mice and humans. This will provide additional input to improve the *in vitro* differentiation protocols by modulating yet unknown signaling cues.

The other growing challenge is the divergence of results between different groups studying the same mutation/genes but using either different hPSC lines and/or different protocols. Indeed, genetic background and culture conditions can have a strong effect on phenotype, thus leading to different experimental outcomes. New hiPSC lines derived from monogenic diabetic patients continue to be reported. A very recent study described the generation of hiPSCs derived from MODY2 patients, but in this case its differentiation into the pancreatic lineage was not evaluated so far (108). Thus, there is a need to develop standard hiPSC lines which could be shared between laboratories. More importantly, the use of "universal" culture conditions to grow and to differentiate hPSC lines would be incredibly useful to allow the comparison of data generated and to precisely establish the importance of genetic background on the phenotype observed *in vitro*. Such standardization implies that culture conditions are fully described and shared between laboratories. The use of isogenic control hPSC lines is also essential and is helping to overcome the limitations related to the variability between lines, especially when compared with the use of family controls, which is inherent to differences in genetic background. Numerous studies have successfully used CRISPR/Cas9 tools to generate isogenic hPSC lines by introducing patient-specific mutations, editing genes in control-hPSC lines to investigate the implication of a single genetic variant on β cell differentiation and function. Here, we revisited the latest advances in the application of *in vitro* pancreatic cell differentiation from hPSCs to model several types of monogenic diabetes. Much work remains to be done to improve the modeling of monogenic diabetes but, as it stems from this review, *in vitro* pancreatic differentiation from hPSCs is definitely gaining momentum.

AUTHOR CONTRIBUTIONS

All authors reviewed the literature, wrote, and edited the paper. JIB prepared the table. LV and SAR-S conceptualized the review topic and contents, and approved the paper. All authors contributed to the article and approved the submitted version.

FUNDING

This work was funded by Office of the Royal Society and the Consejo Nacional de Investigaciones Científicas y Técnicas of Argentina (IEC\R2\181023 to SAR-S and LV). Work in the Rodríguez-Seguí laboratory is supported by grants from the Agencia Nacional de Promoción Científica y Tecnológica of Argentina (ANPCyT: PICT-2015 3605, PICT-2017 2071) and the University of Buenos Aires (UBACyT 20020170200156BA). Work in the Vallier laboratory is supported by grants from European Research Council Grant New-Chol, the Cambridge Hospitals National Institute for Health Research Biomedical Research Center and core support grant from the Wellcome and Medical Research Council to the Wellcome-Medical Research Council Cambridge Stem Cell Institute.

ACKNOWLEDGMENTS

We apologize to our colleagues whose references were omitted owing to space constraints. SAR-S is a career investigator from the Consejo Nacional de Investigaciones Científicas y Técnicas of Argentina (CONICET). JIB is supported by a postdoctoral fellowship from the ANPCyT.

REFERENCES

1. Atkinson MA, Eisenbarth GS, Michels AW. Type 1 Diabetes. *Lancet* (2014) 383:69–82. doi: 10.1016/S0140-6736(13)60591-7
2. Matveyenko AV, Butler PC. Beta-Cell Deficit Due to Increased Apoptosis in the Human Islet Amyloid Polypeptide Transgenic (HIP) Rat Recapitulates the Metabolic Defects Present in Type 2 Diabetes. *Diabetes* (2006) 55:2106–14. doi: 10.2337/db05-1672
3. Butler AE, Janson J, Bonner-Weir S, Ritzel R, Rizza RA, Butler PC. Beta-Cell Deficit and Increased Beta-Cell Apoptosis in Humans With Type 2 Diabetes. *Diabetes* (2003) 52:102–10. doi: 10.2337/diabetes.52.1.102
4. Weyer C, Bogardus C, Mott DM, Pratley RE. The Natural History of Insulin Secretory Dysfunction and Insulin Resistance in the Pathogenesis of Type 2 Diabetes Mellitus. *J Clin Invest* (1999) 104:787–94. doi: 10.1172/JCI7231
5. Malecki MT. The Search for Undiagnosed MODY Patients: What Is the Next Step? *Diabetologia* (2010) 53:2465–7. doi: 10.1007/s00125-010-1908-4
6. Murphy R, Ellard S, Hattersley AT. Clinical Implications of a Molecular Genetic Classification of Monogenic Beta-Cell Diabetes. *Nat Clin Pract Endocrinol Metab* (2008) 4:200–13. doi: 10.1038/ncpendmet0778
7. Rubio-Cabezas O, Hattersley AT, Njolstad PR, Mlynarski W, Ellard S, White N, et al. ISPAD Clinical Practice Consensus Guidelines 2014. The Diagnosis and Management of Monogenic Diabetes in Children and Adolescents. *Pediatr Diabetes* (2014) 15 Suppl 20:47–64. doi: 10.1111/pedi.12192
8. Bansal V, Gassenhuber J, Phillips T, Oliveira G, Harbaugh R, Villarasa N, et al. Spectrum of Mutations in Monogenic Diabetes Genes Identified From High-Throughput DNA Sequencing of 6888 Individuals. *BMC Med* (2017) 15:213–3. doi: 10.1186/s12916-017-0977-3
9. Fujita Y, Tanaka D, Tatsuoka H, Matsubara M, Hyo T, Hamamoto Y, et al. A Novel Splice-Site Mutation of the HNF1B Gene in a Family With Maturity Onset Diabetes of the Young Type 5 (MODY5). *Endocrinol Diabetes Metab Case Rep* (2020) 2020:20-0092. doi: 10.1530/edm-20-0092
10. Gragnoli C, Stanojevic V, Gorini A, Preussenthal G, Thomas M, Habener J. IPF-1/MODY4 Gene Missense Mutation in an Italian Family With Type 2 and Gestational Diabetes. *Metabolism: Clin Exp* (2005) 54:983–8. doi: 10.1016/j.metabol.2005.01.037
11. Patel KA, Kettunen J, Laakso M, Stančáková A, Laver TW, Colclough K, et al. Heterozygous RFX6 Protein Truncating Variants Are Associated With MODY With Reduced Penetrance. *Nat Commun* (2017) 8:888. doi: 10.1038/s41467-017-00895-9
12. Xiao X, Liu L, Xiao Y, Xie Z, Li L, Zhou H, et al. A Novel Frameshift Mutation in the Insulin Gene in a Family With Maturity Onset Diabetes of the Young. *J Diabetes* (2019) 11:83–6. doi: 10.1111/1753-0407.12849
13. Stoffers DA, Zinkin NT, Stanojevic V, Clarke WL, Habener JF. Pancreatic Agenesis Attributable to a Single Nucleotide Deletion in the Human IPF1 Gene Coding Sequence. *Nat Genet* (1997) 15:106–10. doi: 10.1038/ng0197-106
14. Horikawa Y, Iwasaki N, Hara M, Furuta H, Hinokio Y, Cockburn BN, et al. Mutation in Hepatocyte Nuclear Factor 1β Gene (TCF2) Associated With MODY. *Nat Genet* (1997) 17:384–5. doi: 10.1038/ng1297-384
15. Hattersley AT, Patel KA. Precision Diabetes: Learning From Monogenic Diabetes. *Diabetologia* (2017) 60:769–77. doi: 10.1007/s00125-017-4226-2
16. Loo LSW, Lau HH, Jasmen JB, Lim CS, Teo AKK. An Arduous Journey From Human Pluripotent Stem Cells to Functional Pancreatic β Cells. *Diabetes Obes Metab* (2017) 20:3–13. doi: 10.1111/dom.12996
17. Santosa MM, Low BSJ, Pek NMQ, Teo AKK. Knowledge Gaps in Rodent Pancreas Biology: Taking Human Pluripotent Stem Cell-Derived Pancreatic Beta Cells Into Our Own Hands. *Front Endocrinol* (2016) 6:194. doi: 10.3389/fendo.2015.00194
18. Bakhti M, Böttcher A, Lickert H. Modelling the Endocrine Pancreas in Health and Disease. *Nat Rev Endocrinol* (2019) 15:155–71. doi: 10.1038/s41574-018-0132-z
19. Pan FC, Wright C. Pancreas Organogenesis: From Bud to Plexus to Gland. *Dev Dyn* (2011) 240:530–65. doi: 10.1002/dvdy.22584
20. Nair G, Hebrok M. Islet Formation in Mice and Men: Lessons for the Generation of Functional Insulin-Producing Beta-Cells From Human Pluripotent Stem Cells. *Curr Opin Genet Dev* (2015) 32:171–80. doi: 10.1016/j.gde.2015.03.004
21. Jennings RE, Scharfmann R, Staels W. Transcription Factors That Shape the Mammalian Pancreas. *Diabetologia* (2020) 63:1974–80. doi: 10.1007/s00125-020-05161-0
22. Millette K, Georgia S. Gene Editing and Human Pluripotent Stem Cells: Tools for Advancing Diabetes Disease Modeling and Beta-Cell Development. *Curr Diabetes Rep* (2017) 17:116. doi: 10.1007/s11892-017-0947-3
23. Hu M, Cherkaoui I, Misra S, Rutter GA. Functional Genomics in Pancreatic Beta Cells: Recent Advances in Gene Deletion and Genome Editing Technologies for Diabetes Research. *Front Endocrinol* (2020) 11:576632. doi: 10.3389/fendo.2020.576632
24. Petersen MBK, Gonçalves CAC, Kim YH, Grapin-Botton A. Chapter Six - Recapitulating and Deciphering Human Pancreas Development From Human Pluripotent Stem Cells in a Dish. *Curr Topics Dev Biol Acad Press* (2018) 129:143–90. pp. doi: 10.1016/bs.ctdb.2018.02.009
25. Yiangou L, Ross ADB, Goh KJ, Vallier L. Human Pluripotent Stem Cell-Derived Endoderm for Modeling Development and Clinical Applications. *Cell Stem Cell* (2018) 22:485–99. doi: 10.1016/j.stem.2018.03.016
26. Efrat S. Epigenetic Memory: Lessons From Ips Cells Derived From Human β Cells. *Front Endocrinol* (2020) 11:614234. doi: 10.3389/fendo.2020.614234
27. Gaertner B, Carrano AC, Sander M. Human Stem Cell Models: Lessons for Pancreatic Development and Disease. *Genes Dev* (2019) 33:1475–90. doi: 10.1101/gad.331397.119
28. Balboa D, Saarimäki-Vire J, Otonkoski T. Concise Review: Human Pluripotent Stem Cells for the Modeling of Pancreatic β-Cell Pathology. *Stem Cells* (2019) 37:33–41. doi: 10.1002/stem.2913
29. Abdelalim EM. Modeling Different Types of Diabetes Using Human Pluripotent Stem Cells. *Cell Mol Life Sci* (2020) 78:2459–83. doi: 10.1007/s00018-020-03710-9
30. Amirruddin NS, Low BSJ, Lee KO, Tai ES, Teo AKK. New Insights Into Human Beta Cell Biology Using Human Pluripotent Stem Cells. *Semin Cell Dev Biol* (2019) 103:31–40. doi: 10.1016/j.semcdb.2019.11.004
31. Balboa D, Iworima DG, Kieffer TJ. Human Pluripotent Stem Cells to Model Islet Defects in Diabetes. *Front Endocrinol* (2021) 12:642152. doi: 10.3389/fendo.2021.642152
32. Jennings RE, Berry AA, Kirkwood-Wilson R, Roberts NA, Hearn T, Salisbury RJ, et al. Development of the Human Pancreas From Foregut to Endocrine Commitment. *Diabetes* (2013) 62:3514–22. doi: 10.2337/db12-1479
33. Jeon J, Correa-Medina M, Ricordi C, Edlund H, Diez JA. Endocrine Cell Clustering During Human Pancreas Development. *J Histochem Cytochem* (2009) 57:811–24. doi: 10.1369/jhc.2009.953307
34. Salisbury RJ, Blaylock J, Berry AA, Jennings RE, De Krijger R, Piper Hanley K, et al. The Window Period of NEUROGENIN3 During Human Gestation. *Islets* (2014) 6:e954436. doi: 10.4161/19382014.2014.954436
35. Sarkar SA, Kobberup S, Wong R, Lopez AD, Quayum N, Still T, et al. Global Gene Expression Profiling and Histochemical Analysis of the Developing Human Fetal Pancreas. *Diabetologia* (2008) 51:285–97. doi: 10.1007/s00125-007-0880-0
36. Oliver-Krasinski JM, Stoffers DA. On the Origin of the Beta Cell. *Genes Dev* (2008) 22:1998–2021. doi: 10.1101/gad.1670808

37. Zaret KS. Genetic Programming of Liver and Pancreas Progenitors: Lessons for Stem-Cell Differentiation. *Nat Rev Genet* (2008) 9:329–40. doi: 10.1038/nrg2318
38. Byrnes LE, Wong DM, Subramaniam M, Meyer NP, Gilchrist CL, Knox SM, et al. Lineage Dynamics of Murine Pancreatic Development at Single-Cell Resolution. *Nat Commun* (2018) 9:3922. doi: 10.1038/s41467-018-06176-3
39. Scavuzzo MA, Hill MC, Chmielowiec J, Yang D, Teaw J, Sheng K, et al. Endocrine Lineage Biases Arise in Temporally Distinct Endocrine Progenitors During Pancreatic Morphogenesis. *Nat Commun* (2018) 9:3356. doi: 10.1038/s41467-018-05740-1
40. Yu XX, Qiu WL, Yang L, Zhang Y, He MY, Li LC, et al. Defining Multistep Cell Fate Decision Pathways During Pancreatic Development at Single-Cell Resolution. *EMBO J* (2019) 38:e100164. doi: 10.15252/embj.2018100164
41. Bastidas-Ponce A, Tritschler S, Dony L, Scheibner K, Tarquis-Medina M, Salinno C, et al. Comprehensive Single Cell Mrna Profiling Reveals a Detailed Roadmap for Pancreatic Endocrinogenesis. *Development* (2019) 146:dev173849. doi: 10.1242/dev.173849
42. van Gurp L, Muraro MJ, Dielen T, Seneby L, Dharmadhikari G, Gradwohl G, et al. A Transcriptomic Roadmap to Alpha- and Beta-Cell Differentiation in the Embryonic Pancreas. *Development* (2019) 146:dev173716. doi: 10.1242/dev.173716
43. George NM, Day CE, Boerner BP, Johnson RL, Sarvetnick NE. Hippo Signaling Regulates Pancreas Development Through Inactivation of Yap. *Mol Cell Biol* (2012) 32:5116–28. doi: 10.1128/MCB.01034-12
44. Gao T, Zhou D, Yang C, Singh T, Penzo-Mendez A, Maddipati R, et al. Hippo Signaling Regulates Differentiation and Maintenance in the Exocrine Pancreas. *Gastroenterology* (2013) 144:1543–53.e1. doi: 10.1053/j.gastro.2013.02.037
45. Cebola I, Rodriguez-Segui SA, Cho CH, Bessa J, Rovira M, Luengo M, et al. TEAD and YAP Regulate the Enhancer Network of Human Embryonic Pancreatic Progenitors. *Nat Cell Biol* (2015) 17:615–26. doi: 10.1038/ncb3160
46. Rodriguez-Segui SA, Bessa J. Gatekeepers of Pancreas: TEAD and YAP. *Oncotarget* (2015) 6:15736–7. doi: 10.18632/oncotarget.4607
47. Guo T, Landsman L, Li N, Hebrok M. Factors Expressed by Murine Embryonic Pancreatic Mesenchyme Enhance Generation of Insulin-Producing Cells From Hescs. *Diabetes* (2013) 62:1581–92. doi: 10.2337/db12-0167
48. Landsman L, Nijagal A, Whitchurch TJ, Vanderlaan RL, Zimmer WE, Mackenzie TC, et al. Pancreatic Mesenchyme Regulates Epithelial Organogenesis Throughout Development. *PloS Biol* (2011) 9:e1001143. doi: 10.1371/journal.pbio.1001143
49. Sneddon JB, Borowiak M, Melton DA. Self-Renewal of Embryonic-Stem-Cell-Derived Progenitors by Organ-Matched Mesenchyme. *Nature* (2012) 491:765–8. doi: 10.1038/nature11463
50. Cozzitorto C, Mueller L, Ruzittu S, Mah N, Willnow D, Darrigrand J-F, et al. A Specialized Niche in the Pancreatic Microenvironment Promotes Endocrine Differentiation. *Dev Cell* (2020) 55:150–162.e6. doi: 10.1016/j.devcel.2020.08.003
51. D'Amour KA, Agulnick AD, Eliazer S, Kelly OG, Kroon E, Baetge EE. Efficient Differentiation of Human Embryonic Stem Cells to Definitive Endoderm. *Nat Biotechnol* (2005) 23:1534–41. doi: 10.1038/nbt1163
52. D'Amour KA, Bang AG, Eliazer S, Kelly OG, Agulnick AD, Smart NG, et al. Production of Pancreatic Hormone Expressing Endocrine Cells From Human Embryonic Stem Cells. *Nat Biotechnol* (2006) 24:1392–401. doi: 10.1038/nbt1259
53. Kroon E, Martinson LA, Kadoya K, Bang AG, Kelly OG, Eliazer S, et al. Pancreatic Endoderm Derived From Human Embryonic Stem Cells Generates Glucose-Responsive Insulin-Secreting Cells *In Vivo*. *Nat Biotechnol* (2008) 26:443–52. doi: 10.1038/nbt1393
54. Nostro MC, Sarangi F, Ogawa S, Holtzinger A, Corneo B, Li X, et al. Stage-Specific Signaling Through Tgfβ Family Members and WNT Regulates Patterning and Pancreatic Specification of Human Pluripotent Stem Cells. *Development* (2011) 138:861–71. doi: 10.1242/dev.055236
55. Cho CH, Hannan NR, Docherty FM, Docherty HM, Joao Lima M, Trotter MW, et al. Inhibition of Activin/Nodal Signalling Is Necessary for Pancreatic Differentiation of Human Pluripotent Stem Cells. *Diabetologia* (2012) 55:3284–95. doi: 10.1007/s00125-012-2687-x
56. Jennings RE, Berry AA, Strutt JP, Gerrard DT, Hanley NA. Human Pancreas Development. *Development* (2015) 142:3126–37. doi: 10.1242/dev.120063
57. Jennings RE, Berry AA, Gerrard DT, Wearne SJ, Strutt J, Withey S, et al. Laser Capture and Deep Sequencing Reveals the Transcriptomic Programmes Regulating the Onset of Pancreas and Liver Differentiation in Human Embryos. *Stem Cell Rep* (2017) 9:1387–94. doi: 10.1016/j.stemcr.2017.09.018
58. Villani V, Thornton ME, Zook HN, Crook CJ, Grubbs BH, Orlando G, et al. SOX9+/PTF1A+ Cells Define the Tip Progenitor Cells of the Human Fetal Pancreas of the Second Trimester. *Stem Cells Trans Med* (2019) 8:1249–64. doi: 10.1002/sctm.19-0231
59. Ramond C, Beydag-Tasoz BS, Azad A, van de Bunt M, Petersen MBK, Beer NL, et al. Understanding Human Fetal Pancreas Development Using Subpopulation Sorting, RNA Sequencing and Single-Cell Profiling. *Development* (2018) 145:dev165480. doi: 10.1242/dev.165480
60. Ramond C, Glaser N, Berthault C, Ameri J, Kirkegaard JS, Hansson M, et al. Reconstructing Human Pancreatic Differentiation by Mapping Specific Cell Populations During Development. *Elife* (2017) 6:e27564. doi: 10.7554/eLife.27564
61. Pagliuca FW, Millman JR, Gurtler M, Segel M, Van Dervort A, Ryu JH, et al. Generation of Functional Human Pancreatic Beta Cells *In Vitro*. *Cell* (2014) 159:428–39. doi: 10.1016/j.cell.2014.09.040
62. Rezania A, Bruin JE, Arora P, Rubin A, Batushansky I, Asadi A, et al. Reversal of Diabetes With Insulin-Producing Cells Derived *In Vitro* From Human Pluripotent Stem Cells. *Nat Biotechnol* (2014) 32:1121–33. doi: 10.1038/nbt.3033
63. Russ HA, Parent AV, Ringler JJ, Hennings TG, Nair GG, Shveygert M, et al. Controlled Induction of Human Pancreatic Progenitors Produces Functional Beta-Like Cells. *Vitro EMBO J* (2015) 34:1759–72. doi: 10.15252/embj.201591058
64. Amin S, Cook B, Zhou T, Ghazizadeh Z, Lis R, Zhang T.A.-O.X., et al. Discovery of a Drug Candidate for GLIS3-Associated Diabetes. *Nat Commun* (2018) 9:2681. doi: 10.1038/s41467-018-04918-x
65. Sui L, Danzl N, Campbell SR, Viola R, Williams D, Xing Y, et al. β-Cell Replacement in Mice Using Human Type 1 Diabetes Nuclear Transfer Embryonic Stem Cells. *Diabetes* (2017) 67:26–35. doi: 10.2337/db17-0120
66. Veres A, Faust AL, Bushnell HL, Engquist EN, Kenty JH, Harb G, et al. Charting Cellular Identity During Human *In Vitro* Beta-Cell Differentiation. *Nature* (2019) 569:368–73. doi: 10.1038/s41586-019-1168-5
67. Nair GG, Liu JS, Russ HA, Tran S, Saxton MS, Chen R, et al. Recapitulating Endocrine Cell Clustering in Culture Promotes Maturation of Human Stem-Cell-Derived β Cells. *Nat Cell Biol* (2019) 21:263–74. doi: 10.1038/s41556-018-0271-4
68. Velazco-Cruz L, Song J, Maxwell KG, Goedegebuure MM, Augsornworawat P, Hogrebe NJ, et al. Acquisition of Dynamic Function in Human Stem Cell-Derived β Cells. *Stem Cell Rep* (2019) 12:351–65. doi: 10.1016/j.stemcr.2018.12.012
69. Hogrebe NA-O, Augsornworawat PA-O, Maxwell KA-O, Velazco-Cruz L, Millman JA-O. Targeting the Cytoskeleton to Direct Pancreatic Differentiation of Human Pluripotent Stem Cells. *Nat Biotechnol* (2020) 38:460–70. doi: 10.1038/s41587-020-0430-6
70. Mamidi A, Prawiro C, Seymour PA, de Lichtenberg KH, Jackson A, Serup P, et al. Mechanosignalling *via* Integrins Directs Fate Decisions of Pancreatic Progenitors. *Nature* (2018) 564:114–8. doi: 10.1038/s41586-018-0762-2
71. Rosado-Olivieri EA, Anderson K, Kenty JH, Melton DA. YAP Inhibition Enhances the Differentiation of Functional Stem Cell-Derived Insulin-Producing β Cells. *Nat Commun* (2019) 10:1464. doi: 10.1038/s41467-019-09404-6
72. Xie R, Everett LJ, Lim HW, Patel NA, Schug J, Kroon E, et al. Dynamic Chromatin Remodeling Mediated by Polycomb Proteins Orchestrates Pancreatic Differentiation of Human Embryonic Stem Cells. *Cell Stem Cell* (2013) 12:224–37. doi: 10.1016/j.stem.2012.11.023
73. Loh KM, Ang LT, Zhang J, Kumar V, Ang J, Auyeong JQ, et al. Efficient Endoderm Induction From Human Pluripotent Stem Cells by Logically Directing Signals Controlling Lineage Bifurcations. *Cell Stem Cell* (2014) 14:237–52. doi: 10.1016/j.stem.2013.12.007
74. Wang A, Yue F, Li Y, Xie R, Harper T, Patel NA, et al. Epigenetic Priming of Enhancers Predicts Developmental Competence of Hesc-Derived

Endodermal Lineage Intermediates. *Cell Stem Cell* (2015) 16:386–99. doi: 10.1016/j.stem.2015.02.013

75. Li J, Wu X, Zhou Y, Lee M, Guo L, Han W, et al. Decoding the Dynamic DNA Methylation and Hydroxymethylation Landscapes in Endodermal Lineage Intermediates During Pancreatic Differentiation of Hesc. *Nucleic Acids Res* (2018) 46:2883–900. doi: 10.1093/nar/gky063
76. van Arensbergen J, Garcia-Hurtado J, Moran I, Maestro MA, Xu XB, Van de Casteele M, et al. Derepression of Polycomb Targets During Pancreatic Organogenesis Allows Insulin-Producing Beta-Cells to Adopt a Neural Gene Activity Program. *Genome Res* (2010) 20:722–32. doi: 10.1101/gr.101709.109
77. van Arensbergen J, Garcia-Hurtado J, Maestro MA, Correa-Tapia M, Rutter GA, Vidal M, et al. Ring1b Bookmarks Genes in Pancreatic Embryonic Progenitors for Repression in Adult Beta Cells. *Genes Dev* (2013) 27:52–63. doi: 10.1101/gad.206094.112
78. Xu CR, Li LC, Donahue G, Ying L, Zhang YW, Gadue P, et al. Dynamics of Genomic H3k27me3 Domains and Role of EZH2 During Pancreatic Endocrine Specification. *EMBO J* (2014) 33:2157–70. doi: 10.15252/embj.201488671
79. Petersen MBK, Azad A, Ingvorsen C, Hess K, Hansson M, Grapin-Botton A, et al. Single-Cell Gene Expression Analysis of a Human ESC Model of Pancreatic Endocrine Development Reveals Different Paths to Beta-Cell Differentiation. *Stem Cell Rep* (2017) 9:1246–61. doi: 10.1016/j.stemcr.2017.08.009
80. Schaffer AE, Freude KK, Nelson SB, Sander M. Nkx6 Transcription Factors and Ptf1a Function as Antagonistic Lineage Determinants in Multipotent Pancreatic Progenitors. *Dev Cell* (2010) 18:1022–9. doi: 10.1016/j.devcel.2010.05.015
81. Aigha I, Memon B, Elsayed A, Abdelalim E. Differentiation of Human Pluripotent Stem Cells Into Two Distinct NKX6.1 Populations of Pancreatic Progenitors. *Stem Cell Res Ther* (2018) 9:83. doi: 10.1186/s13287-018-0834-0
82. Krentz NAJ, Lee MYY, Xu EE, Sproul SLJ, Maslova A, Sasaki S, et al. Single-Cell Transcriptome Profiling of Mouse and Hesc-Derived Pancreatic Progenitors. *Stem Cell Rep* (2018) 11:1551–64. doi: 10.1016/j.stemcr.2018.11.008
83. Barkas N, Petukhov V, Nikolaeva D, Lozinsky Y, Demharter S, Khodosevich K, et al. Joint Analysis of Heterogeneous Single-Cell RNA-Seq Dataset Collections. *Nat Methods* (2019) 16:695–8. doi: 10.1038/s41592-019-0466-z
84. Korsunsky I, Millard N, Fan J, Slowikowski K, Zhang F, Wei K, et al. Loh, P.-R., and Raychaudhuri, s., Fast, Sensitive and Accurate Integration of Single-Cell Data With Harmony. *Nat Methods* (2019) 16:1289–96. doi: 10.1038/s41592-019-0619-0
85. Helman A, Cangelosi AL, Davis JC, Pham Q, Rothman A, Faust AL, et al. A Nutrient-Sensing Transition at Birth Triggers Glucose-Responsive Insulin Secretion. *Cell Metab* (2020) 31:1004–1016.e5. doi: 10.1016/j.cmet.2020.04.004
86. Hattersley AT, Turner RC, Patel P, O'Rahilly S, Wainscoat JS, Permutt MA, et al. Linkage of Type 2 Diabetes to the Glucokinase Gene. *Lancet* (1992) 339:1307–10. doi: 10.1016/0140-6736(92)91958-B
87. Froguel P, Vaxillaire M, Sun F, Velho G, Zouali H, Butel MO, et al. Close Linkage of Glucokinase Locus on Chromosome 7p to Early-Onset non-Insulin-Dependent Diabetes Mellitus. *Nature* (1992) 356:162–4. doi: 10.1038/356162a0
88. Yamagata K, Oda N, Kaisaki PJ, Menzel S, Furuta H, Vaxillaire M, et al. Mutations in the Hepatocyte Nuclear Factor-1 Alpha Gene in Maturity-Onset Diabetes of the Young (MODY3). *Nature* (1996) 384:455–8. doi: 10.1038/384455a0
89. Yamagata K, Furuta H, Oda N, Kaisaki PJ, Menzel S, Cox NJ, et al. Mutations in the Hepatocyte Nuclear Factor-4 Alpha Gene in Maturity-Onset Diabetes of the Young (MODY1). *Nature* (1996) 384:458–60. doi: 10.1038/384458a0
90. Horikawa Y, Iwasaki N, Hara M, Furuta H, Hinokio Y, Cockburn BN, et al. Mutation in Hepatocyte Nuclear Factor-1 Beta Gene (TCF2) Associated With MODY. *Nat Genet* (1997) 17:384–5. doi: 10.1038/ng1297-384
91. Firdous P, Nissar K, Ali S, Ganai BA, Shabir U, Hassan T, et al. Genetic Testing of Maturity-Onset Diabetes of the Young Current Status and Future Perspectives. *Front Endocrinol* (2018) 9:253. doi: 10.3389/fendo.2018.00253
92. Senee V, Chelala C, Duchatelet S, Feng D, Blanc H, Cossec JC, et al. Mutations in GLIS3 Are Responsible for a Rare Syndrome With Neonatal Diabetes Mellitus and Congenital Hypothyroidism. *Nat Genet* (2006) 38:682–7. doi: 10.1038/ng1802
93. Smith SB, Qu HQ, Taleb N, Kishimoto NY, Scheel DW, Lu Y, et al. Rfx6 Directs Islet Formation and Insulin Production in Mice and Humans. *Nature* (2010) 463:775–80. doi: 10.1038/nature08748
94. Flanagan SE, De Franco E, Lango Allen H, Zerah M, Abdul-Rasoul MM, Edge JA, et al. Analysis of Transcription Factors Key for Mouse Pancreatic Development Establishes NKX2-2 and MNX1 Mutations as Causes of Neonatal Diabetes in Man. *Cell Metab* (2014) 19:146–54. doi: 10.1016/j.cmet.2013.11.021
95. Bonnefond A, Vaillant E, Philippe J, Skrobek B, Lobbens S, Yengo L, et al. Transcription Factor Gene MNX1 Is a Novel Cause of Permanent Neonatal Diabetes in a Consanguineous Family. *Diabetes Metab* (2013) 39:276–80. doi: 10.1016/j.diabet.2013.02.007
96. Rubio-Cabezas O, Jensen JN, Hodgson MI, Codner E, Ellard S, Serup P, et al. Permanent Neonatal Diabetes and Enteric Anendocrinosis Associated With Biallelic Mutations in NEUROG3. *Diabetes* (2011) 60:1349. doi: 10.2337/db10-1008
97. Rubio-Cabezas O, Minton JAL, Kantor I, Williams D, Ellard S, Hattersley AT. Homozygous Mutations in NEUROD1 Are Responsible for a Novel Syndrome of Permanent Neonatal Diabetes and Neurological Abnormalities. *Diabetes* (2010) 59:2326. doi: 10.2337/db10-0011
98. Sellick GS, Barker KT, Stolte-Dijkstra I, Fleischmann C, J Coleman R, Garrett C, et al. Mutations in PTF1A Cause Pancreatic and Cerebellar Agenesis. *Nat Genet* (2004) 36:1301–5. doi: 10.1038/ng1475
99. Dimitri P, Warner JT, Minton JA, Patch AM, Ellard S, Hattersley AT, et al. Novel GLIS3 Mutations Demonstrate an Extended Multisystem Phenotype. *Eur J Endocrinol* (2011) 164:437–43. doi: 10.1530/EJE-10-0893
100. Concepcion JP, Reh CS, Daniels M, Liu X, Paz VP, Ye H, et al. Neonatal Diabetes, Gallbladder Agenesis, Duodenal Atresia, and Intestinal Malrotation Caused by a Novel Homozygous Mutation in RFX6. *Pediatr Diabetes* (2014) 15:67–72. doi: 10.1111/pedi.12063
101. Rieck S, Bankaitis ED, Wright CV. Lineage Determinants in Early Endocrine Development. *Semin Cell Dev Biol* (2012) 23:673–84. doi: 10.1016/j.semcdb.2012.06.005
102. Clissold RL, Hamilton AJ, Hattersley AT, Ellard S, Bingham C. HNF1B-Associated Renal and Extra-Renal Disease-an Expanding Clinical Spectrum. *Nat Rev Nephrol* (2014) 11:102–12. doi: 10.1038/nrneph.2014.232
103. Edghill EL, Bingham C, Slingerland AS, Minton JA, Noordam C, Ellard S, et al. Hepatocyte Nuclear Factor-1 Beta Mutations Cause Neonatal Diabetes and Intrauterine Growth Retardation: Support for a Critical Role of HNF-1beta in Human Pancreatic Development. *Diabetes Med* (2006) 23:1301–6. doi: 10.1111/j.1464-5491.2006.01999.x
104. Bellanne-Chantelot C, Clauin S, Chauveau D, Collin P, Daumont M, Douillard C, et al. Large Genomic Rearrangements in the Hepatocyte Nuclear Factor-1beta (TCF2) Gene Are the Most Frequent Cause of Maturity-Onset Diabetes of the Young Type 5. *Diabetes* (2005) 54:3126–32. doi: 10.2337/diabetes.54.11.3126
105. Vethe H, Bjørlykke Y, Ghila LM, Paulo JA, Scholz H, Gygi SP, et al. Probing the Missing Mature β-Cell Proteomic Landscape in Differentiating Patient Ipsc-Derived Cells. *Sci Rep* (2017) 7:4780. doi: 10.1038/s41598-017-04979-w
106. Braverman-Gross C, Nudel N, Ronen D, Beer NL, McCarthy MI, Benvenisty N. Derivation and Molecular Characterization of Pancreatic Differentiated MODY1-Ipscs. *Stem Cell Res* (2018) 31:16–26. doi: 10.1016/j.scr.2018.06.013
107. Ng NHJ, Jasmen JB, Lim CS, Lau HH, Krishnan VG, Kadiwala J, et al. HNF4A Haploinsufficiency in MODY1 Abrogates Liver and Pancreas Differentiation From Patient-Derived Induced Pluripotent Stem Cells. *iScience* (2019) 16:192–205. doi: 10.1016/j.isci.2019.05.032
108. Aqel YWA, Ali G, Elsayed AK, Al-Khawaga S, Hussain K, Abdelalim EM. Generation of Two Human Ipsc Lines From Patients With Maturity-Onset Diabetes of the Young Type 2 (MODY2) and Permanent Neonatal Diabetes Due to Mutations in the GCK Gene. *Stem Cell Res* (2020) 48:101991. doi: 10.1016/j.scr.2020.101991
109. Cardenas-Diaz FL, Osorio-Quintero C, Diaz-Miranda MA, Kishore S, Leavens K, Jobaliya C, et al. Modeling Monogenic Diabetes Using Human Escs Reveals Developmental and Metabolic Deficiencies Caused by Mutations in HNF1A. *Cell Stem Cell* (2019) 25:273–289.e5. doi: 10.1016/j.stem.2019.07.007

110. Low BSJ, Lim CS, Ding SSL, Tan YS, Ng NHJ, Krishnan VG, et al. Decreased GLUT2 and Glucose Uptake Contribute to Insulin Secretion Defects in MODY3/HNF1A Hipsc-Derived Mutant β Cells. *Nat Commun* (2021) 12:3133. doi: 10.1038/s41467-021-22843-4
111. Zhu Z, Li QV, Lee K, Rosen BP, González F, Soh C-L, et al. Genome Editing of Lineage Determinants in Human Pluripotent Stem Cells Reveals Mechanisms of Pancreatic Development and Diabetes. *Cell Stem Cell* (2016) 18:755–68. doi: 10.1016/j.stem.2016.03.015
112. Wang X, Sterr M, Ansarullah, Burtscher I, Böttcher A, Beckenbauer J, et al. Point Mutations in the PDX1 Transactivation Domain Impair Human β-Cell Development and Function. *Mol Metab* (2019) 24:80–97. doi: 10.1016/j.molmet.2019.03.006
113. Teo AK, Lau HH, Valdez IA, Dirice E, Tjora E, Raeder H, et al. Early Developmental Perturbations in a Human Stem Cell Model of MODY5/HNF1B Pancreatic Hypoplasia. *Stem Cell Rep* (2016) 6:357–67. doi: 10.1016/j.stemcr.2016.01.007
114. Tiyaboonchai A, Cardenas-Diaz FL, Ying L, Maguire JA, Sim X, Jobaliya C, et al. GATA6 Plays an Important Role in the Induction of Human Definitive Endoderm, Development of the Pancreas, and Functionality of Pancreatic β Cells. *Stem Cell Rep* (2017) 8:589–604. doi: 10.1016/j.stemcr.2016.12.026
115. Shi Z-D, Lee K, Yang D, Amin S, Verma N, Li QV, et al. Genome Editing in Hpscs Reveals GATA6 Haploinsufficiency and a Genetic Interaction With GATA4 in Human Pancreatic Development. *Cell Stem Cell* (2017) 20:675–688.e6. doi: 10.1016/j.stem.2017.01.001
116. Chia CY, Madrigal P, Denil SLIJ, Martinez I, Garcia-Bernardo J, El-Khairi R, et al. GATA6 Cooperates With EOMES/SMAD2/3 to Deploy the Gene Regulatory Network Governing Human Definitive Endoderm and Pancreas Formation. *Stem Cell Rep* (2019) 12:57–70. doi: 10.1016/j.stemcr.2018.12.003
117. Kishore S, De Franco E, Cardenas-Diaz FL, Letourneau-Freiberg LR, Sanyoura M, Osorio-Quintero C, et al. A non-Coding Disease Modifier of Pancreatic Agenesis Identified by Genetic Correction in a Patient-Derived Ipsc Line. *Cell Stem Cell* (2020) 27:137–146.e6. doi: 10.1016/j.stem.2020.05.001
118. Trott J, Alpagu Y, Tan EK, Shboul M, Dawood Y, Elsy M, et al. Mitchell-Riley Syndrome Ipscs Exhibit Reduced Pancreatic Endoderm Differentiation Due to a Mutation in RFX6. *Development* (2020) 147:dev194878. doi: 10.1242/dev.194878
119. Shang L, Hua H, Foo K, Martinez H, Watanabe K, Zimmer M, et al. β-Cell Dysfunction Due to Increased ER Stress in a Stem Cell Model of Wolfram Syndrome. *Diabetes* (2014) 63:923. doi: 10.2337/db13-0717
120. Maxwell KG, Augsornworawat P, Velazco-Cruz L, Kim MH, Asada R, Hogrebe NJ, et al. Gene-Edited Human Stem Cell-Derived β Cells From a Patient With Monogenic Diabetes Reverse Preexisting Diabetes in Mice. *Sci Trans Med* (2020) 12:eaax9106. doi: 10.1126/scitranslmed.aax9106
121. Offield MF, Jetton TL, Labosky PA, Ray M, Stein RW, Magnuson MA, et al. PDX-1 Is Required for Pancreatic Outgrowth and Differentiation of the Rostral Duodenum. *Development* (1996) 122:983–95. doi: 10.1242/dev.122.3.983
122. Jonsson J, Carlsson L, Edlund T, Edlund H. Insulin-Promoter-Factor 1 Is Required for Pancreas Development in Mice. *Nature* (1994) 371:606–9. doi: 10.1038/371606a0
123. Ahlgren U, Jonsson J, Edlund H. The Morphogenesis of the Pancreatic Mesenchyme Is Uncoupled From That of the Pancreatic Epithelium in IPF1/PDX1-Deficient Mice. *Development* (1996) 122:1409–16. doi: 10.1242/dev.122.5.1409
124. Johnson JD, Ahmed NT, Luciani DS, Han Z, Tran H, Fujita J, et al. Increased Islet Apoptosis in Pdx1+/- Mice. *J Clin Invest* (2003) 111:1147–60. doi: 10.1172/JCI16537
125. Stoffers DA, Ferrer J, Clarke WL, Habener JF. Early-Onset Type-II Diabetes Mellitus (MODY4) Linked to IPF1. *Nat Genet* (1997) 17:138–9. doi: 10.1038/ng1097-138
126. Brissova M, Blaha M, Spear C, Nicholson W, Radhika A, Shiota M, et al. Reduced PDX-1 Expression Impairs Islet Response to Insulin Resistance and Worsens Glucose Homeostasis. *Am J Physiol-Endocrinol Metab* (2005) 288: E707–14. doi: 10.1152/ajpendo.00252.2004
127. Brissova M, Shiota M, Nicholson WE, Gannon M, Knobel SM, Piston DW, et al. Reduction in Pancreatic Transcription Factor PDX-1 Impairs Glucose-Stimulated Insulin Secretion. *J Biol Chem* (2002) 277:11225–32. doi: 10.1074/jbc.M111272200
128. Zegre Amorim M, Houghton JAL, Carmo S, Salva IS, Pita A, Pereira-da-Silva L. Mitchell-Riley Syndrome: A Novel Mutation in RFX6 Gene. *Case Rep Genet* (2015) 2015:937201. doi: 10.1155/2015/937201
129. Gradwohl G, Dierich A, LeMeur M, Guillemot F. Neurogenin3 Is Required for the Development of the Four Endocrine Cell Lineages of the Pancreas. *Proc Natl Acad Sci* (2000) 97:1607. doi: 10.1073/pnas.97.4.1607
130. Rubio-Cabezas O, Codner E, Flanagan SE, Gómez JL, Ellard S, Hattersley AT. Neurogenin 3 Is Important But Not Essential for Pancreatic Islet Development in Humans. *Diabetologia* (2014) 57:2421–4. doi: 10.1007/s00125-014-3349-y
131. Solorzano-Vargas RS, Bjerknes M, Wang J, Wu SV, Garcia-Careaga MG, Pitukcheewanont P, et al. Null Mutations of NEUROG3 Are Associated With Delayed-Onset Diabetes Mellitus. *JCI Insight* (2020) 5:e127657. doi: 10.1172/jci.insight.127657
132. Zhang X, McGrath PS, Salomone J, Rahal M, McCauley HA, Schweitzer J, et al. A Comprehensive Structure-Function Study of Neurogenin3 Disease-Causing Alleles During Human Pancreas and Intestinal Organoid Development. *Dev Cell* (2019) 50:367–80.e7. doi: 10.1016/j.devcel.2019.05.017
133. McGrath PS, Watson CL, Ingram C, Helmrath MA, Wells JM. The Basic Helix-Loop-Helix Transcription Factor NEUROG3 Is Required for Development of the Human Endocrine Pancreas. *Diabetes* (2015) 64:2497. doi: 10.2337/db14-1412
134. Wang D, Wang J, Bai L, Pan H, Feng H, Clevers H, et al. Long-Term Expansion of Pancreatic Islet Organoids From Resident Procr(+) Progenitors. *Cell* (2020) 180:1198–1211 e19. doi: 10.1016/j.cell.2020.02.048
135. Watanabe N, Hiramatsu K, Miyamoto R, Yasuda K, Suzuki N, Oshima N, et al. A Murine Model of Neonatal Diabetes Mellitus in Glis3-Deficient Mice. *FEBS Lett* (2009) 583:2108–13. doi: 10.1016/j.febslet.2009.05.039
136. Yang Y, Bush SP, Wen X, Cao W, Chan L. Differential Gene Dosage Effects of Diabetes-Associated Gene GLIS3 in Pancreatic β Cell Differentiation and Function. *Endocrinology* (2017) 158:9–20. doi: 10.1210/en.2016-1541
137. Servitja JM, Pignatelli M, Maestro MA, Cardalda C, Boj SF, Lozano J, et al. Hnf1alpha (MODY3) Controls Tissue-Specific Transcriptional Programs and Exerts Opposed Effects on Cell Growth in Pancreatic Islets and Liver. *Mol Cell Biol* (2009) 29:2945–59. doi: 10.1128/MCB.01389-08
138. Pontoglio M, Sreenan S, Roe M, Pugh W, Ostrega D, Doyen A, et al. Defective Insulin Secretion in Hepatocyte Nuclear Factor 1alpha-Deficient Mice. *J Clin Invest* (1998) 101:2215–22. doi: 10.1172/JCI2548
139. Ellard S, Colclough K. Mutations in the Genes Encoding the Transcription Factors Hepatocyte Nuclear Factor 1 Alpha (HNF1A) and 4 Alpha (HNF4A) in Maturity-Onset Diabetes of the Young. *Hum Mutat* (2006) 27:854–69. doi: 10.1002/humu.20357
140. Haliyur R, Tong X, Sanyoura M, Shrestha S, Lindner J, Saunders DC, et al. Human Islets Expressing HNF1A Variant Have Defective β Cell Transcriptional Regulatory Networks. *J Clin Invest* (2019) 129:246–51. doi: 10.1172/jci121994
141. Allen HL, Flanagan SE, Shaw-Smith C, De Franco E, Akerman I, Caswell R, et al. GATA6 Haploinsufficiency Causes Pancreatic Agenesis in Humans. *Nat Genet* (2012) 44:20–2. doi: 10.1038/ng.1035
142. Xuan S, Borok MJ, Decker KJ, Battle MA, Duncan SA, Hale MA, et al. Pancreas-Specific Deletion of Mouse Gata4 and Gata6 Causes Pancreatic Agenesis. *J Clin Invest* (2012) 122:3516–28. doi: 10.1172/JCI63352
143. Carrasco M, Delgado I, Soria B, Martin F, Rojas A. GATA4 and GATA6 Control Mouse Pancreas Organogenesis. *J Clin Invest* (2012) 122:3504–15. doi: 10.1172/JCI63240
144. Thomson JA, Itskovitz-Eldor J, Shapiro SS, Waknitz MA, Swiergiel JJ, Marshall VS, et al. Embryonic Stem Cell Lines Derived From Human Blastocysts. *Science* (1998) 282:1145. doi: 10.1126/science.282.5391.1145
145. Chen WS, Manova K, Weinstein DC, Duncan SA, Plump AS, Prezioso VR, et al. Disruption of the HNF-4 Gene, Expressed in Visceral Endoderm, Leads to Cell Death in Embryonic Ectoderm and Impaired Gastrulation of Mouse Embryos. *Genes Dev* (1994) 8:2466–77. doi: 10.1101/gad.8.20.2466[doi
146. Stoffel M, Duncan SA. The Maturity-Onset Diabetes of the Young (MODY1) Transcription Factor HNF4α Regulates Expression of Genes Required for Glucose Transport and Metabolism. *Proc Natl Acad Sci* (1997) 94:13209. doi: 10.1073/pnas.94.24.13209
147. Shih DQ, Heimesaat M, Kuwajima S, Stein R, Wright CVE, Stoffel M. Profound Defects in Pancreatic Beta-Cell Function in Mice With Combined

Heterozygous Mutations in Pdx-1, Hnf-1alpha, and Hnf-3beta. *Proc Natl Acad Sci* (2002) 99:3818. doi: 10.1073/pnas.062605899

148. Fajans SS, Bell GI, Polonsky KS. Molecular Mechanisms and Clinical Pathophysiology of Maturity-Onset Diabetes of the Young. *New Engl J Med* (2001) 345:971–80. doi: 10.1056/NEJMra002168
149. Krapp A, Knöfler M, Ledermann B, Bürki K, Berney C, Zoerkler N, et al. The Bhlh Protein PTF1-P48 Is Essential for the Formation of the Exocrine and the Correct Spatial Organization of the Endocrine Pancreas. *Genes Dev* (1998) 12:3752–63. doi: 10.1101/gad.12.23.3752
150. Burlison JS, Long Q, Fujitani Y, Wright CV, Magnuson MA. Pdx-1 and Ptf1a Concurrently Determine Fate Specification of Pancreatic Multipotent Progenitor Cells. *Dev Biol* (2008) 316:74–86. doi: 10.1016/j.ydbio.2008.01.011
151. Weedon MN, Cebola I, Patch AM, Flanagan SE, De Franco E, Caswell R, et al. Recessive Mutations in a Distal PTF1A Enhancer Cause Isolated Pancreatic Agenesis. *Nat Genet* (2014) 46:61–4. doi: 10.1038/ng.2826
152. Balboa D, Saarimäki-Vire J, Borshagovski D, Survila M, Pulkkila P, Galli E, et al. Insulin Mutations Impair Beta-Cell Development in a Patient-Derived Ipsc Model of Neonatal Diabetes. *Elife* (2018) 7:e38519. doi: 10.7554/eLife.38519
153. Ma S, Viola R, Sui L, Cherubini V, Barbetti F, Egli D. β Cell Replacement After Gene Editing of a Neonatal Diabetes-Causing Mutation at the Insulin Locus. *Stem Cell Rep* (2018) 11:1407–15. doi: 10.1016/j.stemcr.2018.11.006
154. Saarimäki-Vire J, Balboa D, Russell MA, Saarikettu J, Kinnunen M, Keskitalo S, et al. An Activating STAT3 Mutation Causes Neonatal Diabetes Through Premature Induction of Pancreatic Differentiation. *Cell Rep* (2017) 19:281–94. doi: 10.1016/j.celrep.2017.03.055

9

Human Pluripotent Stem Cells to Model Islet Defects in Diabetes

Diego Balboa [1], Diepiriye G. Iworima [2,3] and Timothy J. Kieffer [2,3,4]*

[1] *Regulatory Genomics and Diabetes, Centre for Genomic Regulation, The Barcelona Institute of Science and Technology, Barcelona, Spain,* [2] *Department of Cellular and Physiological Sciences, University of British Columbia, Vancouver, BC, Canada,* [3] *School of Biomedical Engineering, The University of British Columbia, Vancouver, BC, Canada,* [4] *Department of Surgery, University of British Columbia, Vancouver, BC, Canada*

Correspondence:
Diego Balboa
diego.balboa@crg.eu

Diabetes mellitus is characterized by elevated levels of blood glucose and is ultimately caused by insufficient insulin production from pancreatic beta cells. Different research models have been utilized to unravel the molecular mechanisms leading to the onset of diabetes. The generation of pancreatic endocrine cells from human pluripotent stem cells constitutes an approach to study genetic defects leading to impaired beta cell development and function. Here, we review the recent progress in generating and characterizing functional stem cell-derived beta cells. We summarize the diabetes disease modeling possibilities that stem cells offer and the challenges that lie ahead to further improve these models.

Keywords: diabetes, insulin, modeling, stem cells, genetic defects, insulin secretion

INTRODUCTION

More than 450 million people worldwide are diagnosed with diabetes, a number unfortunately expected to increase dramatically in the next decades (1). Diabetes unfolds when the pancreatic beta cells fail to secrete enough insulin to meet physiological demand, resulting in abnormally high blood glucose levels. Our understanding of the distinct molecular mechanisms that lead to beta cell failure in the different types of diabetes has remarkably improved thanks to progress in the genetic characterization of people with diabetes and the development of animal and cellular models (2). Among these models, the generation of islet cells from human pluripotent stem cells is gaining traction as a useful approach to dissect diabetes molecular mechanisms (3). In this review, we aim to summarize recent progress in diabetes disease modeling using human pluripotent stem cells, discussing current limitations and potential improvements. We particularly focus on advances in functional islet cell generation, and how these cells may be utilized to study beta cell insulin secretory defects.

Beta cell failure leading to diabetes occurs in different ways. While in type 1 diabetes, beta cells are destroyed by cytotoxic T lymphocytes (4), in type 2 diabetes, which represents 90% of all diabetes cases, the beta cells are dysfunctional as a result of maladaptation to elevated demand for insulin secretion, usually in the context of systemic insulin resistance (5, 6). Both type 1 and type 2 diabetes (T2D) result from interactions between a polygenic background and environmental factors like viral infections or obesity (7). Other forms of diabetes that are less frequent result from highly penetrant monogenic mutations that impair beta cell development and/or function. They can manifest at birth, transiently or permanently, in what is known as neonatal diabetes, or in the young adult (10–25 years of age), termed maturity onset diabetes of the young (MODY) (2, 8). While genetic variants in ~30 loci are associated with neonatal diabetes and MODY, over 50% of clinically diagnosed cases remain genetically unexplained despite continuous efforts to find causative genetic variants by using genome sequencing (9, 10).

The characterization of the genetic defects associated with these different types of diabetes has improved the understanding of the molecular mechanisms that trigger or increase the risk for this disease. Genome-wide association studies have so far identified over 400 association signals across ~200 loci associated with T2D. These genetic variants are particularly enriched in coding and non-coding genomic regions characteristic of pancreatic islet cells, highlighting their central role in the development of diabetes (11–13). Interestingly, several genetic variants associated with T2D are in loci of genes that are also mutated in cases of neonatal diabetes and MODY [e.g. *KCNJ11* (14, 15), *HNF1A* (16, 17), *GCK* (18, 19)]. These genes are critical for beta cell function and the severity of the disease is determined by the precise molecular mechanism disrupted by the particular genetic variant and its functional impact. There is a spectrum within diabetes in which the pathogenetic mechanisms might range from protein-truncating mutations causing neonatal diabetes due to pancreas developmental failure (20, 21), to increased T2D risk due to regulatory variants modulating adult islet cell function (22, 23). While genetic studies have identified numerous candidate genetic variants associated with different types of diabetes, functional validation of their impact on glucose homeostasis requires models that recapitulate as faithfully as possible human islet physiology.

Rodent animal models have provided abundant knowledge of pancreatic development and beta cell physiology. The generation of genetically modified mouse models have contributed to understanding the role of genes involved in these processes (21, 24). However, animal models have inherent limitations due to key differences with humans at the genetic and physiological level (25, 26). Primary human islets obtained from the pancreas of cadaveric donors are a valuable research material to study diabetes. They have been used to study particular aspects of human islet physiology (27) and to understand how genetic variation affects islet function (28). However, human islet preparations are scarce and exhibit considerable variability in terms of purity, function, and cell type composition after isolation (29–32). Furthermore, isolated human islets are challenging to keep in culture for extended periods of time, and the ability to use them to study the effect of particular genetic variants is limited by the current capabilities to genetically manipulate them. As an alternative, there have been many attempts to generate immortalized human beta cells resulting in the derivation of several cell lines that are now widely used in research. They constitute a renewable source of beta-like cells that can be used to perform diverse *in vitro* experiments. In particular, EndoC-βH lines have proven to be a particularly useful model since they present glucose-stimulated insulin secretion *in vitro* and are transcriptomically similar to primary beta cells (33, 34). Such lines can be utilized to study the impact of particular genetic variants and perform drug screenings since they are amenable to genetic modification and other perturbations (23, 34). A drawback of these cells is that they are aneuploid, which can be a confounding factor for genetic studies (35). They also proliferate, which compromises the functional characteristics of adult beta cells (36, 37). This has been resolved in conditionally immortalized versions of this cell line where the SV40LT oncogene used to transform them can be removed by inducible genetic recombination (37, 38); these cells continue to be a useful resource for the field.

Differentiated human pluripotent stem cells (hPSCs) represent another source of human beta cells. hPSCs can be derived from human embryos (human embryonic stem cells, hESCs) (39) or from somatic cells *via* nuclear reprogramming (human induced pluripotent stem cells, hiPSCs) (40). Notably, hiPSCs can be obtained from somatic cells of people that carry diabetes-associated genetic variants. By doing so, pluripotent cell lines preserving the donor genetic background can then be differentiated *in vitro* into particular cell types to model the molecular consequences of the genetic variant under study (41). Importantly, hPSCs are amenable to different genome editing approaches, facilitating the correction or introduction of desired genetic variants. This is a useful approach to generate optimal isogenic controls or to create new models when donor sources are not available (42).

Here we discuss the possibilities of using hPSCs to model the impact of diabetes-associated genetic variants on the physiology of the beta cell, focusing on the molecular mechanisms impairing insulin secretion.

BETA CELL INSULIN SECRETION DEFECTS

All forms of diabetes have in common the ultimate dysfunction of the pancreatic beta cells and the consequent inadequate circulating insulin levels. Beta cells constitute about 60% of the cells in the human islets. They are highly intermingled with the other endocrine cells, in particular with glucagon producing alpha cells, the second most abundant type, a configuration that is crucial for the optimal function of the beta cells (43, 44), and somatostatin secreting delta cells that dampen the release of both insulin and glucagon (45). The particular organization of human islet cells is remarkably heterogeneous, with variable islet size and cell type composition across parts of the pancreas, but also showing important variation across individuals and from birth to adulthood (29–31).

In conjunction with glucagon secreting alpha cells, beta cells keep human fasting blood glucose concentrations around 5 mM, normoglycemia, by adjusting their insulin secretion output (44). Beta cells are fine-tuned glucose sensors with an intricate machinery that enables them to respond with exquisite precision to deviations from normoglycemia, such as during meals, to minimize glucose excursions (46, 47). Genetic variants that result in the disruption of these molecular mechanisms impact the capacity of beta cells to secrete insulin in a regulated manner. These can cause reduced insulin secretion, leading to the development of different forms of diabetes, or increased insulin secretion (hyperinsulinism) (20, 48). We discuss some of these in detail below (summarized in **Table 1** and **Figure 1**).

TABLE 1 | Genetic defects leading to dysregulated beta cell insulin secretion.

Mechanism affected	Genes	Impact of genetic defect	Type of disease	References
Glucose import and metabolism	*GCK*	Reduced or increased glucokinase activity results in abnormal glycolytic flux, ATP generation, and insulin secretion	ND, MODY, T2D, CHI	(18, 19, 49)
	G6PC2	Loss of function mutations are associated with reduced fasting glycemia	T2D	(50)
	SLC2A2 (GLUT2)	Loss of function mutations result in impaired glucose uptake	ND, T2D	(11, 51)
	HK1	Abnormal silencing of HK1 in beta-cells results in increased glycolytic flux, ATP generation and insulin secretion	CHI	(52)
	SLC16A1 (MCT1)	Promoter mutations impair SLC16A1 silencing in beta-cells, resulting in abnormal pyruvate uptake, increased ATP generation, and insulin secretion	CHI	(53)
	GLUD1	Gain of function mutations result in increased entrance of glutamate in TCA cycle, increased ATP generation, and insulin secretion	CHI	(54)
	HADH	Loss of function mutations result in abnormal activation of GLUD1, increased glutamate into TCA, ATP generation, and insulin secretion	CHI	(55, 56)
	UCP2	Gain or loss of function mutations alter the mitochondrial uncoupling activity of UCP2, resulting in abnormal ATP generation and insulin secretion	T2D, CHI	(57, 58)
	mtDNA	Mitochondrial DNA mutations impair oxidative phosphorylation, ATP generation, and insulin secretion	–	(59)
Membrane depolarization	*KCNJ11*	Gain or loss of function mutations result in abnormal closure or opening of the channel, altered membrane depolarization, and insulin secretion	ND, MODY, T2D, CHI	(55, 60–62)
	ABCC8	Gain or loss of function mutations result in abnormal closure or opening of the channel, altered membrane depolarization, and insulin secretion	ND, MODY, T2D, CHI	(14, 15, 55, 63, 64)
	KCNQ1	Genetic variants in this locus are associated with T2D risk.	T2D	(11, 65)
Membrane receptors	*GLP1R*	Genetic variants in this locus are associated with lower fasting glucose levels. Altered GLP-1 signaling affects amplification of insulin secretion.	T2D	(66)
	GIPR	Genetic variants associated with reduced GIP signaling, impair incretin-mediated amplification of insulin secretion.	T2D	(67)
	MTNR1B	A genetic variant increasing melatonin signaling lowers cAMP levels, inhibiting insulin secretion.	T2D	(68)
Insulin synthesis and secretion	*INS*	Loss of function mutations disrupt INS protein synthesis, folding, transport or bioactivity.	ND, MODY, T2D	(11, 69)
	SLC30A8 (ZNT8)	Different coding genetic variants increase risk or protect against T2D.	T2D	(70–72)
	ADCY5	Non-coding genetic variant reduces ADCY5 expression, which couples glucose to cAMP generation, increasing T2D risk.	T2D	(73, 74)
ER homeostasis	*WFS1*	Loss of function mutations lead to elevated ER stress and beta cell dysfunction.	ND, T2D	(75)
	CDKAL1	Loss of function mutations induce beta cell ER stress and hypersensitivity to glucotoxicity and lipotoxicity.	T2D	(76)
	THADA	Coding genetic variants associated with increased T2D risk.	T2D	(77)
	MANF	Loss of function mutations cause childhood diabetes and a neurodevelopmental disorder.	ND, T2D	(78, 79)
	YIPF5	Loss of function mutations impaired ER-to-Golgi trafficking leading to increased beta cell ER-stress.	ND	(80)
Transcriptional regulation	*PDX1*	Loss of function mutations impair transcriptional regulation of pancreatic development and adult islet cell function.	ND, MODY, T2D	(81–84)
	RFX6	Loss of function mutations impair transcriptional regulation of pancreatic development and adult islet cell function.	ND, MODY	(85–87)
	NEUROD1	Loss of function mutations impair transcriptional regulation of pancreatic development and adult islet cell function.	ND, MODY	(88–90)
	GLIS3	Coding and non-coding genetic variants impair transcriptional pancreatic development and adult islet cell function	ND, MODY, T2D	(91–93)
	HNF1A	Coding and non-coding genetic variants impair transcriptional pancreatic development and adult islet cell function.	MODY, T2D, CHI	(16, 17, 94)
	HNF1B	Coding and non-coding genetic variants impair transcriptional pancreatic development and adult islet cell function.	ND, MODY, T2D	(95, 96)
	HNF4A	Coding and non-coding genetic variants impair transcriptional pancreatic development and adult islet cell function.	MODY, T2D, CHI	(54, 97)
	TCF7L2	Coding and non-coding genetic variants impair transcriptional pancreatic development and adult islet cell function.	T2D	(98, 99)

ND, Neonatal Diabetes; MODY, Maturity Onset Diabetes of the Young; T2D, Type 2 Diabetes; CHI, Congenital Hyperinsulinism.
Summary of genetic variants that impact on molecular mechanisms involved in insulin secretion by the beta cell, classified by the mechanism affected and detailing the impact of the genetic defect.

Islet cells are profusely vascularized, and this facilitates the sensing of circulating blood glucose levels. Glucose is imported into human beta cells primarily *via* glucose transporters 1 (GLUT1) and 3 (GLUT3). Glucose transporter 2 (GLUT2), the main transporter in rodent beta cells, is expressed at lower levels in human beta cells (100). Interestingly, genetic variants associated with T2D are found in *GLUT2* (51) suggesting an important role in human beta cells.

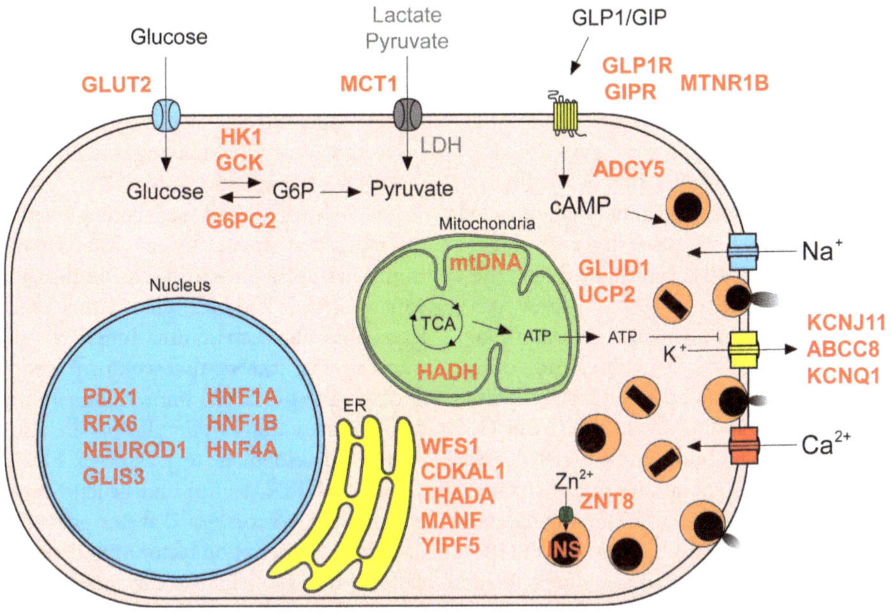

FIGURE 1 | Insulin secretion molecular mechanisms affected in diabetes. Genetic defects can impair different processes involved in regulated insulin secretion (known genes affected in red text): glucose import and metabolism (G6P, glucose 6-phosphate; LDH, lactate dehydrogenase; TCA, tricarboxylic acid cycle; abnormal beta cell metabolism of non-glucose carbon sources due to failure in silencing of disallowed genes depicted in gray text), membrane depolarization, membrane receptors, insulin synthesis and secretion, endoplasmic reticulum (ER) homeostasis, and transcriptional regulation.

Imported glucose is phosphorylated by glucokinase (GCK), a low affinity hexokinase. Diverse genetic defects in *GCK* lead to different kinds of insulin secretion phenotypes, resulting in a range of disease severity, from neonatal diabetes and MODY, to increased T2D risk and congenital hyperinsulinism (18, 19, 49). Other regulators of glucose phosphorylation have also been implicated in insulin secretion dysregulation. For example, glucose-6-phosphatase 2 (*G6PC2*) harbors genetic variants associated with reduced fasting glycemia (50), while the abnormal beta cell expression of "disallowed gene" (101) hexokinase I (*HK1*) has been linked to congenital hyperinsulinism (52).

Once phosphorylated, glucose is retained within the beta cell and it enters the glycolytic pathway to generate pyruvate. Pyruvate is then further oxidized in the mitochondrial tricarboxylic acid cycle, generating abundant chemical energy in the form of ATP, and thus increasing the ATP to ADP ratio. This results in depolarization of the beta cell membrane *via* closure of the ATP-sensitive potassium channels (K^+_{ATP}), triggering insulin secretion. Thus, oxidative metabolism of pyruvate constitutes a crucial coupling process enabling regulated insulin secretion (102). As an alternative to this canonical, one-state model of insulin secretion, recent work by Lewandowski et al. proposes a dynamic model in which beta cells in high glucose conditions oscillate between two states: a biosynthetic state in which conversion of ADP and phosphoenolpyruvate into ATP and pyruvate by pyruvate kinase results in closure of the K^+_{ATP} channels, triggering exocytosis, followed by a state of active oxidative phosphorylation that supports the elevated ATP to ADP ratio sustaining membrane depolarization until exocytosis-associated processes reduce the ATP levels (103, 104). Activators of pyruvate kinase resulted in potentiated GSIS in both rodent and human islets, suggesting that pyruvate kinase may be a potential therapeutic target for T2D (53). Different defects related to the abnormal incorporation of metabolites into the tricarboxylic acid cycle are associated with congenital hyperinsulinism and T2D. Impaired silencing of the pyruvate and lactate transporter *SLC16A1* (*MCT1*), a beta cell disallowed gene, results in congenital hyperinsulinism (53). Gain of function missense mutations in *GLUD1*, glutamate dehydrogenase, or loss of function mutations in hydroxyacyl-coenzyme A dehydrogenase (*HADH*) increase incorporation of glutamate into the TCA cycle leading to congenital hyperinsulinism (54–56). Also, genetic variation in the mitochondrial uncoupler *UCP2* has been associated with T2D and congenital hyperinsulinism (57, 58). In addition, mitochondrial DNA mutations that impair ATP generation cause syndromes that present with diabetes of variable severity (59).

The increase of ATP to ADP ratio triggers the closure of membrane K^+_{ATP} channels, formed by the proteins KCNJ11 and ABCC8 (105). Channel closure leads to depolarization of cell membrane and opening of additional Na^+ and Ca^{2+} channels. Ca^{2+} influx crucially couples membrane depolarization with insulin exocytosis, in a process mediated by the Ca^{2+}-sensing proteins synaptotagmins which trigger the fusion of insulin granules with the plasma membrane (105). Gain of function mutations in *KCNJ11* and *ABCC8* resulting in constant channel opening are the most common cause of neonatal diabetes due to islet physiology defects (15, 60, 63). Genetic variation in these genes can also cause MODY and increased T2D risk (61, 62).

Loss of function mutations that result in constant K_{ATP} closure, or impair its trafficking to the membrane, lead to congenital hyperinsulinism (14, 55). Furthermore, genetic variants in the voltage-gated K^+ channel KCNQ1 are associated with increased T2D risk (65).

Influx of Ca^{2+} ions into depolarized beta cells induces insulin exocytosis by activating the synaptotagmins and SNARE proteins that regulate the fusion of insulin granules with the plasma membrane (106). This exocytosis machinery is not only regulated by the intrinsic pathway triggered by membrane depolarization but is also critically modulated by the intracellular levels of cyclic AMP (cAMP), in what is known as the amplifying pathway (107). Incretin hormones glucagon-like peptide-1 (GLP-1) and glucose dependent insulinotropic polypeptide (GIP), released by intestinal enteroendocrine cells, potentiate insulin secretion upon binding to their cognate G-protein coupled receptors in the membrane of beta cells (108, 109). This binding results in the generation of cAMP and activation of protein kinase A pathway resulting in augmented K^+ channel inhibition, Ca^{2+} influx, and insulin exocytosis (110). GLP-1 also regulates the alpha cells in a glucose-dependent manner, inhibiting glucagon at high glucose levels, and thereby further contributing to glucose homeostasis (111, 112). GIP stimulates glucagon secretion in a glucose-dependent manner in healthy individuals, with enhanced activity at lower glycemia (113). However, GIP stimulates glucagon secretion even in the presence of hyperglycemia in subjects with T2D, and thereby could contribute to the pathogenesis of T2D (113). Genetic variants found in both the incretin receptors genes *GLP1R* and *GIPR* have been associated with increased and decreased risk of T2D (66, 67). Melatonin receptor 1 B (*MTNR1B)*, another G-protein coupled receptor present in the membrane of beta cells, has also been linked to T2D. A genetic variant increasing the expression of *MTNR1B* has been shown to lower cAMP levels in beta cells, leading to reduced insulin secretion (68). Furthermore, a genetic variant that results in reduced expression of the adenyl cyclase five (*ADCY5*), which regulates beta cell cAMP levels, has been associated with increased risk of T2D (73, 74).

Insulin protein is synthesized in remarkable amounts, representing up to 50% of beta cell total protein synthesis (114), and imposes a high demand on the protein folding machinery of the endoplasmic reticulum (ER). These processes are controlled by the unfolded protein response (UPR) pathway, which is highly efficient in beta cells in order to cope with the insulin biosynthesis-induced ER-stress (115, 116). After potassium channel defects, coding mutations in the insulin gene are the second most common cause of neonatal diabetes due to beta cell dysfunction. These missense mutations cause defects in proinsulin translation, folding, or processing, and may induce high levels of ER-stress that leads to dysfunction of the beta cell. Some *INS* coding mutations can also cause MODY (69) and genetic variants in the *INS/IGF2* locus have been associated with T2D increased risk (11). The fine-tuning of ER-stress levels in beta cells is crucial for the proper functioning of these busy insulin factories. Coding mutations in components of the UPR pathway can cause neonatal diabetes or increased risk for T2D (e.g. *WFS1*, *CDKAL*, *THADA*, *MANF*, *YIPF5*) (11, 75–80). Processed proinsulin molecules are tightly packaged as Zn^{2+} complexed crystals in dense core exocytotic granules. Genetic variants in the Zn^{2+} transporter *SLC30A8* (*ZNT8*), present in the membrane of insulin granules, have been associated with T2D susceptibility (70, 71). A rare loss of function mutation in ZNT8 protects against T2D, making this Zn^{2+} transporter a potentially interesting therapeutic target (72).

While coding and non-coding genetic variants linked to diabetes often impact mechanisms regulating insulin secretion from beta cells, some of them perturb the development of the pancreas, islets, and beta cells themselves (23, 48). The expression levels of insulin secretion machinery components is controlled by transcription factors that conform gene regulatory networks governing the beta cell transcriptional program (117). However, many of these transcription factors are also involved in regulating beta cell development (e.g. FOXA2, PDX1, MNX1, NEUROD1, PTF1A, HNF1A, RFX6) and genetic defects in their loci might lead to a wide variety of diabetes phenotypes (20). While highly damaging transcription factor mutations can cause developmental defects leading to pancreatic agenesis and neonatal diabetes, other genetic variants with milder effects might lead to MODY with different clinical features and penetrance (17, 81, 118), increased T2D risk (91, 98), or even congenital hyperinsulinism (54, 94). Epigenetic profiling of human islets has enabled the characterization of their regulatory landscape, showing that dense enhancer areas are enriched in genetic variants associated with T2D risk (13). Furthermore, a recent study characterizing human islet chromatin architecture resulted in the identification of 3D higher-order hubs of enhancers and promoters (23). These regions are enriched for genetic variants that impact on the heritability of islet-cell traits. We summarize in **Table 1** a list of genes that harbor genetic variants specifically linked to dysregulated insulin secretion. The impact of a given genetic variant will depend on how deleterious it is for a particular mechanism controlling insulin secretion, thus determining the diabetes phenotype and the possible therapeutic interventions. Given the wide spectrum in the functional consequences of coding and non-coding genetic variants, we need suitable research models that enable precise dissection of the detailed mechanisms by which these genetic variants impair human islet physiology.

MODELING INSULIN SECRETION DEFECTS USING STEM CELL DERIVED ISLET CELLS

The generation of hPSC-derived beta cells typically relies on differentiation protocols recapitulating the inductive signaling cues that instruct pancreatic development *in vivo*. These protocols have been devised based on knowledge gained from developmental biology, mostly using mouse models, that deciphered the dynamic signaling environment required for pancreas specification, endocrinogenesis, and beta cell formation (119, 120). With this information, different research teams have empirically determined the recipe of recombinant proteins and small molecules that reproduce developmental signals in a stepwise manner. These efforts have crystallized in differentiation protocols that make possible the efficient derivation of beta cells from hPSCs (**Figure 2**).

The first report demonstrating the feasibility of generating insulin-producing cells from human embryonic stem cells *in vitro* relied on an spontaneous differentiation approach (122). The first directed differentiation protocol was reported by D'Amour and colleagues from the company CyThera (now Viacyte Inc.). They devised a multistage, adherent culture differentiation protocol, that relied on a first step of efficient definitive endoderm induction (123), followed by four additional steps to induce primitive gut tube, posterior foregut, pancreatic progenitors, and hormone expressing cells (124). While this pioneer protocol generated relatively few insulin producing cells (about 7%), these cells became functionally mature after implantation into mice. Furthermore, the implanted cells were able to protect against streptozotocin induced diabetes (125).

These results sparked intense effort to develop improved protocols leading to more efficient and robust ways to obtain hPSC-derived beta cells over the next decade. Modulation of additional signaling pathways (e.g. FGF, TGF-beta, EGF, PKC) in a time-wise manner enhanced the differentiation efficiency of pancreatic progenitors and endocrine cells (126–132). However, detailed characterization of the hPSC-derived beta cells showed that these insulin expressing cells were frequently co-expressing other hormones, like glucagon or somatostatin, (usually termed as "polyhormonal" cells) (133–135). Polyhormonal cells had impaired glucose stimulated insulin secretion (134), aberrant epigenetic profiles (136), inappropriate glucose transporter GLUT1 expression, imbalanced $K+_{ATP}$ channel subunit expression (137), and resembled human fetal beta cells at the transcriptomic level (138).

A critical realization was the importance of beta cell programming transcription NKX6-1 for beta cell development and functionality (139). NKX6-1 expression at the pancreatic progenitor stage of the differentiation was shown to be crucial for the generation of "monohormonal" beta cells, expressing insulin together with NKX6-1 (INS+NKX6-1+ beta cells) (140). Delaying NEUROG3 induction to later stages, when PDX1+NKX6-1+ progenitors are more abundant, increased the fraction of insulin^{+} glucagon^{-} beta cells (140, 141). Protocols generating functional beta cells *in vitro* from hPSCs were reported in 2014 (121, 142). Both differentiation protocols have similarities in the length, stages, and signaling cues used, resulting in abundant INS+NKX6-1+ beta cells. Endocrine cell differentiation was induced by a combination of ALK5 (a TGF-beta receptor) and NOTCH signal inhibitors. Thyroid hormone triiodothyronine (T3) was used to induce the expression of MAFA, a beta cell maturation marker (143, 144). The stem cell-derived beta cells secreted insulin in response to high glucose under static conditions, however, a more detailed analysis of dynamic insulin secretion and calcium influx showed the response was minimal compared to human islets (121). In both studies, the implantation of these functional hPSC-derived beta cells rescued diabetes in mice and had increasing levels of human insulin produced by the grafts over time.

These landmark reports demonstrated a viable path towards the generation of glucose-responsive hPSC-derived beta cells *in vitro*, despite the cells not matching the fine-tuned responses of human islets. It is important to recognize that human islets isolated from cadaveric donors, while presently used as the gold-standard control, have the limitation of considerable variability across islet preparations from different donors in terms of purity, cell-type composition, functionality, and expression of important beta cell markers (30, 121, 145).

Subsequent studies have built on these protocols and further refined them to achieve a higher percentage of hPSC-derived beta cells with better functional responses. For example, different studies have demonstrated how NKX6-1 expression can be increased by aggregating the pancreatic progenitors (146, 147) or by adding EGF and Nicotinamide (148). Also, Rho-associated kinase (ROCK) inhibitors were shown to boost the expression levels of NKX6-1 (149) and the numbers and maturation of hPSC-derived beta cells (150). ROCK inhibitor Y-27632 together with TGF-beta ligand Activin A was reported to induce the formation of endocrine cell enriched protrusions in a 3D-aggregate differentiation setup (151). Performing the differentiation in 3D suspension conditions, in an attempt to recapitulate the developing pancreas cytoarchitecture, has improved the generation of pancreatic progenitors and endocrine cells, as well as the reproducibility and scalability of the differentiation (141, 142, 146, 152).

Induction of endocrine cell formation in these differentiation protocols has relied on the modulation of NOTCH (using gamma

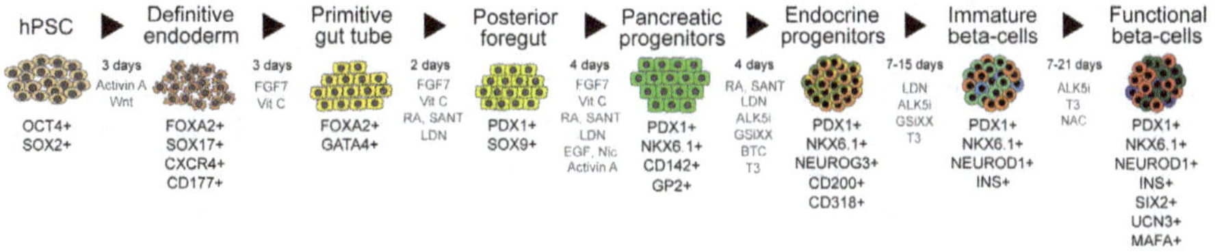

FIGURE 2 | Multistage differentiation protocol to generate functional islet cells from human pluripotent stem cells. Current islet cell differentiation protocols mimic pancreatic developmental stages. Here we represent the commonly used stages [based on (121)], with their usual duration in days, together with cell markers used for the characterization of the differentiated cells (black text) and the cocktails of signaling molecules utilized to induce differentiation (gray text; FGF7, fibroblast growth factor 7; VitC, vitamin C, ascorbic acid; RA, retinoic acid; SANT, SANT-1, a sonic hedgehog signaling inhibitor; LDN, LDN-193189, a BMP inhibitor; EGF, epidermal growth factor; Nic, nicotinamide; ALK5i, a TGF-beta inhibitor; GSiXX, gamma secretase inhibitor used to inhibit Notch signaling; BTC, betacellulin; T3, triiodothyronine; NAC, N-Acetylcysteine).

secretase inhibitors), TGF-beta (ALK5 receptor inhibitors), and EGF (EGF and betacellulin ligands) signaling to trigger NEUROG3 expression. Interestingly, a newly developed 2D planar differentiation protocol generated cells with improved function using latrunculin A to depolymerize the cytoskeleton during endocrine induction, demonstrating that the cytoskeletal state of cells during differentiation can also modulate NEUROG3 expression. These cells were also able to reverse diabetes in STZ treated mice faster than cells generated with a 3D suspension protocol (153). Appropriate timing of NEUROG3 expression is important for beta cell lineage commitment. Its induction in pancreatic progenitor cells expressing PDX1+NKX6-1+ favors the generation of beta cells, while inducing at earlier stages will result in polyhormonal cells that appear to largely resolve into alpha cells (129, 140, 141, 148). Regulatory genomics analyses of embryonic and stem cell derived pancreatic progenitors identified TEAD and YAP as important regulators critical for pancreatic progenitors outgrowth (154). These effectors of the Hippo signaling pathway form part of the gene regulatory network that recruits pancreatic progenitor enhancers and controls their proliferation. Disruption of the TEAD-YAP complex with verteporfin results in reduced proliferation of mouse, zebrafish, and hPSC-derived pancreatic progenitors (154). Additional studies on the role of TEAD-YAP in pancreatic progenitors have shown that cell confinement prevents YAP nuclear accumulation and is a prerequisite for NEUROG3 upregulation (155). In this model, endocrinogenesis is triggered by the disruption of extracellular matrix signaling *via* integrin alpha 5, which maintains the expression of NEUROG3 repressor complex YAP1-TEAD4-HES1. Consistent with this, the use of verteporfin in stem cell-derived pancreatic progenitors resulted in reduced progenitor proliferation, increased NEUROG3 expression, and more C-peptide+ cells (155, 156).

Other approaches to improve the function of stem cell derived beta cells have relied on enrichment steps at various stages and controlling 3D cluster size. For example, enrichment of GP2+ pancreatic progenitors led to the generation of increased numbers of beta cells (157, 158), while enrichment of later differentiation stages using an INS-GFP reporter cell line or magnetic-based enrichment for ITGA1 improved the functionality of the stem-cell derived islet-like aggregates (159, 160). Optimal cluster diameter is important in order to avoid necrosis in the cell cluster core due to hypoxia, maintain a good surface to volume relationship, and is critical for glucose sensing and insulin release dynamics. Across mammalian species with different pancreas sizes, the diameter of islets averages 100–200 μm. The fact that mammalian islets do not scale with the weight of the animal suggests there is an optimal size for the function of these endocrine miniorgans (161, 162). Recapitulating the size of human islets by spontaneous reaggregation (159, 163) or controlled forced aggregation using micropatterned culture plates improves the functionality of stem cell-derived beta cells (160).

The signaling cues required in the later stages of the differentiation protocols to induce maturation are not completely identified. Recent studies have shown that better functioning hPSC-derived beta cells are generated when serum-free media with no added factors is used in the later stages (159, 163). Velazco-Cruz and colleagues reported remarkable acquisition of dynamic glucose stimulated insulin secretion, including robust first and second phases, following the omission of TGF-beta inhibition together with cluster resizing and use of serum-free media during the last stage of the differentiation process. Letting the stem cell-derived islet-like cells establish their own niche and paracrine/autocrine signaling might be a better alternative to achieve more functional cell types (44, 164).

A common problem in the field of hPSC differentiation is the robustness of a given protocol applied to different hPSC lines. In most instances, protocols are optimized specifically for one or few cell lines, and they tend to yield variable differentiation efficiencies when other cell lines are used. In the case of pancreatic differentiation, reports have shown how a particular differentiation protocol results in different percentages of pancreatic progenitors and insulin expressing cell numbers depending on the hPSC line used (148, 165). This is an important obstacle to the wide application of published differentiation protocols, affecting reproducibility. It also complicates the generation of beta cells from diverse patient-derived hiPSCs for disease modeling purposes.

The variability in the efficiency of a particular differentiation protocol has been attributed to the hPSC line genetic background, which can condition its response to inductive cues (166, 167). Recent studies suggest that specific genetic variants in hiPSC lines may alter the differentiation efficiency towards definitive endoderm (168). By using pools of 125 different iPSCs and single-cell RNA sequencing, the authors mapped the population variation during definitive endoderm differentiation stages. They identified several molecular markers predictive of differentiation efficiency, demonstrating that it can be altered by germline genetic variants. Despite intense efforts to improve *in vitro* differentiation protocols, currently they only partially recapitulate the optimal *in vivo* signaling environment. Missing signaling cues are probably better tolerated in some cell lines than in others, explaining this apparent genetic background-determined fitness to respond efficiently to a given protocol. A partial solution to the problem of variability in differentiation efficiency across cell lines is the generation of genome edited cell lines. Genome editing technologies have made possible the introduction and correction of point mutations in hPSCs (169–171). In particular, CRISPR-Cas9 technology has proven particularly useful to efficiently generate isogenic cell line pairs. These can be obtained either by correcting the genetic variant of interest in patient-derived iPSC, or by introducing mutations in a hPSC line that differentiates robustly to the cell type of interest (172, 173). CRISPR can also be used to elucidate which signaling pathways and mechanisms are important to achieve a particular differentiation stage. A recent report illustrates this approach by using a genome-wide CRISPR screening to identify JNK-JUN signaling as a barrier for pluripotency exit and endoderm differentiation (174).

Generation of patient-derived hiPSCs and their differentiations towards the pancreatic lineage has facilitated the generation of cellular models to study diabetes. In combination with genome editing technologies, these approaches make it now feasible to study how a particular genetic variant impacts pancreas development and beta cell physiology. The first diabetes disease modeling studies assessed the ability of patient-derived hiPSCs and healthy donor controls to efficiently differentiate into beta cells (175, 176). CRISPR-Cas9 genome editing has been exploited to correct point mutations associated with diabetes in patient-derived hiPSCs or to generate knockouts (KOs) of critical pancreatic and beta cell genes (177–181). Maxwell and colleagues showed that they were able to generate functional beta cells using a CRISPR-Cas9 edited iPSC line obtained from a patient with *WFS1* mutation. The corrected cells exhibited robust first- and second-phase insulin secretion in response to glucose challenge and restored euglycemia when implanted into diabetic mice, while the unedited controls did not (182).

Diabetes disease modeling studies based on hPSCs have been mostly focused on genes that cause neonatal diabetes, since the expected severe phenotype due to the developmental defect is assumed to be easier to detect. Together with patient-derived hiPSCs, several KO hPSC lines have been genome engineered to study neonatal diabetes disease phenotypes. Several reports have studied the outcomes of disrupting critical pancreatic developmental genes like *NEUROG3* (177), *PDX1* (183), *GLIS3* (184), *RFX6*, *PTF1A*, *MNX1*, *HES1*, *ARX* (178, 185), *GATA4*, *GATA6* (180, 181), or *SIX2* (186). Similar approaches have been exploited to dissect the disease mechanisms behind mutations in *HNF1B* (176) and *HNF4A* (187) that cause MODY or a rare mutation in *STAT3* gene causing neonatal diabetes (179). Genes that harbor genetic variants associated with increased risk of T2D have also been knocked out with CRISPR in hPSCs to study their role in beta cell development and function (e.g. *CDKAL1*, *KCNQ1*, *KCNJ11*) (188, 189).

Beyond genetic defects impairing pancreatic and beta cell development, those directly affecting beta cell insulin secretion are more challenging to study due to the current limitations of the hPSC-based models, in particular the functional immaturity of the derived beta cells. Genetic defects in K^{+}_{ATP} channel genes (64, 188, 190), the insulin gene (191, 192), or the ER-stress related genes *WFS1*, *YIPF5* and *MANF* (79, 80, 175) that cause neonatal diabetes and congenital hyperinsulinism have been modeled with hPSCs using diverse strategies. These include detailed characterization of the *in vitro* obtained pancreatic cell populations, their expression of relevant beta cell markers, their tolerance to different stresses, and their functionality in response to glucose and other secretagogues. Additional characterization with *in vivo* studies allows the assessment of how defective cells respond to systemic environment cues in terms of further differentiation, maturation, and acquisition of regulated insulin secretion. Phenotyping of the implanted cell populations may be particularly useful when the disease mechanisms do not clearly manifest *in vitro*. This is of particular importance when considering the modeling of strict insulin secretion defects, where the functionality of the beta cells generated *in vitro* may be too immature to correctly ascertain a particular phenotype.

In order to model insulin secretion defects reliably, we will need completely functional hPSC-derived beta cells that are as comparable as possible to the ones found in native human islets. Major obstacles in this quest are: i) the lack of differentiation protocols that robustly generate functional beta cells and are widely applicable to any hPSC line; ii) the high variability of human islet preparations, which makes them a problematic gold-standard control to rely upon; iii) the absence of standardized phenotyping methods for hPSC-derived beta cells and human islets which hinders faithful comparison of results across laboratories.

CHARACTERIZATION OF STEM CELL DERIVED BETA CELLS: HOW DO THEY COMPARE TO HUMAN ADULT BETA CELLS?

Beta cell differentiation protocols are technically complex: they have multiple stages, last over a month, and utilize combinations of recombinant proteins and small molecules at different dosages. During the course of any differentiation, many aspects can deviate from the optimal parameters, leading to poor reproducibility and consistency across experiments. In order to minimize experimental variation and minimize costs, laboratories differentiating hPSC usually implement standard operating procedures to prepare culture reagents and execute the differentiation experiments. Current approaches to characterize the outcomes of the hPSC differentiations towards beta cells rely on a battery of methods applied at select stages of the differentiation process. These methods commonly include, but are not limited to, flow cytometry, immunohistochemistry, and RT-qPCR. It is not uncommon for differentiation experiments to fail due to poor definitive endoderm induction, limited expression of pancreatic progenitor markers, or reduced number of INS+ cells. Since the differentiation of one cell type into the next is not 100% efficient, it is critical to address the identity of the cells in the population at given time points. The percentage of cells reaching definitive endoderm stage, the abundance of PDX1+NKX6-1+ pancreatic progenitors, and the fraction of INS+NKX6-1+ cells are examples of common flow cytometry quantifications. They are proxies for the efficiency and quality of the differentiation in terms of achieving bona-fide beta cells. The ultrastructure of hPSC-derived beta cells and human islets has been compared using electron microscopy, using insulin granule morphology as another indicator of beta cell maturity (121, 142, 153, 160).

Current differentiation protocols yield 40–75% INS+ cells in their later stages, although only 20–52% usually represent bona-fide beta cells expressing INS+NKX6-1+ (see an example of differentiation protocol presented in **Figure 2**). Furthermore, the proportions of cell populations can widely vary between experiments and different cell lines (148, 165). One of the important aspects of human pancreatic development that is still poorly understood is the fate allocation of the different endocrine cell types. The timing of NEUROG3 upregulation seems to influence the fate selection of the endocrine precursors (193), which have been shown to be unipotent (194). Endocrine cell fate selection is likely determined by heterogeneous spatiotemporal signals present in the niche of the trunk domain endocrine progenitors. For example, different ligands of the EGF family can modulate the cell-fate selection: betacellulin was reported to promote differentiation into the beta cell lineage when added to mouse embryonic explant cultures (195). This effect was later shown be mediated *via* EGFR-PI3K/AKT-RAC1 signaling resulting in apical polarity inhibition, NOTCH signaling reduction, and induction of NEUROG3 expression (196).

In comparison with the stem cell differentiation outcomes, *in vivo* pancreatic development is also a highly heterogeneous process. Human islet endocrine cell composition varies depending on the islet size and location (161). It is also highly variable across individuals (30) and ages (197). This heterogeneity probably reflects the complexity of endocrine cell fate allocation during development and the plasticity of the pancreas to adapt during the life of the individual to the different metabolic needs.

At the transcriptomic level, gene expression profiling of stem cell-derived islet cells is determined by bulk RT-qPCR or RNAseq at different stages. Sorting of antibody-stained or INS-GFP reporter lines have been used to study beta cell transcriptomes. Several reports have compared the transcriptome of stem cell-derived islet and beta cells with human islets (121, 141, 142, 153, 160, 163, 198). Hrvatin and colleagues compared the transcriptome of stem cell-derived beta cells to both fetal and adult human beta cells (138). RNAseq analyses showed that the INS+ cells generated with that differentiation protocol were transcriptionally closer to fetal beta cells than to adult ones. They had reduced expression levels of genes associated with the functionality and maturation of the beta cell like *PDX1, NKX6-1, MAFA, GLI3,* and *MNX1.* Recent reports describing the generation of hPSC-derived beta cells with dynamic glucose stimulated insulin secretion (GSIS) have curiously shown that some important mature beta cell markers associated with functionality are expressed at much lower levels compared to adult beta cells (i.e. MAFA, UCN3, SIX3) (159, 163, 186, 198, 199). It is therefore unclear what should be considered a reliable marker of functional maturity for hPSC-derived beta cells, especially when the expression levels of some of these markers are age-dependent, being low in functional juvenile islets and taking years to increase (200).

Arising single cell technologies are generating a new important source of knowledge that can be utilized in the quest of generating better beta cells. Single cell transcriptomics, mass spectrometry, and epigenomics are providing new insights on the development and physiology of pancreas, islets, and beta cells (30, 201–208). Single cell transcriptomics has proven particularly useful to investigate the differentiation of hPSCs into beta cells by providing novel information about the heterogeneity of the stage-specific populations, their differentiation trajectories, the role of putative regulators of fate decisions, as well as a mean to assess the impact of diabetes-associated genetic variants. Using single-cell RT-qPCR, Petersen and colleagues explored the trajectories of stem cell derived pancreatic progenitors differentiating towards beta cells (209). They identified two pancreatic progenitor populations that give rise to "monohormonal" beta cells. This suggests the existence of alternative differentiation routes toward beta cells, *via* a progenitor stage that expresses NKX6-1 before or after NEUROG3 upregulation. Single-cell RNA sequencing approaches enable the transcriptional profiling of thousands of cells simultaneously. Krentz and colleagues used this approach to characterize mouse and hPSC-derived endocrine progenitors (210). Exploiting fluorescent reporter mouse strains and hPSC lines labeling Neurog3 lineages, they described and compared the heterogeneity of the mouse and human endocrine progenitor populations and the gene markers they express. scRNA-seq can also aid in the interrogation of the molecular mechanisms behind mutations causing neonatal diabetes (182, 192). Veres et al. used scRNA-seq to chart the differentiation trajectories of stem cell-derived populations, showing the presence of different endocrine and non-endocrine cell populations (159). Single cell transcriptomic technologies have been exploited to identify surface markers like ITGA1, which can be used to enrich for beta cells (159), or CD9 which can be used as a negative marker of functional beta cells (211). Similarly, single cell RNA sequencing analysis led to the observation that WNT signaling is reduced in endocrine cells compared to pancreatic progenitors. Chemical inhibition of WNT signaling in hPSC-derived progenitors induced differentiation to endocrine cells (212). By performing scRNA-seq on *in vitro* and grafted stem cell derived islets, Augsornworawat and colleagues were able to show that 6-month grafted cells undergo important transcriptomic changes, acquiring a gene expression profile more similar to human adult islets (199). scRNA-seq technologies thus offer a new window into the understanding of how transcriptomic regulation determines cell state. Part of its potential for the development of stem cell differentiation approaches relies on the direct comparison of the *in vitro* cells with their *in vivo* "real" counterpart. Enterprises like the human cell atlas are yielding body-wide datasets of single cell transcriptomic that are being used to benchmark *in vitro* stem cell differentiated cells (213).

The assessment of hPSC-derived beta cell functionality relies on methods established to characterize human islets. The most conventional method is the evaluation of insulin secretion in response to high glucose, either in a static setup or in a dynamic fashion using a perifusion setup. Additionally, different secretagogues can be used to probe the different insulin secretion mechanisms in place: K^{+}_{ATP} channel blockers (Tolbutamide), cAMP level modulators (Forskolin, IBMX), GLP1R ligands (Exendin-4, liraglutide), voltage dependent calcium channel agonists (Bay K 8644), non-glucose metabolic substrates (pyruvate, glutamine, leucine), and forced membrane depolarization (arginine, KCl).

The stimulation index (fold increase in insulin secretion from low to high glucose) of hPSC-derived islet cells in static GSIS reported by most studies ranges from 2 to 3, while human islet indexes have a median of about 7 (32). While static glucose stimulated insulin secretion is seemingly a straight-forward assay, there is a wide range of protocols used in the field for both hPSC-derived cells and human islets. They diverge in important critical points: the concentrations of low and high glucose used, the length of the stimulation, the number of cells/

aggregates/islets used for the test, the composition of the stimulation buffer, the washing steps, the length of the equilibration period, and the glucose concentration used during that time. Also, the selection of the samples for this assay is not always clearly reported: it is a common practice to hand-pick human islets of homogeneous aspect and size to perform GSIS, which might yield better results than randomly sampled islets. Many of these parameters and details may seem trivial but can introduce important systematic biases that make comparison of results across labs difficult. These comparisons could greatly benefit from the adoption of standardized practices in the functional assessment and phenotyping of both hPSC-derived cells and human islets. Furthermore, a stable positive control could be used in each GSIS test to have a reference point between experiments, but this is usually not possible due to the scarcity and high variability of human islets.

An important characteristic of human islets is their fine-tuned secretion of insulin in response to glucose. This can be clearly observed in dynamic GSIS assays using perifusion systems, where a robust first insulin secretion phase is followed by sustained second phase of insulin secretion with lower output (145). The acquisition of this biphasic insulin secretion pattern does not occur in human islets until birth (214). The levels of basal insulin secretion are also a good indicator of beta cell function. Immature beta cells have reduced glucose threshold for insulin secretion which leads to higher basal insulin secretion levels and a relatively low stimulation index (138, 215). Robust dynamic glucose stimulated insulin secretion of hPSC-derived beta cells has been only recently reported (160, 163). It seems to depend on a combination of abundant insulin positive cells in the aggregates, achieved either by high differentiation efficiencies involving late stage reaggregation in media containing no additional signaling cues and the expression of beta cell maturation marker SIX2 (163, 186), enrichment using fluorescent reporter lines (160) or surface marker antigens (159), followed by forced or spontaneous reaggregation, respectively. Interesting, Hogrebe et al. also showed that dynamic glucose stimulated insulin was achieved in beta cells generated using their planar differentiation protocol (153). A remaining challenge faced with hPSC-derived beta cell GSIS is the lower magnitude of insulin secretion in comparison to human islets. Davis and colleagues demonstrated that the disparity may be due to a metabolic bottleneck in the glycolytic pathway that can be ameliorated when glyceraldehyde 3-phosphate dehydrogenase (GAPDH) and phosphoglycerate kinase (PGK1) activities are bypassed (216).

Additional functional characterization has relied on surrogate indicators of insulin secretion, like the measurements of Ca^{2+} influx into the cytoplasm in response to different stimuli. Calcium imaging can be performed on dispersed individual cells or on whole aggregates/islets. It has been used to assess the function of hPSC-derived beta cells in some studies, showing that although calcium dynamics might be similar in a small fraction of cells, they are not as robust as in primary human islet cells (121, 142, 160). Electrophysiological studies of human beta cells using patch-clamp technique have demonstrated the electrical properties of their membranes in response to different stimuli (105, 217). In a recent study, Camunas-Soler and colleagues exploited Patch-seq technology to generate healthy and diseased human islet single cell transcriptomic profiles linked with their electrophysiological characteristics. This valuable dataset enabled them to generate predictive sets of genes that reliably linked gene expression to beta cell function and identify transcriptional programs that contribute to beta cell dysfunction in type 1 and type 2 diabetes (218). Basford and colleagues examined the electrophysiological properties of beta cells derived from an INS-GFP+ reporter stem cell. Compared to human adult beta cells, stem cell-derived beta cells presented heterogeneous K_{ATP} (45% of the cells) and Ca^{2+} (42% of the cells) channel currents and no Na^{+} channel currents (134). To the best of our knowledge, there are presently no reports of stem cell derived beta cells demonstrating electrophysiological properties identical to those of primary human adult beta cells.

Glycolysis coupled with efficient mitochondrial respiration is required for normal insulin secretion (219). During beta cell maturation, active DNA methylation silences the expression of disallowed genes (e.g. *HK1*, *LDHA*) that interfere with the glucose-secretion coupling (220). Neonatal acquisition of aerobic oxidative metabolism is a crucial step for the maturing beta cell, a process shown to be induced by non-canonical WNT4 signaling and estrogen related receptor gamma (ESRRG) (221, 222). Rates of O_2 consumption and CO_2 production can be used to evaluate the respiratory capacity of islets and hPSC-derived cells (223, 224), and serve as both a functionality and maturation surrogate marker in the efforts of making better beta cells (160, 221). Enrichment and reaggregation of hPSC-derived beta cells induced mitochondrial metabolic maturation, and the ultrastructure of mitochondria showed increased folding and stacking of cristae (160).

All hPSC-derived beta cell characterization approaches are ultimately benchmarked against human islets. Unfortunately, human islets typically demonstrate wide phenotypic variability across batches (30, 121). This particular point is frequently not suitably acknowledged, and it is particularly problematic when batches of poorly performing islets are used for the comparison to hPSC-derived beta cell preparations. Systematic evaluation of human islets batches at different levels (cell composition, functionality, transcriptomics, etc.) is a step in the right direction to highlight this variability and define a canonical human islet response (**Table 2**). This is illustrated by the Alberta Diabetes Institute IsletCore database collaborative initiative spearheaded by the MacDonald laboratory (32), where traceable phenotypes of over 300 human islets batches demonstrate the remarkable variation in functionality.

MODELING DIABETES WITH IMPLANTED STEM CELL DERIVED ISLET CELLS

An alternative approach to derive functionally mature hPSC-derived beta cells is to implant their precursors into immunocompromised host rodents. The first report describing this approach showed that a few months after implantation the grafts secreted human insulin in response to systemic glucose

TABLE 2 | Key characteristics of human islets.

Morphology: spheroid
Number of endocrine cells/islet: ~1500
Mean islet size: ~150 µm
Endocrine architecture and composition:
* Beta cells ~50–60% interspersed throughout the islet
* Alpha cells ~40% interspersed throughout the islet
* Delta cells ~10%
* Gamma and PP cells <1%

Stimulated insulin secretion
* High glucose-stimulated insulin challenge (static GSIS) stimulation index: ~7-fold
* High glucose-stimulated insulin challenge (dynamic GSIS): rapid biphasic response
* Potentiated secretion in response to cAMP modulators (e.g. Forskolin, IBMX)
* Potentiated secretion in response to incretin hormones (e.g. GIP, GLP-1)
* Increased secretion in response to membrane depolarization (e.g. KCl)
* Increased secretion in response to K_{ATP} channel activators (e.g. tolbutamide)
* Increased secretion in response to calcium channel agonists (e.g. Bay K 8644)
* Increased secretion in response to non-glucose nutrients (e.g. palmitate, leucine)

Key transcription factors and maturation markers: SIX2, SIX3, UCN3, MAFA, NKX6.1, INS, PDX1, GLIS3, MNX1
Morphology of mature insulin granules: dense core vesicles
Dithizone staining: brick red color
Respiratory capacity: primarily mitochondrial
Calcium signaling: increased pulsatile signaling in response to high glucose

List of various features of primary isolated human islets that can be taken into consideration for comparison to stem cell derived islet cells.

administration (125). Since then, multiple implantation sites (subcutaneous, intramuscular, renal subcapsular space, epididymal fat pad, pancreas) (121, 125, 225, 226) and several animal models (SCID-beige, NSG, NOG, NRG-Akita mice; nude rats) (125, 142, 227) have been employed with variable success.

Implantation in the renal subcapsular space is one of the preferred approaches since it is relatively easy to implant and retrieve the cells months later *via* survival nephrectomy. Upon implantation, cells become vascularized and interestingly their cytoarchitecture can undergo rearrangement (131), concomitantly with an increase in the functional maturation (121, 227). Endogenous pancreatic beta cells in recipient animals can be largely eliminated by administration of a beta cell toxin (e.g. alloxan or streptozotocin) either before or after implantation of hPSC-derived cells. The doses used are typically relatively harmless to the hPSC-derived cells owing to species differences in toxin sensitivity (228, 229). Graft function is monitored by measuring circulating human C-peptide levels (using assays that can distinguish the graft derived human *versus* recipient's endogenous C-peptide owing to sequence differences), and the response to glucose can be determined with intraperitoneal, intravenous, or oral glucose tolerance tests. Also, hPSC-derived pancreatic progenitors or endocrine cells can be transplanted within macro- or micro-encapsulation devices (132, 230–232).

An interesting phenomenon is the functional maturation of implanted beta cells with time. Rezania and colleagues described the progressive increase in circulating C-peptide levels for several weeks after implanting pancreatic progenitor cells or more matured cells (121, 131), a phenomenon which has also been observed by others (160, 192, 199). Furthermore, Rezania et al. observed faster diabetes recovery and achieved higher circulating C-peptide levels sooner when the implanted cells were further along in their differentiation prior to implant (121). It remains unclear what factors promote the apparently successful maturation of differentiated hPSCs post implant. One possibility is that immature hPSC-derived beta cells require a critical niche and systemic factors including vascularization and proper oxygenation to acquire full functionality (233). Interestingly, maturation of hPSC-derived pancreatic progenitors is accelerated in rats compared with mice (228), something that the authors correlated with increased levels of thyroid hormone in the rats, in line with the fact that thyroid hormone promotes beta cell maturation in rats (143) and in differentiating hPSCs (121). Sex hormones may also influence *in vivo* maturation of pancreatic progenitors since following implant, glucose-stimulated insulin secretion was observed in female mice before males (234). The systemic environment also provides periodic signals, which might entrain the circadian clock of the implanted stem cell-derived islet cells, leading to their functional maturation (235–238).

Several studies reporting the generation of hPSC-based diabetes disease models have relied on implanting mice with hPSC-derived beta cells. This constitutes a practical solution to study the function of the beta cells in a systemic environment, especially when the disease phenotype is not apparent *in vitro* (80, 175, 188, 191, 192). It is also useful to investigate the impairment of development caused by mutations perturbing critical regulators of islet cell development (177, 179, 180). Once grafts have matured, these models offer the possibility to dissect the effect of particular mutations on insulin secretion by carefully examining their responses to different stimuli.

An important aspect after implantation of hPSC-derived beta cells, is their capacity to survive the hypoxic environment of the implantation site until they become vascularized. This is a critical stress period that may result in the apoptosis of the most differentiated endocrine cells (239). Faleo and colleagues reported that this might be partially overcome by acclimatizing the cells to hypoxic conditions before implantation. In this regard, the format of the implanted cells likely also plays a critical role in successful engraftment, with smaller aggregates probably benefiting from faster vascularization kinetics as shown for engineered pseudoislets (240).

Another concern with stem cell derived islet cell implants, especially when using them on diabetic rodents, is the "pellet" effect, in which the non-regulated basal insulin secretion coming from immature beta cells might be enough to rescue hyperglycemia. This brings up the question of how many cells should be implanted to achieve an optimal working graft, which obviously will depend on the stage of differentiation and quality of the cells in terms of functionality (121). The composition and format (e.g. aggregate size) of the implanted cell population is also likely critical for a successful outcome. In fact, the formation of pancreatic progenitor cell aggregates prior to implant was shown to be essential for the formation endocrine cells (132). In order to investigate the *in vivo* maturation process of hPSC-derived beta cells, novel *in vivo* imaging technologies could be exploited. Radio tracer-based imaging of beta cell mass and function could prove to be particularly useful in this regard (241, 242).

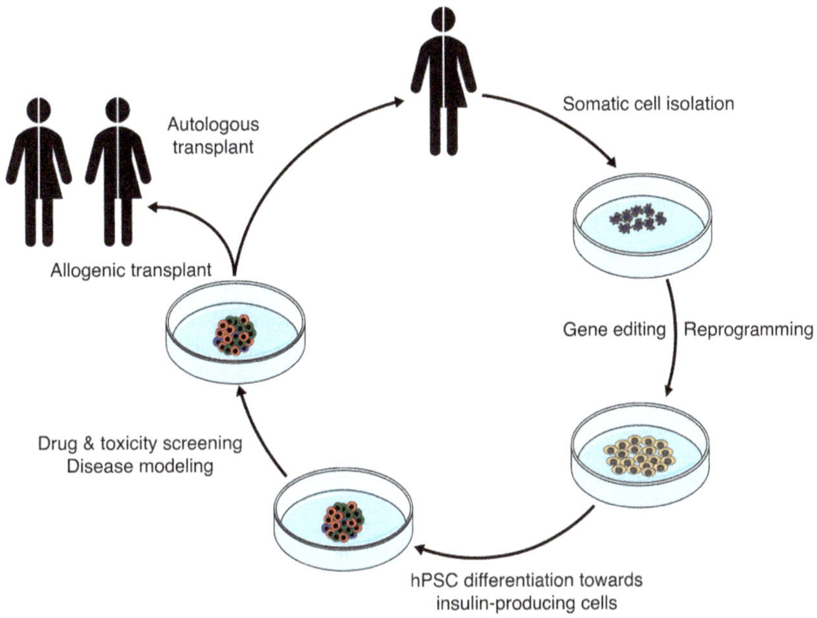

FIGURE 3 | Schematic showing the potential application of stem cell-derived insulin-producing cells for the treatment of diabetes.

Although intra islet paracrine signaling between the different endocrine cell types is crucial for fine-tuned insulin secretion (44, 45, 243), different reports have shown that diabetes can be rescued with nearly pure populations of islet beta cells with different efficiency (244, 245). Nair and colleagues described the implantation of 90% enriched hPSC-derived beta cells using an INS-GFP reporter cell line. While the grafts presented a few polyhormonal (INS+GCG+) cells 3 days after implantation, 8 weeks later, there was clear presence of GCG+ and SST+ monohormonal cells together with the beta cells, suggesting that the polyhormonal cells gave rise to monohormonal alpha and delta cells that likely mediate paracrine signaling contributing to optimal insulin secretion (160). The proportion of endocrine cell types in the implant to achieve the best glycemic control possible is an interesting question that requires further investigation.

Further understanding of the factors playing an important role in the functional maturation of implanted hPSC-derived beta cells will pave the way to the generation of better humanized mouse models to study insulin secretion. Ultimately, optimal control of implantation parameters will reduce the associated variability of these experiments enabling the careful assessment of the impact of genetic variants on insulin secretion.

CONCLUSION

Derivation of endocrine islet cells from hPSCs has become an attractive possibility to model diabetes disease and screen for new treatments (**Figure 3**). The progress in the last decade has made it feasible to obtain cells *in vitro* that closely resemble the native adult counterpart. Arising technologies like CRISPR-Cas9 genome editing and single cell transcriptomics are aiding in the generation of more reliable stem cell models, the refinement of the differentiation protocols and the characterization of the resulting differentiated islet cells. There are still important remaining questions in the quest for more functionally mature beta cells: how can we determine and increase beta cell specification? What are reliable mature beta cell markers and the key triggers of functional maturation? Together with detailed single cell transcriptomic characterization, improved characterization of the metabolism, proteomics and functional genomics of the hPSC-derived islets cells, and their comparison with human islets, will certainly pave the way forward. In this common effort, consensus in standardized characterization of the resulting hPSC progeny, the development of robust and reproducible differentiation protocols, and open dissemination of results, will enable prompt replication and speed up the implementation of successful strategies for beta cell generation and diabetes disease modeling.

AUTHOR CONTRIBUTIONS

Writing—original draft preparation, DB. Writing—review and editing, DB, DI, and TK. Funding acquisition, TK. All authors contributed to the article and approved the submitted version.

ACKNOWLEDGMENTS

DB gratefully acknowledges funding support by EMBO long-term fellowship (ALTF 295-2019). TK gratefully acknowledges lab support by the NIH, Stem Cell Network, STEMCELL Technologies, JDRF, and Canadian Institutes of Health Research. DI gratefully acknowledges funding support by the Natural Science and Engineering Research Council, and Canadian Federation of University Women.

REFERENCES

1. International Diabetes Federation. *IDF Diabetes Atlas, 9th edn*. International Diabetes Federation (2019). Available at: https://www.diabetesatlas.org.
2. Ashcroft FMM, Rorsman P. Diabetes Mellitus and the β Cell: The Last Ten Years. *Cell* (2012) 148:1160–71. doi: 10.1016/j.cell.2012.02.010
3. Krentz NAJ, Gloyn AL. Insights into pancreatic islet cell dysfunction from type 2 diabetes mellitus genetics. *Nat Rev Endocrinol* (2020) 16:202–12. doi: 10.1038/s41574-020-0325-0
4. Tuomilehto J. The emerging global epidemic of type 1 diabetes. *Curr Diabetes Rep* (2013) 13:795–804. doi: 10.1007/s11892-013-0433-5
5. Tuomi T, Santoro N, Caprio S, Cai M, Weng J, Groop L. The many faces of diabetes: a disease with increasing heterogeneity. *Lancet (London England)* (2014) 383:1084–94. doi: 10.1016/S0140-6736(13)62219-9
6. Alberti KGMM, Zimmet PZ. Definition, diagnosis and classification of diabetes mellitus and its complications. Part 1: Diagnosis and classification of diabetes mellitus. Provisional report of a WHO consultation. *Diabetes Med* (1998) 15:539–53. doi: 10.1002/(SICI)1096-9136(199807)15:7<539::AID-DIA668>3.0.CO;2-S
7. Redondo MJ, Hagopian WA, Oram R, Steck AK, Vehik K, Weedon M, et al. The clinical consequences of heterogeneity within and between different diabetes types. *Diabetologia* (2020) 63:2040–8. doi: 10.1007/s00125-020-05211-7
8. Hattersley AT, Patel KA. Precision diabetes: learning from monogenic diabetes. *Diabetologia* (2017) 60:769–77. doi: 10.1007/s00125-017-4226-2
9. Besser REJ, Flanagan SE, Mackay DGJ, Temple IK, Shepherd MH, Shields BM, et al. Prematurity and Genetic Testing for Neonatal Diabetes. *Pediatrics* (2016) 138:e20153926–e20153926. doi: 10.1542/peds.2015-3926
10. Kwak SH, Jung C, Ahn CH, Park J, Chae J, Jung HS, et al. Clinical whole exome sequencing in early onset diabetes patients. *Diabetes Res Clin Pract* (2016) 122:71–7. doi: 10.1016/j.diabres.2016.10.005
11. Mahajan A, Taliun D, Thurner M, Robertson NR, Torres JM, Rayner NW, et al. Fine-mapping type 2 diabetes loci to single-variant resolution using high-density imputation and islet-specific epigenome maps. *Nat Genet* (2018) 50:1505–13. doi: 10.1038/s41588-018-0241-6
12. Bonàs-Guarch S, Guindo-Martínez M, Miguel-Escalada I, Grarup N, Sebastian D, Rodriguez-Fos E, et al. Re-analysis of public genetic data reveals a rare X-chromosomal variant associated with type 2 diabetes. *Nat Commun* (2018) 9:321. doi: 10.1038/s41467-017-02380-9
13. Pasquali L, Gaulton KJ, Rodríguez-Seguí SA, Mularoni L, Miguel-Escalada I, Akerman İ, et al. Pancreatic islet enhancer clusters enriched in type 2 diabetes risk-associated variants. *Nat Genet* (2014) 46:136–43. doi: 10.1038/ng.2870
14. Thomas PM, Cote GJ, Hallman DM, Mathew PM. Homozygosity mapping, to chromosome 11p, of the gene for familial persistent hyperinsulinemic hypoglycemia of infancy. *Am J Hum Genet* (1995) 56:416–21.
15. Ellard S, Flanagan SE, Girard CA, Patch A-M, Harries LW, Parrish A, et al. Permanent Neonatal Diabetes Caused by Dominant, Recessive, or Compound Heterozygous SUR1 Mutations with Opposite Functional Effects. *Am J Hum Genet* (2007) 81:375–82. doi: 10.1086/519174
16. Yamagata K, Oda N, Kaisaki PJ, Menzel S, Furuta H, Vaxillaire M, et al. Mutations in the hepatocyte nuclear factor-1α gene in maturity-onset diabetes of the young (MODY3). *Nature* (1996) 384:455–8. doi: 10.1038/384455a0
17. Estrada K, Aukrust I, Bjørkhaug L, Burtt NP, Mercader JM, García-Ortiz H, et al. Association of a Low-Frequency Variant in *HNF1A* With Type 2 Diabetes in a Latino Population. *JAMA* (2014) 311:2305. doi: 10.1001/jama.2014.6511
18. Osbak KK, Colclough K, Saint-Martin C, Beer NL, Bellanné-Chantelot C, Ellard S, et al. Update on mutations in glucokinase (*GCK*), which cause maturity-onset diabetes of the young, permanent neonatal diabetes, and hyperinsulinemic hypoglycemia. *Hum Mutat* (2009) 30:1512–26. doi: 10.1002/humu.21110
19. Njølstad PR, Søvik O, Cuesta-Muñoz A, Bjørkhaug L, Massa O, Barbetti F, et al. Neonatal Diabetes Mellitus Due to Complete Glucokinase Deficiency. *N Engl J Med* (2001) 344:1588–92. doi: 10.1056/NEJM200105243442104
20. Flannick J, Johansson S, Njølstad PR. Common and rare forms of diabetes mellitus: towards a continuum of diabetes subtypes. *Nat Rev Endocrinol* (2016) 12:394–406. doi: 10.1038/nrendo.2016.50
21. Flanagan SE, De Franco E, Lango Allen H, Zerah M, Abdul-Rasoul MM, Edge JA, et al. Analysis of Transcription Factors Key for Mouse Pancreatic Development Establishes NKX2-2 and MNX1 Mutations as Causes of Neonatal Diabetes in Man. *Cell Metab* (2014) 19:146–54. doi: 10.1016/j.cmet.2013.11.021
22. Gaulton KJ, Nammo T, Pasquali L, Simon JM, Giresi PG, Fogarty MP, et al. A map of open chromatin in human pancreatic islets. *Nat Genet* (2010) 42:255–9. doi: 10.1038/ng.530
23. Miguel-Escalada I, Bonàs-Guarch S, Cebola I, Ponsa-Cobas J, Mendieta-Esteban J, Atla G, et al. Human pancreatic islet three-dimensional chromatin architecture provides insights into the genetics of type 2 diabetes. *Nat Genet* (2019) 51:1137–48. doi: 10.1038/s41588-019-0457-0
24. Collombat P, Hecksher-Sørensen J, Serup P, Mansouri A. Specifying pancreatic endocrine cell fates. *Mech Dev* (2006) 123:501–12. doi: 10.1016/j.mod.2006.05.006
25. Balboa D, Otonkoski T. Human pluripotent stem cell based islet models for diabetes research. *Best Pract Res Clin Endocrinol Metab* (2015) 29:899–909. doi: 10.1016/j.beem.2015.10.012
26. Cabrera O, Berman DM, Kenyon NS, Ricordi C, Berggren P-O, Caicedo A. The unique cytoarchitecture of human pancreatic islets has implications for islet cell function. *Proc Natl Acad Sci USA* (2006) 103:2334–9. doi: 10.1073/pnas.0510790103
27. Henquin J-C, Dufrane D, Nenquin M. Nutrient control of insulin secretion in isolated normal human islets. *Diabetes* (2006) 55:3470–7. doi: 10.2337/db06-0868
28. Fadista J, Vikman P, Laakso EO, Mollet IG, Esguerra JL, Taneera J, et al. Global genomic and transcriptomic analysis of human pancreatic islets reveals novel genes influencing glucose metabolism. *Proc Natl Acad Sci* (2014) 111:13924–9. doi: 10.1073/pnas.1402665111
29. Ionescu-Tirgoviste C, Gagniuc PA, Gubceac E, Mardare L, Popescu I, Dima S. Militaru M. A 3D map of the islet routes throughout the healthy human pancreas. *Sci Rep* (2015) 5:14634. doi: 10.1038/srep14634
30. Wang YJ, Golson ML, Schug J, Traum D, Liu C, Vivek K, et al. Single-Cell Mass Cytometry Analysis of the Human Endocrine Pancreas. *Cell Metab* (2016) 24:616–26. doi: 10.1016/j.cmet.2016.09.007
31. Brissova M, Fowler MJ, Nicholson WE, Chu A, Hirshberg B, Harlan DM, et al. Assessment of Human Pancreatic Islet Architecture and Composition by Laser Scanning Confocal Microscopy. *J Histochem Cytochem* (2005) 53:1087–97. doi: 10.1369/jhc.5C6684.2005
32. Database ADII. *Alberta Diabetes Institute IsletCore Database* (2020). Available at: http://www.isletcore.ca (Accessed December 1, 2020).
33. Ravassard P, Hazhouz Y, Pechberty S, Bricout-Neveu E, Armanet M, Czernichow P, et al. A genetically engineered human pancreatic β cell line exhibiting glucose-inducible insulin secretion. *J Clin Invest* (2011) 121:3589–97. doi: 10.1172/JCI58447
34. Tsonkova VG, Sand FW, Wolf XA, Grunnet LG, Kirstine Ringgaard A, Ingvorsen C, et al. The EndoC-βH1 cell line is a valid model of human beta cells and applicable for screenings to identify novel drug target candidates. *Mol Metab* (2018) 8:144–57. doi: 10.1016/j.molmet.2017.12.007
35. Lawlor N, Márquez EJ, Orchard P, Narisu N, Shamim MS, Thibodeau A, et al. Multiomic Profiling Identifies cis-Regulatory Networks Underlying Human Pancreatic β Cell Identity and Function. *Cell Rep* (2019) 26:788–801.e6. doi: 10.1016/J.CELREP.2018.12.083
36. Puri S, Roy N, Russ HA, Leonhardt L, French EK, Roy R, et al. Replication confers β cell immaturity. *Nat Commun* (2018) 9:485. doi: 10.1038/s41467-018-02939-0
37. Scharfmann R, Pechberty S, Hazhouz Y, von Bülow M, Bricout-Neveu E, Grenier-Godard M, et al. Development of a conditionally immortalized human pancreatic β cell line. *J Clin Invest* (2014) 124:2087–98. doi: 10.1172/JCI72674
38. Benazra M, Lecomte M-J, Colace C, Müller A, Machado C, Pechberty S, et al. A human beta cell line with drug inducible excision of immortalizing transgenes. *Mol Metab* (2015) 4:916–25. doi: 10.1016/j.molmet.2015.09.008
39. Thomson JA, Itskovitz-Eldor J, Shapiro SS, Waknitz MA, Swiergiel JJ, Marshall VS, et al. Embryonic stem cell lines derived from human blastocysts. *Science* (1998) 282:1145–7. doi: 10.1126/science.282.5391.1145
40. Takahashi K, Tanabe K, Ohnuki M, Narita M, Ichisaka T, Tomoda K, et al. Induction of Pluripotent Stem Cells from Adult Human Fibroblasts by Defined Factors. *Cell* (2007) 131:861–72. doi: 10.1016/j.cell.2007.11.019

41. Tiscornia G, Vivas EL, Belmonte JCI. Diseases in a dish: modeling human genetic disorders using induced pluripotent cells. *Nat Med* (2011) 17:1570–6. doi: 10.1038/nm.2504
42. González F. CRISPR/Cas9 genome editing in human pluripotent stem cells: Harnessing human genetics in a dish. *Dev Dyn* (2016) 245:788–806. doi: 10.1002/dvdy.24414
43. Roscioni SS, Migliorini A, Gegg M, Lickert H. Impact of islet architecture on β-cell heterogeneity, plasticity and function. *Nat Rev Endocrinol* (2016) 12:695–709. doi: 10.1038/nrendo.2016.147
44. Rodriguez-Diaz R, Molano RD, Weitz JR, Abdulreda MH, Berman DM, Leibiger B, et al. Paracrine Interactions within the Pancreatic Islet Determine the Glycemic Set Point. *Cell Metab* (2018) 27:549–558.e4. doi: 10.1016/j.cmet.2018.01.015
45. Huising MO. Paracrine regulation of insulin secretion. *Diabetologia* (2020) 63:2057–63. doi: 10.1007/s00125-020-05213-5
46. Nicholls DG. The Pancreatic β-Cell: A Bioenergetic Perspective. *Physiol Rev* (2016) 96:1385–447. doi: 10.1152/physrev.00009.2016
47. German MS. Glucose sensing in pancreatic islet beta cells: The key role of glucokinase and the glycolytic intermediates. *Proc Natl Acad Sci USA* (1993) 90:1781–5. doi: 10.1073/pnas.90.5.1781
48. Flannick J, Mercader JM, Fuchsberger C, Udler MS, Mahajan A, Wessel J, et al. Exome sequencing of 20,791 cases of type 2 diabetes and 24,440 controls. *Nature* (2019) 570:71–6. doi: 10.1038/s41586-019-1231-2
49. Glaser B, Kesavan P, Heyman M, Davis E, Cuesta A, Buchs A, et al. Familial Hyperinsulinism Caused by an Activating Glucokinase Mutation. *N Engl J Med* (1998) 338:226–30. doi: 10.1056/NEJM199801223380404
50. Mahajan A, Sim X, Ng HJ, Manning A, Rivas MA, Highland HM, et al. Identification and functional characterization of G6PC2 coding variants influencing glycemic traits define an effector transcript at the G6PC2-ABCB11 locus. *PloS Genet* (2015) 11:e1004876. doi: 10.1371/journal.pgen.1004876
51. Sansbury FH, Flanagan SE, Houghton JAL, Shuixian Shen FL, Al-Senani AMS, Habeb AM, et al. SLC2A2 mutations can cause neonatal diabetes, suggesting GLUT2 may have a role in human insulin secretion. *Diabetologia* (2012) 55:2381–5. doi: 10.1007/s00125-012-2595-0
52. Henquin J-C, Sempoux C, Marchandise J, Godecharles S, Guiot Y, Nenquin M, et al. Congenital Hyperinsulinism Caused by Hexokinase I Expression or Glucokinase-Activating Mutation in a Subset of -Cells. *Diabetes* (2013) 62:1689–96. doi: 10.2337/db12-1414
53. Otonkoski T, Kaminen N, Ustinov J, Lapatto R, Meissner T, Mayatepek E, et al. Physical Exercise-Induced Hyperinsulinemic Hypoglycemia Is an Autosomal-Dominant Trait Characterized by Abnormal Pyruvate-Induced Insulin Release. *Diabetes* (2003) 52:199–204. doi: 10.2337/diabetes.52.1.199
54. Flanagan SE, Kapoor RR, Mali G, Cody D, Murphy N, Schwahn B, et al. Diazoxide-responsive hyperinsulinemic hypoglycemia caused by HNF4A gene mutations. *Eur J Endocrinol* (2010) 162:987–92. doi: 10.1530/EJE-09-0861
55. Kapoor RR, Flanagan SE, Arya VB, Shield JP, Ellard S, Hussain K. Clinical and molecular characterisation of 300 patients with congenital hyperinsulinism. *Eur J Endocrinol* (2013) 168:557–64. doi: 10.1530/EJE-12-0673
56. Li C, Chen P, Palladino A, Narayan S, Russell LK, Sayed S, et al. Mechanism of Hyperinsulinism in Short-chain 3-Hydroxyacyl-CoA Dehydrogenase Deficiency Involves Activation of Glutamate Dehydrogenase. *J Biol Chem* (2010) 285:31806–18. doi: 10.1074/jbc.M110.123638
57. González-Barroso MM, Giurgea I, Bouillaud F, Anedda A, Bellanné-Chantelot C, Hubert L, et al. Mutations in UCP2 in Congenital Hyperinsulinism Reveal a Role for Regulation of Insulin Secretion. *PloS One* (2008) 3:e3850. doi: 10.1371/journal.pone.0003850
58. Sesti G, Cardellini M, Marini MA, Frontoni S, D'Adamo M, Del Guerra S, et al. A Common Polymorphism in the Promoter of UCP2 Contributes to the Variation in Insulin Secretion in Glucose-Tolerant Subjects. *Diabetes* (2003) 52:1280–3. doi: 10.2337/diabetes.52.5.1280
59. Maassen JA, 'T Hart LM, Van Essen E, Heine RJ, Nijpels G, Jahangir Tafrechi RS, et al. Mitochondrial diabetes: molecular mechanisms and clinical presentation. *Diabetes* (2004) 53 Suppl 1:S103–9. doi: 10.2337/diabetes.53.2007.s103
60. Gloyn AL, Pearson ER, Antcliff JF, Proks P, Bruining GJ, Slingerland AS, et al. Activating Mutations in the Gene Encoding the ATP-Sensitive Potassium-Channel Subunit Kir6.2 and Permanent Neonatal Diabetes. *N Engl J Med* (2004) 350:1838–49. doi: 10.1056/NEJMoa032922
61. Bonnefond A, Philippe J, Durand E, Dechaume A, Huyvaert M, Montagne L, et al. Whole-Exome Sequencing and High Throughput Genotyping Identified KCNJ11 as the Thirteenth MODY Gene. *PloS One* (2012) 7: e37423. doi: 10.1371/journal.pone.0037423
62. Gloyn AL, Weedon MN, Owen KR, Turner MJ, Knight BA, Hitman G, et al. Large-Scale Association Studies of Variants in Genes Encoding the Pancreatic -Cell KATP Channel Subunits Kir6.2 (KCNJ11) and SUR1 (ABCC8) Confirm That the KCNJ11 E23K Variant Is Associated With Type 2 Diabetes. *Diabetes* (2003) 52:568–72. doi: 10.2337/diabetes.52.2.568
63. Huopio H, Otonkoski T, Vauhkonen I, Reimann F, Ashcroft FM, Laakso M. A new subtype of autosomal dominant diabetes attributable to a mutation in the gene for sulfonylurea receptor 1. *Lancet* (2003) 361:301–7. doi: 10.1016/S0140-6736(03)12325-2
64. Lithovius V, Saarimäki-Vire J, Balboa D, Ibrahim H, Montaser H, Barsby T, et al. SUR1-mutant iPS cell-derived islets recapitulate the pathophysiology of congenital hyperinsulinism. *Diabetologia* (2021) 64:630–40. doi: 10.1007/s00125-020-05346-7
65. Travers ME, Mackay DJG, Dekker Nitert M, Morris AP, Lindgren CM, Berry A, et al. Insights into the molecular mechanism for type 2 diabetes susceptibility at the KCNQ1 locus from temporal changes in imprinting status in human islets. *Diabetes* (2013) 62:987–92. doi: 10.2337/db12-0819
66. Kwak SH, Chae J, Lee S, Choi S, Koo BK, Yoon JW, et al. Nonsynonymous Variants in PAX4 and GLP1R Are Associated With Type 2 Diabetes in an East Asian Population. *Diabetes* (2018) 67:1892–902. doi: 10.2337/db18-0361
67. Saxena R, Hivert M-F, Langenberg C, Tanaka T, Pankow JS, Vollenweider P, et al. Genetic variation in GIPR influences the glucose and insulin responses to an oral glucose challenge. *Nat Genet* (2010) 42:142–8. doi: 10.1038/ng.521
68. Tuomi T, Nagorny CL, Singh P, Bennet H, Yu Q, Alenkvist I, et al. Increased Melatonin Signaling Is a Risk Factor for Type 2 Diabetes. *Cell Metab* (2016) 23:1067–77. doi: 10.1016/j.cmet.2016.04.009
69. Støy J, Steiner DF, Park S-Y, Ye H, Philipson LH, Bell GI. Clinical and molecular genetics of neonatal diabetes due to mutations in the insulin gene. *Rev Endocr Metab Disord* (2010) 11:205–15. doi: 10.1007/s11154-010-9151-3
70. Sladek R, Rocheleau G, Rung J, Dina C, Shen L, Serre D, et al. A genome-wide association study identifies novel risk loci for type 2 diabetes. *Nature* (2007) 445:881–5. doi: 10.1038/nature05616
71. Flannick J, Thorleifsson G, Beer NL, Jacobs SBR, Grarup N, Burtt NP, et al. Loss-of-function mutations in SLC30A8 protect against type 2 diabetes. *Nat Genet* (2014) 46:357–63. doi: 10.1038/ng.2915
72. Dwivedi OP, Lehtovirta M, Hastoy B, Chandra V, Krentz NAJJ, Kleiner S, et al. Loss of ZnT8 function protects against diabetes by enhanced insulin secretion. *Nat Genet* (2019) 51:1596–606. doi: 10.1038/s41588-019-0513-9
73. Roman TS, Cannon ME, Vadlamudi S, Buchkovich ML, Wolford BN, Welch RP, et al. A Type 2 Diabetes–Associated Functional Regulatory Variant in a Pancreatic Islet Enhancer at the *ADCY5* Locus. *Diabetes* (2017) 66:2521–30. doi: 10.2337/db17-0464
74. Hodson DJ, Mitchell RK, Marselli L, Pullen TJ, Gimeno Brias S, Semplici F, et al. ADCY5 Couples Glucose to Insulin Secretion in Human Islets. *Diabetes* (2014) 63:3009–21. doi: 10.2337/db13-1607
75. Inoue H, Tanizawa Y, Wasson J, Behn P, Kalidas K, Bernal-Mizrachi E, et al. A gene encoding a transmembrane protein is mutated in patients with diabetes mellitus and optic atrophy (Wolfram syndrome). *Nat Genet* (1998) 20:143–8. doi: 10.1038/2441
76. Steinthorsdottir V, Thorleifsson G, Reynisdottir I, Benediktsson R, Jonsdottir T, Walters GB, et al. A variant in CDKAL1 influences insulin response and risk of type 2 diabetes. *Nat Genet* (2007) 39:770–5. doi: 10.1038/ng2043
77. Thomsen SK, Ceroni A, van de Bunt M, Burrows C, Barrett A, Scharfmann R, et al. Systematic Functional Characterization of Candidate Causal Genes for Type 2 Diabetes Risk Variants. *Diabetes* (2016) 65:3805–11. doi: 10.2337/db16-0361
78. Yavarna T, Al-Dewik N, Al-Mureikhi M, Ali R, Al-Mesaifri F, Mahmoud L, et al. High diagnostic yield of clinical exome sequencing in Middle Eastern patients with Mendelian disorders. *Hum Genet* (2015) 134:967–80. doi: 10.1007/s00439-015-1575-0

79. Montaser H, Patel KA, Balboa D, Ibrahim H, Lithovius V, Näätänen A, et al. Loss of MANF causes childhood onset syndromic diabetes due to increased endoplasmic reticulum stress. *Diabetes* (2021), db201174. doi: 10.2337/db20-1174
80. De Franco E, Lytrivi M, Ibrahim H, Montaser H, Wakeling MN, Fantuzzi F, et al. YIPF5 mutations cause neonatal diabetes and microcephaly through endoplasmic reticulum stress. *J Clin Invest* (2020) 130:6338–53. doi: 10.1172/jci141455
81. Staffers DA, Ferrer J, Clarke WL, Habener JF. Early-onset type-ll diabetes mellitus (MODY4) linked to IPF1. *Nat Genet* (1997) 17:138–9. doi: 10.1038/ng1097-138
82. Stoffers DA, Zinkin NT, Stanojevic V, Clarke WL, Habener JF. Pancreatic agenesis attributable to a single nucleotide deletion in the human IPF1 gene coding sequence. *Nat Genet* (1997) 15:106–10. doi: 10.1038/ng0197-106
83. Macfarlane WM, Frayling TM, Ellard S, Evans JC, Allen LIS, Bulman MP, et al. Missense mutations in the insulin promoter factor-1 gene predispose to type 2 diabetes. *J Clin Invest* (1999) 104:R33–9. doi: 10.1172/JCI7449
84. Hani EH, Stoffers DA, Chèvre J-C, Durand E, Stanojevic V, Dina C, et al. Defective mutations in the insulin promoter factor-1 (IPF-1) gene in late-onset type 2 diabetes mellitus. *J Clin Invest* (1999) 104:R41–8. doi: 10.1172/JCI7469
85. Smith SB, Qu H-Q, Taleb N, Kishimoto NY, Scheel DW, Lu Y, et al. Rfx6 directs islet formation and insulin production in mice and humans. *Nature* (2010) 463:775–80. doi: 10.1038/nature08748
86. Sansbury FH, Kirel B, Caswell R, Lango Allen H, Flanagan SE, Hattersley AT, et al. Biallelic RFX6 mutations can cause childhood as well as neonatal onset diabetes mellitus. *Eur J Hum Genet* (2015) 23:1744–8. doi: 10.1038/ejhg.2015.161
87. Patel KA, Kettunen J, Laakso M, Stančáková A, Laver TW, Colclough K, et al. Heterozygous RFX6 protein truncating variants are associated with MODY with reduced penetrance. *Nat Commun* (2017) 8:888. doi: 10.1038/s41467-017-00895-9
88. Rubio-Cabezas O, Minton JAL, Kantor I, Williams D, Ellard S, Hattersley AT. Homozygous mutations in NEUROD1 are responsible for a novel syndrome of permanent neonatal diabetes and neurological abnormalities. *Diabetes* (2010) 59:2326–31. doi: 10.2337/db10-0011
89. Malecki MT, Jhala US, Antonellis A, Fields L, Doria A, Orban T, et al. Mutations in NEUROD1 are associated with the development of type 2 diabetes mellitus. *Nat Genet* (1999) 23:323–8. doi: 10.1038/15500
90. Fajans SS, Bell GI, Polonsky KS. Molecular Mechanisms and Clinical Pathophysiology of Maturity-Onset Diabetes of the Young. *N Engl J Med* (2001) 345:971–80. doi: 10.1056/NEJMra002168
91. Dooley J, Tian L, Schonefeldt S, Delghingaro-Augusto V, Garcia-Perez JE, Pasciuto E, et al. Genetic predisposition for beta cell fragility underlies type 1 and type 2 diabetes. *Nat Genet* (2016) 48:519–27. doi: 10.1038/ng.3531
92. Dimitri P, Warner JT, Minton JAL, Patch AM, Ellard S, Hattersley AT, et al. Novel GLIS3 mutations demonstrate an extended multisystem phenotype. *Eur J Endocrinol* (2011) 164:437–43. doi: 10.1530/EJE-10-0893
93. Senée V, Chelala C, Duchatelet S, Feng D, Blanc H, Cossec J-C, et al. Mutations in GLIS3 are responsible for a rare syndrome with neonatal diabetes mellitus and congenital hypothyroidism. *Nat Genet* (2006) 38:682–7. doi: 10.1038/ng1802
94. Snider KE, Becker S, Boyajian L, Shyng S-L, MacMullen C, Hughes N, et al. Genotype and Phenotype Correlations in 417 Children With Congenital Hyperinsulinism. *J Clin Endocrinol Metab* (2013) 98:E355–63. doi: 10.1210/jc.2012-2169
95. Haldorsen IS, Vesterhus M, Rder H, Jensen DK, Svik O, Molven A, et al. Lack of pancreatic body and tail in *HNF1B* mutation carriers. *Diabetes Med* (2008) 25:782–7. doi: 10.1111/j.1464-5491.2008.02460.x
96. Horikawa Y, Iwasaki N, Hara M, Furuta H, Hinokio Y, Cockburn BN, et al. Mutation in hepatocyte nuclear factor–1β gene (TCF2) associated with MODY. *Nat Genet* (1997) 17:384–5. doi: 10.1038/ng1297-384
97. Yamagata K, Furuta H, Oda N, Kaisaki PJ, Menzel S, Cox NJ, et al. Mutations in the hepatocyte nuclear factor-4α gene in maturity-onset diabetes of the young (MODY1). *Nature* (1996) 384:458–60. doi: 10.1038/384458a0
98. Florez JC, Jablonski KA, Bayley N, Pollin TI, de Bakker PIW, Shuldiner AR, et al. Diabetes Prevention Program Research Group. *TCF7L2* Polymorphisms and Progression to Diabetes in the Diabetes Prevention Program. *N Engl J Med* (2006) 355:241–50. doi: 10.1056/NEJMoa062418
99. Grant SFA, Thorleifsson G, Reynisdottir I, Benediktsson R, Manolescu A, Sainz J, et al. Variant of transcription factor 7-like 2 (TCF7L2) gene confers risk of type 2 diabetes. *Nat Genet* (2006) 38:320–3. doi: 10.1038/ng1732
100. McCulloch LJ, van de Bunt M, Braun M, Frayn KN, Clark A, Gloyn AL. GLUT2 (SLC2A2) is not the principal glucose transporter in human pancreatic beta cells: implications for understanding genetic association signals at this locus. *Mol Genet Metab* (2011) 104:648–53. doi: 10.1016/j.ymgme.2011.08.026
101. Quintens R, Hendrickx N, Lemaire K, Schuit F. Why expression of some genes is disallowed in beta-cells. *Biochem Soc Trans* (2008) 36:300–5. doi: 10.1042/BST0360300
102. Rutter GAA, Pullen TJJ, Hodson DJJ, Martinez–Sanchez A, Martinez-Sanchez A. Martinez–Sanchez A. Pancreatic β-cell identity, glucose sensing and the control of insulin secretion. *Biochem J* (2015) 466:203–18. doi: 10.1042/BJ20141384
103. Lewandowski SL, Cardone RL, Foster HR, Ho T, Potapenko E, Poudel C, et al. Pyruvate Kinase Controls Signal Strength in the Insulin Secretory Pathway. *Cell Metab* (2020) 32:736–50. doi: 10.1016/j.cmet.2020.10.007
104. Abulizi A, Cardone RL, Stark R, Lewandowski SL, Zhao X, Hillion J, et al. Multi-Tissue Acceleration of the Mitochondrial Phosphoenolpyruvate Cycle Improves Whole-Body Metabolic Health. *Cell Metab* (2020) 32:751–766.e11. doi: 10.1016/j.cmet.2020.10.006
105. Rorsman P, Ashcroft FM. Pancreatic β-Cell Electrical Activity and Insulin Secretion: Of Mice and Men. *Physiol Rev* (2018) 98:117–214. doi: 10.1152/physrev.00008.2017
106. Gaisano HY. Here come the newcomer granules, better late than never. *Trends Endocrinol Metab* (2014) 25:381–8. doi: 10.1016/j.tem.2014.03.005
107. Henquin JC. Triggering and amplifying pathways of regulation of insulin secretion by glucose. *Diabetes* (2000) 49:1751–60. doi: 10.2337/diabetes.49.11.1751
108. Drucker DJ. Incretin action in the pancreas: Potential promise, possible perils, and pathological pitfalls. *Diabetes* (2013) 62:3316–23. doi: 10.2337/db13-0822
109. Campbell JE, Drucker DJ. Pharmacology, Physiology, and Mechanisms of Incretin Hormone Action. *Cell Metab* (2013) 17:819–37. doi: 10.1016/J.CMET.2013.04.008
110. Hodson DJ, Tarasov AI, Gimeno Brias S, Mitchell RK, Johnston NR, Haghollahi S, et al. Incretin-Modulated Beta Cell Energetics in Intact Islets of Langerhans. *Mol Endocrinol* (2014) 28:860–71. doi: 10.1210/me.2014-1038
111. Dunning BE, Foley JE, Ahrén B. Alpha cell function in health and disease: Influence of glucagon-like peptide-1. *Diabetologia* (2005) 48:1700–13. doi: 10.1007/s00125-005-1878-0
112. Zhang Y, Parajuli KR, Fava GE, Gupta R, Xu W, Nguyen LU, et al. GLP-1 receptor in pancreatic A-cells regulates glucagon secretion in a glucose-dependent bidirectional manner. *Diabetes* (2019) 68:34–44. doi: 10.2337/db18-0317
113. El K, Campbell JE. The role of GIP in α-cells and glucagon secretion. *Peptides* (2020) 125:170213. doi: 10.1016/j.peptides.2019.170213
114. Schuit FC, In't Veld PA, Pipeleers DG. Glucose stimulates proinsulin biosynthesis by a dose-dependent recruitment of pancreatic beta cells. *Proc Natl Acad Sci USA* (1988) 85:3865–9. doi: 10.1073/pnas.85.11.3865
115. Cnop M, Toivonen S, Igoillo-Esteve M, Salpea P. Endoplasmic reticulum stress and eIF2α phosphorylation: The Achilles heel of pancreatic β cells. *Mol Metab* (2017) 6:1024–39. doi: 10.1016/j.molmet.2017.06.001
116. Szabat M, Page MM, Panzhinskiy E, Skovsø S, Mojibian M, Fernandez-Tajes J, et al. Reduced Insulin Production Relieves Endoplasmic Reticulum Stress and Induces β Cell Proliferation. *Cell Metab* (2016) 23:179–93. doi: 10.1016/j.cmet.2015.10.016
117. Arda HE, Benitez CM, Kim SK. Gene regulatory networks governing pancreas development. *Dev Cell* (2013) 25:5–13. doi: 10.1016/j.devcel.2013.03.016
118. Patel KA, Kettunen K, Laakso M, Stančáková A, Laver TW, Colclough K, et al. Heterozygous RFX6 protein truncating variants cause Maturity-Onset Diabetes of the Young (MODY) with reduced penetrance. *Nat Commun* (2017) 8:888. doi: 10.1038/s41467-017-00895-9
119. Larsen HL, Grapin-Botton A. The molecular and morphogenetic basis of pancreas organogenesis. *Semin Cell Dev Biol* (2017) 66:51–68. doi: 10.1016/J.SEMCDB.2017.01.005

120. Pan FC, Wright C. Pancreas organogenesis: From bud to plexus to gland. *Dev Dyn* (2011) 240:530–65. doi: 10.1002/dvdy.22584
121. Rezania A, Bruin JE, Arora P, Rubin A, Batushansky I, Asadi A, et al. Reversal of diabetes with insulin-producing cells derived in vitro from human pluripotent stem cells. *Nat Biotechnol* (2014) 32:1121–33. doi: 10.1038/nbt.3033
122. Assady S, Maor G, Amit M, Itskovitz-Eldor J, Skorecki KL, Tzukerman M. Insulin Production by Human Embryonic Stem Cells. *Diabetes* (2001) 50:1691–7. doi: 10.2337/diabetes.50.8.1691
123. D'Amour KA, Agulnick AD, Eliazer S, Kelly OG, Kroon E, Baetge EE. Efficient differentiation of human embryonic stem cells to definitive endoderm. *Nat Biotechnol* (2005) 23:1534–41. doi: 10.1038/nbt1163
124. D'Amour KA, Bang AG, Eliazer S, Kelly OG, Agulnick AD, Smart NG, et al. Production of pancreatic hormone-expressing endocrine cells from human embryonic stem cells. *Nat Biotechnol* (2006) 24:1392–401. doi: 10.1038/nbt1259
125. Kroon E, Martinson LA, Kadoya K, Bang AG, Kelly OG, Eliazer S, et al. Pancreatic endoderm derived from human embryonic stem cells generates glucose-responsive insulin-secreting cells in vivo. *Nat Biotechnol* (2008) 26:443–52. doi: 10.1038/nbt1393
126. Cho YM, Lim JM, Yoo DH, Kim JH, Chung SS, Park SG, et al. Betacellulin and nicotinamide sustain PDX1 expression and induce pancreatic beta-cell differentiation in human embryonic stem cells. *Biochem Biophys Res Commun* (2008) 366:129–34. doi: 10.1016/j.bbrc.2007.11.112
127. Zhang D, Jiang W, Liu M, Sui X, Yin X, Chen S, et al. Highly efficient differentiation of human ES cells and iPS cells into mature pancreatic insulin-producing cells. *Cell Res* (2009) 19:429–38. doi: 10.1038/cr.2009.28
128. Mfopou JK, Chen B, Mateizel I, Sermon K, Bouwens L. Noggin, retinoids, and fibroblast growth factor regulate hepatic or pancreatic fate of human embryonic stem cells. *Gastroenterology* (2010) 138:2233–45, 2245.e1–14. doi: 10.1053/j.gastro.2010.02.056
129. Nostro MC, Sarangi F, Ogawa S, Holtzinger A, Corneo B, Li X, et al. Stage-specific signaling through TGFβ family members and WNT regulates patterning and pancreatic specification of human pluripotent stem cells. *Development* (2011) 138:861–71. doi: 10.1242/dev.055236
130. Kunisada Y, Tsubooka-Yamazoe N, Shoji M, Hosoya M. Small molecules induce efficient differentiation into insulin-producing cells from human induced pluripotent stem cells. *Stem Cell Res* (2012) 8:274–84. doi: 10.1016/j.scr.2011.10.002
131. Rezania A, Bruin JE, Riedel MJ, Mojibian M, Asadi A, Xu J, et al. Maturation of Human Embryonic Stem Cell-Derived Pancreatic Progenitors Into Functional Islets Capable of Treating Pre-existing Diabetes in Mice. *Diabetes* (2012) 61:2016–29. doi: 10.2337/db11-1711
132. Bruin JE, Rezania A, Xu J, Narayan K, Fox JK, O'Neil JJ, et al. Maturation and function of human embryonic stem cell-derived pancreatic progenitors in macroencapsulation devices following transplant into mice. *Diabetologia* (2013) 56:1987–98. doi: 10.1007/s00125-013-2955-4
133. Kelly OG, Chan MY, Martinson LA, Kadoya K, Ostertag TM, Ross KG, et al. Cell-surface markers for the isolation of pancreatic cell types derived from human embryonic stem cells. *Nat Biotechnol* (2011) 29:750–6. doi: 10.1038/nbt.1931
134. Basford CL, Prentice KJ, Hardy a B, Sarangi F, Micallef SJ, Li X, et al. The functional and molecular characterisation of human embryonic stem cell-derived insulin-positive cells compared with adult pancreatic beta cells. *Diabetologia* (2012) 55:358–71. doi: 10.1007/s00125-011-2335-x
135. Micallef SJ, Li X, Schiesser JV, Hirst CE, Yu QC, Lim SM, et al. INS GFP/w human embryonic stem cells facilitate isolation of in vitro derived insulin-producing cells. *Diabetologia* (2012) 55:694–706. doi: 10.1007/s00125-011-2379-y
136. Xie R, Everett LJ, Lim H-W, Patel NA, Schug J, Kroon E, et al. Dynamic Chromatin Remodeling Mediated by Polycomb Proteins Orchestrates Pancreatic Differentiation of Human Embryonic Stem Cells. *Cell Stem Cell* (2013) 12:224–37. doi: 10.1016/j.stem.2012.11.023
137. Bruin JE, Erener S, Vela J, Hu X, Johnson JD, Kurata HT, et al. Characterization of polyhormonal insulin-producing cells derived in vitro from human embryonic stem cells. *Stem Cell Res* (2014) 12:194–208. doi: 10.1016/j.scr.2013.10.003
138. Hrvatin S, O'Donnell CW, Deng F, Millman JR, Pagliuca FW, Diiorio P, et al. Differentiated human stem cells resemble fetal, not adult, β cells. *Proc Natl Acad Sci U S A* (2014) 111:3038–43. doi: 10.1073/pnas.1400709111
139. Schaffer AE, Taylor BL, Benthuysen JR, Liu J, Thorel F, Yuan W, et al. Nkx6.1 controls a gene regulatory network required for establishing and maintaining pancreatic Beta cell identity. *PloS Genet* (2013) 9:e1003274. doi: 10.1371/journal.pgen.1003274
140. Rezania A, Bruin JE, Xu J, Narayan K, Fox JK, O'Neil JJ, et al. Enrichment of human embryonic stem cell-derived NKX6.1-expressing pancreatic progenitor cells accelerates the maturation of insulin-secreting cells in vivo. *Stem Cells* (2013) 31:2432–42. doi: 10.1002/stem.1489
141. Russ HA, Parent AV, Ringler JJ, Hennings TG, Nair GG, Shveygert M, et al. Controlled induction of human pancreatic progenitors produces functional beta-like cells in vitro. *EMBO J* (2015) 34:1–14. doi: 10.15252/embj.201591058
142. Pagliuca FW, Millman JR, Gürtler M, Segel M, Van Dervort A, Ryu JH, et al. Generation of Functional Human Pancreatic β Cells In Vitro. *Cell* (2014) 159:428–39. doi: 10.1016/j.cell.2014.09.040
143. Aguayo-Mazzucato C, Zavacki AM, Marinelarena A, Hollister-Lock J, El Khattabi I, Marsili A, et al. Thyroid hormone promotes postnatal rat pancreatic β-cell development and glucose-responsive insulin secretion through MAFA. *Diabetes* (2013) 62:1569–80. doi: 10.2337/db12-0849
144. Aguayo-Mazzucato C, DiIenno A, Hollister-Lock J, Cahill C, Sharma A, Weir G, et al. MAFA and T 3 Drive Maturation of Both Fetal Human Islets and Insulin-Producing Cells Differentiated From hESC. *J Clin Endocrinol Metab* (2015) 100:3651–9. doi: 10.1210/jc.2015-2632
145. Henquin J-C, Dufrane D, Kerr-Conte J, Nenquin M. Dynamics of glucose-induced insulin secretion in normal human islets. *Am J Physiol Metab* (2015) 309:E640–50. doi: 10.1152/ajpendo.00251.2015
146. Toyoda T, Mae S-I, Tanaka H, Kondo Y, Funato M, Hosokawa Y, et al. Cell aggregation optimizes the differentiation of human ESCs and iPSCs into pancreatic bud-like progenitor cells. *Stem Cell Res* (2015) 14:185–97. doi: 10.1016/j.scr.2015.01.007
147. Tran R, Moraes C, Hoesli CA. Controlled clustering enhances PDX1 and NKX6.1 expression in pancreatic endoderm cells derived from pluripotent stem cells. *Sci Rep* (2020) 10:1190. doi: 10.1038/s41598-020-57787-0
148. Nostro MCC, Sarangi F, Yang C, Holland A, Elefanty AGG, Stanley EGG, et al. Efficient Generation of NKX6-1+ Pancreatic Progenitors from Multiple Human Pluripotent Stem Cell Lines. *Stem Cell Rep* (2015) 4:591–604. doi: 10.1016/j.stemcr.2015.02.017
149. Toyoda T, Kimura A, Tanaka H, Ameku T, Mima A, Hirose Y, et al. Rho-Associated Kinases and Non-muscle Myosin IIs Inhibit the Differentiation of Human iPSCs to Pancreatic Endoderm. *Stem Cell Rep* (2017) 9:419–28. doi: 10.1016/j.stemcr.2017.07.005
150. Ghazizadeh Z, Kao D-I, Amin S, Cook B, Rao S, Zhou T, et al. ROCKII inhibition promotes the maturation of human pancreatic beta-like cells. *Nat Commun* (2017) 8:298. doi: 10.1038/s41467-017-00129-y
151. Sharon N, Chawla R, Mueller J, Vanderhooft J, Whitehorn LJ, Rosenthal B, et al. A Peninsular Structure Coordinates Asynchronous Differentiation with Morphogenesis to Generate Pancreatic Islets. *Cell* (2019) 176:790–804. doi: 10.1016/J.CELL.2018.12.003
152. Schulz TC, Young HY, Agulnick AD, Babin MJ, Baetge EE, Bang AG, et al. A Scalable System for Production of Functional Pancreatic Progenitors from Human Embryonic Stem Cells. *PloS One* (2012) 7:e37004. doi: 10.1371/journal.pone.0037004
153. Hogrebe NJ, Augsornworawat P, Maxwell KG, Velazco-Cruz L, Millman JR. Targeting the cytoskeleton to direct pancreatic differentiation of human pluripotent stem cells. *Nat Biotechnol* (2020) 38:460–70. doi: 10.1038/s41587-020-0430-6
154. Cebola I, Rodríguez-Seguí SA, Cho CH-H, Bessa J, Rovira M, Luengo M, et al. TEAD and YAP regulate the enhancer network of human embryonic pancreatic progenitors. *Nat Cell Biol* (2015) 17:615–26. doi: 10.1038/ncb3160
155. Mamidi A, Prawiro C, Seymour PA, de Lichtenberg KH, Jackson A, Serup P, et al. Mechanosignalling via integrins directs fate decisions of pancreatic progenitors. *Nature* (2018) 564:114–8. doi: 10.1038/s41586-018-0762-2
156. Rosado-Olivieri EA, Anderson K, Kenty JH, Melton DA. YAP inhibition enhances the differentiation of functional stem cell-derived insulin-producing β cells. *Nat Commun* (2019) 10:1464. doi: 10.1038/s41467-019-09404-6
157. Cogger KF, Sinha A, Sarangi F, McGaugh EC, Saunders D, Dorrell C, et al. Glycoprotein 2 is a specific cell surface marker of human pancreatic progenitors. *Nat Commun* (2017) 8:331. doi: 10.1038/s41467-017-00561-0

158. Ameri J, Borup R, Prawiro C, Ramond C, Schachter KA, Scharfmann R, et al. Efficient Generation of Glucose-Responsive Beta Cells from Isolated GP2 +Human Pancreatic Progenitors. *Cell Rep* (2017) 19:36–49. doi: 10.1016/j.celrep.2017.03.032
159. Veres A, Faust AL, Bushnell HL, Engquist EN, Kenty JH-R, Harb G, et al. Charting cellular identity during human in vitro β-cell differentiation. *Nature* (2019) 569:368–73. doi: 10.1038/s41586-019-1168-5
160. Nair GG, Liu JS, Russ HA, Tran S, Saxton MS, Chen R, et al. Recapitulating endocrine cell clustering in culture promotes maturation of human stem-cell-derived β cells. *Nat Cell Biol* (2019) 21:263–74. doi: 10.1038/s41556-018-0271-4
161. In't Veld P, Smeets S. "Microscopic Anatomy of the Human Islet of Langerhans,". In: *Islets of Langerhans.* Dordrecht: Springer Netherlands (2015). p. 19–38. doi: 10.1007/978-94-007-6686-0_1
162. Henderson JR. WHY ARE THE ISLETS OF LANGERHANS? *Lancet* (1969) 294:469–70. doi: 10.1016/S0140-6736(69)90171-8
163. Velazco-Cruz L, Song J, Maxwell KG, Goedegebuure MM, Augsornworawat P, Hogrebe NJ, et al. Acquisition of Dynamic Function in Human Stem Cell-Derived β Cells. *Stem Cell Rep* (2019) 12:351–65. doi: 10.1016/j.stemcr.2018.12.012
164. Gylfe E, Tengholm A. Neurotransmitter control of islet hormone pulsatility. *Diabetes Obes Metab* (2014) 16:102–10. doi: 10.1111/dom.12345
165. Sui L, Danzl N, Campbell SR, Viola R, Williams D, Xing Y, et al. β-Cell Replacement in Mice Using Human Type 1 Diabetes Nuclear Transfer Embryonic Stem Cells. *Diabetes* (2018) 67:26–35. doi: 10.2337/db17-0120
166. Choi J, Lee S, Mallard W, Clement K, Tagliazucchi GM, Lim H, et al. A comparison of genetically matched cell lines reveals the equivalence of human iPSCs and ESCs. *Nat Biotechnol* (2015) 33:1173–81. doi: 10.1038/nbt.3388
167. Kyttälä A, Moraghebi R, Valensisi C, Kettunen J, Andrus C, Pasumarthy KK, et al. Genetic Variability Overrides the Impact of Parental Cell Type and Determines iPSC Differentiation Potential. *Stem Cell Rep* (2016) 6:200–12. doi: 10.1016/j.stemcr.2015.12.009
168. Cuomo ASE, Seaton DD, McCarthy DJ, Martinez I, Bonder MJ, Garcia-Bernardo J, et al. Single-cell RNA-sequencing of differentiating iPS cells reveals dynamic genetic effects on gene expression. *Nat Commun* (2020) 11:810. doi: 10.1038/s41467-020-14457-z
169. Hockemeyer D, Wang H, Kiani S, Lai CS, Gao Q, Cassady JP, et al. Genetic engineering of human pluripotent cells using TALE nucleases. *Nat Biotechnol* (2011) 29:731–4. doi: 10.1038/nbt.1927
170. Hockemeyer D, Soldner F, Beard C, Gao Q, Mitalipova M, DeKelver RC, et al. Efficient targeting of expressed and silent genes in human ESCs and iPSCs using zinc-finger nucleases. *Nat Biotechnol* (2009) 27:851–7. doi: 10.1038/nbt.1562
171. Mali P, Yang L, Esvelt KM, Aach J, Guell M, DiCarlo JE, et al. RNA-Guided Human Genome Engineering via Cas9. *Sci (80-)* (2013) 339:823–6. doi: 10.1126/science.1232033
172. Hotta A, Yamanaka S. From Genomics to Gene Therapy: Induced Pluripotent Stem Cells Meet Genome Editing. *Annu Rev Genet* (2015) 49:47–70. doi: 10.1146/annurev-genet-112414-054926
173. Mojica FJM, Montoliu L. On the Origin of CRISPR-Cas Technology: From Prokaryotes to Mammals. *Trends Microbiol* (2016) 24:811–20. doi: 10.1016/j.tim.2016.06.005
174. Li QV, Dixon G, Verma N, Rosen BP, Gordillo M, Luo R, et al. Genome-scale screens identify JNK–JUN signaling as a barrier for pluripotency exit and endoderm differentiation. *Nat Genet* (2019) 51:999–1010. doi: 10.1038/s41588-019-0408-9
175. Shang L, Hua H, Foo K, Martinez H, Watanabe K, Zimmer M, et al. β-cell dysfunction due to increased ER stress in a stem cell model of Wolfram syndrome. *Diabetes* (2014) 63:923–33. doi: 10.2337/db13-0717
176. Teo AKK, Lau HH, Valdez IA, Dirice E, Tjora E, Raeder H, et al. Early Developmental Perturbations in a Human Stem Cell Model of MODY5/HNF1B Pancreatic Hypoplasia. *Stem Cell Rep* (2016) 6:357–67. doi: 10.1016/j.stemcr.2016.01.007
177. McGrath PS, Watson CL, Ingram C, Helmrath MA, Wells JM. The Basic Helix-Loop-Helix Transcription Factor NEUROG3 Is Required for Development of the Human Endocrine Pancreas. *Diabetes* (2015) 64:2497–505. doi: 10.2337/db14-1412
178. Zhu Z, Li QV, Lee K, Rosen BP, González F, Soh C-L, et al. Genome Editing of Lineage Determinants in Human Pluripotent Stem Cells Reveals Mechanisms of Pancreatic Development and Diabetes Article Genome Editing of Lineage Determinants in Human Pluripotent Stem Cells Reveals Mechanisms of Pancreatic Development an. *Cell Stem Cell* (2016) 18:755–68. doi: 10.1016/j.stem.2016.03.015
179. Saarimäki-Vire J, Balboa D, Russell MA, Saarikettu J, Kinnunen M, Keskitalo S, et al. An Activating STAT3 Mutation Causes Neonatal Diabetes through Premature Induction of Pancreatic Differentiation. *Cell Rep* (2017) 19:281–94. doi: 10.1016/j.celrep.2017.03.055
180. Shi Z, Lee K, Yang D, Amin S, Verma N, Li QV, et al. Genome Editing in hPSCs Reveals GATA6 Haploinsufficiency and a Genetic Interaction with GATA4 in Human Pancreatic Development. *Cell Stem Cell* (2017) 20:675–688.e6. doi: 10.1016/j.stem.2017.01.001
181. Tiyaboonchai A, Cardenas-Diaz FL, Ying L, Maguire JA, Sim X, Jobaliya C, et al. GATA6 Plays an Important Role in the Induction of Human Definitive Endoderm, Development of the Pancreas, and Functionality of Pancreatic β Cells. *Stem Cell Rep* (2017) 8:589–604. doi: 10.1016/j.stemcr.2016.12.026
182. Maxwell KG, Augsornworawat P, Velazco-Cruz L, Kim MH, Asada R, Hogrebe NJ, et al. Gene-edited human stem cell-derived β cells from a patient with monogenic diabetes reverse preexisting diabetes in mice. *Sci Transl Med* (2020) 12:eaax9106. doi: 10.1126/scitranslmed.aax9106
183. Wang X, Sterr M, Ansarullah, Burtscher I, Böttcher A, Beckenbauer J, et al. Point mutations in the PDX1 transactivation domain impair human β-cell development and function. *Mol Metab* (2019) 24:80–97. doi: 10.1016/J.MOLMET.2019.03.006
184. Amin S, Cook B, Zhou T, Ghazizadeh Z, Lis R, Zhang T, et al. Discovery of a drug candidate for GLIS3-associated diabetes. *Nat Commun* (2018) 9:2681. doi: 10.1038/s41467-018-04918-x
185. Gage BK, Asadi A, Baker RK, Webber TD, Wang R, Itoh M, et al. The Role of ARX in Human Pancreatic Endocrine Specification. *PloS One* (2015) 10:e0144100. doi: 10.1371/journal.pone.0144100
186. Velazco-cruz L, Goedegebuure MM, Maxwell KG, Augsornworawat P, Hogrebe NJ, Millman JR, et al. SIX2 Regulates Human b Cell Differentiation from Stem Cells and Functional Maturation In Vitro ll ll SIX2 Regulates Human b Cell Differentiation from Stem Cells and Functional Maturation In Vitro. *CellReports* (2020) 31:107687. doi: 10.1016/j.celrep.2020.107687
187. Vethe H, Bjørlykke Y, Ghila LM, Paulo JA, Scholz H, Gygi SP, et al. Probing the missing mature β-cell proteomic landscape in differentiating patient iPSC-derived cells. *Sci Rep* (2017) 7:1–14. doi: 10.1038/s41598-017-04979-w
188. Zeng H, Guo M, Zhou T, Tan L, Chong CN, Zhang T, et al. An Isogenic Human ESC Platform for Functional Evaluation of Genome-wide-Association-Study-Identified Diabetes Genes and Drug Discovery. *Cell Stem Cell* (2016) 0:1660–9. doi: 10.1016/j.stem.2016.07.002
189. Guo M, Zhang T, Dong X, Xiang JZ, Lei M, Evans T, et al. Using hESCs to Probe the Interaction of the Diabetes-Associated Genes CDKAL1 and MT1E. *Cell Rep* (2017) 19:1512–21. doi: 10.1016/j.celrep.2017.04.070
190. Guo D, Liu H, Ruzi A, Gao G, Nasir A, Liu Y, et al. Modeling Congenital Hyperinsulinism with ABCC8-Deficient Human Embryonic Stem Cells Generated by CRISPR/Cas9. *Sci Rep* (2017) 7:3156. doi: 10.1038/s41598-017-03349-w
191. Ma S, Viola R, Sui L, Cherubini V, Barbetti F, Egli D. β Cell Replacement after Gene Editing of a Neonatal Diabetes-Causing Mutation at the Insulin Locus. *Stem Cell Rep* (2018) 11:1407–15. doi: 10.1016/j.stemcr.2018.11.006
192. Balboa D, Saarimäki-Vire J, Borshagovski D, Survila M, Lindholm P, Galli E, et al. Insulin mutations impair beta-cell development in a patient-derived iPSC model of neonatal diabetes. *Elife* (2018) 7. doi: 10.7554/eLife.38519
193. Johansson KA, Dursun U, Jordan N, Gu G, Beermann F, Gradwohl G, et al. Temporal control of neurogenin3 activity in pancreas progenitors reveals competence windows for the generation of different endocrine cell types. *Dev Cell* (2007) 12:457–65. doi: 10.1016/j.devcel.2007.02.010
194. Desgraz R, Herrera PL. Pancreatic neurogenin 3-expressing cells are unipotent islet precursors. *Development* (2009) 136:3567–74. doi: 10.1242/dev.039214
195. Huotari MA, Miettinen PJ, Palgi J, Koivisto T, Ustinov J, Harari D, et al. ErbB signaling regulates lineage determination of developing pancreatic islet cells in embryonic organ culture. *Endocrinology* (2002) 143:4437–46. doi: 10.1210/en.2002-220382

196. Löf-Öhlin ZM, Nyeng P, Bechard ME, Hess K, Bankaitis E, Greiner TU, et al. EGFR signalling controls cellular fate and pancreatic organogenesis by regulating apicobasal polarity. *Nat Cell Biol* (2017) 19:1313–25. doi: 10.1038/ncb3628
197. Mizukami H, Takahashi K, Inaba W, Osonoi S, Kamata K, Tsuboi K, et al. Age-associated changes of islet endocrine cells and the effects of body mass index in Japanese. *J Diabetes Investig* (2014) 5:38–47. doi: 10.1111/jdi.12118
198. Mahaddalkar PU, Scheibner K, Pfluger S, Ansarullah, Sterr M, Beckenbauer J, et al. Generation of pancreatic β cells from CD177+ anterior definitive endoderm. *Nat Biotechnol* (2020) 38:1061–72. doi: 10.1038/s41587-020-0492-5
199. Augsornworawat P, Maxwell KG, Velazco-Cruz L, Millman JR. Single-Cell Transcriptome Profiling Reveals β Cell Maturation in Stem Cell-Derived Islets after Transplantation. *Cell Rep* (2020) 32:108067. doi: 10.1016/j.celrep.2020.108067
200. Arda HE, Li L, Tsai J, Powers AC, Chang HY, Kim SK, et al. Age-Dependent Pancreatic Gene Regulation Reveals Mechanisms Governing Human b Cell Function. *Cell Metab* (2016) 23:1–12. doi: 10.1016/j.cmet.2016.04.002
201. Zeng C, Mulas F, Sui Y, Guan T, Miller N, Tan Y, et al. Pseudotemporal Ordering of Single Cells Reveals Metabolic Control of Postnatal β Cell Proliferation. *Cell Metab* (2017) 25:1160–75. doi: 10.1016/j.cmet.2017.04.014
202. Qiu W-L, Zhang Y-W, Feng Y, Li L-C, Yang L, Xu C-R. Deciphering Pancreatic Islet β Cell and α Cell Maturation Pathways and Characteristic Features at the Single-Cell Level. *Cell Metab* (2017) 25:1194–205. doi: 10.1016/j.cmet.2017.04.003
203. Dorrell C, Schug J, Canaday PS, Russ HA, Tarlow BD, Grompe MT, et al. Human islets contain four distinct subtypes of [beta] cells. *Nat Commun* (2016) 7:11756. doi: 10.1038/NCOMMS11756
204. Segerstolpe Å, Palasantza A, Eliasson P, Andersson E, Andréasson A-C, Sun X, et al. Single-Cell Transcriptome Profiling of Human Pancreatic Islets in Health and Type 2 Diabetes. *Cell Metab* (2016) 24:593–607. doi: 10.1016/j.cmet.2016.08.020
205. Xin Y, Dominguez Gutierrez G, Okamoto H, Kim J, Lee A-H, Adler C, et al. Pseudotime Ordering of Single Human β-Cells Reveals States of Insulin Production and Unfolded Protein Response. *Diabetes* (2018) 67:1783–94. doi: 10.2337/db18-0365
206. Baron M, Veres A, Wolock SLL, Faust ALL, Gaujoux R, Vetere A, et al. A Single-Cell Transcriptomic Map of the Human and Mouse Pancreas Reveals Inter- and Intra-cell Population Structure. *Cell Syst* (2016) 3:1–15. doi: 10.1016/j.cels.2016.08.011
207. Muraro MJ, Dharmadhikari G, Grün D, Groen N, Dielen T, Jansen E, et al. A Single-Cell Transcriptome Atlas of the Human Pancreas. *Cell Syst* (2016) 3:385–94. doi: 10.1016/j.cels.2016.09.002
208. Ramond C, Beydag-Tasöz BS, Azad A, van de Bunt M, Petersen MBK, Beer NL, et al. Understanding human fetal pancreas development using subpopulation sorting, RNA sequencing and single-cell profiling. *Development* (2018) 145:dev165480. doi: 10.1242/dev.165480
209. Petersen MBK, Azad A, Ingvorsen C, Hess K, Hansson M, Grapin-Botton A, et al. Single-Cell Gene Expression Analysis of a Human ESC Model of Pancreatic Endocrine Development Reveals Different Paths to β-Cell Differentiation. *Stem Cell Rep* (2017) 9:1246–61. doi: 10.1016/j.stemcr.2017.08.009
210. Krentz NA, Lee MYY, Xu EE, Sproul SLJ, Maslova A, Sasaki S, et al. Single-Cell Transcriptome Profiling of Mouse and hESC-Derived Pancreatic Progenitors. *Stem Cell Rep* (2018) 11:1551–64. doi: 10.1016/j.stemcr.2018.11.008
211. Li X, Yang KY, Chan VW, Leung KT, Zhang X-B, Wong AS, et al. Single-Cell RNA-Seq Reveals that CD9 Is a Negative Marker of Glucose-Responsive Pancreatic β-like Cells Derived from Human Pluripotent Stem Cells. *Stem Cell Rep* (2020) 15:1111–26. doi: 10.1016/j.stemcr.2020.09.009
212. Sharon N, Vanderhooft J, Straubhaar J, Mueller J, Chawla R, Zhou Q, et al. Wnt Signaling Separates the Progenitor and Endocrine Compartments during Pancreas Development. *Cell Rep* (2019) 27:2281–2291.e5. doi: 10.1016/j.celrep.2019.04.083
213. Regev A, Teichmann SA, Lander ES, Amit I, Benoist C, Birney E, et al. The Human Cell Atlas. *Elife* (2017) 6:e27041. doi: 10.7554/eLife.27041
214. Otonkoski T, Andersson S, Knip M, Simell O. Maturation of insulin response to glucose during human fetal and neonatal development. Studies with perifusion of pancreatic isletlike cell clusters. *Diabetes* (1988) 37:286–91. doi: 10.2337/diabetes.37.3.286
215. Blum B, Hrvatin S, Schuetz C, Bonal C, Rezania A. Melton D a. Functional beta-cell maturation is marked by an increased glucose threshold and by expression of urocortin 3. *Nat Biotechnol* (2012) 30:261–4. doi: 10.1038/nbt.2141
216. Davis JC, Alves TC, Helman A, Chen JC, Kenty JH, Cardone RL, et al. Glucose Response by Stem Cell-Derived β Cells In Vitro Is Inhibited by a Bottleneck in Glycolysis. *Cell Rep* (2020) 31:107623. doi: 10.1016/j.celrep.2020.107623
217. Hastoy B, Godazgar M, Clark A, Nylander V, Spiliotis I, van de Bunt M, et al. Electrophysiological properties of human beta-cell lines EndoC-βH1 and -βH2 conform with human beta-cells. *Sci Rep* (2018) 8:16994. doi: 10.1038/s41598-018-34743-7
218. Camunas-Soler J, Dai X, Hang Y, Bautista A, Lyon J, Suzuki K, et al. Patch-Seq Links Single-Cell Transcriptomes to Human Islet Dysfunction in Diabetes Resource Patch-Seq Links Single-Cell Transcriptomes to Human Islet Dysfunction in Diabetes. *Cell Metab* (2020) 31:1017–31.e4. doi: 10.1016/j.cmet.2020.04.005
219. Henquin J-C. Metabolic control of the potassium permeability in pancreatic islet cells. *Biochem J* (1980) 186:541–50. doi: 10.1042/bj1860541
220. Dhawan S, Tschen S-I, Zeng C, Guo T, Hebrok M, Matveyenko A, et al. DNA methylation directs functional maturation of pancreatic β cells. *J Clin Invest* (2015) 125:2851–60. doi: 10.1172/JCI79956
221. Yoshihara E, Wei Z, Lin CS, Fang S, Ahmadian M, Kida Y, et al. ERRγ Is Required for the Metabolic Maturation of Therapeutically Functional Glucose-Responsive β Cells. *Cell Metab* (2016) 23:622–34. doi: 10.1016/j.cmet.2016.03.005
222. Yoshihara E, O'Connor C, Gasser E, Wei Z, Oh TG, Tseng TW, et al. Immune-evasive human islet-like organoids ameliorate diabetes. *Nature* (2020) 586:606–11. doi: 10.1038/s41586-020-2631-z
223. Andersson LE, Valtat B, Bagge A, Sharoyko VV, Nicholls DG, Ravassard P, et al. Characterization of stimulus-secretion coupling in the human pancreatic EndoC-βH1 beta cell line. *PloS One* (2015) 10:e0120879. doi: 10.1371/journal.pone.0120879
224. Wikstrom JD, Sereda SB, Stiles L, Elorza A, Allister EM, Neilson A, et al. A novel high-throughput assay for islet respiration reveals uncoupling of rodent and human islets. *PloS One* (2012) 7:e33023. doi: 10.1371/journal.pone.0033023
225. Ma H, Wert KJ, Shvartsman D, Melton DA, Jaenisch R. Establishment of human pluripotent stem cell-derived pancreatic β-like cells in the mouse pancreas. *Proc Natl Acad Sci* (2018) 115:3924–9. doi: 10.1073/pnas.1702059115
226. Sui L, Mfopou JK, Chen B, Sermon K, Bouwens L. Transplantation of Human Embryonic Stem Cell-Derived Pancreatic Endoderm Reveals a Site-Specific Survival, Growth, and Differentiation. *Cell Transplant* (2013) 22:821–30. doi: 10.3727/096368912X636812
227. Bruin JE, Asadi A, Fox JK, Erener S, Rezania A, Kieffer TJ. Accelerated Maturation of Human Stem Cell-Derived Pancreatic Progenitor Cells into Insulin-Secreting Cells in Immunodeficient Rats Relative to Mice. *Stem Cell Rep* (2015) 5:1081–96. doi: 10.1016/j.stemcr.2015.10.013
228. Yang H, Wright JR. Human beta cells are exceedingly resistant to streptozotocin in vivo. *Endocrinology* (2002) 143:2491–5. doi: 10.1210/endo.143.7.8901
229. Hosokawa M, Dolci W, Thorens B. Differential sensitivity of GLUT1- and GLUT2-expressing beta cells to streptozotocin. *Biochem Biophys Res Commun* (2001) 289:1114–7. doi: 10.1006/bbrc.2001.6145
230. Agulnick AD, Ambruzs DM, Moorman MA, Bhoumik A, Cesario RM, Payne JK, et al. Insulin-Producing Endocrine Cells Differentiated In Vitro From Human Embryonic Stem Cells Function in Macroencapsulation Devices In Vivo. *Stem Cells Transl Med* (2015) 4:1214–22. doi: 10.5966/sctm.2015-0079
231. Robert T, De Mesmaeker I, Stangé GM, Suenens KG, Ling Z, Kroon EJ, et al. Functional Beta Cell Mass from Device-Encapsulated hESC-Derived Pancreatic Endoderm Achieving Metabolic Control. *Stem Cell Rep* (2018) 10:739–50. doi: 10.1016/j.stemcr.2018.01.040
232. Vegas AJ, Veiseh O, Gürtler M, Millman JR, Pagliuca FW, Bader AR, et al. Long-term glycemic control using polymer-encapsulated human stem cell–derived beta cells in immune-competent mice. *Nat Med* (2016) 22:306–11. doi: 10.1038/nm.4030

233. Heinis M, Simon M-T, Ilc K, Mazure NM, Pouysségur J, Scharfmann R, et al. Oxygen tension regulates pancreatic beta-cell differentiation through hypoxia-inducible factor 1alpha. *Diabetes* (2010) 59:662–9. doi: 10.2337/db09-0891
234. Saber N, Bruin JE, O'Dwyer S, Schuster H, Rezania A, Kieffer TJ. Sex Differences in Maturation of Human Embryonic Stem Cell–Derived β Cells in Mice. *Endocrinology* (2018) 159:1827–41. doi: 10.1210/en.2018-00048
235. Rakshit K, Qian J, Satish Gaonkar K, Dhawan S, Colwell CS, Matveyenko AV, et al. Postnatal ontogenesis of the islet circadian clock plays a contributory role in β-cell maturation process(2018). Available at: http://diabetes.diabetesjournals.org/content/diabetes/early/2018/03/01/db17-0850.full.pdf (Accessed March 16, 2018).
236. Perelis M, Marcheva B, Moynihan Ramsey K, Schipma MJ, Hutchison AL, Taguchi A, et al. Pancreatic cell enhancers regulate rhythmic transcription of genes controlling insulin secretion. *Sci (80-)* (2015) 350:aac4250–aac4250. doi: 10.1126/science.aac4250
237. Alvarez-Dominguez JR, Donaghey J, Rasouli N, Kenty JHR, Helman A, Charlton J, et al. Circadian Entrainment Triggers Maturation of Human In Vitro Islets. *Cell Stem Cell* (2020) 26:108–22.e10. doi: 10.1016/j.stem.2019.11.011
238. Mochida T, Ueno H, Tsubooka-Yamazoe N, Hiyoshi H, Ito R, Matsumoto H, et al. Insulin-Deficient Diabetic Condition Upregulates Insulin Secreting Capacity of Human iPSC-Derived Pancreatic Endocrine Progenitor Cells After Implantation in Mice. *Diabetes* (2020) 69:634–46. doi: 10.2337/db19-0728
239. Faleo G, Russ HA, Wisel S, Parent AV, Nguyen V, Nair GG, et al. Mitigating Ischemic Injury of Stem Cell-Derived Insulin-Producing Cells after Transplant. *Stem Cell Rep* (2017) 9:807–19. doi: 10.1016/j.stemcr.2017.07.012
240. Yu Y, Gamble A, Pawlick R, Pepper AR, Salama B, Toms D, et al. Bioengineered human pseudoislets form efficiently from donated tissue, compare favourably with native islets in vitro and restore normoglycaemia in mice. *Diabetologia* (2018) 61:2016–29. doi: 10.1007/s00125-018-4672-5
241. Gotthardt M, Eizirik DL, Cnop M, Brom M. Beta cell imaging - a key tool in optimized diabetes prevention and treatment. *Trends Endocrinol Metab* (2014) 25:375–7. doi: 10.1016/j.tem.2014.02.002
242. Eter WA, Van Der Kroon I, Andralojc K, Buitinga M, Willekens SMA, Frielink C, et al. Non-invasive in vivo determination of viable islet graft volume by 111 In-exendin-3. *Sci Rep* (2017) 7:3–8. doi: 10.1038/s41598-017-07815-3
243. van der Meulen T, Donaldson CJ, Cáceres E, Hunter AE, Cowing-Zitron C, Pound LD, et al. Urocortin3 mediates somatostatin-dependent negative feedback control of insulin secretion. *Nat Med* (2015) 21:769–76. doi: 10.1038/nm.3872
244. King AJF, Fernandes JR, Hollister-Lock J, Nienaber CE, Bonner-Weir S, Weir GC. Normal relationship of beta- and non-beta-cells not needed for successful islet transplantation. *Diabetes* (2007) 56:2312–8. doi: 10.2337/db07-0191
245. Keymeulen B, Anselmo J, Pipeleers D. Length of metabolic normalization after rat islet cell transplantation depends on endocrine cell composition of graft and on donor age. *Diabetologia* (1997) 40:1152–8. doi: 10.1007/s001250050800

10

Lessons from Human Islet Transplantation Inform Stem Cell-Based Approaches in the Treatment of Diabetes

Taylor M. Triolo[1*] *and Melena D. Bellin*[2]

[1] The Barbara Davis Center for Diabetes, School of Medicine, University of Colorado Denver, Aurora, CO, United States,
[2] Department of Pediatrics, School of Medicine, University of Minnesota, Minneapolis, MN, United States

Correspondence:
Taylor M. Triolo
taylor.triolo@cuanschutz.edu

Diabetes mellitus is characterized by the body's inability to control blood glucose levels within a physiological range due to loss and/or dysfunction of insulin producing beta cells. Progressive beta cell loss leads to hyperglycemia and if untreated can lead to severe complications and/or death. Treatments at this time are limited to pharmacologic therapies, including exogenous insulin or oral/injectable agents that improve insulin sensitivity or augment endogenous insulin secretion. Cell transplantation can restore physiologic endogenous insulin production and minimize hyper- and hypoglycemic excursions. Islet isolation procedures and management of transplant recipients have advanced over the last several decades; both tight glycemic control and insulin independence are achievable. Research has been conducted in isolating islets, monitoring islet function, and mitigating the immune response. However, this procedure is still only performed in a small minority of patients. One major barrier is the scarcity of human pancreatic islet donors, variation in donor pancreas quality, and variability in islet isolation success. Advances have been made in generation of glucose responsive human stem cell derived beta cells (sBCs) and islets from human pluripotent stem cells using directed differentiation. This is an emerging promising treatment for patients with diabetes because they could potentially serve as an unlimited source of functional, glucose-responsive beta cells. Challenges exist in their generation including long term survival of grafts, safety of transplantation, and protection from the immune response. This review focuses on the progress made in islet allo- and auto transplantation and how these advances may be extrapolated to the sBC context.

Keywords: islet transplantation, type 1 diabetes, diabetes treatment, total pancreatectomy with islet autotransplant (TPIAT), allotransplantation, type 2 diabetes, beta cell transplantation, stem cell differentiation

INTRODUCTION

Patients With Insulin Deficient Forms of Diabetes May Be Considered for Islet Transplantation

Diabetes is characterized by the body's inability to control blood glucose within a tight physiological range, due to insulin deficiency from beta cell loss or dysfunction and/or insulin resistance. Prolonged or recurrent hyperglycemia can lead to macro and microvascular complications, associated with substantial morbidity and early mortality (1, 2).

Type 1 diabetes is an autoimmune attack on a patient's own insulin producing beta cells. If left untreated, severe insulin deficiency leads to hyperglycemia, diabetic ketoacidosis and potentially death. Treatment options at this time are limited to exogenous insulin; although progress has been made in the precision of delivery of insulin and blood glucose monitoring (2), patients are at risk for life-threatening hypoglycemia (3). While therapies targeted specifically at risk relatives have shown some promise in delaying onset of disease (4) there is still no cure for type 1 diabetes. Allogeneic islet transplantation may be considered in highly selected type 1 diabetes patients with either repeated severe hypoglycemic, significant glycemic variability, or microvascular complications (usually renal failure necessitating kidney transplant) (5). In the United States islet allotransplant is considered investigational and only performed in the context of a research study but is offered as standard clinical care in areas of Canada, Europe, and elsewhere. However, only a limited supply of suitable cadaveric donor pancreases are available for islet isolation and transplant.

In contrast, type 2 diabetes is a condition of insulin resistance and progressive beta cell decline (6). Patients are often treated with oral or injectable agents that improve endogenous insulin function. In some cases, patients with type 2 diabetes are also treated with insulin. Because the large islet mass needed to overcome insulin resistance is unlikely to be obtained with isolated islets, patients with type 2 diabetes are generally not considered candidates for islet transplantation. However, a renewable cell source could overcome this barrier of insufficient islet mass.

While type 1 and 2 diabetes are the more common causes of glucose dysfunction, beta cell loss and dysfunction can also occur in the setting of persistent inflammation and stress within the pancreas due to chronic pancreatitis (7). This is a painful and disabling condition that can be treated with analgesics, procedural interventions, and sometimes removal of the pancreas, rendering the patient without endocrine or exocrine pancreatic function. Total pancreatectomy and intraportal islet cell autotransplantation (TPIAT) can provide pain relief and sustained islet graft function in these patients. In this procedure, patients receive their own islets and therefore donor pancreases are not required. However, because of the damage and fibrosis from pancreatitis, these individuals usually have a sub-optimal islet mass and only about 1 out of every 3 achieves insulin independence.

In these etiologies of dysglycemia, replacement of beta cell function is a potential treatment to alleviate glycemic variability and reduce risk for the complications associated with long term hyperglycemia. A common challenge is obtaining a sufficient number of islets to successfully treat individual patients and offer cell therapy to a larger number of patients with diabetes. Here we review the uses of allo- and auto- islet transplantation and how stem cell derived beta cells (sBCs) or islets may overcome barriers and limitations currently inherent in islet transplantation (**Table 1**).

Islet Allotransplantation

Allogenic transplantation of cadaveric islets as a functional source of beta cells has become a treatment for patients with type 1 diabetes, particularly those with either life-threatening hypoglycemia or diabetes-related kidney failure requiring kidney transplantation. Although whole organ pancreas transplants can also be performed for these individuals (8–10), the appeal of islet transplantation is the lack of major surgery and very low risk for procedural complications. There have been improvements in isolation of islets and the procedure is considered minimally invasive. Initial transplants of pancreatic islets were trialed as early as the 1970s, with initially low rates of success (11). A turning point came with the introduction of the Edmonton Protocol in 2000 (12)—by introducing glucocorticoid-free immunosuppression and using multiple donors to increase islet mass, all 7 patients transplanted in the initial Edmonton trial achieved insulin independence. Subsequent refinements in immunosuppression protocols have improved the longevity of insulin independence (5, 13).

Even when insulin independence is not achieved, islet transplantation is highly successful in preventing severe hypoglycemia, if islet graft function is maintained (5, 14, 15). While success of islet transplants have generally improved over the last 20 years (5), variability in achieving insulin independence and concern for immunosuppression impact on kidney function (16) remain a concern. In addition, donor tissue availability

TABLE 1 | Considerations in islet transplantation.

	Allotransplantation	Autotransplantation	Stem cell transplantation
Patient Population	Patients with type 1 diabetes and severe hypoglycemia, glycemic variability or microvascular complications	Patients with chronic pancreatitis undergoing pancreatectomy	Investigational; type 1 diabetes with potential application to other forms of diabetes
Source of Islet Material	Cadaveric islets	Autologous transplant	Human pluripotent stem cells (induced pluripotent stem cells or embryonic stem cells)
Limitations	- Immunosuppression required - Limited supply of donors	- Limited cell mass, from one's own diseased pancreas	-Remains investigational - Need to scale up to sufficient functional mass to reverse diabetes in humans

continues to be the limiting factor of allogenic islet transplantation as a treatment for diabetes as 2–3 donors are typically required to obtain the necessary beta cell mass required for transplantation (17).

Despite improving long-term outcomes after islet allotransplantation, challenges remain around the longevity of insulin independence. Transplanted islets are subjected to non-immune attrition, and at risk for alloimmune rejection and recurrent autoimmunity (18–20). Immunosuppressive drugs necessary for islet allotransplant also carry risk for beta cell toxicity (21). Although intraportally transplanted islets are rarely accessible for study, limited histopathology of intraportal islet allografts have shown amyloid deposition, postulated due to over-stimulation of insulin production from a marginal islet mass or immunosuppressive drug toxicity (22); a recent report documenting absence of islet amyloid in an islet autograft patient with marginal islet mass suggests drug toxicity as a more likely culprit (23). More recently de-differentiation of the mature beta cell phenotype was observed in two islet allotransplant recipients, possibly consequences of hypoxia and metabolic stress (23). Innate immune destruction of islets stimulated upon intraportal infusion of islets has led to study of alternate sites for transplant, including omentum, bone marrow, intramuscular, and subcutaneous sites, though none has yet established the same efficacy as the liver (24–27).

As with any organ transplant, alloimmune rejection can occur in islet transplant and is more common in patients exhibiting high levels of HLA-sensitization pre-transplant (28–31). Unfortunately, immune rejection is difficult to treat due to limitations in early detection and lack of effective treatment strategies (10, 18, 32–34). Genetically engineered human beta cell lines can be used *in vivo* to augment the immune response to evaluate immune interactions and perhaps protect transplanted beta cells from immune destruction (35). Recurrent autoimmunity has been associated with positive autoantibodies, but the presence of autoreactive T-cell studies is more strongly associated with islet graft failure (36–39). Potential strategies to address these immunologic losses include encapsulation and use of bioengineered scaffold devices with enhanced vascularization and/or local drug release.

Islet Autotransplantation

TPIAT is a treatment option for patients suffering from intractable abdominal pain from chronic pancreatitis. Total pancreatectomy provides pain relief by removing the primary source of chronic pain but results in complete exocrine insufficiency and insulin deficient diabetes. By combining total pancreatectomy with islet transplantation, patients can maintain some beta cell mass with insulin secretory capacity, in order to mitigate the severity of post-operative diabetes (40). Unlike allotransplantation, TPIAT does not require immunosuppression and patients serve as their own islet donors (41). Rather the challenge with islet transplantation is obtaining a sufficient number or mass of islets from a diseased pancreas.

TPIAT was first performed in the 1970s at the University of Minnesota (42) and since then has been adapted to many centers worldwide as a treatment for chronic pancreatitis (43–46). Because of the limitations in obtaining sufficient islet mass, only around one out of every three individuals is insulin independent after the procedure, but the majority preserve some endogenous insulin secretion benefiting glycemic control (41). Younger age and higher islet mass transplantation are predictors for functional graft survival (47) and normal preoperative glucose status can also improve post-operative graft success (48). Improvements have been made in the isolation of the islets and minimization of ischemia to the pancreatic islets (41). Work has been done to minimize risk of ischemia by avoiding prolonged cold ischemia to the isolated tissue but length of time has not been shown to have a detrimental effect on islet isolation and location of isolation (remote or onsite) does not affect insulin independence (49). Although some patients have maintained insulin independence for >10 years after TPIAT, as seen with Considerations in islet transplantation, insulin independence, and islet graft function wane over time (50). It is possible that metabolic strain due to glucotoxicity, exposure to toxins and medications (51) or the inability for islet neogenesis to occur in the liver (52) may contribute to the observed decline in islet graft function. An autologous renewable cell source could address the diabetes challenges after TPIAT by increasing islet mass and providing potential to "redose" islets later to address the apparent slow loss of islet mass over time after the procedure. For obvious reasons, autologous transplantation can only occur once in a patient's lifetime and cadaveric islets require a deceased donor, therefore a source of renewable islet sources could benefit patients with islet dysfunction, either due to diabetes or chronic pancreatitis.

Moving Beyond Allo and Autotransplantation

There is clearly a need for access to a renewable cell source for allotransplantation, and for re-transplant after autotransplantation or in chronic pancreatitis. There are some key lessons to be learned from allo- and autotransplantation for the future of cell therapy. From islet autotransplantation, we have functional data to establish a dose-response in the absence of targeted immunity– Islet graft function (C-peptide positivity) is nearly universal when a minimum threshold of 5,000 IEQ/kg is transplanted in the autograft setting, suggesting this may be an appropriate minimum "dose" target for a stem cell-derived therapy [particularly if immune barriers are fully addressed (41)]. Islet attrition occurs due to immune and non-immune stressors, and thus engineering the proper microenviroment for renewable cell sources may enhance the potential for long-term benefit. Encapsulation, engineered scaffolds, and alternate transplant sites are particularly relevant to stem cell therapy, where encapsulation may also both immunoprotect and to "contain" the cell product and sites outside the liver may be desired for safety. Addressing auto and alloimmunity, such as through encapsulation approaches, will continue to be a need for stem cell derived therapy.

A Future for Stem Cell Derived Islets

Given the limitations of donor availability, the generation of glucose responsive human sBCs and islets from human pluripotent stem cells (hPSCs) are a potential future treatment for those with diabetes. Both human embryonic stem cells (hESCs) and induced pluripotent stem cells (iPSCs) are potential sources of hPSCs from which sBCs can be generated. hESCs are have been derived from blastocysts (53, 54). iPSCs are somatic cells that can be taken from a patient blood sample or fibroblast and reprogrammed with defined factors to the pluripotent state (55). Both hESCs and iPSCs are able to undergo differentiation and self-renewal to generate an unlimited source of potentially therapeutic cells. Much work has been done to direct the differentiation from hPSCs to the pancreatic lineage through stepwise differentiation protocols (56–59) (**Figure 1**). These cells are functionally mature (60) and display insulin secretory properties similar to human islets (61–63). Transplantation of stem cell derived pancreatic endoderm can mature to functional islets *in vivo* in rodents (58). Further work has been done to make this process functional and scalable (64). While in their infancy, current and future studies are underway in humans to investigate safety and efficacy of hPSC derived islets (NCT02239354, NCT03163511, and NCT02939118). These advances are the initial steps to providing renewable, functional islets to patients with beta cell dysfunction (65).

While great progress has been made in the development of these sBCs, there are several challenges and factors to consider. One factor to consider is the presence of off-target or undifferentiated cells that could interfere with the functional sBCs or be tumorigenic. Current and future clinical approaches using partially mature (pancreatic endoderm) or fully mature islets for implantation may reduce this risk, but close clinical follow up will be needed.

A second consideration is the role of immunogenicity of the sBCs or stem-cell derived islets. hESC derived cells, in the absence of genome editing, are subject to risk for alloimmune rejection which may require immunosuppression or encapsulation. Using iPSCs subverts this risk of alloimmunity and would be ideal for a cell source in TPIAT. Autologous sBCs have been successfully derived from patients with type 1 diabetes (62), but in this setting would remain at risk for autoimmune attack by autoreactive T-cells against pancreatic beta cells characteristic of type 1 diabetes. Therapeutic strategies for delaying this autoimmune attack have shown promise in at-risk relatives (4) but there is not yet an accepted therapy for halting this immune response. Even in the setting of allogeneic islet or simultaneous pancreas-kidney transplantation, where multi-drug immunosuppression is administered, islet autoimmunity can recur (66). Transplanted islet exosome profiling can be used as a way to monitor for evidence of recurrent autoimmunity (67). This can be tracked from a peripheral blood sample from a patient and may be a marker of beta cell injury patients who have undergone islet transplantation. Plasma detection of glutamate decarboxylase (GAD-65) can serve as a marker of beta cell loss after transplantation (68). Although hPSCs can evade allogenic response (69) once fully differentiated, these cells lose their immunologic privilege (70). Additionally, strategies being explored include the use of genetically engineered immune silent cells. Advances in genome engineering using CRISPR/Cas9 allows for modification of hPSCs (71) and can knock out HLA surface molecules implicated in autoimmunity (72, 73).

Pluripotent stem cells may be better poised to overcome the immunologic challenges of allotransplantation when combined with genetic engineering, encapsulation, or scaffolding technology. Bioengineered scaffolds offer novel opportunity to improve islet vascularization and optimize the islet microenvironment to protect grafts (27, 74, 75). Macroencapsulation of sBCs have been explored as a way to protect sBCs *in vivo* (76, 77) which could block transplanted sBCs from an immune attack but provide an environment to allow the survival of transplanted tissue. Trials are underway of to encapsulate sBCs for transplantation (78). These considerations will be important aspects to consider prior to considering transplantation of sBCs.

DISCUSSION

While much progress has been made in transplantation functional islet tissue as a treatment for diabetes, there are still many aspects that must be faced. While sBCs and islets can be generated on a large scale, there are still challenges to ensure

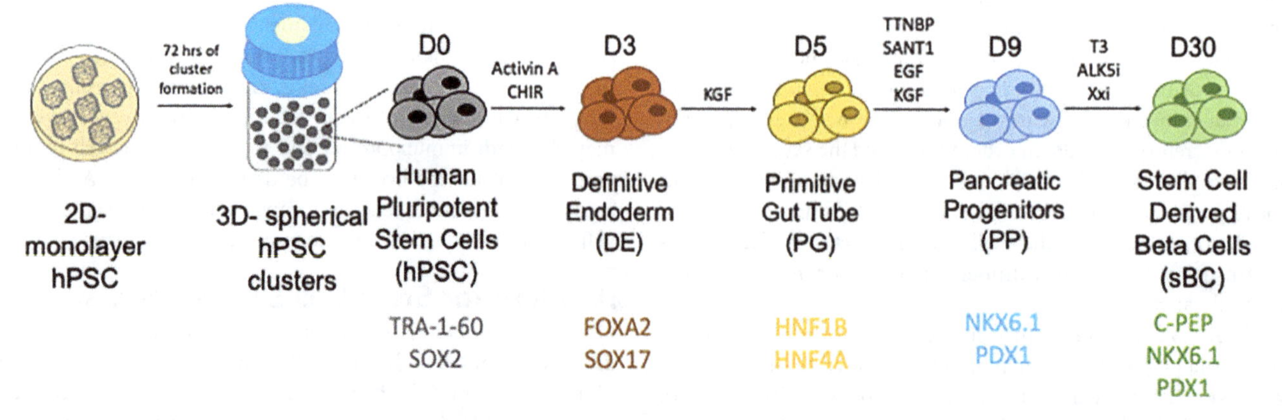

FIGURE 1 | Generation of stem cell derived beta cells from human pluripotent stem cells through a stepwise differentiation protocol. Courtesy of Dr. Holger Russ.

protection from rejection, continued functionality and assurance of safety. Lessons learned from allo and auto islet transplantation will be helpful to apply in the sBC context.

AUTHOR CONTRIBUTIONS

TT is the guarantor of this work and, as such, takes responsibility for the integrity of the data and the accuracy of the data analysis. TT researched data and wrote the manuscript. MB researched data and reviewed/edited manuscript. All authors contributed to the article and approved the submitted version.

REFERENCES

1. Diabetes Control and Complications Trial Research Group, Nathan DM, Genuth S, Lachin J, Cleary P, Crofford O, et al. The effect of intensive treatment of diabetes on the development and progression of long-term complications in insulin-dependent diabetes mellitus. *N Engl J Med* (1993) 329(14):977–86. doi: 10.1056/NEJM199309303291401
2. Diabetes Control and Complications Trial (DCCT)/Epidemiology of Diabetes Interventions and Complications (EDIC) Research Group, Lachin JM, White NH, Hainsworth DP, Sun W, Cleary PA. Effect of intensive diabetes therapy on the progression of diabetic retinopathy in patients with type 1 diabetes: 18 years of follow-up in the DCCT/EDIC. *Diabetes* (2015) 64(2):631–42. doi: 10.2337/db14-0930
3. Cryer PE. Mechanisms of hypoglycemia-associated autonomic failure in diabetes. *N Engl J Med* (2013) 369(4):362–72. doi: 10.1056/NEJMra1215228
4. An Anti-CD3 Antibody, Teplizumab, in Relatives at Risk for Type 1 Diabetes. *N Engl J Med* (2020) 382(6):586. doi: 10.1056/NEJMx190033
5. Barton FB, Rickels MR, Alejandro R, Hering BJ, Wease S, Naziruddin B, et al. Improvement in outcomes of clinical islet transplantation: 1999-2010. *Diabetes Care* (2012) 35(7):1436–45. doi: 10.2337/dc12-0063
6. Ferrannini E. Insulin resistance versus insulin deficiency in non-insulin-dependent diabetes mellitus: problems and prospects. *Endocr Rev* (1998) 19 (4):477–90. doi: 10.1210/edrv.19.4.0336
7. Sasikala M, Talukdar R, Pavan kumar P, Radhika G, Rao GV, Pradeep R, et al. beta-Cell dysfunction in chronic pancreatitis. *Dig Dis Sci* (2012) 57(7):1764–72. doi: 10.1007/s10620-012-2086-7
8. Bosi E, Bottazzo GF, Secchi A, Pozza G, Shattock M, Saunders A, et al. Islet cell autoimmunity in type I diabetic patients after HLA-mismatched pancreas transplantation. *Diabetes* (1989) 38(Suppl 1):82–4. doi: 10.2337/diab.38.1.S82
9. Braghi S, Bonifacio E, Secchi A, Di Carlo V, Pozza G, Bosi E. Modulation of humoral islet autoimmunity by pancreas allotransplantation influences allograft outcome in patients with type 1 diabetes. *Diabetes* (2000) 49 (2):218–24. doi: 10.2337/diabetes.49.2.218
10. Vendrame F, Pileggi A, Laughlin E, Allende G, Martin-Pagola A, Molano RD, et al. Recurrence of type 1 diabetes after simultaneous pancreas-kidney transplantation, despite immunosuppression, is associated with autoantibodies and pathogenic autoreactive CD4 T-cells. *Diabetes* (2010) 59 (4):947–57. doi: 10.2337/db09-0498
11. Scharp DW, Lacy PE, Santiago JV, McCullough CS, Weide LG, Falqui L, et al. Insulin independence after islet transplantation into type I diabetic patient. *Diabetes* (1990) 39(4):515–8. doi: 10.2337/diabetes.39.4.515
12. Shapiro AM, Lakey JR, Ryan EA, Korbutt GS, Toth E, Warnock GL, et al. Islet transplantation in seven patients with type 1 diabetes mellitus using a glucocorticoid-free immunosuppressive regimen. *N Engl J Med* (2000) 343 (4):230–8. doi: 10.1056/NEJM200007273430401
13. Bellin MD, Barton FB, Heitman A, Harmon JV, Kandaswamy R, Balamurugan AN, et al. Potent induction immunotherapy promotes long-term insulin independence after islet transplantation in type 1 diabetes. *Am J Transplant* (2012) 12(6):1576–83. doi: 10.1111/j.1600-6143.2011.03977.x
14. Hering BJ, Clarke WR, Bridges ND, Eggerman TL, Alejandro R, Bellin MD, et al. Phase 3 Trial of Transplantation of Human Islets in Type 1 Diabetes Complicated by Severe Hypoglycemia. *Diabetes Care* (2016) 39(7):1230–40. doi: 10.2337/dc15-1988
15. Ryan EA, Shandro T, Green K, Paty BW, Senior PA, Bigam D, et al. Assessment of the Severity of Hypoglycemia and Glycemic Lability in Type 1 Diabetic Subjects Undergoing Islet Transplantation. *Diabetes* (2004) 53 (4):955–62. doi: 10.2337/diabetes.53.4.955
16. Senior PA, Zeman M, Paty BW, Ryan EA, Shapiro AM. Changes in renal function after clinical islet transplantation: four-year observational study. *Am J Transplant* (2007) 7(1):91–8. doi: 10.1111/j.1600-6143.2006.01573.x
17. Ryan EA, Lakey JRT, Rajotte RV, Korbutt GS, Kin T, Imes S, et al. Clinical Outcomes and Insulin Secretion After Islet Transplantation With the Edmonton Protocol. *Diabetes* (2001) 50(4):710–9. doi: 10.2337/diabetes.50.4.710
18. Huurman VA, Hilbrands R, Pinkse GG, Gillard P, Duinkerken G, van de Linde P, et al. Cellular islet autoimmunity associates with clinical outcome of islet cell transplantation. *PloS One* (2008) 3(6):e2435. doi: 10.1371/journal.pone.0002435
19. Roelen DL, Huurman VA, Hilbrands R, Gillard P, Duinkerken G, van der Meer-Prins PW, et al. Relevance of cytotoxic alloreactivity under different immunosuppressive regimens in clinical islet cell transplantation. *Clin Exp Immunol* (2009) 156(1):141–8. doi: 10.1111/j.1365-2249.2008.03812.x
20. Qi M, Kinzer K, Danielson KK, Martellotto J, Barbaro B, Wang Y, et al. Five-year follow-up of patients with type 1 diabetes transplanted with allogeneic islets: the UIC experience. *Acta Diabetol* (2014) 51(5):833–43. doi: 10.1007/s00592-014-0627-6
21. Triñanes J, Rodriguez-Rodriguez AE, Brito-Casillas Y, Wagner A, De Vries APJ, Cuesto G, et al. Deciphering Tacrolimus-Induced Toxicity in Pancreatic β Cells. *Am J Transplant* (2017) 17(11):2829–40. doi: 10.1111/ajt.14323
22. Westermark GT, Westermark P, Berne C, Korsgren O. Widespread amyloid deposition in transplanted human pancreatic islets. *N Engl J Med* (2008) 359 (9):977–9. doi: 10.1056/NEJMc0802893
23. Generette GS, Bachul PJ, Boylan KE, Yassan LJ, Hart J, Pyda JS, et al. Neither amyloid depositions nor hepatic steatosis are associated with marginal islet mass early after autotransplantation. *Am J Transplant* (2020) 0:1–3. doi: 10.1111/ajt.16406
24. Maffi P, Nano R, Monti P, Melzi R, Sordi V, Mercalli A, et al. Islet Allotransplantation in the Bone Marrow of Patients With Type 1 Diabetes: A Pilot Randomized Trial. *Transplantation* (2019) 103(4):839–51. doi: 10.1097/TP.0000000000002416
25. Bertuzzi F, Colussi G, Lauterio A, De Carlis L. Intramuscular islet allotransplantation in type 1 diabetes mellitus. *Eur Rev Med Pharmacol Sci* (2018) 22(6):1731–6. doi: 10.26355/eurrev_201803_14588
26. Baidal DA, Ricordi C, Berman DM, Alvarez A, Padilla N, Ciancio G, et al. Bioengineering of an Intraabdominal Endocrine Pancreas. *N Engl J Med* (2017) 376(19):1887–9. doi: 10.1056/NEJMc1613959
27. Carlsson PO, Espes D, Sedigh A, Rotem A, Zimerman B, Grinberg H, et al. Transplantation of macroencapsulated human islets within the bioartificial pancreas βAir to patients with type 1 diabetes mellitus. *Am J Transplant* (2018) 18(7):1735–44. doi: 10.1111/ajt.14642
28. Campbell PM, Salam A, Ryan EA, Senior P, Paty BW, Bigam D, et al. Pretransplant HLA antibodies are associated with reduced graft survival after clinical islet transplantation. *Am J Transplant* (2007) 7(5):1242–8. doi: 10.1111/j.1600-6143.2007.01777.x
29. Piemonti L, Everly MJ, Maffi P, Scavini M, Poli F, Nano R, et al. Alloantibody and autoantibody monitoring predicts islet transplantation outcome in

human type 1 diabetes. *Diabetes* (2013) 62(5):1656–64. doi: 10.2337/db12-1258
30. Mohanakumar T, Narayanan K, Desai N, Ramachandran S, Shenoy S, Jendrisak M, et al. A significant role for histocompatibility in human islet transplantation. *Transplantation* (2006) 82(2):180–7. doi: 10.1097/01.tp.0000226161.82581.b2
31. Olack BJ, Swanson CJ, Flavin KS, Phelan D, Brennan DC, White NH, et al. Sensitization to HLA antigens in islet recipients with failing transplants. *Transplant Proc* (1997) 29(4):2268–9. doi: 10.1016/S0041-1345(97)00327-8
32. Hilbrands R, Huurman VA, Gillard P, Velthuis JH, De Waele M, Mathieu C, et al. Differences in baseline lymphocyte counts and autoreactivity are associated with differences in outcome of islet cell transplantation in type 1 diabetic patients. *Diabetes* (2009) 58(10):2267–76. doi: 10.2337/db09-0160
33. Laughlin E, Burke G, Pugliese A, Falk B, Nepom G. Recurrence of autoreactive antigen-specific CD4+ T cells in autoimmune diabetes after pancreas transplantation. *Clin Immunol* (2008) 128(1):23–30. doi: 10.1016/j.clim.2008.03.459
34. Monti P, Scirpoli M, Maffi P, Ghidoli N, De Taddeo F, Bertuzzi F, et al. Islet transplantation in patients with autoimmune diabetes induces homeostatic cytokines that expand autoreactive memory T cells. *J Clin Invest* (2008) 118 (5):1806–14. doi: 10.1172/JCI35197
35. van der Torren CR, Zaldumbide A, Roelen DL, Duinkerken G, Brand-Schaaf SH, Peakman M, et al. Innate and adaptive immunity to human beta cell lines: implications for beta cell therapy. *Diabetologia* (2016) 59(1):170–5. doi: 10.1007/s00125-015-3779-1
36. Pinkse GG, Tysma OH, Bergen CA, Kester MG, Ossendorp F, van Veelen PA, et al. Autoreactive CD8 T cells associated with beta cell destruction in type 1 diabetes. *Proc Natl Acad Sci U S A* (2005) 102(51):18425–30. doi: 10.1073/pnas.0508621102
37. Unger WW, Velthuis J, Abreu JR, Laban S, Quinten E, Kester MG, et al. Discovery of low-affinity preproinsulin epitopes and detection of autoreactive CD8 T-cells using combinatorial MHC multimers. *J Autoimmun* (2011) 37 (3):151–9. doi: 10.1016/j.jaut.2011.05.012
38. Velthuis JH, Unger WW, van der Slik AR, Duinkerken G, Engelse M, Schaapherder AF, et al. Accumulation of autoreactive effector T cells and allo-specific regulatory T cells in the pancreas allograft of a type 1 diabetic recipient. *Diabetologia* (2009) 52(3):494–503. doi: 10.1007/s00125-008-1237-z
39. Velthuis JH, Unger WW, Abreu JR, Duinkerken G, Franken K, Peakman M, et al. Simultaneous detection of circulating autoreactive CD8+ T-cells specific for different islet cell-associated epitopes using combinatorial MHC multimers. *Diabetes* (2010) 59(7):1721–30. doi: 10.2337/db09-1486
40. McEachron KR, Bellin MD. Total pancreatectomy and islet autotransplantion for chronic and recurrent acute pancreatitis. *Curr Opin Gastroenterol* (2018) 34(5):367–73. doi: 10.1097/MOG.0000000000000458
41. Sutherland DE, Radosevich DM, Bellin MD, Hering BJ, Beilman GJ, Dunn TB, et al. Total pancreatectomy and islet autotransplantation for chronic pancreatitis. *J Am Coll Surg* (2012) 214(4):409–24; discussion 24-6. doi: 10.1016/j.jamcollsurg.2011.12.040
42. Najarian JS, Sutherland DE, Matas AJ, Goetz FC. Human islet autotransplantation following pancreatectomy. *Transplant Proc* (1979) 11 (1):336–40.
43. Morgan KA, Lancaster WP, Owczarski SM, Wang H, Borckardt J, Adams DB. Patient Selection for Total Pancreatectomy with Islet Autotransplantation in the Surgical Management of Chronic Pancreatitis. *J Am Coll Surg* (2018) 226 (4):446–51. doi: 10.1016/j.jamcollsurg.2017.12.018
44. Wilson GC, Sutton JM, Abbott DE, Smith MT, Lowy AM, Matthews JB, et al. Long-Term Outcomes After Total Pancreatectomy and Islet Cell Autotransplantation: Is It a Durable Operation? *Ann Surg* (2014) 260 (4):659–67. doi: 10.1097/SLA.0000000000000920
45. Garcea G, Weaver J, Phillips J, Pollard CA, Ilouz SC, Webb MA, et al. Total pancreatectomy with and without islet cell transplantation for chronic pancreatitis: a series of 85 consecutive patients. *Pancreas* (2009) 38(1):1–7. doi: 10.1097/MPA.0b013e3181825c00
46. Takita M, Lara LF, Naziruddin B, Shahbazov R, Lawrence MC, Kim PT, et al. Effect of the Duration of Chronic Pancreatitis on Pancreas Islet Yield and Metabolic Outcome Following Islet Autotransplantation. *J Gastrointest Surg* (2015) 19(7):1236–46. doi: 10.1007/s11605-015-2828-x
47. Chinnakotla S, Beilman GJ, Dunn TB, Bellin MD, Freeman ML, Radosevich DM, et al. Factors Predicting Outcomes After a Total Pancreatectomy and Islet Autotransplantation Lessons Learned From Over 500 Cases. *Ann Surg* (2015) 262(4):610–22. doi: 10.1097/SLA.0000000000001453
48. Quartuccio M, Hall E, Singh V, Makary MA, Hirose K, Desai N, et al. Glycemic Predictors of Insulin Independence After Total Pancreatectomy With Islet Autotransplantation. *J Clin Endocrinol Metab* (2017) 102(3):801–9. doi: 10.1210/jc.2016-2952
49. Kesseli SJ, Smith KD, Jung MK, Lin YK, Walsh RM, Hatipoglu B, et al. Islet Cell Yield Following Remote Total Pancreatectomy With Islet Autotransplant is Independent of Cold Ischemia Time. *Pancreas* (2017) 46(3):380–4. doi: 10.1097/MPA.0000000000000792
50. Bellin MD, Beilman GJ, Sutherland DER, Ali H, Petersen A, Mongin S, et al. How Durable Is Total Pancreatectomy and Intraportal Islet Cell Transplantation for Treatment of Chronic Pancreatitis? *J Am Coll Surg* (2019) 228(4):329–39. doi: 10.1016/j.jamcollsurg.2018.12.019
51. Ngo A, Sutherland DE, Beilman GJ, Bellin MD. Deterioration of glycemic control after corticosteroid administration in islet autotransplant recipients: a cautionary tale. *Acta Diabetol* (2014) 51(1):141–5. doi: 10.1007/s00592-011-0315-8
52. Soltani SM, O'Brien TD, Loganathan G, Bellin MD, Anazawa T, Tiwari M, et al. Severely fibrotic pancreases from young patients with chronic pancreatitis: evidence for a ductal origin of islet neogenesis. *Acta Diabetol* (2013) 50(5):807–14. doi: 10.1007/s00592-011-0306-9
53. Thomson JA, Itskovitz-Eldor J, Shapiro SS, Waknitz MA, Swiergiel JJ, Marshall VS, et al. Embryonic stem cell lines derived from human blastocysts. *Science* (1998) 282(5391):1145–7. doi: 10.1126/science.282.5391.1145
54. Cowan CA, Klimanskaya I, McMahon J, Atienza J, Witmyer J, Zucker JP, et al. Derivation of Embryonic Stem-Cell Lines from Human Blastocysts. *N Engl J Med* (2004) 350(13):1353–6. doi: 10.1056/NEJMsr040330
55. Takahashi K, Yamanaka S. Induction of Pluripotent Stem Cells from Mouse Embryonic and Adult Fibroblast Cultures by Defined Factors. *Cell* (2006) 126 (4):663–76. doi: 10.1016/j.cell.2006.07.024
56. D'Amour KA, Agulnick AD, Eliazer S, Kelly OG, Kroon E, Baetge EE. Efficient differentiation of human embryonic stem cells to definitive endoderm. *Nat Biotechnol* (2005) 23(12):1534–41. doi: 10.1038/nbt1163
57. Kroon E, Martinson LA, Kadoya K, Bang AG, Kelly OG, Eliazer S, et al. Pancreatic endoderm derived from human embryonic stem cells generates glucose-responsive insulin-secreting cells in vivo. *Nat Biotechnol* (2008) 26 (4):443–52. doi: 10.1038/nbt1393
58. D'Amour KA, Bang AG, Eliazer S, Kelly OG, Agulnick AD, Smart NG, et al. Production of pancreatic hormone–expressing endocrine cells from human embryonic stem cells. *Nat Biotechnol* (2006) 24(11):1392–401. doi: 10.1038/nbt1259
59. Pagliuca Felicia W, Millman Jeffrey R, Gürtler M, Segel M, Van Dervort A, Ryu Jennifer H, et al. Generation of Functional Human Pancreatic β Cells In Vitro. *Cell* (2014) 159(2):428–39. doi: 10.1016/j.cell.2014.09.040
60. Velazco-Cruz L, Song J, Maxwell KG, Goedegebuure MM, Augsornworawat P, Hogrebe NJ, et al. Acquisition of Dynamic Function in Human Stem Cell-Derived β Cells. *Stem Cell Rep* (2019) 12(2):351–65. doi: 10.1016/j.stemcr.2018.12.012
61. Robert T, De Mesmaeker I, Stangé GM, Suenens KG, Ling Z, Kroon EJ, et al. Functional Beta Cell Mass from Device-Encapsulated hESC-Derived Pancreatic Endoderm Achieving Metabolic Control. *Stem Cell Rep* (2018) 10(3):739–50. doi: 10.1016/j.stemcr.2018.01.040
62. Millman JR, Xie C, Van Dervort A, Gürtler M, Pagliuca FW, Melton DA. Generation of stem cell-derived β-cells from patients with type 1 diabetes. *Nat Commun* (2016) 7:11463. doi: 10.1038/ncomms11463
63. Nair GG, Liu JS, Russ HA, Tran S, Saxton MS, Chen R, et al. Recapitulating endocrine cell clustering in culture promotes maturation of human stem-cell-derived beta cells. *Nat Cell Biol* (2019) 21(2):263–74. doi: 10.1038/s41556-018-0271-4
64. Schulz TC, Young HY, Agulnick AD, Babin MJ, Baetge EE, Bang AG, et al. A scalable system for production of functional pancreatic progenitors from human embryonic stem cells. *PloS One* (2012) 7(5):e37004. doi: 10.1371/journal.pone.0037004

65. Henry RR, Pettus J, Wilensky J, Shapiro AMJ, Senior PA, Roep B, et al. Initial Clinical Evaluation of VC-01TM Combination Product—A Stem Cell-Derived Islet Replacement for Type 1 Diabetes (T1D). *Diabetes* (2018) 67 (Supplement 1):138–OR. doi: 10.2337/db18-138-OR
66. Burke GW,3, Vendrame F, Virdi SK, Ciancio G, Chen L, Ruiz P, et al. Lessons From Pancreas Transplantation in Type 1 Diabetes: Recurrence of Islet Autoimmunity. *Curr Diabetes Rep* (2015) 15(12):121. doi: 10.1007/s11892-015-0691-5
67. Korutla L, Rickels MR, Hu RW, Freas A, Reddy S, Habertheuer A, et al. Noninvasive diagnosis of recurrent autoimmune type 1 diabetes after islet cell transplantation. *Am J Transplant* (2019) 19(6):1852–8. doi: 10.1111/ajt.15322
68. Pipeleers D, Robert T, De Mesmaeker I, Ling Z. Concise Review: Markers for Assessing Human Stem Cell-Derived Implants as β-Cell Replacement in Type 1 Diabetes. *Stem Cells Transl Med* (2016) 5(10):1338–44. doi: 10.5966/sctm.2015-0187
69. English K, Wood KJ. Immunogenicity of embryonic stem cell-derived progenitors after transplantation. *Curr Opin Organ Transplant* (2011) 16 (1):90–5. doi: 10.1097/MOT.0b013e3283424faa
70. van der Torren CR, Zaldumbide A, Duinkerken G, Brand-Schaaf SH, Peakman M, Stangé G, et al. Immunogenicity of human embryonic stem cell-derived beta cells. *Diabetologia* (2017) 60(1):126–33. doi: 10.1007/s00125-016-4125-y
71. Lin S, Staahl BT, Alla RK, Doudna JA. Enhanced homology-directed human genome engineering by controlled timing of CRISPR/Cas9 delivery. *Elife* (2014) 3:e04766. doi: 10.7554/eLife.04766
72. Wang D, Quan Y, Yan Q, Morales JE, Wetsel RA. Targeted Disruption of the β2-Microglobulin Gene Minimizes the Immunogenicity of Human Embryonic Stem Cells. *Stem Cells Transl Med* (2015) 4(10):1234–45. doi: 10.5966/sctm.2015-0049
73. Han X, Wang M, Duan S, Franco PJ, Kenty JH, Hedrick P, et al. Generation of hypoimmunogenic human pluripotent stem cells. *Proc Natl Acad Sci U S A* (2019) 116(21):10441–6. doi: 10.1073/pnas.1902566116
74. Nalbach L, Roma LP, Schmitt BM, Becker V, Körbel C, Wrublewsky S, et al. Improvement of islet transplantation by the fusion of islet cells with functional blood vessels. *EMBO Mol Med* (2020) 13(1):e12616. doi: 10.15252/emmm.202012616
75. Elizondo DM, Brandy NZD, da Silva RLL, de Moura TR, Ali J, Yang D, et al. Pancreatic islets seeded in a novel bioscaffold forms an organoid to rescue insulin production and reverse hyperglycemia in models of type 1 diabetes. *Sci Rep* (2020) 10(1):4362. doi: 10.1038/s41598-020-60947-x
76. Motte E, Szepessy E, Suenens K, Stange G, Bomans M, Jacobs-Tulleneers-Thevissen D, et al. Composition and function of macroencapsulated human embryonic stem cell-derived implants: comparison with clinical human islet cell grafts. *Am J Physiol Endocrinol Metab* (2014) 307(9):E838–46. doi: 10.1152/ajpendo.00219.2014
77. Haller C, Piccand J, De Franceschi F, Ohi Y, Bhoumik A, Boss C, et al. Macroencapsulated Human iPSC-Derived Pancreatic Progenitors Protect against STZ-Induced Hyperglycemia in Mice. *Stem Cell Rep* (2019) 12 (4):787–800. doi: 10.1016/j.stemcr.2019.02.002
78. Agulnick AD, Ambruzs DM, Moorman MA, Bhoumik A, Cesario RM, Payne JK, et al. Insulin-Producing Endocrine Cells Differentiated In Vitro From Human Embryonic Stem Cells Function in Macroencapsulation Devices In Vivo. *Stem Cells Transl Med* (2015) 4(10):1214–22. doi: 10.5966/sctm.2015-0079

11

Epigenetic Memory: Lessons from iPS Cells Derived from Human β Cells

Shimon Efrat[*]

Department of Human Molecular Genetics and Biochemistry, Sackler School of Medicine, Tel Aviv University, Tel Aviv, Israel

***Correspondence:**
Shimon Efrat
sefrat@tauex.tau.ac.il

Incomplete reprogramming of somatic cells into induced pluripotent stem cells (iPSCs) may be responsible for the heterogeneity in differentiation capacity observed among iPSC lines. It remains unclear whether it results from stochastic reprogramming events, or reflects consistent genetic or cell-of-origin differences. Some evidence suggests that epigenetic memory predisposes iPSCs to enhanced differentiation into the parental cell type. We investigated iPSCs reprogrammed from human pancreatic islet β cells (BiPSCs), as a step in development of a robust differentiation protocol for generation of β-like cells. BiPSCs derived from multiple human donors manifested enhanced and reproducible spontaneous and induced differentiation towards insulin-producing cells, compared with iPSCs derived from isogenic non-β-cell types and fibroblast-derived iPSCs (FiPSCs). Genome-wide analyses of open chromatin in BiPSCs and FiPSCs identified thousands of differential open chromatin sites (DOCs) between the two iPSC types. DOCs more open in BiPSCs (Bi-DOCs) were significantly enriched for known regulators of endodermal development, including bivalent and weak enhancers, and FOXA2 binding sites. Bi-DOCs were associated with genes related to pancreas development and β-cell function. These studies provide evidence for reproducible epigenetic memory in BiPSCs. Bi-DOCs may provide clues to genes and pathways involved in the differentiation process, which could be manipulated to increase the efficiency and reproducibility of differentiation of pluripotent stem cells from non-β-cell sources.

Keywords: Epigenetic memory, Islet β Cells, pluripotent stem cell differentiation, ATAC-seq, Foxa2

INCOMPLETE REPROGRAMMING INTO IPSCS

Reprogramming of somatic cells into induced pluripotent stem cells (iPSCs) using Yamanaka's combination of four transcription factors (OCT4, SOX2, KLF4, and c-MYC, together termed OSKM) (1) has opened new avenues for in-vitro generation of multiple human differentiated cell types for disease modeling, drug screening, and regenerative medicine. Initially iPSCs derived from multiple cell types have been expected to manifest similar differentiation capacities, resembling those of human embryonic stem cells (ESCs). However, accumulated experience has shown a great heterogeneity in differentiation capacity among iPSC lines. This heterogeneity is thought to result from incomplete reprogramming.

The mechanisms activated by ectopic expression of OSKM in somatic cells, which result in reprogramming to pluripotency, are only partly understood. This prolonged process, lasting several weeks, involves multiple events, including silencing of somatic cell genes and activation of pluripotency-associated genes. The low efficiency of OSKM-mediated reprogramming is thought to reflect the stochastic nature of these complex events. Only a small fraction of cells acquire pluripotency, as judged by accepted assays (e.g. differentiation into cells from the three embryonic germ layers in embryoid bodies (EB), and teratoma formation assay). OSKM factors function by binding to chromatin regions and inducing their remodeling, thereby activating or repressing gene expression. Evidence suggests that broad epigenetic changes are among early key events of the reprogramming process (2).

The epigenetic landscape, shaped by DNA methylation and histone modifications, is critical for maintaining cell identity. Erasing cell-specific patterns of epigenetic modifications, and replacing them with pluripotency patterns, are central to cell reprogramming to pluripotency. Ample evidence supports the incomplete and varying erasure of the original epigenetic marks in both mouse and human iPSC lines. However, it remains unclear whether these variations represent consistent patterns, based on the cell type of origin, or genetic differences among donors, or reflect stochastic differences caused by low efficiency of the reprogramming mechanisms (**Table 1**). While epigenetic memory may not necessarily affect gene expression patterns in iPSCs (3), likely due to missing transcription factors, it is expected to affect differentiation capacity towards specific cell fates. This may limit some applications of iPSCs, but at the same time may predispose iPSCs to enhanced differentiation into the parental cell type, thereby facilitating generation of cells for specific uses.

IPSC DIFFERENCES RELATED TO THE CELL TYPE OF ORIGIN

Mouse iPSCS

Early comparison of mouse iPSCs derived from bone marrow progenitor cells, dermal fibroblasts, and neural progenitor cells, identified residual DNA methylation signatures characteristic of the somatic tissue of origin (4). These differences favored iPSC differentiation towards the donor cell type, while restricting alternative cell fates. These results were contrasted to the methylation and differentiation patterns of nuclear-transfer-derived pluripotent stem cells, which were more similar to those of ESCs. Similar findings were reported in early-passage iPSCs obtained from mouse fibroblasts, hematopoietic and myogenic cells, which exhibited distinct epigenetic patterns (5). These patterns were reflected in different transcriptional profiles of the iPSCs, and in their differentiation efficiency into embryoid bodies (EBs) and hematopoietic cell types. These differences were eroded with iPSC passaging, suggesting that epigenetic memory was a transient phenomenon. iPSCs derived from mouse neonatal cardiomyocytes (CMs) were also shown to differentiate toward CMs more efficiently than fibroblast-derived iPSCs (FiPSCs) or mouse ESCs (6).

TABLE 1 | Possible sources of iPSC variability.

Possible source of iPSC variability	Expected phenotype
Stochastic reprogramming events	Differences among clones from a single cell type of origin from a single donor
Donor-related	Similarity among clones from multiple cell types of origin from a single donor; differences among iPSCs from multiple donors
Cell type of origin	Similarity among iPSCs from the same cell type of origin from multiple donors

Human iPSCs

A number of reports documented similar epigenetic memory in human iPSCs. Kim et al. compared DNA methylation profiles and differentiation potential of iPSCs derived from human umbilical cord blood and neonatal keratinocytes (7). They identified distinct genome-wide DNA methylation patterns in iPSCs derived from each cell type, resulting from both incomplete erasure of tissue-specific methylation and aberrant de-novo methylation. These differences did not disappear upon extended passaging. Ohi et al. observed that human iPSCs generated from hepatocytes, skin fibroblasts, and melanocytes, retained some transcriptional characteristics of the original cells at low passages, which could be partially explained by incomplete promoter DNA methylation (8). They noticed that incompletely silenced genes tended to be isolated from other genes that were repressed during reprogramming, indicating that silencing of isolated genes may be less efficient. Global DNA methylation analyses of iPSCs reprogrammed from human cornea limbal epithelial stem cells (LESC) showed gene methylation patterns similar to those of the parental cells, and different from those of FiPSCs (9). Upon differentiation, LESC-derived iPSCs expressed higher levels of LESC markers, compared with FiPSCs.

IPSC DIFFERENCES RELATED TO DONORS AND STOCHASTIC VARIABILITY

In contrast to these findings, which associated epigenetic and differentiation differences among iPSCs with their cell type of origin, other studies supported a donor-related or stochastic basis for these differences. In one such study (10), whole-genome profiling of DNA methylation in five human iPSC lines derived from adipocytes and fibroblasts (including 3 subclones of a single line) identified over 1,000 differentially methylated sites, most of them associated with CG islands and genes, indicating stochastic interclone reprogramming variability. Two other studies (11, 12) surveyed 16–18 iPSC lines each, derived from fibroblasts and peripheral blood cells of four human donors in each study. Both studies concluded that the majority of transcriptional and DNA methylome differences among iPSCs, as well as differences in their differentiation capacity towards the hematopoietic cell lineage, could be attributed to the donor, rather than the cell type of origin, indicating that genetic differences among donors can result in reproducible reprogramming differences. Finally, two studies, which analyzed differences in gene expression patterns in human iPSCs, concluded that genetic differences between individual donors were the major cause of transcriptional variation between lines. One of these studies (13) compared 25 iPSC lines from four donors and three tissues. The second study compared two human ESC lines with genetically matched iPSCs derived from fibroblasts differentiated from each ESC line (14). Their findings showed reproducible gene expression patterns among each ESC line and three iPSC clones derived from it, and variations compared with the other ESC line and its iPSC derivatives. The study concluded that ESCs and iPSCs exhibited similar gene expression patterns, and that the donor genetic background was responsible for transcriptional variations among pluripotent stem cell lines.

The donor genetic background may also influence the differentiation capacity of ESC lines. However, in contrast to the relatively large numbers of iPSC lines that have been generated by multiple laboratories, data on differentiation of ESC lines accumulated to date is based on a small number of established ESC lines commonly used by all research groups. This limited number, which is a result of ethical barriers to generation of new ESC lines, does not allow proper comparisons of variations among ESC lines in differentiation potential into specific cell lineages.

EPIGENETIC MEMORY IN IPSCS DERIVED FROM HUMAN β CELLS

Directed in-vitro differentiation of iPSCs into β-like cells is a promising approach for generation of abundant insulin-producing cells for cell therapy of diabetes, disease modeling and drug screening. Despite significant progress (15–21), current differentiation protocols result in cells with heterogeneous and immature phenotype, and suffer from low efficiency and variability among iPSC lines. In a step towards developing a more robust differentiation protocol, we investigated iPSCs reprogrammed from human pancreatic islet β cells (BiPSCs). This approach depended on lineage tracing of human β-cell-derived (BCD) cells within the heterogeneous cell population in cultures of isolated human islets (22). The stable genetic label allowed positive identification of the β-cell origin of individual iPSC clones, that otherwise would be difficult to distinguish from iPSCs derived from non-β-cell types present in the expanded islet cell culture. Initial studies of four BiPSC lines derived from β cells of three nondiabetic human donors established their pluripotency, as judged by standard assays (23). Nevertheless, chromatin immunoprecipitation (ChIP) analyses showed that the levels of histone H3 acetylation, a hallmark of open chromatin structure, at the *INS* and *PDX1* loci in BiPSCs were similar to those found in human islets and BCD cells, and significantly higher than those in FiPSCs, as well as in two iPSC lines derived from isogenic islet non-β cells (termed PiPSCs) from two of the donors. BiPSCs also exhibited significantly lower DNA methylation levels, characteristic of transcribed genes, in genes expressed in human islet cells, such as *PDX1*, compares with FiPSCs. Despite the open chromatin marks, β-cell genes were not expressed in BCD and BiPS cells.

The epigenetic memory of BiPSCs was associated with higher expression levels of *PDX1*, *FOXA2*, and *INS* transcripts following spontaneous differentiation into EBs and teratomas, compared with those derived from FiPSCs and PiPSCs, and an enhanced induced differentiation capacity into insulin-producing cells in mice transplanted with BiPSC-derived endocrine progenitors, following the protocol of Kroon et al. (24). The epigenetic phenotype and differentiation capacity of BiPSCs were reproducible among the four lines obtained from three human donors, and appeared stable within the passage range analyzed (passages 10–20).

To identify genes and pathways, which may be responsible for the enhanced and reproducible differentiation capacity of BiPSCs, we performed a global analysis of chromatin sites differentially open in BiPSCs, compared with FiPSCs, using an Assay for Transposase Accessible Chromatin with high-throughput sequencing (ATAC-seq) (25). For this analysis we generated five new BiPSC lines from three nondiabetic donors, which were compared to five FiPSC lines from two nondiabetic donors. All these lines passed pluripotency assays, and EBs generated from BiPSCs at passages 9-12 showed enhanced spontaneous expression of *FOXA2*, *PDX1*, and *INS*, compared with those derived from FiPSCs (25), similarly to the four BiPSC lines in the initial study.

Despite high overall similarity in open chromatin between the two iPSC types, the ATAC-seq analysis identified thousands of significantly differential open chromatin (DOC) sites between BiPSCs and FiPSCs, most of which were more open in BiPSCs (Bi-DOCs). Bi-DOCS overlapped gene regulatory elements known to be involved in development, such as weak enhancers (marked by H3K4me1) and bivalent enhancers and promoters (marked by H3K27me3), especially near genes involved in endodermal development, such as *FOXA2* and its target genes, and pancreas development, such as *PDX1*, and *NKX2-2*, as well as genes expressed in mature β cells, such as *INS*. The ATAC-seq data for these four genes (*FOXA2*, *PDX1*, *NKX2-2*, and *INS*) was validated by H3K4me3 ChIP analysis, which found higher levels of this open chromatin mark in the promoter regions of all four genes in BiPSCs, compares with FiPSCs. These findings could explain the enhanced expression of these genes in EBs generated from BiPSCs, compared with those derived from FiPSCs.

The relevance of Bi-DOCs to differentiation towards islet cells was analyzed by comparing directed differentiation of BiPSCs and FiPSCs into definitive endoderm (DE) and pancreatic progenitor (PP) cells according to the protocol of Rezania et al. (16). Global transcriptome analyses by RNA-seq identified 567 protein-coding genes expressed at higher levels in BiPSC-derived DE, compared with FiPSC-derived DE, and 181 genes expressed at higher levels in BiPSC-derived PP, compared with FiPSC-derived PP (25). These genes were significantly enriched for genes mapping near Bi-DOCs. Among genes expressed at higher levels in BiPSC at both stages, the most prominent was estrogen receptor 1 (*ESR1*; 3.5-fold and 53.4-fold higher in DE and PP, respectively) and several of its target genes. 17β-estradiol (E2) has been shown to protect mouse β cells from apoptosis by signaling through estrogen receptor (ER) α encoded by *ESR1* (26). ERα activity has been reported to increase *Neurog3* expression and β-cell proliferation in a mouse model of pancreas partial duct ligation, and during mouse islet development (27). It has been suggested to regulate endocrine lineage specification through downregulation of NOTCH signaling. Thus, inclusion of E2 in the culture medium at key stages of the in-vitro differentiation protocol may increase its efficiency and reproducibility.

CONCLUSION

Overall, the analyses of Bi-DOCs support the existence of reproducible epigenetic memory in BiPSCs. The association between Bi-DOCs and gene expression levels at early stages of the in-vitro differentiation protocol provides a plausible explanation for the enhanced differentiation capacity of BiPSCs into the β-cell lineage, compared with pluripotent stem cells from a non-β-cell source. Both chromatin structure and differentiation capacity were reproducible in a combined number of nine BiPSC lines from six human islet donors in two separate studies. Bi-DOCs may provide clues to genes and pathways involved in the differentiation process, which could be manipulated to increase the efficiency and reproducibility of differentiation of pluripotent stem cells from other sources. Such manipulations could include activation of candidate genes using CRISPR-on approaches or small-molecule compounds.

AUTHOR CONTRIBUTIONS

The author confirms being the sole contributor of this work and has approved it for publication.

ACKNOWLEDGMENTS

This research in my laboratory was supported by the Juvenile Diabetes Research Foundation (JDRF), the Israel Science Foundation, and the Innovative Medicines Initiative of the European Union Seventh Framework Program. Human islets for these studies were provided through the JDRF ECIT Islets for Basic Research Program and the Integrated Islet Distribution Program.

REFERENCES

1. Takahashi K, Yamanaka S. Induction of pluripotent stem cells from mouse embryonic and adult fibroblast cultures by defined factors. *Cell* (2006) 126:663–76. doi: 10.1016/j.cell.2006.07.024
2. Watanabe A, Yamada Y, Yamanaka S. Epigenetic regulation in pluripotent stem cells: a key to breaking the epigenetic barrier. *Philos Trans R Soc Lond B Biol Sci* (2013) 368:20120292. doi: 10.1098/rstb.2012.0292
3. Cahan P, Daley GQ. Origins and implications of pluripotent stem cell variability and heterogeneity. *Nat Rev Mol Cell Biol* (2013) 14:357–68. doi: 10.1038/nrm3584
4. Kim K, Doi A, Wen B, Ng K, Zhao R, Cahan P, et al. Epigenetic memory in induced pluripotent stem cells. *Nature* (2010) 467:285–90. doi: 10.1038/nature09342
5. Polo JM, Liu S, Figueroa ME, Kulalert W, Eminli S, Tan KY, et al. Cell type of origin influences the molecular and functional properties of mouse induced pluripotent stem cells. *Nat Biotechnol* (2010) 28:848–55. doi: 10.1038/nbt.1667
6. Rizzi R, Di Pasquale E, Portararo P, Papait R, Cattaneo P, Latronico MV, et al. Post-natal cardiomyocytes can generate iPS cells with an enhanced capacity toward cardiomyogenic re-differentation. *Cell Death Differ* (2012) 19:1162–74. doi: 10.1038/cdd.2011.205
7. Kim K, Zhao R, Doi A, Ng K, Unternaehrer J, Cahan P, et al. Donor cell type can influence the epigenome and differentiation potential of human induced pluripotent stem cells. *Nat Biotechnol* (2011) 29:1117–9. doi: 10.1038/nbt.2052
8. Ohi Y, Qin H, Hong C, Blouin L, Polo JM, Guo T, et al. Incomplete DNA methylation underlies a transcriptional memory of somatic cells in human iPS cells. *Nat Cell Biol* (2011) 13:541–9. doi: 10.1038/ncb2239
9. Sareen D, Saghizadeh M, Ornelas L, Winkler MA, Narwani K, Sahabian A, et al. Differentiation of human limbal-derived induced pluripotent stem cells into limbal-like epithelium. *Stem Cells Transl Med* (2014) 3:1002–12. doi: 10.5966/sctm.2014-0076
10. Lister R, Pelizzola M, Kida YS, Hawkins RD, Nery JR, Hon G, et al. Hotspots of aberrant epigenomic reprogramming in human induced pluripotent stem cells. *Nature* (2011) 471:68–73. doi: 10.1038/nature09798
11. Kyttälä A, Moraghebi R, Valensisi C, Kettunen J, Andrus C, Pasumarthy KK, et al. Genetic Variability Overrides the Impact of Parental Cell Type and Determines iPSC Differentiation Potential. *Stem Cell Reports* (2016) 6:200–12. doi: 10.1016/j.stemcr.2015.12.009
12. Burrows CK, Banovich NE, Pavlovic BJ, Patterson K, Gallego Romero I, Pritchard JK, et al. Genetic Variation, Not Cell Type of Origin, Underlies the Majority of Identifiable Regulatory Differences in iPSCs. *PloS Genet* (2016) 12: e1005793. doi: 10.1371/journal.pgen.1005793
13. Rouhani F, Kumasaka N, de Brito MC, Bradley A, Vallier L, Gaffney D. Genetic background drives transcriptional variation in human induced pluripotent stem cells. *PloS Genet* (2014) 10:e1004432. doi: 10.1371/journal.pgen.1004432
14. Choi J, Lee S, Mallard W, Clement K, Tagliazucchi GM, Lim H, et al. A comparison of genetically matched cell lines reveals the equivalence of human iPSCs and ESCs. *Nat Biotechnol* (2015) 33:1173–81. doi: 10.1038/nbt.3388
15. Pagliuca FW, Millman JR, Gürtler M, Segel M, Van Dervort A, Ryu JH, et al. Generation of functional human pancreatic β cells in vitro. *Cell* (2014) 159:428–39. doi: 10.1016/j.cell.2014.09.040
16. Rezania A, Bruin JE, Arora P, Rubin A, Batushansky I, Asadi A, et al. Reversal of diabetes with insulin-producing cells derived in vitro from human pluripotent stem cells. *Nat Biotechnol* (2014) 32:1121–33. doi: 10.1038/nbt.3033
17. Russ HA, Parent AV, Ringler JJ, Hennings TG, Nair GG, Shveygert M, et al. Controlled induction of human pancreatic progenitors produces functional beta-like cells in vitro. *EMBO J* (2015) 34:1759–72. doi: 10.15252/embj.201591058
18. Nair GG, Liu JS, Russ HA, Tran S, Saxton MS, Chen R, et al. Recapitulating endocrine cell clustering in culture promotes maturation of human stem-cell-derived β cells. *Nat Cell Biol* (2019) 21:263–74. doi: 10.1038/s41556-018-0271-4
19. Veres A, Faust AL, Bushnell HL, Engquist EN, Kenty JH, Harb G, et al. Charting cellular identity during human in vitro β-cell differentiation. *Nature* (2019) 569:368–73. doi: 10.1038/s41586-019-1168-5
20. Velazco-Cruz L, Song J, Maxwell KG, Goedegebuure MM, Augsornworawat P, Hogrebe NJ, et al. Acquisition of Dynamic Function in Human Stem Cell-Derived β Cells. *Stem Cell Reports* (2019) 12:351–65. doi: 10.1016/j.stemcr.2018.12.012
21. Hogrebe NJ, Augsornworawat P, Maxwell KG, Velazco-Cruz L, Millman JR. Targeting the cytoskeleton to direct pancreatic differentiation of human

pluripotent stem cells. *Nat Biotechnol* (2020) 38:460–70. doi: 10.1038/s41587-020-0430-6
22. Russ HA, Bar Y, Ravassard P, Efrat S. In vitro proliferation of cells derived from adult human beta cells revealed by cell-lineage tracing. *Diabetes* (2008) 57:1575–83. doi: 10.2337/db07-1283
23. Bar-Nur O, Russ HA, Efrat S, Benvenisty N. Epigenetic memory and preferential lineage-specific differentiation in induced pluripotent stem cells derived from human pancreatic islet beta cells. *Cell Stem Cell* (2011) 9:17–23. doi: 10.1016/j.stem.2011.06.007
24. Kroon E, Martinson LA, Kadoya K, Bang AG, Kelly OG, Eliazer S, et al. Pancreatic endoderm derived from human embryonic stem cells generates glucose-responsive insulin-secreting cells in vivo. *Nat Biotechnol* (2008) 26:443–52. doi: 10.1038/nbt1393
25. Thurner M, Shenhav L, Wesolowska-Andersen A, Bennett AJ, Barrett A, Gloyn AL, et al. Genes associated with pancreas development and function maintain open chromatin structure in iPSCs generated from human pancreatic beta cells. *Stem Cell Rep* (2017) 9:1395–405. doi: 10.1016/j.stemcr.2017.09.020
26. Le May C, Chu K, Hu M, Ortega CS, Simpson ER, Korach KS, et al. Estrogens protect pancreatic beta-cells from apoptosis and prevent insulin-deficient diabetes mellitus in mice. *Proc Natl Acad Sci USA* (2006) 103:9232–37. doi: 10.1073/pnas.0602956103
27. Yuchi Y, Cai Y, Legein B, De Groef S, Leuckx G, Coppens V, et al. Estrogen receptor α regulates β-cell formation during pancreas development and following injury. *Diabetes* (2015) 64:3218–28. doi: 10.2337/db14-1798

12

Human Pluripotent Stem Cells go Diabetic: A Glimpse on Monogenic Variants

Sandra Heller[1], Michael Karl Melzer[1,2], Ninel Azoitei[1], Cécile Julier[3*] and Alexander Kleger[1*]*

[1] *Department of Internal Medicine I, Ulm University Hospital, Ulm, Germany,* [2] *Department of Urology, Ulm University Hospital, Ulm, Germany,* [3] *Université de Paris, Institut Cochin, INSERM U1016, CNRS UMR-8104, Paris, France*

***Correspondence:**
Sandra Heller
sandra.heller@uni-ulm.de
Cécile Julier
cecile.julier@inserm.fr
Alexander Kleger
alexander.kleger@uni-ulm.de

Diabetes, as one of the major diseases in industrial countries, affects over 350 million people worldwide. Type 1 (T1D) and type 2 diabetes (T2D) are the most common forms with both types having invariable genetic influence. It is accepted that a subset of all diabetes patients, generally estimated to account for 1–2% of all diabetic cases, is attributed to mutations in single genes. As only a subset of these genes has been identified and fully characterized, there is a dramatic need to understand the pathophysiological impact of genetic determinants on β-cell function and pancreatic development but also on cell replacement therapies. Pluripotent stem cells differentiated along the pancreatic lineage provide a valuable research platform to study such genes. This review summarizes current perspectives in applying this platform to study monogenic diabetes variants.

Keywords: pluripotent stem cells, diabetes, monogenic variants, Maturity Onset of Diabetes in the Young, type 2 diabetes, type 1 diabetes

INTRODUCTION

Diabetes, as one of the major diseases in industrial countries, affects over 350 million people worldwide. Type 1 (T1D) and type 2 diabetes (T2D) are the most common forms. T2D accounts for most diabetes cases and is a multifactorial metabolic disease where insulin deficiency is caused by insulin resistance in target organs and pancreatic β-cell failure. The current diabetes classifications are insufficient to explain the large clinical and biological variability of diabetes, suggesting an unrecognized level of heterogeneity (1). T1D is described as a chronic autoimmune disease against insulin-producing β-cells leading to hyperglycemia. T1D results from the combination of multiple factors, including environment, genes, and a prominent role of the immune system. Genetic studies have long recognized that mutations of the human leukocyte antigens (HLAs) within the Major Histocompatibility Complex (MHC) represent major genetic risk factors in T1D (2, 3). More recently, genome-wide association studies (GWAS) and candidate gene approaches have identified

---

Abbreviations: CRISPR, Clustered regularly interspaced short palindromic repeats; ER, Endoplasmatic reticulum; GWAS, Genome-wide association studies; hESC, Human embryonic stem cell; HLA, Human leucocyte antigen; hPSC, Human pluripotent stem cell; iPSC, Induced pluripotent stem cell; JOD, Juvenile-onset diabetes; KO, Knockout; MHC, Major histocompatibility complex; MODY, Maturity Onset of Diabetes in the Young; PNDM, Permanent neonatal diabetes mellitus; SNP, Single nucleotide polymorphism; T1D, Type 1 diabetes; T2D, Type 2 diabetes.

more than 50 other loci contributing to T1D risk, including *INS, PTPN22, CTLA4*, and *GLIS3* genes (4, 5). In addition, mutations in several genes, such as *AIRE, FOXP3, and STAT* (6, 7), may cause rare monogenic forms of autoimmune diabetes.

Similarly, genetic studies of T2D identified many single nucleotide polymorphisms (SNPs) associated with T2D risk, which are located near several functionally relevant genes such as *PPARG* (8), *WFS1* (9), *KCNJ11* (10), *KLF14* (9), *ANK1* (11), *INS* (12), *HNF1A* (9), *HNF1B* (13), and *GLIS3* (14). In addition to genetic predisposition, environmental factors and epigenetic changes are influencing the pathophysiology of T2D, which may contribute to the additional variance in susceptibility.

Overall, genetic studies of T1D and T2D resulted in the identification of many disease-associated variants, most of which, with the exception of the HLA locus for T1D, contribute to a small increase in disease risk (5, 15, 16). These studies have provided valuable information on putative genes and the mechanisms involved in diabetes. For example, many genes identified by T2D GWAS are expressed in human islets (17) and may regulate β-cell mass and function (18). While a large proportion of T1D susceptibility genes are surprisingly not related to the immune system (19), studies from D. Eizirik's group have also shown that >60% of these genes are expressed in β-cells, and their expression is affected upon exposure to cytokines, viruses, and double-stranded RNA, a by-product of viral infection, in human and rodent β-cells (19–21). Altogether, these studies suggest an essential role of mechanisms acting at the level of β-cells in the etiology of both T1D and T2D (19–21).

Interestingly, several of these genes are involved in both monogenic diabetes (rare variants) and multifactorial diabetes (frequent variants). This is the case of the insulin gene (*INS*, monogenic neonatal diabetes, Maturity Onset of Diabetes in the Young (MODY) and multifactorial T1D), as well as *KCNJ11, WFS1, HNF1A, HNF1B* (neonatal, syndromic, or MODY monogenic diabetes, and multifactorial T2D) and *GLIS3* (neonatal monogenic diabetes and multifactorial T1D and T2D). This plethora of genes involved in common multifactorial and rare monogenic forms of diabetes suggests that some disease mechanisms and biological pathways may be shared between different forms of diabetes. The identification and detailed study of genes responsible for monogenic diabetes are therefore extremely valuable to investigate important genes and pathways involved in both monogenic diabetes and common forms of diabetes. Noteworthy, it has become evident most recently that a subset of all diabetes patients, generally estimated to account for 1–5% of all diabetic cases, is attributed to mutations in single genes (22, 23).

As only a subset of these genes has been identified and fully characterized, there is a dramatic need to understand the pathophysiological impact of genetic determinants on β-cell function and pancreatic development but also on cell replacement therapies. Although islet transplantation can lead to insulin-independency of diabetic patients for 5 years or longer, this therapeutic option is only accessible for a rare number of patients due to the limited number of cadaveric human islets and complex handling (24). On the other hand, the use of human pluripotent stem cells [hPSCs, induced pluripotent (iPSC) and embryonic stem cells (hESC)] may bypass this need by generating mature β-cells *in vitro* upon improving the current protocols of β-cell generation.

PSCs have been used as a relevant model system to elucidate pathophysiological mechanisms in diseases such as diabetes, blood disorders, defined neurological disorders, and genetic liver disease (25–27). Induced pluripotent stem cells (iPSCs) allow dissecting monogenic human disease mechanisms (28) as well as mechanisms of genetically complex human disorders such as schizophrenia (29). This opens promising perspectives in both regenerative medicine but also in drug development to screen for innovative, "druggable" targets (30) and to develop *ex vivo* gene-targeting therapies (28). Given the still high intra- and interpatient variability of patient-derived iPSCs, controls are the key for a precise analysis (31). Recent advances in the development of genomic editing tools such as the Zinc-finger or clustered regularly interspaced short palindromic repeats (CRISPR)/Cas9 technology have further revolutionized this research field. Now researchers can precisely modify a human pluripotent stem cell genome with (i) high efficiency, (ii) on a single-base resolution, (iii) without altering the pluripotent capacity, and (iv) with negligible off-target effects to provide isogenic controls and to facilitate data interpretation. In turn, these recent tools represent novel state-of-the-art disease-in-a-dish models and will pioneer research fields aiming to understand also the mechanisms underlying monogenic diseases (32–35). Human pancreatic disease modeling is highly dependent on reliable and efficient differentiation protocols for human PSCs. We and others have recently challenged the currently existing protocols (36), first, to optimize the step toward pancreatic progenitor cells (37), second, to drive maturation in a 3D environment (38–40), and third, to increase yields of true monohormonal β-cells (41–44). In turn, optimized differentiation platforms now allow for appropriately modeling complex pancreatic diseases such as diabetes (45). A schematic overview of currently available disease modeling tools for diabetes employing hPSCs is presented in **Figure 1**.

MATURITY-ONSET DIABETES OF THE YOUNG

So far, 14 subtypes of maturity-onset diabetes of the young (MODY) have been described to be caused by mainly heterozygous dominant mutations in genes for pancreas-specific transcription factors as well as enzymes, hormones, and ion channels (46, 47). These mutations impair endocrine function at various levels ranging from alterations in development, glucose sensing, synthesis, and storage of insulin to inappropriate secretion of insulin in β-cells. The most frequently identified mutations are located in the *HNF4A* gene (MODY1) (48) with a frequency of 4–10% (49–51), in the *GCK* gene (MODY2) (52) with 30–60% (49–51, 53) and in the *HNF1A* gene (MODY3) (54) with 30–50% (49–51) depending on the study population. MODY1 patients are particularly characterized by defective glucose-stimulated insulin secretion possibly caused by disrupted gene expression playing a role in glucose transport

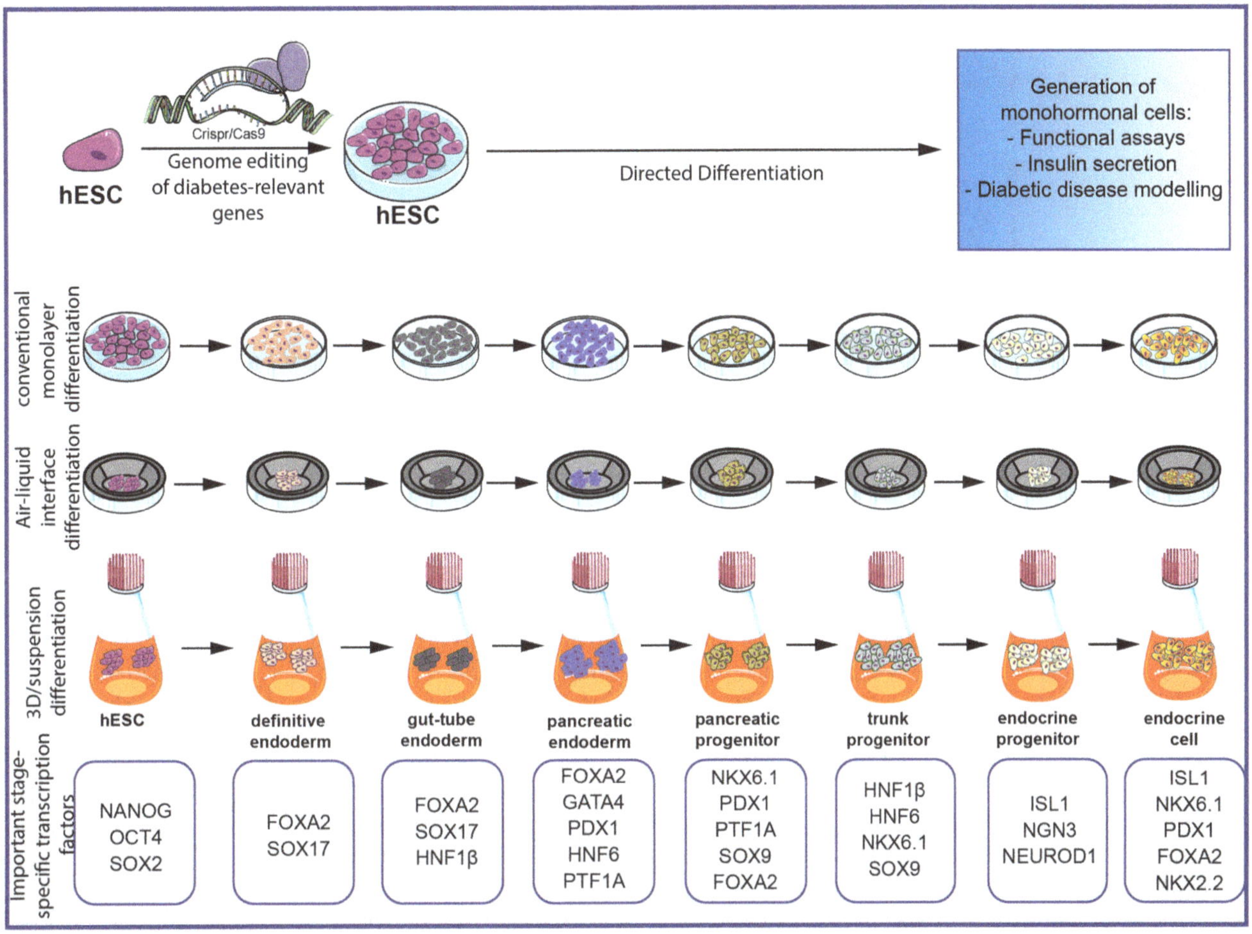

FIGURE 1 | Schematic overview of currently existing protocols for the investigation of diabetes on an hPSC-based platform. Gene editing with CRISPR/Cas9 allows precise editing of diabetes-relevant genes and generation of hESC for further differentiation experiments. Different differentiation protocols allow the generation of monohormonal cells by passing through different milestones during embryonic development. Important stage-specific transcription factors are indicated below the schematics. Subsequent analysis of monohormonal β-cells, including insulin secretions assays, can be performed and generate hypotheses about the influence of specific genetic variants.

and glycolysis (48, 55, 56). MODY2 due to glucokinase deficiency often results in mild hyperglycemia during early life (57). These patients have a defect in glucose-stimulated insulin secretion caused by impaired glucose sensitivity in β-cells. MODY3 patients develop β-cell dysfunction and hyperglycemia caused by impairment of glucose-dependent insulin secretion (58).

In addition, rare MODY cases [accounting for up to 6% of all MODY forms (49)] have been diagnosed with mutations in *PDX1* (59), *HNF1B* (60), *NEUROD1* (61, 62), *KLF11* (63), *CEL* (64), *PAX4* (65), *INS* (66), *BLK* (67), *ABCC8* (68), *KCNJ11* (69), and *APPL1* (70) (known as MODY4-14).

mpaired functions of proteins caused by pathogenic variants can vary depending on the nature of the mutation, therefore causing a spectrum of clinical manifestations. Patients harboring heterozygous *HNF1B* mutations suffer from MODY, but may also feature pancreas exocrine dysfunction as well as kidney and liver abnormalities (71) and vaginal and uterine malformation (72). Few specified cases of *NEUROD1* mutations are characterized mainly by early onset diabetes (61, 62, 73), but patients with neurological defects such as pituitary gland hypoplasia, growth hormone deficiency, epilepsy, and intellectual disability have also been described (74, 75). *CEL* mutations cause early onset diabetes associated with exocrine pancreatic dysfunction and chronic pancreatitis (76, 77). In addition to the diabetic phenotype in patients with *PAX4* mutations, diabetic complications such as retinopathy and nephropathy have been observed (65, 78). MODY-causing *INS* mutations have been associated with early onset diabetes as well as ketoacidosis in some cases (79), whereas rare cases with *BLK* mutations have also been associated with overweight (67).

No other clear clinical manifestation besides diabetes has been described for patients with heterozygous mutations in *KLF11* (MODY7), *ABCC8 (*MODY12), *KCNJ11 (MODY13*), and APPL1 (MODY14). A summary of confirmed MODY-causing mutations as well as of other prominent clinical features is presented in **Table 1**.

TABLE 1 | Different MODY forms, including their frequencies, affected genes, and potential other prominent clinical manifestations are presented.

MODY form	Affected gene	Frequency	Potential prominent additional clinical manifestations besides diabetes and its complications	Affected gene investigated using hESC	Affected gene investigated using hiPSC
MODY1	*HNF4A*	4–10%	Not relevant	No	Yes (80–82)
MODY2	*GCK*	30–60%	Not relevant	No	Yes (80, 83)
MODY3	*HNF1A*	30–50%	Not relevant	Yes (84, 85)	Yes (80)
MODY4	*PDX1*	Rare	Pancreatic agenesis and miscarriages	Yes (86)	Yes (87–89)
MODY5	*HNF1B*	Rare	Exocrine pancreatic dysfunction, kidney and liver abnormalities, vaginal aplasia, and uterus hypoplasia	No	Yes (80, 90, 91)
MODY6	*NEUROD1*	Rare	Neurological defects including pituitary hypoplasia, growth hormone deficiency, epilepsy, and intellectual disability	No	No
MODY7	*KLF11*	Rare	Nothing else described	No	No
MODY8	*CEL*	Rare	Exocrine pancreatic dysfunction, chronic pancreatitis	No	Yes (80)
MODY9	*PAX4*	Rare	Not relevant	Yes (92)	No
MODY10	*INS*	Rare	Not relevant	No	Yes (93)
MODY11	*BLK*	Rare	Overweight	No	No
MODY12	*ABCC8*	Rare	Nothing else described	Yes (94, 95)	No
MODY13	*KCNJ11*	Rare	Nothing else described	Yes (96)	Yes (97)
MODY14	*APPL1*	Rare	Nothing else described	No	No

Furthermore, a statement about the current research of the respective mutations or MODY forms, including hESC and hiPSC, is included.

Noteworthy, the homozygous status for mutations in several MODY genes has been found to lead to extreme clinical presentations, contrasting with the less severe early onset diabetes observed in heterozygous carriers. For example, homozygous mutations in the *PDX1* gene result in early onset diabetes associated with pancreatic agenesis and maternal miscarriages (59, 98).

A subset of patients with MODY-like phenotype doesn't carry any mutation in the known MODY genes, suggesting the involvement of additional genes. The identification of these additional genes responsible for rare MODY forms is now facilitated by the availability of large databases of diabetic cases and control cohorts that enable increased efficiency to detect novel genes with rare contributing variants (including MODY-like effects) (99), compared to earlier studies with smaller sample size (100). In addition, the availability of large databases of control subjects (*e.g.*, gnomAD, TOPMED) provides now the possibility to estimate the frequency of rare coding variants in candidate genes, hence allowing for efficient association and burden-testing for rare monogenic contributions, such as MODY. Consequently, recent studies identified *RFX6* as a novel MODY gene (101) and *WFS1, PPARG*, and *GLIS3* have recently been proposed as potential candidates for these rare MODY forms (101–103).

Taken into consideration the overlap in genes involved in common multifactorial and rare monogenic forms of diabetes, the specific analysis of monogenic pathogenic variants can therefore reveal novel interaction partners and gene targets that might be helpful to better understand the mechanisms involved in the onset of T1D and T2D.

PERMANENT NEONATAL DIABETES MELLITUS

Permanent neonatal diabetes mellitus (PNDM) is the second form of monogenic diabetes. It is characterized by hyperglycemia and partial or complete insulin deficiency in patients in the first 6 months postnatal (104, 105). Moreover, patients with PNDM may suffer from intrauterine growth retardation, glycosuria, ketoacidosis, failure to thrive as well as various clinical features depending on the gene. Mutations in more than 20 genes with monogenic contribution important for β-cell development have been identified to cause PNDM (46). Treatment of PNDM includes oral sulfonylureas or insulin therapy and may require pancreatic enzyme replacement for infants with pancreatic aplasia or hypoplasia.

Some genes, including *ABCC8* (106), *GCK* (107), *INS* (108), *KCNJ11* (109), and *PDX1* (110) may alternatively cause PNMD or MODY, with various severity and clinical features depending on the gene, nature of the mutation, and genotype (homozygous or heterozygous). Common variants in these genes may also be associated with multifactorial T1D or T2D.

In addition to diabetes, *KCNJ11*-PNMD patients may also have neurological features such as developmental delay and epilepsy (DEND syndrome) (111). Similarly, pathogenic homozygous *PDX1*-PNDM patients have pancreatic agenesis and pancreatic hypoplasia leading to exocrine pancreatic insufficiency (110, 112). Pancreatic agenesis is furthermore caused by homozygous mutations in another pancreatic transcription factor, *PTF1A* (113). Here, PNDM patients additionally suffer from severe intrauterine growth retardation, cerebellar agenesis, and neurological dysfunction.

In addition, PNDM may manifest in the context of specific syndromes. Homozygous mutations in *EIF2AK3* cause Wolcott–Rallison syndrome, characterized by PNDM, exocrine pancreas dysfunction, and abnormalities such as liver failure, developmental delay, and epiphyseal dysplasia (114). Inactivating *GATA4* variants can induce pancreatic agenesis or hypoplasia, causing PNDM but also lead to extrapancreatic symptoms such as cardiac and neurodevelopmental abnormalities (115). Similarly, *GATA6* mutations cause pancreatic agenesis leading to PNDM, together with abnormalities of the heart, biliary tract, and gut development

(116). Homozygous mutations in *GLIS3* cause PNDM together with congenital hypothyroidism associated with congenital glaucoma, hepatic fibrosis, and polycystic kidneys (117). In addition to neonatal diabetes, *NEUROD1* mutations cause cerebellar hypoplasia, sensorineural deafness, and visual impairment (118), whereas *NEUROG3* mutations affect intestinal development leading to congenital malabsorptive diarrhea (119–121). Mutations in *PAX6*, encoding a transcription factor involved in *β*-cell development as well as eye and brain development, cause neonatal diabetes combined with abnormalities of the central nervous system and visual system (*e.g.* microencephaly, optic nerve defects, microphthalmia) (122). In addition to PNDM, mutations in the transcription factor *RFX6* cause pancreatic hypoplasia, intestinal atresia, and gall bladder hypoplasia (123, 124). Patients with Wolfram syndrome caused by mutations in *WFS1* suffer from early onset diabetes as well as optic atrophy, deafness, ataxia, and dementia (125). Other neonatal diabetes syndromes have been described for mutations in *SLC19A2* (associated with thiamin-responsive megaloblastic anemia, neurological disorders, cardiac abnormalities, and deafness) (126, 127), *MNX1* (associated with growth retardation, delayed central nervous system development, hypoplastic lungs, renal maldevelopment, skeletal dysplasia) (128), *NKX2.2* (further leading to growth retardation, delayed central nervous system development, constipation) (128), and *IER3IP1* (additional microcephaly, CNS maldevelopment) (129). Furthermore, some mutations in the glucose transporter *SLC2A2* can cause neonatal diabetes prior to the Fanconi–Bickel syndrome associated with glycosuria, galactosemia, aminoaciduria, proteinuria, hepatomegaly, as well as glucose and galactose intolerance (130, 131). Franco et al. recently showed that mutations in *YIPF5* cause neonatal diabetes associated with microcephaly and epilepsy (132).

Interestingly, an overlapping phenotype between PNDM and autoimmune T1D was observed for a patient with an activating mutation in the *STAT3* gene (133). Although the autoimmune-mediated destruction of *β*-cells was prominent Saarimäki-Vire et al. revealed an additional mechanism (PNDM) due to the observed pancreatic hypoplasia (134).

An overview of all described genes leading to the development of PNDM if affected by mutations is presented in **Table 2**.

MODELING PANCREATIC ENDOCRINE DEVELOPMENT

The detailed pathomechanisms of monogenic diabetes are not yet fully understood since mouse models do not completely recapitulate the human disease phenotype (121, 144, 145), and patient samples such as *β*-cells have very limited availability. Moreover, animal models with a specific knockout of MODY genes show species-specific differences that do not entirely recapitulate the patient phenotype (146–150). Therefore, even more suitable disease models are crucial to develop an adequate therapy.

In the recent years, human pluripotent stem cells have been deployed as a suitable human model system. On the one hand, human embryonic stem cells (hESCs) can be subjected to a directed differentiation protocol to investigate different mechanisms during differentiation of mature pancreatic *β*-cells. Additionally, patient-specific induced pluripotent stem cells (iPSCs) can be generated from materials such as fibroblasts and keratinocytes, allowing to address various genetic backgrounds of patients. Of note, gene-mutated iPSCs show high heterogeneity in terms of differentiation efficiency and are best controlled with isogenic, repaired lines. Furthermore, patient-specific iPSCs are a useful tool for biobanking because of their unlimited expansion capacity. Subsequently, these cells are differentiated *in vitro* into disease-relevant cell types such as pancreatic endocrine cells or β-cells.

To better understand the importance of certain genes in the maturation of *β*-cells, genetic engineering may be performed with hESCs and iPSCs. Here, state-of-the-art gene-editing tools such as zinc-finger nucleases (ZFNs), transcription activator-like effector nucleases (TALENs), and the more recent clustered regularly interspaced short palindromic repeats (CRISPRs)/Cas allow the generation of specific point mutations or gene knockouts (KOs). A potential option of genetic engineering might involve gene correction in iPSCs. This allows the generation of autologous *β*-cells for transplantation that may circumvent immune reaction and donor scarcity.

The generation of pancreatic endocrine cells is achieved by different differentiation protocols. Established protocols try to mimic signaling pathways of *in vivo* embryonic developmental stages by involving different/various combinations of growth factors, cytokines, and small molecules known/reported to guide the stem cells through stages of definitive endoderm, gut-tube endoderm, pancreatic endoderm, pancreatic progenitors, and endocrine progenitors to finally yield mature mono-hormonal endocrine cells (**Figure 1**).

Overall differentiation models of hESCs and hiPSCs provide a versatile tool to study the influence of genetic disorders on *β*-cell development in the human pancreas as well as the embryonic development of the human pancreas itself. Different published protocols whose concepts are described below allow the investigation of multiple facets of *β*-cell maturation during different steps of embryonic development. However, the procedure itself may have a huge influence on the phenotype of differentiated cells. This might lead to the bias that a "good" differentiation protocol can overcome the inherent genotypic features, which would rather underestimate the phenotypic features of a certain genotype. This is, for example, the case for a *GATA6*-mutant iPSC cell line with a severe phenotypic loss of endoderm and *β*-cell differentiation capacity in a differentiation condition involving low levels of retinoic acid. On the other hand, high levels of retinoic acid mask this phenotype (136). The same principle might apply to a "poor" differentiation protocol that may overestimate the phenotypic properties of a certain genotype.

GENERATION OF MATURE PANCREATIC *β*-CELLS REQUIRES COMPLEX AND SOPHISTICATED DIFFERENTIATION PROTOCOLS

During the last decade, differentiation protocols have been adapted to achieve a more mature state of *in vitro* differentiated *β*-cells. Since

TABLE 2 | Overview of mutations in genes that can lead to PNDM.

Affected gene in PNDM	Affected gene also described for MODY	Part of a syndromic phenotype	Potential prominent additional clinical manifestations	Affected gene investigated in hESC	Affected gene investigated in hiPSC
ABCC8	MODY12	No	Nothing else described	Yes (94, 95)	No
EIF2AK3	No	Yes, Wolcott-Rallison syndrome	Exocrine dysfunction, acute liver failure, developmental delay, epiphyseal dysplasia	No	No
GATA4	No	Yes	Pancreatic agenesis, cardiac, and neurodevelopmental abnormalities	No	No
GATA6	No	Yes	Pancreatic agenesis, abnormalities of heart, biliary tract, and gut development	Yes (135, 136)	Yes (135, 136)
GCK	MODY2	No	Not relevant	No	Yes (80, 83)
GLIS3	Potentially	Yes	Congenital hypothyroidism, congenital glaucoma, hepatic fibrosis, polycystic kidneys, pancreatic exocrine insufficiency, kidney, liver, and biliary dysfunction	Yes (86)	No
IER3IP1	No	Yes	Microcephaly, CNS maldevelopment	No	No
INS	MODY10	No	Not relevant	No	Yes (93, 137, 138)
KCNJ11	MODY13	No	Nothing else described	Yes (96)	Yes (97)
MNX1	No	Yes	Growth retardation, delayed central nervous system development, hypoplastic lungs, renal maldevelopment, skeletal dysplasia	Yes (86)	No
NEUROD1	MODY6	Yes	Cerebellar hypoplasia, sensorineural deafness, visual impairment	No	No
NEUROG3/ NGN3	No	Yes	Intestinal maldevelopment with malabsorptive diarrhea	Yes (86, 139)	No
NKX2.2	No	Yes	growth retardation, delayed central nervous system development, constipation	No	No
PAX6	No	Yes	Abnormalities of the central nervous system and visual system including microencephaly, optic nerve defects, microphthalmia	No	No
PDX1	MODY4	Yes	Pancreatic agenesis and miscarriages	Yes (86)	Yes (87–89)
PTF1A	No	Yes	Intrauterine growth retardation, pancreatic agenesis, cerebellar agenesis, and neurological dysfunction	Yes (86)	No
RFX6	Potentially	Yes, Mitchell–Riley syndrome	Pancreatic hypoplasia, intestinal atresia, and gall bladder hypoplasia	Yes (86)	Yes (140)
SLC19A2	No	Yes, Thiamine-responsive megaloblastic anemia	Megaloblastic anemia, hearing loss, neurological disorders, cardiac abnormalities	No	No
SLC2A2	No	Yes, Fanconi–Bickel syndrome	Glycosuria, galactosemia, aminoaciduria, proteinuria, hepatomegaly, glucose intolerance, galactose intolerance (130, 131)	No	Yes (141)
STAT3	No	Potentially correlated to autoimmune diabetes	Strong autoimmune component of diabetes, pancreatic hypoplasia	No	Yes (134)
YIPF5	No	Yes	Microcephaly, epilepsy	Yes (132)	Yes (132)
WFS1	Potentially	Yes, Wolfram syndrome	Optic atrophy, deafness, ataxia, and dementia	No	Yes (142, 143)

Additional information of mutations leading to syndromic phenotypes and their characteristics is included. Further statements regarding the relationship to MODY forms and feasibility of hPSC to model the disease development and progression in terms of PNDM are included.

earlier pancreatic endocrine differentiation protocols in monolayers yield mainly an immature or heterogeneous population of polyhormonal cells lacking robust insulin secretion in response to glucose stimulation, a prerequisite of β-cells (36, 151), novel *in vitro* approaches including different culture conditions have been established. Recent protocols include a transition to 3D culture using a suspension culture system with spinner flasks and orbital shaker generating endocrine spheres (41, 42). Alternatively, a switch from the initial culture in monolayer to an air–liquid interphase culture stage promoting basal and apical cell polarity generates even more functionally mature β-cells (152) (**Figure 1**).

Air–liquid interphase culture systems require spotting the cells from the pancreatic endoderm stage on filters. Upon formation of small cell clusters, the differentiation is further improved as measured by *NGN3* and insulin expression (43). This transition might help to mimic the natural 3D environment and cell orientation within the developing tissue, thus promoting *in vitro* differentiation (153).

Further progress in β-cell maturation was also achieved by reaggregating immature cells into enriched β-cell clusters using an insulin-driven fluorescence reporter (154). Veres et al. combined cellular reaggregation and β-cell purification using CD49a to enrich endocrine cells and promote functional maturation of β-cells able to maintain their identity for several weeks in culture (155).

An alternative approach is to enrich precursor cells in the differentiation process. A recent study demonstrated that enrichment of anterior definitive endoderm with CD177 results in a more homogenous pancreatic progenitor population and subsequent better functional maturation (156).

Moreover, besides optimizing technical conditions for differentiation, the modulation of signaling pathways and cytoskeleton is a promising mean to increase the β-cell yield. Inhibition of certain pathways such as ROCKII and WNT using specific inhibitors promoted maturation (157, 158). Hogrebe et al. investigated the role of the actin cytoskeleton in promoting the expression of pancreas-specific transcription factors such as NEUROG3 during differentiation (159). The manipulation of actin polymerization during early developmental stages influences the expression of transcription factors important for the specification of lineage fate in pancreatic progenitors. Depolymerization of the cytoskeleton during endocrine induction further improved the functionality of derived β-cells, also allowing for a planar differentiation protocol.

Another strategy to improve differentiation efficacy is to modulate the basic content of cell culture media. Two studies explored metabolic changes during β-cell maturation (160, 161). Adaptation of nutrient-sensing *via* mTORC1 signaling during the transition from fetal to adult pancreatic islets can be recapitulated by reduced amino acid content in differentiation media, further advancing cellular insulin content and glucose-stimulated insulin secretion (160). Similarly, epigenomic characterization of primary and *in vitro* differentiated pancreatic cells revealed that entrainment to cycles of fasting and feeding leads to circadian control of genes important for energy and insulin metabolism, further improving β-cell function (161).

Protocol improvements resulted in more mature β-cells and faster reversal of diabetes after transplantation in mice. However, manifold successful approaches show that regulation of human pancreas development is still not fully understood, and various adaptations to endocrine differentiation protocols are difficult to compare because they use different cell lines and culture methods as well as slightly different functional assays. Therefore, more research is necessary to determine the appropriate combination of culture methods, cytokine and small molecule cocktails, purification markers, and metabolic modifications, generating a protocol that robustly produces the desired pancreatic cell types.

For potential clinical use, several questions/issues regarding the composition of transplanted cells containing only β-cells or more than one endocrine cell type, the best transplantation site, and whether transplanted cells benefit from co-transplantation with other cell types such as mesenchymal stem cells need to be answered. Moreover, long-term survival and functionality of transplanted cells exceeding the life span of mice have to be addressed. Additionally, the possibility of teratoma formation from the remaining progenitor cells even after prolonged time has to be eliminated. Encapsulation of cells in suitable biomaterials such as alginates or synthetic polymer hydrogels might not only reduce the risk of tumor formation but also provide protection from the immune rejection of the host, removing the need for lifelong immunosuppression.

Although the latest research significantly improved our knowledge about transcriptional regulation, signaling pathways as well as metabolic adaptation during *in vitro* differentiation and maturation of β-cells and paved the way for future clinical use, more research is necessary until *in vitro*-generated pancreatic endocrine cells can be used as potential diabetes therapy.

PLURIPOTENT STEM CELL MODELS TO UNDERSTAND MONOGENIC DIABETES

Although iPSC could be successfully generated from T1D and T2D patients, complex autoimmune reactions, environmental influence, as well as multifactorial genetic factors hampered the intimate recapitulation of pathogenesis (162–164). Despite the complexity of T1D and T2D pathogenesis, some recent approaches have been performed to model T1D using pluripotent stem cells. Co-culture studies of iPSCs derived from T1D patients together with immune cells are one such way to model the mechanisms of T1D *in vitro* (165). Yet, it has to be kept in mind that this kind of model requires additional prerequisites such as environmental factors and complex composition of different immune cell types as recently reviewed by Joshi et al. (166). Modeling T2D *in vitro* is far more complex as many more different pathogenic mechanisms can cause or even interact to promote T2D development, including multiple genetic and environmental factors. This fact makes it even harder to investigate T2D by pluripotent stem cell-based approaches solely *in vitro*.

Whereas modeling T1D and T2D using PSCs remains challenging due to their complex nature, monogenic alterations leading to a MODY or PNMD diabetes phenotype are ideal to be investigated by PSC-based approaches. The role of specific variants of the respective genes has already been investigated using pancreatic differentiation of pluripotent stem cells. Compared to various genetic and environmental aspects contributing to other diabetes types, single mutations present in monogenic diabetes allow tighter control of the observed phenotypes. Deciphered mechanisms for the development of MODY, which were uncovered using hPSC-based systems, are presented in **Figure 2** (MODY1, 2, 3, 4, 5, 10, 13).

Teo et al. showed that karyotypically normal iPSC expressing pluripotency markers and able to differentiate in all three germ layers could be derived from different MODY patients (MODY1, MODY2, MODY3, MODY5, and MODY8) (80) and can serve as a tool to study the role of the respective genes in pancreatic development. A more detailed study of these MODY1 iPSCs with premature HNF4A protein truncation revealed impaired foregut and hepatopancreatic progenitor development. These events were associated with HNF4A mislocalization and reduced expression of target genes such as the *FOXA* gene family, *HNF1B*, *PDX1*, *GATA4*, and *RFX6* (81). In turn, impaired activation of target genes disturbs β-cell gene signatures. Prior to that study, iPSCs from MODY1 patients with a nonsense mutation were characterized (82). Here, the patient phenotype was caused by a reduction in levels of functional HNF4A accompanied by increased expression of pancreatic transcription factors and pancreatic hormones as a compensatory mechanism (82).

Understanding the role of transcription factor HNF1A in MODY3, human ESC lacking one or both alleles have been differentiated to study endocrine development (84). HNF1A deficiency increased expression of markers for α-cells but reduced expression of β-cell markers suggesting a role in endocrine hormone expression. In addition, HNF1A is required for insulin secretion, in line with hyperglycemia observed in patients. Moreover, mutated

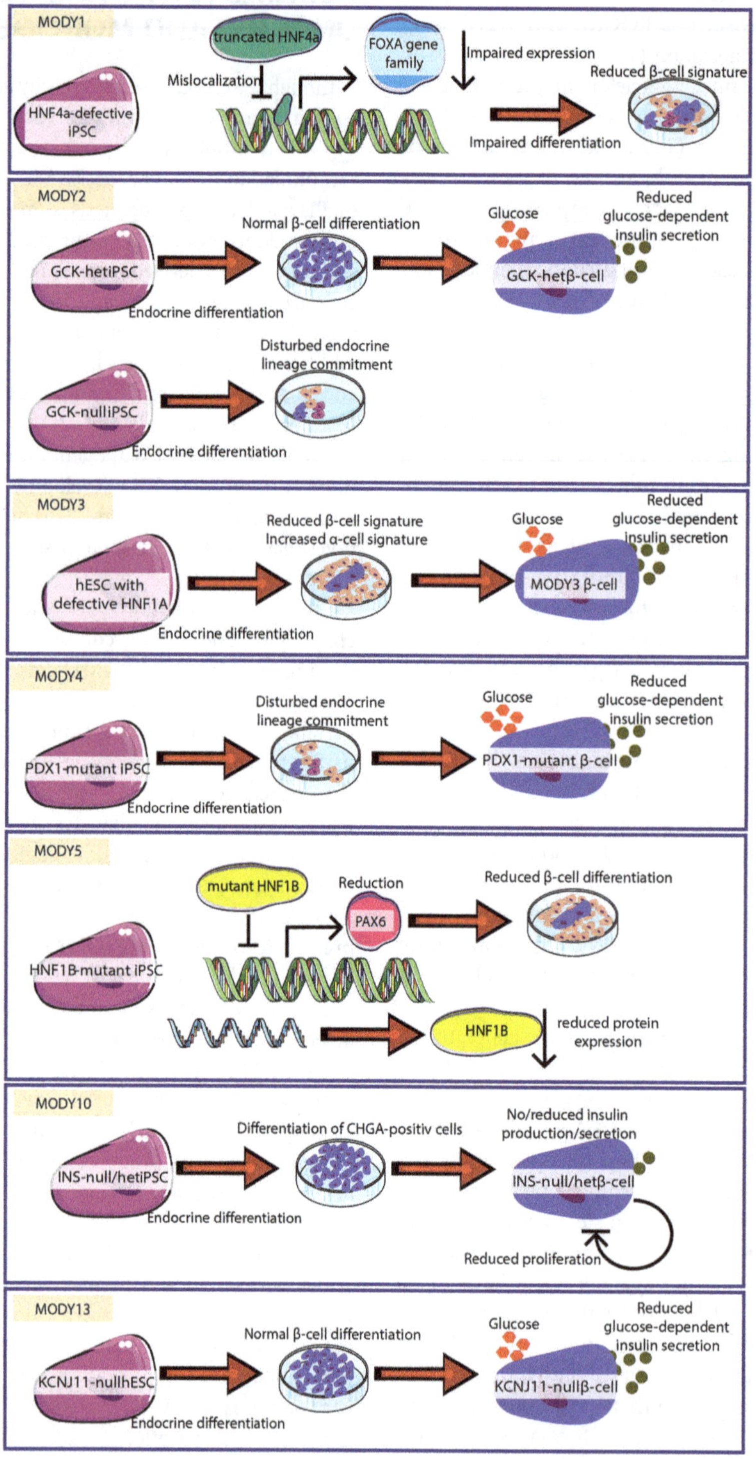

FIGURE 2 | Successful uncovering of pathomechanisms for different MODY forms. Different MODY forms were modeled by employing hESC/hiPSC with respective mutations. Mechanisms leading to monogenic diabetes could be delineated/characterized. In MODY1, mutated HNF4a leads to reduced FOXA gene family expression and impaired *β*-cell signature. MODY2 is characterized by reduced differentiation or reduced glucose-dependent insulin secretion. MODY3 is caused by reduced *β*-cell differentiation and insulin secretion. MODY4 shows reduced endocrine lineage entrance and impaired insulin secretion. MODY5 is caused by diminished *β*-cell differentiation. MODY10 is highlighted by lacking production and secretion of insulin. MODY13 is characterized by impaired glucose-dependent insulin secretion.

cells show metabolic defects in glycolysis and mitochondrial respiration, also typical for T2D. This observation, together with the finding that frequent *HNF1A* variants are associated with T2D, suggests a link between mechanisms identified in MODY and in common T2D. Furthermore, iPSCs from a patient harboring heterozygous *HNF1A* mutation generated by non-integrative viral transduction, show a normal karyotype, and express pluripotency factors (85). These cells can be used for further functional analysis of this specific mutation.

Relevant defects in the *GATA6* gene disrupt the endoderm differentiation by decreased cell survival. Later, the pancreas specification and β-cell function were identified using iPSC and genome-edited ESC (135, 136). By circumventing the developmental block at the endoderm stage, cell lines with *GATA6* mutations were differentiated with low dose retinoic acid to mimic severe patient phenotype. These cells failed to show normal insulin secretion after glucose stimulation and harbor defective insulin processing (136).

Fibroblasts from patients with heterozygous point mutations in the *PDX1* transactivation domain were successfully reprogrammed to iPSCs and can be used to study diabetes-associated pathomechanisms (87–89). Further pancreatic differentiation reveals that mutations in the *PDX1* transactivation domain disturb the pancreatic endocrine lineage development and result in impairment of the glucose-responsive function of β-cells (88).

The analysis of patient-derived iPSCs with *HNF1B* mutations (MODY5) suffering from early onset diabetes and pancreatic hypoplasia revealed a compensatory increase in markers of definitive endoderm and pancreatic transcription factor expression such as PDX1 (90). Additionally, downregulation of transcription factor PAX6, important for islet development (167), may result in the observed patient phenotype (90). Furthermore, iPSCs were generated from a Japanese MODY5 patient with a truncated *HNF1B* variant in order to account for differences in insulin sensitivity and insulin response depending on the genetic background (91). Yabe et al. compared iPSCs derived from healthy and patient skin fibroblasts and detected degradation of mutant mRNA by the nonsense-mediated decay pathway in differentiated patient-derived iPSCs (91).

A more systematic analysis of pancreatic transcription factors *PDX1*, *RFX6*, *PTF1A*, *GLIS3*, *MNX1*, *NGN3*, *HES1*, and *ARX*, partly identified in monogenic variants of MODY and PNDM, characterized the transcriptional control and corresponding defects at several developmental stages (86). This study highlights especially the role of RFX6 in controlling pancreatic progenitor numbers and differences of NEUROG3 requirement in humans and mice. Mutations in *WFS1*, causing Wolfram syndrome, lead to chronic endoplasmic reticulum stress activating the unfolded protein response, which impairs survival of β-cells and neurons (168–170). This could be recapitulated using iPSC with *WFS1* variants (142). A recent publication from Maxwell et al. also characterized iPSC from a Wolfram syndrome patient (143). Patient-specific iPSCs harboring a pathogenic variant of *WFS1* were corrected using CRISPR/Cas9 technology. This study used a differentiation protocol with cytoskeletal modification, which significantly improved differentiation efficiency compared to previously tested suspension culture in these cell lines (159). Corrected cells showed higher WFS1 expression and robust insulin secretion, probably benefiting from reduced ER stress and improved mitochondrial respiratory capacity in endocrine cells (143). In turn, further maturation of *in vitro* generated β-cells allows better identification of effects also in later stages, additionally providing potential use for β-cell replacement therapy.

Homozygous mutations in the insulin gene (*INS*) are known to lead to PNDM. Pancreatic differentiation of patient iPSC results in CHGA-positive endocrine cells expressing β-cell markers NKX6.1, PDX1, and MAFA but lacking insulin expression (93). Gene correction rescued the phenotype and prevented diabetes in a streptozotocin mouse model, providing a future tool for patient cell therapy. Another study involved iPSCs generated from patients with neonatal diabetes and heterozygous insulin mutations disturbing proper proinsulin folding (137). Patient-derived iPSCs show normal pancreatic differentiation comparable to corrected isogenic iPSC but have reduced insulin expression. Moreover, *INS* mutation increases ER stress and hampers proliferation of β-cells but without increased apoptosis promoting diabetes development in patients. In addition, fibroblasts from a PNDM patient harboring an intronic *INS* mutation have been efficiently generated and may serve as a diabetes model to characterize the expected aberrant splicing (138).

Another study characterized iPSCs from MODY2 patients with a heterozygous *GCK* mutation (83). Similar to control cells, these iPSCs differentiated into insulin-producing β-cells but showed reduced insulin secretion in response to glucose stimulation. In addition, iPSCs with two inactive *GCK* alleles also showed reduced differentiation efficiency recapitulating the functional impairment observed in patients and mouse models.

After the identification of *YIPF5* mutations causing a novel PNDM syndrome, Franco et al. characterized *in vitro* differentiated patient-derived iPSCs harboring a homozygous *YIPF5* mutation as well as genome-edited ESC in addition to a β-cell line (132). Functional impairment of YIPF5, responsible for trafficking between the endoplasmic reticulum and the Golgi apparatus, caused proinsulin retention at the ER resulting in ER stress-induced apoptosis and β-cell failure and, thus, diabetes.

In order to understand the role of activating mutations in *STAT3* during pancreatic development, iPSCs derived from patient fibroblasts were subjected to pancreatic differentiation and revealed a premature endocrine differentiation later preferentially forming α-cells which is in line with the observed phenotype of the patient (134). This defect results from enhanced nuclear localization of the mutant protein and NEUROG3 activation and could be rescued by correction of the *STAT3* mutation.

Inactivating mutations in *ABCC8* resulting in excess insulin secretion have been successfully employed for modeling congenital hyperinsulinism using ABCC8-deficient ESC (94, 95). In contrast, activating mutations in *ABCC8* have been described in diabetes (171). Thus, using the hPSC-systems, a better characterization of the components of the β-cell ATP-sensitive potassium channel may be obtained to understand the function of β-cells and associated pathomechanisms such as diabetes (caused by activating mutations) or the *vice-versa* effect of congenital hyperinsulinism (caused by inactivating mutations).

In the context of T2D susceptibility genes identified in GWAS, Zeng et al. generated ESC with null alleles for *KCNJ11*, also associated with MODY13 (96). Although the loss of KCNJ11 does not affect *in vitro* differentiation towards *β*-like cells and insulin production, these cells show impaired glucose-stimulated insulin secretion (96). In addition to hESCs with *KCNJ11* mutations, iPSCs have been generated from peripheral blood mononuclear cells with a heterozygous activating mutation in *KCNJ11* and are available for mechanistic studies as well as drug testing in differentiated pancreatic cells (97).

Skin fibroblasts from patients with Michell–Riley syndrome were used to generate iPSCs harboring a homozygous nonsense mutation in *RFX6* (140). Pancreatic differentiation revealed an impaired formation of pancreatic endoderm and thus, supports the impaired formation of endocrine cells in the pancreas in line with the patient phenotype.

In order to better understand the impact of different pathogenic *NEUROG3* variants, Zhang et al. expressed NEUROG3 mutant proteins at physiological levels in *NEUROG3* knockout ESC during pancreatic and intestinal differentiation determining the ability to rescue the generation of endocrine cells (139). Depending on the variant, expression resulted either in the decreased or abolished formation of pancreatic endocrine cells recapitulating the respective patient phenotype. Moreover, these effects could be retraced to be caused by impairment of NEUROG3 protein stability, DNA-binding affinity, and protein dimerization. Those features can differ in various tissues, a fact that emphasizes the importance of considering the relationship between protein structure and function.

Adenoviral *PAX4* overexpression during pancreatic differentiation of ESCs results in decreased glucagon-positive cells promoting the formation of monohormonal insulin-positive cells supporting its role in cell fate specification (92). This suggests a crucial role of intact *PAX4* in the development of healthy monohormonal insulin-positive cells.

A recent study reported the generation of iPSCs with a homozygous mutation in the *SLC2A2* gene (141). Peripheral mononuclear blood cells from a patient suffering from Fanconi–Bickel syndrome accompanied by early onset diabetes were reprogrammed using a non-integrating Sendai virus vector. These cells can be used to study the pathogenesis associated with defects in the GLUT2 glucose transporter in pancreatic *β*-cells.

In addition to KO mouse models, genome-engineered and patient-specific hPSCs have helped to get more insight into developmental and mechanistic processes as well as transcriptional networks (81). Not only do they provide additional information but sometimes even highlight the species-specific differences that make them even more crucial for a better understanding of monogenic diabetes.

A conditional *Hnf4a* KO in mice did not result in a diabetic phenotype but revealed that expression of the potassium channel subunit Kir6.2 regulating insulin secretion is promoted by *Hnf4a* (148). Patient-specific iPSCs inform about changes in transcriptional network (81). Similarly, *Hnf1a* mouse models with heterozygous KO present without developing diabetes, whereas a homozygous KO impairs *β*-cell function by reducing the insulin secretion. Stem cell-based models provided further details characterizing the developmental, transcriptional, and metabolic role of HNF1a (84). Furthermore, loss of *Wfs1* in mice showed impairment of *β*-cell development and function with a mild diabetic phenotype (149). Patient-specific iPSCs provided deeper insight into endoplasmatic reticulum stress and were used to test a possible therapeutic approach (142). Since GATA6 haploinsufficiency resulting in pancreatic agenesis in patients cannot be recapitulated in mice (172, 173), effects of GATA6 gene dosage on pancreatic differentiation *in vitro* helped in understanding the clinical presentation of different patients (135). Furthermore, hPSC models facilitate the characterization of disease-specific variants. For example, a *Hnf1b* heterozygous KO in mice is not associated with a diabetic phenotype (150), but patient-specific iPSC carrying a heterozygous variant of *HNF1B* helped to explain the MODY5 phenotype (90). In mice, *Neurog3* is essential for the development of the endocrine pancreas (144), but the disease phenotype slightly differs in humans (121). Expression of different disease-associated *NEUROG3* variants during *in vitro* differentiation helped in explaining various phenotypes in patients (139).

Taken together, a comparison of human and mouse model systems can provide further insight into the role of specific genes but also highlights the species-specific differences concerning, for example, transcription factor activity. That explains why human PSC-based models are crucially needed to compensate for those specific differences.

Therefore, genome editing in PSC or even patient-specific hiPSC provide a versatile approach to study developmental and functional effects of selected diabetes genes and variants and complement or even contradict data obtained from mouse models.

So far, many stem-cell-based models exist that characterize monogenic mutations resulting in early-onset diabetes. These models nicely elucidate the diabetic patient phenotype and help in understanding the common pathways in *β*-cell development and function. Altogether, improved screening for pathogenic variants in combination with thorough functional analysis will be the first step to precision medicine in diabetes therapy.

CONCLUSION/OUTLOOK

In vitro pancreatic differentiation of pluripotent stem cells is a powerful tool to better understand pancreatic development and the specific role of the involved transcription factors. Identification and characterization of specific variants in monogenic diabetes help in characterizing the complex transcriptional network and in overcoming phenotypic differences between patients and corresponding mouse models. In addition, these model systems provide the basis for drug development and testing that could benefit both patients of monogenic and multifactorial diabetes. Ideally, iPSCs of genetically disordered persons could be repaired and serve as a major source for tissue engineering and regeneration, *e.g. β*-cells in the case of monogenic diabetes.

Yet, some major roadblocks need to be kept in mind before translating genetically repaired iPSCs into clinics. First, epigenetic modifications of iPSCs which are derived from their originating tissue, might reduce their differentiation capacity and subsequent function as well as immune tolerance after autologous transplantation (174). Furthermore, genetic aberrations after reprogramming might bear tumorigenic potential and thus provoke carcinogenesis in the transplanted iPSC-derived tissues (174). Additionally, iPSC-derived tissues need to be manufactured according to SOPs and GMP guidelines which need lots of effort to implement those prerequisites into standard clinical care (175).

Altogether, these studies further urge the involvement of pluripotent stem cells in deciphering the underlying pathomechanisms as well as the affected genes, particularly when monogenic diabetes displays discrete clinical phenotypes and needs specific treatment depending on the subtype.

AUTHOR CONTRIBUTIONS

All authors wrote the paper. MM constructed figures. All authors contributed to the article and approved the submitted version.

FUNDING

ANR-DFG collaborative research project (ANR-18-CE92-0031, DFG KL 2544/5-1) to CJ and AK; DFG funding with the identifiers KL 2544/6-1 ("Heisenberg-Programm"), KL 2544/7-1 ("Sachbeihilfe") and KL 2544/1-2 (HEIST RTG) to AK; Agence Nationale pour la Recherche (ANR-09-GENO-021), the European Foundation for the Study of Diabetes/JDRF/Novo Nordisk, the Assistance Publique-Hôpitaux de Paris Programme Hospitalier de Recherche Clinique, and France Génomique (project DIAPED) to CJ. Novo Nordisk was not involved in the study design, collection, analysis, interpretation of data, the writing of this article or the decision to submit it for publication. MM received funding from Ulm University in the Clinician Scientist Program.

REFERENCES

1. Merger SR, Leslie RD, Boehm BO. The Broad Clinical Phenotype of Type 1 Diabetes At Presentation. *Diabetes Med* (2013) 30:170–8. doi: 10.1111/dme.12048
2. Nejentsev S, Howson JM, Walker NM, Szeszko J, Field SF, Stevens HE, et al. Localization of Type 1 Diabetes Susceptibility to the MHC Class I Genes HLA-B and HLA-A. *Nature* (2007) 450:887–92. doi: 10.1038/nature06406
3. Nerup J, Platz P, Andersen OO, Christy M, Lyngsøe J, Poulsen J, et al. HL-a Antigens and Diabetes Mellitus. *Lancet* (1974) 304:864–6. doi: 10.1016/S0140-6736(74)91201-X
4. Onengut-Gumuscu S, Chen WM, Burren O, Cooper NJ, Quinlan AR, Mychaleckyj JC, et al. Fine Mapping of Type 1 Diabetes Susceptibility Loci and Evidence for Colocalization of Causal Variants With Lymphoid Gene Enhancers. *Nat Genet* (2015) 47:381–6. doi: 10.1038/ng.3245
5. Barrett JC, Clayton DG, Concannon P, Akolkar B, Cooper JD, Erlich HA, et al. Genome-Wide Association Study and Meta-Analysis Find That Over 40 Loci Affect Risk of Type 1 Diabetes. *Nat Genet* (2009) 41:703–7. doi: 10.1038/ng.381
6. Fabbri M, Frixou M, Degano M, Fousteri G. Type 1 Diabetes in STAT Protein Family Mutations: Regulating the Th17/Treg Equilibrium and Beyond. *Diabetes* (2019) 68:258–65. doi: 10.2337/db18-0627
7. Hattersley AT, Patel KA. Precision Diabetes: Learning From Monogenic Diabetes. *Diabetologia* (2017) 60:769–77. doi: 10.1007/s00125-017-4226-2
8. Mahajan A, Go MJ, Zhang W, Below JE, Gaulton KJ, Ferreira T, et al. Genome-Wide Trans-Ancestry Meta-Analysis Provides Insight Into the Genetic Architecture of Type 2 Diabetes Susceptibility. *Nat Genet* (2014) 46:234–44. doi: 10.1038/ng.2897
9. Voight BF, Scott LJ, Steinthorsdottir V, Morris AP, Dina C, Welch RP, et al. Twelve Type 2 Diabetes Susceptibility Loci Identified Through Large-Scale Association Analysis. *Nat Genet* (2010) 42:579. doi: 10.1038/ng.609
10. Zeggini E, Weedon MN, Lindgren CM, Frayling TM, Elliott KS, Lango H, et al. Replication of Genome-Wide Association Signals in UK Samples Reveals Risk Loci for Type 2 Diabetes. *Science* (2007) 316:1336–41. doi: 10.1126/science.1142364
11. Imamura M, Maeda S, Yamauchi T, Hara K, Yasuda K, Morizono T, et al. A Single-Nucleotide Polymorphism in ANK1 is Associated With Susceptibility to Type 2 Diabetes in Japanese Populations. *Hum Mol Genet* (2012) 21:3042–9. doi: 10.1093/hmg/dds113
12. Ng MC, Shriner D, Chen BH, Li J, Chen W-M, Guo X, et al. Meta-Analysis of Genome-Wide Association Studies in African Americans Provides Insights Into the Genetic Architecture of Type 2 Diabetes. *PloS Genet* (2014) 10:e1004517. doi: 10.1371/journal.pgen.1004517
13. Li H, Gan W, Lu L, Dong X, Han X, Hu C, et al. A Genome-Wide Association Study Identifies GRK5 and RASGRP1 as Type 2 Diabetes Loci in Chinese Hans. *Diabetes* (2013) 62:291–8. doi: 10.2337/db12-0454
14. Cho YS, Chen C-H, Hu C, Long J, Ong RTH, Sim X, et al. Meta-Analysis of Genome-Wide Association Studies Identifies Eight New Loci for Type 2 Diabetes in East Asians. *Nat Genet* (2012) 44:67. doi: 10.1038/ng.1019
15. Dupuis J, Langenberg C, Prokopenko I, Saxena R, Soranzo N, Jackson AU, et al. New Genetic Loci Implicated in Fasting Glucose Homeostasis and Their Impact on Type 2 Diabetes Risk. *Nat Genet* (2010) 42:105–16. doi: 10.1038/ng.520
16. Morris JP4, Cano DA, Sekine S, Wang SC, Hebrok M. Beta-Catenin Blocks Kras-Dependent Reprogramming of Acini Into Pancreatic Cancer Precursor Lesions in Mice. *J Clin Invest* (2010) 120:508–20. doi: 10.1172/JCI40045
17. Cnop M, Abdulkarim B, Bottu G, Cunha DA, Igoillo-Esteve M, Masini M, et al. RNA Sequencing Identifies Dysregulation of the Human Pancreatic Islet Transcriptome by the Saturated Fatty Acid Palmitate. *Diabetes* (2014) 63:1978–93. doi: 10.2337/db13-1383
18. Florez JC. Newly Identified Loci Highlight Beta Cell Dysfunction as a Key Cause of Type 2 Diabetes: Where are the Insulin Resistance Genes? *Diabetologia* (2008) 51:1100–10. doi: 10.1007/s00125-008-1025-9
19. Pociot F, Akolkar B, Concannon P, Erlich HA, Julier C, Morahan G, et al. Genetics of Type 1 Diabetes: What's Next? *Diabetes* (2010) 59:1561–71. doi: 10.2337/db10-0076
20. Eizirik DL, Sammeth M, Bouckenooghe T, Bottu G, Sisino G, Igoillo-Esteve M, et al. The Human Pancreatic Islet Transcriptome: Expression of Candidate Genes for Type 1 Diabetes and the Impact of Pro-Inflammatory Cytokines. *PloS Genet* (2012) 8:e1002552. doi: 10.1371/journal.pgen.1002552
21. Santin I, Moore F, Colli ML, Gurzov EN, Marselli L, Marchetti P, et al. PTPN2, a Candidate Gene for Type 1 Diabetes, Modulates Pancreatic Beta-Cell Apoptosis Via Regulation of the BH3-only Protein Bim. *Diabetes* (2011) 60:3279–88. doi: 10.2337/db11-0758
22. Schwitzgebel VM. Many Faces of Monogenic Diabetes. *J Diabetes Invest* (2014) 5:121–33. doi: 10.1111/jdi.12197
23. Shepherd M, Shields B, Hammersley S, Hudson M, McDonald TJ, Colclough K, et al. Systematic Population Screening, Using Biomarkers and Genetic Testing,

Identifies 2.5% of the U.K. Pediatric Diabetes Population With Monogenic Diabetes. *Diabetes Care* (2016) 39:1879–88. doi: 10.2337/dc16-0645

24. Bellin MD, Barton FB, Heitman A, Harmon JV, Kandaswamy R, Balamurugan AN, et al. Potent Induction Immunotherapy Promotes Long-Term Insulin Independence After Islet Transplantation in Type 1 Diabetes. *Am J Transplant* (2012) 12:1576–83. doi: 10.1111/j.1600-6143.2011.03977.x
25. Maehr R, Chen S, Snitow M, Ludwig T, Yagasaki L, Goland R, et al. Generation of Pluripotent Stem Cells From Patients With Type 1 Diabetes. *Proc Natl Acad Sci USA* (2009) 106:15768–73. doi: 10.1073/pnas.0906894106
26. Raya A, Rodriguez-Piza I, Guenechea G, Vassena R, Navarro S, Barrero MJ, et al. Disease-Corrected Haematopoietic Progenitors From Fanconi Anaemia Induced Pluripotent Stem Cells. *Nature* (2009) 460:53–9. doi: 10.1038/nature08129
27. Rashid ST, Corbineau S, Hannan N, Marciniak SJ, Miranda E, Alexander G, et al. Modeling Inherited Metabolic Disorders of the Liver Using Human Induced Pluripotent Stem Cells. *J Clin Invest* (2010) 120:3127–36. doi: 10.1172/JCI43122
28. Liu GH, Barkho BZ, Ruiz S, Diep D, Qu J, Yang SL, et al. Recapitulation of Premature Ageing With iPSCs From Hutchinson-Gilford Progeria Syndrome. *Nature* (2011) 472:221–5. doi: 10.1038/nature09879
29. Brennand KJ, Simone A, Jou J, Gelboin-Burkhart C, Tran N, Sangar S, et al. Modelling Schizophrenia Using Human Induced Pluripotent Stem Cells. *Nature* (2011) 473(7346):221–5. doi: 10.1038/nature09915
30. Itzhaki I, Maizels L, Huber I, Zwi-Dantsis L, Caspi O, Winterstern A, et al. Modelling the Long QT Syndrome With Induced Pluripotent Stem Cells. *Nature* (2011) 471:225–9. doi: 10.1038/nature09747
31. Thatava T, Kudva YC, Edukulla R, Squillace K, De Lamo JG, Khan YK, et al. Intrapatient Variations in Type 1 Diabetes-Specific iPS Cell Differentiation Into Insulin-Producing Cells. *Mol Ther J Am Soc Gene Ther* (2013) 21:228–39. doi: 10.1038/mt.2012.245
32. Maetzel D, Sarkar S, Wang H, Abi-Mosleh L, Xu P, Cheng AW, et al. Genetic and Chemical Correction of Cholesterol Accumulation and Impaired Autophagy in Hepatic and Neural Cells Derived From Niemann-Pick Type C Patient-Specific iPS Cells. *Stem Cell Rep* (2014) 2:866–80. doi: 10.1016/j.stemcr.2014.03.014
33. Sarkar S, Carroll B, Buganim Y, Maetzel D, Ng AH, Cassady JP, et al. Impaired Autophagy in the Lipid-Storage Disorder Niemann-Pick Type C1 Disease. *Cell Rep* (2013) 5:1302–15. doi: 10.1016/j.celrep.2013.10.042
34. Ryan SD, Dolatabadi N, Chan SF, Zhang X, Akhtar MW, Parker J, et al. Isogenic Human iPSC Parkinson's Model Shows Nitrosative Stress-Induced Dysfunction in MEF2-PGC1alpha Transcription. *Cell* (2013) 155:1351–64. doi: 10.1016/j.cell.2013.11.009
35. Reinhardt P, Schmid B, Burbulla LF, Schondorf DC, Wagner L, Glatza M, et al. Genetic Correction of a LRRK2 Mutation in Human iPSCs Links Parkinsonian Neurodegeneration to ERK-dependent Changes in Gene Expression. *Cell Stem Cell* (2013) 12:354–67. doi: 10.1016/j.stem.2013.01.008
36. D'Amour KA, Bang AG, Eliazer S, Kelly OG, Agulnick AD, Smart NG, et al. Production of Pancreatic Hormone–Expressing Endocrine Cells From Human Embryonic Stem Cells. *Nat Biotechnol* (2006) 24:1392–401. doi: 10.1038/nbt1259
37. Nostro MC, Sarangi F, Yang C, Holland A, Elefanty AG, Stanley EG, et al. Efficient Generation of NKX6-1+ Pancreatic Progenitors From Multiple Human Pluripotent Stem Cell Lines. *Stem Cell Rep* (2015) 4:591–604. doi: 10.1016/j.stemcr.2015.02.017
38. Hohwieler M, Illing A, Hermann PC, Mayer T, Stockmann M, Perkhofer L, et al. Human Pluripotent Stem Cell-Derived Acinar/Ductal Organoids Generate Human Pancreas Upon Orthotopic Transplantation and Allow Disease Modelling. *Gut* (2017) 66:473–86. doi: 10.1136/gutjnl-2016-312423
39. Hohwieler M, Perkhofer L, Liebau S, Seufferlein T, Muller M, Illing A, et al. Stem Cell-Derived Organoids to Model Gastrointestinal Facets of Cystic Fibrosis. *U Eur Gastroenterol J* (2017) 5:609–24. doi: 10.1177/2050640616670565
40. Hohwieler M, Renz S, Liebau S, Lin Q, Lechel A, Klaus J, et al. "Miniguts" From Plucked Human Hair Meet Crohn's Disease. *Z Gastroenterol* (2016) 54:748–59. doi: 10.1055/s-0042-105520
41. Russ HA, Parent AV, Ringler JJ, Hennings TG, Nair GG, Shveygert M, et al. Controlled Induction of Human Pancreatic Progenitors Produces Functional Beta-Like Cells In Vitro. *EMBO J* (2015) 34:1759–72. doi: 10.15252/embj.201591058
42. Pagliuca FW, Millman JR, Gürtler M, Segel M, Van Dervort A, Ryu JH, et al. Generation of Functional Human Pancreatic β Cells In Vitro. *Cell* (2014) 159:428–39. doi: 10.1016/j.cell.2014.09.040
43. Rezania A, Bruin JE, Arora P, Rubin A, Batushansky I, Asadi A, et al. Reversal of Diabetes With Insulin-Producing Cells Derived In Vitro From Human Pluripotent Stem Cells. *Nat Biotechnol* (2014) 32:1121–33. doi: 10.1038/nbt.3033
44. Cheng X, Ying L, Lu L, Galvao AM, Mills JA, Lin HC, et al. Self-Renewing Endodermal Progenitor Lines Generated From Human Pluripotent Stem Cells. *Cell Stem Cell* (2012) 10:371–84. doi: 10.1016/j.stem.2012.02.024
45. Hua H, Shang L, Martinez H, Freeby M, Gallagher MP, Ludwig T, et al. iPSC-derived Beta Cells Model Diabetes Due to Glucokinase Deficiency. *J Clin Invest* (2013) 123:3146–53. doi: 10.1172/JCI67638
46. Carmody D, Støy J, Greeley SAW, Bell GI, Philipson LH. *A Clinical Guide to Monogenic Diabetes*. Genetic Diagnosis of Endocrine Disorders: Elsevier (2016) p. 21–30. doi: 10.1016/B978-0-12-800892-8.00002-6
47. Bishay RH, Greenfield JR. A Review of Maturity Onset Diabetes of the Young (MODY) and Challenges in the Management of Glucokinase-MODY. *Med J Aust* (2016) 205:480–5. doi: 10.5694/mja16.00458
48. Yamagata K, Furuta H, Oda N, Kaisaki PJ, Menzel S, Cox NJ, et al. Mutations in the Hepatocyte Nuclear Factor-4α Gene in Maturity-Onset Diabetes of the Young (MODY1). *Nature* (1996) 384:458–60. doi: 10.1038/384458a0
49. Shields B, Hicks S, Shepherd M, Colclough K, Hattersley AT, Ellard S. Maturity-Onset Diabetes of the Young (MODY): How Many Cases are We Missing? *Diabetologia* (2010) 53:2504–8. doi: 10.1007/s00125-010-1799-4
50. Søvika O, Irgens HU, Molnes J, Sagena JV, Bjørkhaug L, Ræder H, et al. Monogenic diabetes mellitus in Norway. *Norsk Epidemiol* (2013) 23. doi: 10.5324/nje.v23i1.1603
51. Schober E, Rami B, Grabert M, Thon A, Kapellen T, Reinehr T, et al. Phenotypical Aspects of Maturity-Onset Diabetes of the Young (MODY Diabetes) in Comparison With Type 2 Diabetes Mellitus (T2DM) in Children and Adolescents: Experience From a Large Multicentre Database. *Diabetic Med* (2009) 26:466–73. doi: 10.1111/j.1464-5491.2009.02720.x
52. Sakura H, Eto K, Kadowaki H, Simokawa K, Ueno H, Koda N, et al. Structure of the Human Glucokinase Gene and Identification of a Missense Mutation in a Japanese Patient With Early-Onset non-Insulin-Dependent Diabetes Mellitus. *J Clin Endocrinol Metab* (1992) 75:1571–3. doi: 10.1210/jcem.75.6.1464666
53. Froguel P, Zouali H, Vionnet N, Velho G, Vaxillaire M, Sun F, et al. Familial Hyperglycemia Due to Mutations in Glucokinase–Definition of a Subtype of Diabetes Mellitus. *New Engl J Med* (1993) 328:697–702. doi: 10.1056/NEJM199303113281005
54. Yamagata K, Oda N, Kaisaki PJ, Menzel S, Furuta H, Vaxillaire M, et al. Mutations in the Hepatocyte Nuclear factor-1alpha Gene in Maturity-Onset Diabetes of the Young (MODY3). *Nature* (1996) 384:455–8. doi: 10.1038/384455a0
55. Stoffel M, Duncan SA. The Maturity-Onset Diabetes of the Young (MODY1) Transcription Factor HNF4α Regulates Expression of Genes Required for Glucose Transport and Metabolism. *Proc Natl Acad Sci* (1997) 94:13209–14. doi: 10.1073/pnas.94.24.13209
56. Byrne MM, Sturis J, Fajans SS, Ortiz FJ, Stoltz A, Stoffel M, et al. Altered Insulin Secretory Responses to Glucose in Subjects With a Mutation in the MODY1 Gene on Chromosome 20. *Diabetes* (1995) 44:699–704. doi: 10.2337/diabetes.44.6.699
57. Velho G, Blanche H, Vaxillaire M, Bellanne-Chantelot C, Pardini V, Timsit J, et al. Identification of 14 New Glucokinase Mutations and Description of the Clinical Profile of 42 MODY-2 Families. *Diabetologia* (1997) 40:217–24. doi: 10.1007/s001250050666
58. Byrne MM, Sturis J, Menzel S, Yamagata K, Fajans SS, Dronsfield MJ, et al. Altered Insulin Secretory Responses to Glucose in Diabetic and Nondiabetic Subjects With Mutations in the Diabetes Susceptibility Gene MODY3 on Chromosome 12. *Diabetes* (1996) 45:1503–10. doi: 10.2337/diab.45.11.1503
59. Staffers DA, Ferrer J, Clarke WL, Habener JF. Early-Onset Type-Ll Diabetes Mellitus (MODY4) Linked to IPF1. *Nat Genet* (1997) 17:138–9. doi: 10.1038/ng1097-138
60. Horikawa Y, Iwasaki N, Hara M, Furuta H, Hinokio Y, Cockburn BN, et al. Mutation in Hepatocyte Nuclear Factor-1β Gene (TCF2) Associated With MODY. *Nat Genet* (1997) 17:384–5. doi: 10.1038/ng1297-384

61. Malecki MT, Jhala US, Antonellis A, Fields L, Doria A, Orban T, et al. Mutations in NEUROD1 are Associated With the Development of Type 2 Diabetes Mellitus. *Nat Genet* (1999) 23:323–8. doi: 10.1038/15500
62. Szopa M, Ludwig-Galezowska AH, Radkowski P, Skupien J, Machlowska J, Klupa T, et al. A Family With the Arg103Pro Mutation in the NEUROD1 Gene Detected by Next-Generation Sequencing–Clinical Characteristics of Mutation Carriers. *Eur J Med Genet* (2016) 59:75–9. doi: 10.1016/j.ejmg.2016.01.002
63. Neve B, Fernandez-Zapico ME, Ashkenazi-Katalan V, Dina C, Hamid YH, Joly E, et al. Role of Transcription Factor KLF11 and its Diabetes-Associated Gene Variants in Pancreatic Beta Cell Function. *Proc Natl Acad Sci* (2005) 102:4807–12. doi: 10.1073/pnas.0409177102
64. Torsvik J, Johansson S, Johansen A, Ek J, Minton J, Ræder H, et al. Mutations in the VNTR of the Carboxyl-Ester Lipase Gene (CEL) are a Rare Cause of Monogenic Diabetes. *Hum Genet* (2010) 127:55. doi: 10.1007/s00439-009-0740-8
65. Plengvidhya N, Kooptiwut S, Songtawee N, Doi A, Furuta H, Nishi M, et al. PAX4 Mutations in Thais With Maturity Onset Diabetes of the Young. *J Clin Endocrinol Metab* (2007) 92:2821–6. doi: 10.1210/jc.2006-1927
66. Haneda M, Chan SJ, Kwok S, Rubenstein AH, Steiner DF. Studies on Mutant Human Insulin Genes: Identification and Sequence Analysis of a Gene Encoding [SerB24] Insulin. *Proc Natl Acad Sci* (1983) 80:6366–70. doi: 10.1073/pnas.80.20.6366
67. Borowiec M, Liew CW, Thompson R, Boonyasrisawat W, Hu J, Mlynarski WM, et al. Mutations At the BLK Locus Linked to Maturity Onset Diabetes of the Young and β-Cell Dysfunction. *Proc Natl Acad Sci* (2009) 106:14460–5. doi: 10.1073/pnas.0906474106
68. Bowman P, Flanagan S, Edghill E, Damhuis A, Shepherd M, Paisey R, et al. Heterozygous ABCC8 Mutations are a Cause of MODY. *Diabetologia* (2012) 55:123–7. doi: 10.1007/s00125-011-2319-x
69. Bonnefond A, Philippe J, Durand E, Dechaume A, Huyvaert M, Montagne L, et al. Whole-Exome Sequencing and High Throughput Genotyping Identified KCNJ11 as the Thirteenth MODY Gene. *PloS One* (2012) 7: e37423. doi: 10.1371/journal.pone.0037423
70. Prudente S, Jungtrakoon P, Marucci A, Ludovico O, Buranasupkajorn P, Mazza T, et al. Loss-of-Function Mutations in APPL1 in Familial Diabetes Mellitus. *Am J Hum Genet* (2015) 97:177–85. doi: 10.1016/j.ajhg.2015.05.011
71. Dubois-Laforgue D, Cornu E, Saint-Martin C, Coste J, Bellanné-Chantelot C, Timsit J. Diabetes, Associated Clinical Spectrum, Long-Term Prognosis, and Genotype/Phenotype Correlations in 201 Adult Patients With Hepatocyte Nuclear Factor 1B (HNF1B) Molecular Defects. *Diabetes Care* (2017) 40:1436–43. doi: 10.2337/dc16-2462
72. Lindner TH, Njolstad PR, Horikawa Y, Bostad L, Bell GI, Sovik O. A Novel Syndrome of Diabetes Mellitus, Renal Dysfunction and Genital Malformation Associated With a Partial Deletion of the pseudo-POU Domain of Hepatocyte Nuclear Factor-1beta. *Hum Mol Genet* (1999) 8:2001–8. doi: 10.1093/hmg/8.11.2001
73. Abreu G, Tarantino RM, Cabello PH, Zembrzuski VM, da Fonseca ACP, Rodacki M, et al. The First Case of NEUROD1-MODY Reported in Latin America. *Mol Genet Genomic Med* (2019) 7:e989. doi: 10.1002/mgg3.989
74. Ağladıoğlu SY, Aycan Z, Çetinkaya S, Baş VN, Önder A, Kendirci HNP, et al. Maturity Onset Diabetes of Youth (MODY) in Turkish Children: Sequence Analysis of 11 Causative Genes by Next Generation Sequencing. *J Pediatr Endocrinol Metab* (2016) 29:487–96. doi: 10.1515/jpem-2015-0039
75. Horikawa Y, Enya M, Mabe H, Fukushima K, Takubo N, Ohashi M, et al. NEUROD1-Deficient Diabetes (MODY6): Identification of the First Cases in Japanese and the Clinical Features. *Pediatr Diabetes* (2018) 19:236–42. doi: 10.1111/pedi.12553
76. Fjeld K, Weiss FU, Lasher D, Rosendahl J, Chen J-M, Johansson BB, et al. A Recombined Allele of the Lipase Gene CEL and its Pseudogene CELP Confers Susceptibility to Chronic Pancreatitis. *Nat Genet* (2015) 47:518–22. doi: 10.1038/ng.3249
77. Ræder H, Johansson S, Holm PI, Haldorsen IS, Mas E, Sbarra V, et al. Mutations in the CEL VNTR Cause a Syndrome of Diabetes and Pancreatic Exocrine Dysfunction. *Nat Genet* (2006) 38:54–62. doi: 10.1038/ng1708
78. Jo W, Endo M, Ishizu K, Nakamura A, Tajima T. A Novel PAX4 Mutation in a Japanese Patient With Maturity-Onset Diabetes of the Young. *Tohoku J Exp Med* (2011) 223:113–8. doi: 10.1620/tjem.223.113
79. Ushijima K, Fukami M, Ayabe T, Narumi S, Okuno M, Nakamura A, et al. Comprehensive Screening for Monogenic Diabetes in 89 Japanese Children With Insulin-Requiring Antibody-Negative Type 1 Diabetes. *Pediatr Diabetes* (2018) 19:243–50. doi: 10.1111/pedi.12544
80. Teo AK, Windmueller R, Johansson BB, Dirice E, Njolstad PR, Tjora E, et al. Derivation of Human Induced Pluripotent Stem Cells From Patients With Maturity Onset Diabetes of the Young. *J Biol Chem* (2013) 288:5353–6. doi: 10.1074/jbc.C112.428979
81. Ng NHJ, Jasmen JB, Lim CS, Lau HH, Krishnan VG, Kadiwala J, et al. HNF4A Haploinsufficiency in MODY1 Abrogates Liver and Pancreas Differentiation From Patient-Derived Induced Pluripotent Stem Cells. *Iscience* (2019) 16:192–205. doi: 10.1016/j.isci.2019.05.032
82. Braverman-Gross C, Nudel N, Ronen D, Beer NL, McCarthy MI, Benvenisty N. Derivation and Molecular Characterization of Pancreatic Differentiated MODY1-Ipscs. *Stem Cell Res* (2018) 31:16–26. doi: 10.1016/j.scr.2018.06.013
83. Hua H, Shang L, Martinez H, Freeby M, Gallagher MP, Ludwig T, et al. iPSC-derived Beta Cells Model Diabetes Due to Glucokinase Deficiency. *J Clin Invest* (2017) 127:1115. doi: 10.1172/JCI92775
84. Cardenas-Diaz FL, Osorio-Quintero C, Diaz-Miranda MA, Kishore S, Leavens K, Jobaliya C, et al. Modeling Monogenic Diabetes Using Human ESCs Reveals Developmental and Metabolic Deficiencies Caused by Mutations in HNF1A. *Cell Stem Cell* (2019) 25:273–89.e5. doi: 10.1016/j.stem.2019.07.007
85. Griscelli F, Ezanno H, Soubeyrand M, Feraud O, Oudrhiri N, Bonnefond A, et al. Generation of an Induced Pluripotent Stem Cell (iPSC) Line From a Patient With Maturity-Onset Diabetes of the Young Type 3 (MODY3) Carrying a Hepatocyte Nuclear Factor 1-Alpha (HNF1A) Mutation. *Stem Cell Res* (2018) 29:56–9. doi: 10.1016/j.scr.2018.02.017
86. Zhu Z, Li QV, Lee K, Rosen BP, González F, Soh C-L, et al. Genome Editing of Lineage Determinants in Human Pluripotent Stem Cells Reveals Mechanisms of Pancreatic Development and Diabetes. *Cell Stem Cell* (2016) 18:755–68. doi: 10.1016/j.stem.2016.03.015
87. Wang X, Chen S, Burtscher I, Sterr M, Hieronimus A, Machicao F, et al. Generation of a Human Induced Pluripotent Stem Cell (iPSC) Line From a Patient Carrying a P33T Mutation in the PDX1 Gene. *Stem Cell Res* (2016) 17:273–6. doi: 10.1016/j.scr.2016.08.004
88. Wang X, Sterr M, Burtscher I, Böttcher A, Beckenbauer J, Siehler J, et al. Point Mutations in the PDX1 Transactivation Domain Impair Human β-Cell Development and Function. *Mol Metab* (2019) 24:80–97. doi: 10.1016/j.molmet.2019.03.006
89. Wang X, Chen S, Burtscher I, Sterr M, Hieronimus A, Machicao F, et al. Generation of a Human Induced Pluripotent Stem Cell (iPSC) Line From a Patient With Family History of Diabetes Carrying a C18R Mutation in the PDX1 Gene. *Stem Cell Res* (2016) 17:292–5. doi: 10.1016/j.scr.2016.08.005
90. Teo AKK, Lau HH, Valdez IA, Dirice E, Tjora E, Raeder H, et al. Early Developmental Perturbations in a Human Stem Cell Model of MODY5/HNF1B Pancreatic Hypoplasia. *Stem Cell Rep* (2016) 6:357–67. doi: 10.1016/j.stemcr.2016.01.007
91. Yabe SG, Iwasaki N, Yasuda K, Hamazaki TS, Konno M, Fukuda S, et al. Establishment of Maturity-Onset Diabetes of the Young-Induced Pluripotent Stem Cells From a Japanese Patient. *J Diabetes Invest* (2015) 6:543–7. doi: 10.1111/jdi.12334
92. Gage BK, Baker RK, Kieffer TJ. Overexpression of PAX4 Reduces Glucagon Expression in Differentiating Hescs. *Islets* (2014) 6:e29236. doi: 10.4161/isl.29236
93. Ma S, Viola R, Sui L, Cherubini V, Barbetti F, Egli D. β Cell Replacement After Gene Editing of a Neonatal Diabetes-Causing Mutation At the Insulin Locus. *Stem Cell Rep* (2018) 11:1407–15. doi: 10.1016/j.stemcr.2018.11.006
94. Guo D, Liu H, Gao G, Ruzi A, Wang K, Wu H, et al. Generation of an Abcc8 Heterozygous Mutation Human Embryonic Stem Cell Line Using CRISPR/Cas9. *Stem Cell Res* (2016) 17:670–2. doi: 10.1016/j.scr.2016.11.014
95. Guo D, Liu H, Ruzi A, Gao G, Nasir A, Liu Y, et al. Modeling Congenital Hyperinsulinism With ABCC8-deficient Human Embryonic Stem Cells Generated by CRISPR/Cas9. *Sci Rep* (2017) 7:1–8. doi: 10.1038/s41598-017-03349-w
96. Zeng H, Guo M, Zhou T, Tan L, Chong CN, Zhang T, et al. An Isogenic Human ESC Platform for Functional Evaluation of Genome-Wide-Association-Study-Identified Diabetes Genes and Drug Discovery. *Cell Stem Cell* (2016) 19:326–40. doi: 10.1016/j.stem.2016.07.002

97. Griscelli F, Feraud O, Ernault T, Oudrihri N, Turhan AG, Bonnefond A, et al. Generation of an Induced Pluripotent Stem Cell (iPSC) Line From a Patient With Maturity-Onset Diabetes of the Young Type 13 (MODY13) With a the Potassium Inwardly-Rectifying Channel, Subfamily J, Member 11 (KCNJ11) Mutation. *Stem Cell Res* (2017) 23:178–81. doi: 10.1016/j.scr.2017.07.023
98. Gragnoli C, Stanojevic V, Gorini A, Von Preussenthal GM, Thomas MK, Habener JF. Ipf-1/MODY4 Gene Missense Mutation in an Italian Family With Type 2 and Gestational Diabetes. *Metabolism* (2005) 54:983–8. doi: 10.1016/j.metabol.2005.01.037
99. Mahajan A, Wessel J, Willems SM, Zhao W, Robertson NR, Chu AY, et al. Refining the Accuracy of Validated Target Identification Through Coding Variant Fine-Mapping in Type 2 Diabetes. *Nat Genet* (2018) 50:559–71. doi: 10.1038/s41588-018-0084-1
100. Fuchsberger C, Flannick J, Teslovich TM, Mahajan A, Agarwala V, Gaulton KJ, et al. The Genetic Architecture of Type 2 Diabetes. *Nature* (2016) 536:41–7. doi: 10.1038/nature18642
101. Patel KA, Kettunen J, Laakso M, Stančáková A, Laver TW, Colclough K, et al. Heterozygous RFX6 Protein Truncating Variants are Associated With MODY With Reduced Penetrance. *Nat Commun* (2017) 8:1–8. doi: 10.1038/s41467-017-00895-9
102. Shim YJ, Kim JE, Hwang S-K, Choi BS, Choi BH, Cho E-M, et al. Identification of Candidate Gene Variants in Korean MODY Families by Whole-Exome Sequencing. *Hormone Res Paediatr* (2015) 83:242–51. doi: 10.1159/000368657
103. Johansson S, Irgens H, Chudasama KK, Molnes J, Aerts J, Roque FS, et al. Exome Sequencing and Genetic Testing for MODY. *PloS One* (2012) 7: e38050. doi: 10.1371/journal.pone.0038050
104. Fösel S. Transient and Permanent Neonatal Diabetes. *Eur J Pediatr* (1995) 154:944–8. doi: 10.1007/BF01958635
105. Hattersley A, Bruining J, Shield J, Njolstad P, Donaghue KC. The Diagnosis and Management of Monogenic Diabetes in Children and Adolescents. *Pediatr Diabetes* (2009) 10:33–42. doi: 10.1111/j.1399-5448.2009.00571.x
106. Babenko AP, Polak M, Cavé H, Busiah K, Czernichow P, Scharfmann R, et al. Activating Mutations in the ABCC8 Gene in Neonatal Diabetes Mellitus. *New Engl J Med* (2006) 355:456–66. doi: 10.1056/NEJMoa055068
107. Njølstad PR, Søvik O, Cuesta-Muñoz A, Bjørkhaug L, Massa O, Barbetti F, et al. Neonatal Diabetes Mellitus Due to Complete Glucokinase Deficiency. *New Engl J Med* (2001) 344:1588–92. doi: 10.1056/NEJM200105243442104
108. Støy J, Edghill EL, Flanagan SE, Ye H, Paz VP, Pluzhnikov A, et al. Insulin Gene Mutations as a Cause of Permanent Neonatal Diabetes. *Proc Natl Acad Sci* (2007) 104:15040–4. doi: 10.1073/pnas.0707291104
109. Flanagan S, Edghill E, Gloyn A, Ellard S, Hattersley A. Mutations in KCNJ11, Which Encodes Kir6. 2, are a Common Cause of Diabetes Diagnosed in the First 6 Months of Life, With the Phenotype Determined by Genotype. *Diabetologia* (2006) 49:1190–7. doi: 10.1007/s00125-006-0246-z
110. Stoffers DA, Zinkin NT, Stanojevic V, Clarke WL, Habener JF. Pancreatic Agenesis Attributable to a Single Nucleotide Deletion in the Human IPF1 Gene Coding Sequence. *Nat Genet* (1997) 15:106–10. doi: 10.1038/ng0197-106
111. Gloyn AL, Diatloff-Zito C, Edghill EL, Bellanné-Chantelot C, Nivot S, Coutant R, et al. KCNJ11 Activating Mutations are Associated With Developmental Delay, Epilepsy and Neonatal Diabetes Syndrome and Other Neurological Features. *Eur J Hum Genet* (2006) 14:824–30. doi: 10.1038/sj.ejhg.5201629
112. Schwitzgebel VM, Mamin A, Brun T, Ritz-Laser B, Zaiko M, Maret A, et al. Agenesis of Human Pancreas Due to Decreased Half-Life of Insulin Promoter Factor 1. *J Clin Endocrinol Metab* (2003) 88:4398–406. doi: 10.1210/jc.2003-030046
113. Sellick GS, Barker KT, Stolte-Dijkstra I, Fleischmann C, Coleman RJ, Garrett C, et al. Mutations in PTF1A Cause Pancreatic and Cerebellar Agenesis. *Nat Genet* (2004) 36:1301–5. doi: 10.1038/ng1475
114. Delépine M, Nicolino M, Barrett T, Golamaully M, Lathrop GM, Julier C. EIF2AK3, Encoding Translation Initiation Factor 2-α Kinase 3, is Mutated in Patients With Wolcott-Rallison Syndrome. *Nat Genet* (2000) 25:406–9. doi: 10.1038/78085
115. D'Amato E, Giacopelli F, Giannattasio A, d'Annunzio G, Bocciardi R, Musso M, et al. Genetic Investigation in an Italian Child With an Unusual Association of Atrial Septal Defect, Attributable to a New Familial GATA4 Gene Mutation, and Neonatal Diabetes Due to Pancreatic Agenesis. *Diabetic Med* (2010) 27:1195–200. doi: 10.1111/j.1464-5491.2010.03046.x
116. Allen HL, Flanagan SE, Shaw-Smith C, De Franco E, Akerman I, Caswell R, et al. GATA6 Haploinsufficiency Causes Pancreatic Agenesis in Humans. *Nat Genet* (2012) 44:20–2. doi: 10.1038/ng.1035
117. Senée V, Chelala C, Duchatelet S, Feng D, Blanc H, Cossec J-C, et al. Mutations in GLIS3 are Responsible for a Rare Syndrome With Neonatal Diabetes Mellitus and Congenital Hypothyroidism. *Nat Genet* (2006) 38:682–7. doi: 10.1038/ng1802
118. Rubio-Cabezas O, Minton JA, Kantor I, Williams D, Ellard S, Hattersley AT. Homozygous Mutations in NEUROD1 are Responsible for a Novel Syndrome of Permanent Neonatal Diabetes and Neurological Abnormalities. *Diabetes* (2010) 59:2326–31. doi: 10.2337/db10-0011
119. Pinney SE, Oliver-Krasinski J, Ernst L, Hughes N, Patel P, Stoffers DA, et al. Neonatal Diabetes and Congenital Malabsorptive Diarrhea Attributable to a Novel Mutation in the Human Neurogenin-3 Gene Coding Sequence. *J Clin Endocrinol Metab* (2011) 96:1960–5. doi: 10.1210/jc.2011-0029
120. Rubio-Cabezas O, Jensen JN, Hodgson MI, Codner E, Ellard S, Serup P, et al. Permanent Neonatal Diabetes and Enteric Anendocrinosis Associated With Biallelic Mutations in NEUROG3. *Diabetes* (2011) 60:1349–53. doi: 10.2337/db10-1008
121. Wang J, Cortina G, Wu SV, Tran R, Cho J-H, Tsai M-J, et al. Mutant Neurogenin-3 in Congenital Malabsorptive Diarrhea. *New Engl J Med* (2006) 355:270–80. doi: 10.1056/NEJMoa054288
122. Solomon BD, Pineda-Alvarez DE, Balog JZ, Hadley D, Gropman AL, Nandagopal R, et al. Compound Heterozygosity for Mutations in PAX6 in a Patient With Complex Brain Anomaly, Neonatal Diabetes Mellitus, and Microophthalmia. *Am J Med Genet Part A* (2009) 149:2543–6. doi: 10.1002/ajmg.a.33081
123. Spiegel R, Dobbie A, Hartman C, de Vries L, Ellard S, Shalev SA. Clinical Characterization of a Newly Described Neonatal Diabetes Syndrome Caused by RFX6 Mutations. *Am J Med Genet Part A* (2011) 155:2821–5. doi: 10.1002/ajmg.a.34251
124. Smith SB, Qu H-Q, Taleb N, Kishimoto NY, Scheel DW, Lu Y, et al. Rfx6 Directs Islet Formation and Insulin Production in Mice and Humans. *Nature* (2010) 463:775–80. doi: 10.1038/nature08748
125. Bonnycastle LL, Chines PS, Hara T, Huyghe JR, Swift AJ, Heikinheimo P, et al. Autosomal Dominant Diabetes Arising From a Wolfram Syndrome 1 Mutation. *Diabetes* (2013) 62:3943–50. doi: 10.2337/db13-0571
126. Olsen BS, Hahnemann JM, Schwartz M, Østergaard E. Thiamine-Responsive Megaloblastic Anaemia: A Cause of Syndromic Diabetes in Childhood. *Pediatr Diabetes* (2007) 8:239–41. doi: 10.1111/j.1399-5448.2007.00251.x
127. Shaw-Smith C, Flanagan SE, Patch AM, Grulich-Henn J, Habeb AM, Hussain K, et al. Recessive SLC19A2 Mutations are a Cause of Neonatal Diabetes Mellitus in Thiamine-Responsive Megaloblastic Anaemia. *Pediatr Diabetes* (2012) 13:314–21. doi: 10.1111/j.1399-5448.2012.00855.x
128. Flanagan SE, De Franco E, Allen HL, Zerah M, Abdul-Rasoul MM, Edge JA, et al. Analysis of Transcription Factors Key for Mouse Pancreatic Development Establishes NKX2-2 and MNX1 Mutations as Causes of Neonatal Diabetes in Man. *Cell Metab* (2014) 19:146–54. doi: 10.1016/j.cmet.2013.11.021
129. Poulton CJ, Schot R, Kia SK, Jones M, Verheijen FW, Venselaar H, et al. Microcephaly With Simplified Gyration, Epilepsy, and Infantile Diabetes Linked to Inappropriate Apoptosis of Neural Progenitors. *Am J Hum Genet* (2011) 89:265–76. doi: 10.1016/j.ajhg.2011.07.006
130. Sansbury F, Flanagan S, Houghton J, Shen FS, Al-Senani A, Habeb A, et al. SLC2A2 Mutations can Cause Neonatal Diabetes, Suggesting GLUT2 may Have a Role in Human Insulin Secretion. *Diabetologia* (2012) 55:2381–5. doi: 10.1007/s00125-012-2595-0
131. Santer R, Schneppenheim R, Dombrowski A, Götze H, Steinmann B, Schaub J. Mutations in GLUT2, the Gene for the Liver-Type Glucose Transporter, in Patients With Fanconi-Bickel Syndrome. *Nat Genet* (1997) 17:324–6. doi: 10.1038/ng1197-324
132. De Franco E, Lytrivi M, Ibrahim H, Montaser H, Wakeling MN, Fantuzzi F, et al. YIPF5 Mutations Cause Neonatal Diabetes and Microcephaly Through Endoplasmic Reticulum Stress. *J Clin Invest* (2020) 130(12):6338–53. doi: 10.1172/JCI141455

133. Flanagan SE, Haapaniemi E, Russell MA, Caswell R, Allen HL, De Franco E, et al. Activating Germline Mutations in STAT3 Cause Early-Onset Multi-Organ Autoimmune Disease. *Nat Genet* (2014) 46:812–4. doi: 10.1038/ng.3040
134. Saarimäki-Vire J, Balboa D, Russell MA, Saarikettu J, Kinnunen M, Keskitalo S, et al. An Activating STAT3 Mutation Causes Neonatal Diabetes Through Premature Induction of Pancreatic Differentiation. *Cell Rep* (2017) 19:281–94. doi: 10.1016/j.celrep.2017.03.055
135. Shi Z-D, Lee K, Yang D, Amin S, Verma N, Li QV, et al. Genome Editing in hPSCs Reveals GATA6 Haploinsufficiency and a Genetic Interaction With GATA4 in Human Pancreatic Development. *Cell Stem Cell* (2017) 20:675–88.e6. doi: 10.1016/j.stem.2017.01.001
136. Tiyaboonchai A, Cardenas-Diaz FL, Ying L, Maguire JA, Sim X, Jobaliya C, et al. GATA6 Plays an Important Role in the Induction of Human Definitive Endoderm, Development of the Pancreas, and Functionality of Pancreatic β Cells. *Stem Cell Rep* (2017) 8:589–604. doi: 10.1016/j.stemcr.2016.12.026
137. Balboa D, Saarimäki-Vire J, Borshagovski D, Survila M, Lindholm P, Galli E, et al. Insulin Mutations Impair Beta-Cell Development in a Patient-Derived iPSC Model of Neonatal Diabetes. *Elife* (2018) 7:e38519. doi: 10.7554/eLife.38519
138. Panova AV, Klementieva NV, Sycheva AV, Goliusova DV, Khokhlov NV, Zubkova NA, et al. Generation of an Induced Pluripotent Stem Cell Line MNDINSi001-A From a Patient With Neonatal Diabetes Caused by a Heterozygous INS Mutation. *Stem Cell Res* (2020) 47:101929. doi: 10.1016/j.scr.2020.101929
139. Zhang X, McGrath PS, Salomone J, Rahal M, McCauley HA, Schweitzer J, et al. A Comprehensive Structure-Function Study of Neurogenin3 Disease-Causing Alleles During Human Pancreas and Intestinal Organoid Development. *Dev Cell* (2019) 50:367–80.e7. doi: 10.1016/j.devcel.2019.05.017
140. Trott J, Alpagu Y, Tan EK, Shboul M, Dawood Y, Elsy M, et al. Mitchell-Riley Syndrome iPSCs Exhibit Reduced Pancreatic Endoderm Differentiation Due to a Mutation in RFX6. *Development* (2020) 147(21):dev194878. doi: 10.1242/dev.194878
141. Elsayed AK, Aghadi M, Al-Khawaga S, Hussain K, Abdelalim EM. Derivation of a Human Induced Pluripotent Stem Cell Line (Qbrii007-A) From a Patient Carrying a Homozygous Intronic Mutation (C. 613-7T> G) in the SLC2A2 Gene. *Stem Cell Res* (2020) 44:101736. doi: 10.1016/j.scr.2020.101736
142. Shang L, Hua H, Foo K, Martinez H, Watanabe K, Zimmer M, et al. β-Cell Dysfunction Due to Increased ER Stress in a Stem Cell Model of Wolfram Syndrome. *Diabetes* (2014) 63:923–33. doi: 10.2337/db13-0717
143. Maxwell KG, Augsornworawat P, Velazco-Cruz L, Kim MH, Asada R, Hogrebe NJ, et al. Gene-Edited Human Stem Cell–Derived β Cells From a Patient With Monogenic Diabetes Reverse Preexisting Diabetes in Mice. *Sci Trans Med* (2020) 12(540):eaax9106. doi: 10.1126/scitranslmed.aax9106
144. Gradwohl G, Dierich A, LeMeur M, Guillemot F. Neurogenin3 is Required for the Development of the Four Endocrine Cell Lineages of the Pancreas. *Proc Natl Acad Sci* (2000) 97:1607–11. doi: 10.1073/pnas.97.4.1607
145. Rodriguez-Seguí S, Akerman I, Ferrer J. GATA Believe it: New Essential Regulators of Pancreas Development. *J Clin Invest* (2012) 122:3469–71. doi: 10.1172/JCI65751
146. Garcia-Gonzalez MA, Carette C, Bagattin A, Chiral M, Makinistoglu MP, Garbay S, et al. A Suppressor Locus for MODY3-Diabetes. *Sci Rep* (2016) 6:33087. doi: 10.1038/srep35697
147. Pontoglio M, Sreenan S, Roe M, Pugh W, Ostrega D, Doyen A, et al. Defective Insulin Secretion in Hepatocyte Nuclear Factor 1alpha-Deficient Mice. *J Clin Invest* (1998) 101:2215–22. doi: 10.1172/JCI2548
148. Gupta RK, Vatamaniuk MZ, Lee CS, Flaschen RC, Fulmer JT, Matschinsky FM, et al. The MODY1 Gene HNF-4α Regulates Selected Genes Involved in Insulin Secretion. *J Clin Invest* (2005) 115:1006–15. doi: 10.1172/JCI200522365
149. Riggs A, Bernal-Mizrachi E, Ohsugi M, Wasson J, Fatrai S, Welling C, et al. Mice Conditionally Lacking the Wolfram Gene in Pancreatic Islet Beta Cells Exhibit Diabetes as a Result of Enhanced Endoplasmic Reticulum Stress and Apoptosis. *Diabetologia* (2005) 48:2313–21. doi: 10.1007/s00125-005-1947-4
150. Haumaitre C, Barbacci E, Jenny M, Ott M, Gradwohl G, Cereghini S. Lack of TCF2/vHNF1 in Mice Leads to Pancreas Agenesis. *Proc Natl Acad Sci* (2005) 102:1490–5. doi: 10.1073/pnas.0405776102
151. Nostro MC, Sarangi F, Ogawa S, Holtzinger A, Corneo B, Li X, et al. Stage-Specific Signaling Through Tgfβ Family Members and WNT Regulates Patterning and Pancreatic Specification of Human Pluripotent Stem Cells. *Development* (2011) 138:861–71. doi: 10.1242/dev.055236
152. Rezania A, Bruin JE, Riedel MJ, Mojibian M, Asadi A, Xu J, et al. Maturation of Human Embryonic Stem Cell–Derived Pancreatic Progenitors Into Functional Islets Capable of Treating Pre-Existing Diabetes in Mice. *Diabetes* (2012) 61:2016–29. doi: 10.2337/db11-1711
153. Cortijo C, Gouzi M, Tissir F, Grapin-Botton A. Planar Cell Polarity Controls Pancreatic Beta Cell Differentiation and Glucose Homeostasis. *Cell Rep* (2012) 2:1593–606. doi: 10.1016/j.celrep.2012.10.016
154. Nair GG, Liu JS, Russ HA, Tran S, Saxton MS, Chen R, et al. Recapitulating Endocrine Cell Clustering in Culture Promotes Maturation of Human Stem-Cell-Derived β Cells. *Nat Cell Biol* (2019) 21:263–74. doi: 10.1038/s41556-018-0271-4
155. Veres A, Faust AL, Bushnell HL, Engquist EN, Kenty JH-R, Harb G, et al. Charting Cellular Identity During Human In Vitro β-Cell Differentiation. *Nature* (2019) 569:368–73. doi: 10.1038/s41586-019-1168-5
156. Mahaddalkar PU, Scheibner K, Pfluger S, Sterr M, Beckenbauer J, Irmler M, et al. Generation of Pancreatic β Cells From CD177+ Anterior Definitive Endoderm. *Nat Biotechnol* (2020) 38:1061–72. doi: 10.1038/s41587-020-0492-5
157. Ghazizadeh Z, Kao D-I, Amin S, Cook B, Rao S, Zhou T, et al. ROCKII Inhibition Promotes the Maturation of Human Pancreatic Beta-Like Cells. *Nat Commun* (2017) 8:1–12. doi: 10.1038/s41467-017-00129-y
158. Vethe H, Ghila L, Berle M, Hoareau L, Haaland ØA, Scholz H, et al. The Effect of Wnt Pathway Modulators on Human iPSC-derived Pancreatic Beta Cell Maturation. *Front Endocrinol* (2019) 10:293. doi: 10.3389/fendo.2019.00293
159. Hogrebe NJ, Augsornworawat P, Maxwell KG, Velazco-Cruz L, Millman JR. Targeting the Cytoskeleton to Direct Pancreatic Differentiation of Human Pluripotent Stem Cells. *Nat Biotechnol* (2020) 38:460–70. doi: 10.1038/s41587-020-0430-6
160. Helman A, Cangelosi AL, Davis JC, Pham Q, Rothman A, Faust AL, et al. A Nutrient-Sensing Transition At Birth Triggers Glucose-Responsive Insulin Secretion. *Cell Metab* (2020) 31:1004–16.e5. doi: 10.1016/j.cmet.2020.04.004
161. Alvarez-Dominguez JR, Donaghey J, Rasouli N, Kenty JH, Helman A, Charlton J, et al. Circadian Entrainment Triggers Maturation of Human In Vitro Islets. *Cell Stem Cell* (2020) 26:108–22.e10. doi: 10.1016/j.stem.2019.11.011
162. Kim MJ, Lee EY, You Y-H, Yang HK, Yoon K-H, Kim J-W. Generation of iPSC-derived Insulin-Producing Cells From Patients With Type 1 and Type 2 Diabetes Compared With Healthy Control. *Stem Cell Res* (2020) 48:101958. doi: 10.1016/j.scr.2020.101958
163. Millman JR, Xie C, Van Dervort A, Gürtler M, Pagliuca FW, Melton DA. Generation of Stem Cell-Derived β-Cells From Patients With Type 1 Diabetes. *Nat Commun* (2016) 7:1–9. doi: 10.1038/ncomms11463
164. Hosokawa Y, Toyoda T, Fukui K, Baden MY, Funato M, Kondo Y, et al. Insulin-Producing Cells Derived From 'Induced Pluripotent Stem Cells' of Patients With Fulminant Type 1 Diabetes: Vulnerability to Cytokine Insults and Increased Expression of Apoptosis-Related Genes. *J Diabetes Invest* (2018) 9:481–93. doi: 10.1111/jdi.12727
165. Leite NC, Sintov E, Meissner TB, Brehm MA, Greiner DL, Harlan DM, et al. Modeling Type 1 Diabetes In Vitro Using Human Pluripotent Stem Cells. *Cell Rep* (2020) 32:107894. doi: 10.1016/j.celrep.2020.107894
166. Joshi K, Cameron F, Tiwari S, Mannering SI, Elefanty AG, Stanley EG. Modeling Type 1 Diabetes Using Pluripotent Stem Cell Technology. *Front Endocrinol* (2021) 12:260. doi: 10.3389/fendo.2021.635662
167. Sander M, Neubüser A, Kalamaras J, Ee HC, Martin GR, German MS. Genetic Analysis Reveals That PAX6 is Required for Normal Transcription of Pancreatic Hormone Genes and Islet Development. *Genes Dev* (1997) 11:1662–73. doi: 10.1101/gad.11.13.1662
168. Inoue H, Tanizawa Y, Wasson J, Behn P, Kalidas K, Bernal-Mizrachi E, et al. A Gene Encoding a Transmembrane Protein is Mutated in Patients With Diabetes Mellitus and Optic Atrophy (Wolfram Syndrome). *Nat Genet* (1998) 20:143–8. doi: 10.1038/2441
169. Fonseca SG, Fukuma M, Lipson KL, Nguyen LX, Allen JR, Oka Y, et al. WFS1 is a Novel Component of the Unfolded Protein Response and Maintains Homeostasis of the Endoplasmic Reticulum in Pancreatic β-Cells. *J Biol Chem* (2005) 280:39609–15. doi: 10.1074/jbc.M507426200

170. Yamada T, Ishihara H, Tamura A, Takahashi R, Yamaguchi S, Takei D, et al. WFS1-Deficiency Increases Endoplasmic Reticulum Stress, Impairs Cell Cycle Progression and Triggers the Apoptotic Pathway Specifically in Pancreatic β-Cells. *Hum Mol Genet* (2006) 15:1600–9. doi: 10.1093/hmg/ddl081
171. Flanagan SE, Clauin S, Bellanné-Chantelot C, de Lonlay P, Harries LW, Gloyn AL, et al. Update of Mutations in the Genes Encoding the Pancreatic Beta-Cell KATP Channel Subunits Kir6. 2 (KCNJ11) and Sulfonylurea Receptor 1 (ABCC8) in Diabetes Mellitus and Hyperinsulinism. *Hum Mutat* (2009) 30:170–80. doi: 10.1002/humu.20838
172. Carrasco M, Delgado I, Soria B, Martín F, Rojas A. GATA4 and GATA6 Control Mouse Pancreas Organogenesis. *J Clin Invest* (2012) 122:3504–15. doi: 10.1172/JCI63240
173. Xuan S, Borok MJ, Decker KJ, Battle MA, Duncan SA, Hale MA, et al. Pancreas-Specific Deletion of Mouse Gata4 and Gata6 Causes Pancreatic Agenesis. *J Clin Invest* (2012) 122:3516–28. doi: 10.1172/JCI63352
174. Garreta E, Sanchez S, Lajara J, Montserrat N, Belmonte JCI. Roadblocks in the Path of iPSC to the Clinic. *Curr Transplant Rep* (2018) 5:14–8. doi: 10.1007/s40472-018-0177-x
175. Rančić N, Raščanin S, Miljković M, Jovanović M. Induced Pluripotent Stem Cells: Where We are Currently? *Halo 194* (2020) 26:153–61. doi: 10.5937/halo26-27861

13

Harnessing Proliferation for the Expansion of Stem Cell-Derived Pancreatic Cells: Advantages and Limitations

Amanda Oakie[1] *and Maria Cristina Nostro*[1,2*]

[1] *McEwen Stem Cell Institute, University Health Network, Toronto, ON, Canada,* [2] *Department of Physiology, University of Toronto, Toronto, ON, Canada*

Correspondence:
Maria Cristina Nostro
cristina.nostro@utoronto.ca

Restoring the number of glucose-responsive β-cells in patients living with diabetes is critical for achieving normoglycemia since functional β-cells are lost during the progression of both type 1 and 2 diabetes. Stem cell-derived β-cell replacement therapies offer an unprecedented opportunity to replace the lost β-cell mass, yet differentiation efficiencies and the final yield of insulin-expressing β-like cells are low when using established protocols. Driving cellular proliferation at targeted points during stem cell-derived pancreatic progenitor to β-like cell differentiation can serve as unique means to expand the final cell therapeutic product needed to restore insulin levels. Numerous studies have examined the effects of β-cell replication upon functionality, using primary islets *in vitro* and mouse models *in vivo*, yet studies that focus on proliferation in stem cell-derived pancreatic models are only just emerging in the field. This mini review will discuss the current literature on cell proliferation in pancreatic cells, with a focus on the proliferative state of stem cell-derived pancreatic progenitors and β-like cells during their differentiation and maturation. The benefits of inducing proliferation to increase the final number of β-like cells will be compared against limitations associated with driving replication, such as the blunted capacity of proliferating β-like cells to maintain optimal β-cell function. Potential strategies that may bypass the challenges induced by the up-regulation of cell cycle-associated factors during β-cell differentiation will be proposed.

Keywords: human pluripotent stem cell, beta cell, proliferation, islet, diabetes, in vitro differentiation

INTRODUCTION

Major advances in stem cell differentiation protocols for β-cell commitment have taken place throughout the past decade. Early reports that demonstrated the production of insulin$^+$ cells using *in vitro* differentiation protocols found that these cells were polyhormonal and did not form monohormonal β-cells *in vivo* (1–3). Subsequent studies have established that tightly-regulated sequential steps in differentiation that mimic human fetal islet development are required and that NKX6-1 expression is essential to generate pancreatic progenitors (PPs) that form monohormonal cells expressing both C-peptide and NKX6-1, which will be referred to as β-like cells in this review

(4–9). Stem cell-derived β-like cells demonstrate similar, yet not identical, characteristics to primary adult β-cells, including limited glucose-responsive insulin release, calcium dynamics, and mitochondria-driven oxidative phosphorylation (5, 8, 10–12). A defining feature of primary β-cells is their consistently low proliferation rate throughout adulthood, and the guided differentiation of human pluripotent stem cells (hPSCs) towards monohormonal β-like cells mimics this loss of replication, limiting the final number of cells available for further *in vitro* characterization or *in vivo* transplantation analyses. Promoting cell proliferation *in vitro* to generate higher yields of β-like cells can serve as a strategy to create a sufficient number of β-like cells for the reversal of hyperglycemia in diabetic patients. The following report will review current literature on the proliferation capacity of β-cells from primary human islets and in hPSCs differentiated to PPs and β-like cells. The balance between replication and maturation in stem cell-derived β-like cells and methods to increase total β-like cell yield for their potential application in therapeutics will be discussed.

CURRENT EVIDENCE OF HUMAN β-CELL PROLIFERATION

Replication During Fetal Development

β-cell proliferation rates during human islet development have been observed in multiple reports, but the limited availability of human fetal pancreatic tissue samples prevents the level of examination established with rodent models. What is evident from the current information available is that fetal pancreatic samples have comparatively high rates of replication in insulin-expressing cells when assessed against adult tissue samples. β-cell replication has been reported to be retained at ~3% from 10–23 weeks of gestational age in fetal pancreata, but is typically reduced to less than 1% in children less than 2 years old, although high variation in the percentage of proliferating β-cells has been noted during the first year of infancy (up to 5.28% $KI67^+$ β-cells) (13–18). Early stage (7.5 to 9.5 weeks of gestational age) human fetal pancreata transplanted under the kidney capsule of SCID mice demonstrated increased proliferation in progenitor PDX1 pancreatic cells and in early $insulin^+$ endocrine cells, but not in $NGN3^+$ endocrine-committed cells, which may suggest that the maturation status of endocrine cells is inversely related with cell replication (19). Further assessment of the correlation between proliferating insulin-expressing cells and their expression of maturation markers is required to conclude this.

Replication in Mature β-Cells

Low replication is found throughout all endocrine cell types of the adult islet, with approximately 0.38% of islet cells replicating in adults at one time (17). The human β-cell has been reported to have low (less than 0.5%) to no replication occurring under physiological conditions (17, 20). However, a recent report identified subsets of β-cells from donors that are highly proliferative (21). $KI67^+$ β-cells expressed distinct cell surface markers (CD9, CD44, CD49F, PDGFRA) and up-regulated signaling through ERK1/2, STAT3, and STAT5, potentially providing novel markers for identifying proliferating β-cells. Differences in β-cell replication rates within the developing and mature islet are likely due to the maturation status of cells. Fetal β-cells demonstrate a functionally immature phenotype when compared to adult cells and favour cell proliferation over glucose metabolism (10, 22, 23). The changes in fetal versus adult β-cell replication may also be driven by differing mechanisms of shared pathways. For example, activation of Glucagon-Like Peptide 1 Receptor (GLP-1R), which classically potentiates insulin release, also drives proliferation in adolescent, but not adult, β-cells through calcineurin -regulated transcription of cell cycle-associated genes (24). Further examination of developing β-cells can therefore reveal unique systems for driving replication.

β-Cell Proliferation in Diabetes and in Pregnancy

Reports on β-cell replication in patients with diabetes have contrasting findings. Islets from patients with type 1 diabetes do not display increased β-cell proliferation when compared to proliferation in control patient β-cells (16). However, patients with "recent onset" of type 1 diabetes, defined as less than 18 months since diagnosis, demonstrated a 10 times higher rate of β-cell replication compared to control patient β-cells, and this was not observed in patients with long-term type 1 diabetes or in patients with type 2 diabetes (25). This study found that islets from donors with recent onset type 1 diabetes, presenting with insulitis, had higher replication than islets without insulitis, suggesting that inflammation may drive proliferation during this short time frame. β-cells from patients with type 2 diabetes demonstrate unchanged proliferation when compared to control patients, yet proliferation is up-regulated in nondiabetic obese patients (26). This may mark a compensatory mechanism associated with balancing potential obesity stresses present in the period before gradual β-cell failure and apoptosis in type 2 islets. Recent reporting also suggests that human islets undergo adaptive mechanisms under pregnancy to increase the β-cell compartment, and that cadaveric sections from pregnant patients at later gestational ages (32–40 weeks) had increased β-cell proliferation when compared to β-cells from control patients (27). $Insulin^+$ staining near duct cells has been found in obese and pregnant patients and suggests a potential target for β-cell neogenesis, but examining proliferation or neogenesis in humans is limited due to their low detection and the limited availability of samples (18, 27–29). However, it does remain that inducing β-cell expansion can serve as a therapeutic for insulin restoration in patients with β-cell dysfunction.

Comparison of β-Cell Proliferation Between Rodents and Humans

Though rodent models of diabetes are often used to better understand the *in vivo* function of β-cells, evidence in the literature suggests that human β-cells do not demonstrate the same range of proliferation. The application of multiple techniques to induce proliferation in sorted primary β-cells *in vitro*—such as extracellular matrix supplementation and growth factor treatment—revealed that these techniques readily drove rodent β-cell replication but not human replication (30).

Screening primary human β-cells for mitogens that had been previously identified to initiate proliferation in rodent β-cells revealed that many factors were unable to initiate significant replication, with the dual-specificity tyrosine phosphorylation regulated kinase 1A (DYRK1A)-inhibiting compound harmine being the only mitogen that induced significantly increased replication with treatment (31). Indeed, harmine is unique in that it has been repeatedly reported in the literature to induce β-cell and other pancreatic cell replication (32–37). Differing effects on β-cell replication through shared signaling mechanisms between rodent and human β-cells may be due to additional cellular factors that inhibit proliferation, such as the high accumulation of p16 in adult human β-cells (38–40). Thus, careful interpretation of factors that induce rodent β-cell expansion should be taken when examining human β-cells.

PROLIFERATION RATES DURING hPSC DIFFERENTIATION TO β-LIKE CELLS

Pancreatic and Endocrine Progenitor Replication

Sources of hPSCs, such as embryonic or induced pluripotent stem cells, maintain the unique characteristic of self-renewal prior to lineage-specific differentiation. In order to give rise to the monohormonal β-like cell type that resembles primary β-cells, hPSCs are differentiated to pancreatic progenitor cells that demonstrate co-expression of transcription factors PDX1 and NKX6-1. These cells contain differentiation capacity towards all major pancreatic cell lineages (exocrine, ductal, and endocrine) and bypass the polyhormonal cell fate observed in C-peptide-expressing cells that have no detectable NKX6-1 (41–44). A significant portion of end-stage pancreatic progenitor cells have been found to still contain the replication marker KI67 (~30%–50%) (5, 45, 46). Although KI67 expression is present, pancreatic progenitors demonstrate up-regulation of the cyclin dependent kinase (CDK) inhibitors p21 and p16, suggesting that these cells may be exiting the cell cycle (47). The specification and proliferative rates of pancreatic progenitors can be affected by factors controlling differentiation. YAP, a factor required for Hippo signaling, has been shown to decrease during the differentiation from posterior foregut-like cells (PDX1/NKX6-1$^-$) towards PDX1/NKX6-1$^+$ pancreatic progenitors (48). However, chemical inhibition of YAP signaling during this developmental stage decreased both the percentage of PDX1/NKX6-1$^+$ cells and of KI67$^+$ cells, demonstrating that proper commitment towards pancreatic progenitors is also accompanied by proliferation of this compartment (48).

As pancreatic progenitors proceed to differentiate towards more specified endocrine progenitors, characterized by Chromogranin A$^+$/NKX6-1$^+$ expression, the expression of proliferation markers continues to decrease (47). In the same study that examined the effects of YAP signaling modification on pancreatic progenitor commitment and proliferation, it was found that promoting YAP activity during endocrine progenitor commitment through to β-like cell differentiation led to increased proliferation in cultured cells (48). However, this reduced the final percentage of monohormonal β-like cells, while inhibiting YAP during these stages increased the percentage of monohormonal β-like cells. The authors of this study tested the CDK inhibitor roscovitine during endocrine progenitor differentiation and found that endocrine progenitor commitment was not affected by CDK inhibition (48). In contrast, disrupting the cell cycle using the compound aphidicolin throughout endocrine progenitor and β-like cell commitment, which arrested endocrine progenitor cells at G1 and inhibited the completion of S phase, improved the differentiation of endocrine progenitor cells to β-like cells, but this effect was not seen in the same degree with CDK inhibition alone (49). This may suggest that compounds that disrupt cell cycle progression in endocrine progenitors enhance differentiation to end-stage β-like cells.

β-Like Cell Replication

At the stage where monohormonal hPSC-derived β-like cells emerge, only a small subset of end-stage β-like cells are actively replicating (~1%), as demonstrated using common markers of proliferation such as KI67 or 5-ethynyl-2'-deoxyuridine (EdU) incorporation assays (5, 32, 48, 50, 51). As mentioned in the preceding section, modifications to the proliferation of pancreatic or endocrine progenitors can impact the final population of β-like cells, which demonstrates a balance that must be maintained between successful β-like cell production and progenitor cell expansion. Although low proliferation rates in β-like cells follows the events seen during primary β-cell development and maturation, it limits the expansion of hPSC-derived β-like cells for further *in vitro* analyses and *in vivo* transplantation studies. For this reason, the search for methods to target β-like cell-specific replication *in vitro* is under continuous investigation. A notable report from the Melton group demonstrated that the leukemia inhibitory factor (LIF) drove proliferation in β-like cells expressing the corresponding receptor (LIFR). Activation of this pathway induced the expansion of β-like cells *in vitro* and was also able to enhance harmine-induced proliferation in treated β-like cells (32). To improve specificity and minimize the off-target effects from small molecule treatments, a recent study developed a zinc-binding prodrug for the deployment of compounds selectively in zinc-rich β-like cells. The study utilized β-like cells in an *in vitro* 3D platform to screen for molecules that targeted β-like cell replication. This novel platform allowed for the development of a harmine-carrying zinc-binding prodrug that was more effective than harmine alone at expanding β-like cells (37). Therefore, further development of compounds that induce targeted proliferation provide promising therapeutic avenues for both inducing and regulating β-like cell replication.

CAVEATS LINKED TO PROMOTING PROLIFERATION DURING hPSC DIFFERENTIATION TO β-LIKE CELLS

Inducible control over replication during the differentiation of hPSC-derived β-like cells is a potent strategy for restoring the number of β-cells for patients that have insufficient β-cell mass.

Before this can be pursued for basic research and potential clinical use, all challenges associated with cell cycle manipulation must be addressed.

A general issue that arises with the use of stem cell-derived cell sources is ensuring that proliferative capacity within end-stage β-like cells is tightly regulated to avoid uncontrolled cell growth. While cellular outgrowths have been occasionally detected following transplantation of cultures containing pancreatic progenitors (3, 7), endocrine progenitors (10), or end-stage cultures that failed to commit to the endocrine lineages (52), no outgrowths have been reported from transplantation of end-stage populations sorted for insulin-expressing cells (10). These findings from insulin-purified cell sorts are not surprising as human embryonic stem cell (hESC)-derived endocrine cells are typically post-mitotic, and there has been evidence that expanded pancreatic progenitors do not induce outgrowth when transplanted to mice (44). However, if proliferation is induced within β-like cells during *in vitro* differentiation, blocking proliferation prior to transplantation would become necessary to eliminate the risk of outgrowths.

The induction of proliferation during hPSC pancreatic differentiation to expand a transplantable population of β-like cells is challenged by reports that have documented inhibited commitment and maturation of pancreatic progenitors to end-stage β-like cells. Attempts made at expanding the pancreatic progenitor population have found that PDX1 cells generated do not necessarily express NKX6-1 and that pancreatic marker expression varied with passage (44, 46). The inverse relationship between proliferation and functional maturation of β-like cells was further supported in a recent study examining the cell surface CD9 marker (52). Depleting the stem cell-derived population of $CD9^+$ β-like cells removed the population of immature β-like cells, leaving cells that demonstrated higher expression of genes associated with β-cell maturity and insulin secretion. This study also established that human fetal pancreatic C-$peptide^+$/$KI67^+$ cells expressed higher frequencies of $CD9^+$ compared to the C-$peptide^+$/$KI67^-$ cells, confirming the higher proliferative status of the $CD9^+$ fraction *in vivo*. As mentioned previously, recent findings from the Melton lab identified YAP as one factor that could drive proliferation of β-like cells while reducing their maturation when up-regulated during their differentiation (32, 48). This group had also identified that WNT signaling was found in replicating epithelial progenitor cells and is down-regulated in mature endocrine-like cells (53). The TGF-β receptor inhibitor SB431542 has been shown to induce replication in adult human β-cells (54–56). However, SMAD2/3 inhibition through TGF-β can be detrimental in inducing monohormonal β-like cells when applied during pancreatic progenitor differentiation, instead leading towards early lineage polyhormonal endocrine cells that give rise to glucagon-expressing cells (1, 42). Prolonged TGF-β inhibition through ALK5 can also reduce β-like cell glucose-responsiveness when extended beyond endocrine progenitor commitment during β-like cell differentiation (12). Thus, exit from the cell cycle appears to enhance stem cell-derived cultures towards mature β-like cells *in vitro*. The inverse relationship between the maturation and proliferation of β-like cells can be further supported when looking at studies from primary β-cells in human and rodent islets. Signaling pathways such as NOTCH and mTOR have all demonstrated up-regulation in proliferating β-cells at the cost of their functionality (57–59). These findings present a challenge for successfully promoting proliferation in mature β-like cells since there are valid concerns with proliferation inducing characteristics of immature β-like cells and promoting unregulated expansion of transplanted grafts.

POTENTIAL STRATEGIES AND FUTURE DIRECTIONS

The end goal of stem cell-derived β-like cell research is to develop a transplantable system of cells that fully replicate β-cell function. Ideal tactics to achieve this would be to optimize current differentiation protocols, such as through manual purification of mature populations or treatments that increase final β-like cells (10, 12, 41, 52, 60, 61), and to improve *in vivo* site support for transplanted β-like cells (62–64). Despite the limitations listed above, expanding β-like cells or their progenitors *in vitro* still presents a promising strategy to further support these approaches. The final yield of end-stage β-like cells generated using current hPSC-based differentiation protocols is relatively poor and time-consuming, taking 23 days to 1 month to reach β-like cells (5, 10, 12, 53, 60, 65). Therefore, identifying strategies that promote controlled pancreas-specified cell proliferation *in vitro* could facilitate the manufacturing processes necessary to move forward with translational options.

Previous studies have identified methods to expand definitive endoderm, foregut endoderm, and pancreatic progenitor populations (**Figure 1**) (44, 46, 66, 67). While these approaches demonstrated the feasibility to induce self-renewal without impacting the developmental potential of hPSC-derived cells, these methods proved difficult to scale up. As a result of this, the generation of pancreatic cells for clinical purposes focuses on the expansion of undifferentiated hESCs, which can be grown and expanded in large bioreactors in the absence of special matrices or stromal cells (8, 68, 69). However, this approach is far from perfect, as a combination of proliferative and apoptotic events occurring during the differentiation process lead to a net outcome of 1 β-like cell for every 2 hESCs seeded (5). Therefore, being able to prevent these losses or expand cells at specific developmental stages will be key to increasing the β-cell yield. Strategies such as throughput screening, previously used to establish proliferative factors in primary β-cells, can be adapted for hESC-derived endocrine progenitor and β-like cell screening (31, 35, 70). With this tool, it is possible to screen for chemical or genetic regulators that provide control over proliferation in pancreatic cells. In addition to using compound-based treatments previously mentioned, such as harmine (34, 35, 37), another method for inducing replication could be through viral delivery of controllable systems (70–72). Once proliferative drivers have been identified, rigorous examination of the expanded population must confirm

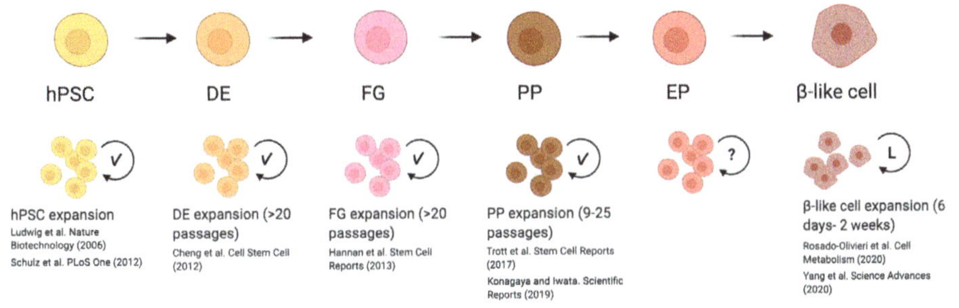

FIGURE 1 | Reported mechanisms and caveats for expanding cells during β-like cell differentiation. Previously reported methods for inducing replication of human pluripotent stem cells (hPSC) to end-stage β-like cells demonstrated that hPSC, definitive endoderm (DE), foregut (FG), and pancreatic progenitor (PP) stages are able to undergo significant expansion with the development of optimized culture conditions (indicated with check mark). These expanded cells are able to maintain their developmental potential with serial passaging. Although proliferation rates have been reported for endocrine progenitors (EP), methods to induce their replication *in vitro* are not well-studied (indicated with question mark). Inducing replication in β-like cells has been successfully initiated. However, β-like cells were only examined after a limited period of expansion *in vitro* (indicated by L), and the long-term effects of driving proliferation are not known. This figure was created using BioRender.com.

that no aberrant mutations have been acquired during the process and that expanded cells retain the ability to form β-like cells. Importantly, if proliferation is induced in β-like cells, one would need to show that cell cycle progression can be blocked and confirm that β-like cells retain maturation features, such as the ability to recapitulate first and second phases of insulin secretion in response to glucose challenge and other secretagogues, and that these cells express molecular markers indicative of bona fide β-cells (10, 12, 60).

Ultimately, the successful establishment of controlled proliferation of hESC-derived differentiating pancreatic cells will allow for the large-scale production of glucose-regulating cells and reduce the high costs associated with growing large batches of undifferentiated hPSCs from the initial steps of differentiation. Strict guidelines for the proposed protocols must ensure that the end product population generate islet-like cells with the ability to normalize glycemia without the risk of teratoma or outgrowth formation. The combination of transient expansion with other strategies to optimize β-like cell differentiation will inform the development of an accessible cellular bank for patients, and ideally contribute to their permanent halt of exogenous therapies for glucose management.

AUTHOR CONTRIBUTIONS

AO and MCN wrote the manuscript. All authors contributed to the article and approved the submitted version.

ACKNOWLEDGMENTS

The authors would like to thank Dr. Rangarajan Sambathkumar and Emily McGaugh for their invaluable help in reviewing this manuscript and providing critical feedback.

REFERENCES

1. Nostro MC, Sarangi F, Ogawa S, Holtzinger A, Corneo B, Li X, et al. Stage-specific signaling through TGFβ family members and WNT regulates patterning and pancreatic specification of human pluripotent stem cells. *Development* (2011) 138(5):861–71. doi: 10.1242/dev.055236
2. Basford CL, Prentice KJ, Hardy AB, Sarangi F, Micallef SJ, Li X, et al. The functional and molecular characterisation of human embryonic stem cell-derived insulin-positive cells compared with adult pancreatic beta cells. *Diabetologia* (2012) 55(2):358–71. doi: 10.1007/s00125-011-2335-x
3. Kelly OG, Chan MY, Martinson LA, Kadoya K, Ostertag TM, Ross KG, et al. Cell-surface markers for the isolation of pancreatic cell types derived from human embryonic stem cells. *Nat Biotechnol* (2011) 29(8):750–6. doi: 10.1038/nbt.1931
4. Rezania A, Bruin JE, Riedel MJ, Mojibian M, Asadi A, Xu J, et al. Maturation of human embryonic stem cell-derived pancreatic progenitors into functional islets capable of treating pre-existing diabetes in mice. *Diabetes* (2012) 61 (8):2016–29. doi: 10.2337/db11-1711
5. Rezania A, Bruin JE, Arora P, Rubin A, Batushansky I, Asadi A, et al. Reversal of diabetes with insulin-producing cells derived in vitro from human pluripotent stem cells. *Nat Biotechnol* (2014) 32(11):1121–33. doi: 10.1038/nbt.3033
6. Rezania A, Bruin JE, Xu J, Narayan K, Fox JK, O'Neil JJ, et al. Enrichment of human embryonic stem cell-derived NKX6.1-expressing pancreatic progenitor cells accelerates the maturation of insulin-secreting cells in vivo. *Stem Cells* (2013) 31(11):2432–42. doi: 10.1002/stem.1489
7. Kroon E, Martinson LA, Kadoya K, Bang AG, Kelly OG, Eliazer S, et al. Pancreatic endoderm derived from human embryonic stem cells generates glucose-responsive insulin-secreting cells in vivo. *Nat Biotechnol* (2008) 26 (4):443–52. doi: 10.1038/nbt1393
8. Pagliuca FW, Millman JR, Gürtler M, Segel M, Van Dervort A, Ryu JH, et al. Generation of functional human pancreatic β cells in vitro. *Cell* (2014) 159 (2):428–39. doi: 10.1016/j.cell.2014.09.040

9. Nostro MC, Sarangi F, Yang C, Holland A, Elefanty AG, Stanley EG, et al. Efficient generation of NKX6-1+ pancreatic progenitors from multiple human pluripotent stem cell lines. *Stem Cell Rep* (2015) 4(4):591–604. doi: 10.1016/j.stemcr.2015.02.017
10. Nair GG, Liu JS, Russ HA, Tran S, Saxton MS, Chen R, et al. Recapitulating endocrine cell clustering in culture promotes maturation of human stem-cell-derived β cells. *Nat Cell Biol* (2019) 21(2):263–74. doi: 10.1038/s41556-018-0271-4
11. Augsornworawat P, Maxwell KG, Velazco-Cruz L, Millman JR. Single-Cell Transcriptome Profiling Reveals β Cell Maturation in Stem Cell-Derived Islets after Transplantation. *Cell Rep* (2020) 32(8):108067. doi: 10.1016/j.celrep.2020.108067
12. Velazco-Cruz L, Song J, Maxwell KG, Goedegebuure MM, Augsornworawat P, Hogrebe NJ, et al. Acquisition of Dynamic Function in Human Stem Cell-Derived β Cells. *Stem Cell Rep* (2019) 12(2):351–65. doi: 10.1016/j.stemcr.2018.12.012
13. Lyttle BM, Li J, Krishnamurthy M, Fellows F, Wheeler MB, Goodyer CG, et al. Transcription factor expression in the developing human fetal endocrine pancreas. *Diabetologia* (2008) 51(7):1169–80. doi: 10.1007/s00125-008-1006-z
14. Bouwens L, Lu WG, De Krijger R. Proliferation and differentiation in the human fetal endocrine pancreas. *Diabetologia* (1997) 40(4):398–404. doi: 10.1007/s001250050693
15. Kassem SA, Ariel I, Thornton PS, Scheimberg I, Glaser B. Beta-cell proliferation and apoptosis in the developing normal human pancreas and in hyperinsulinism of infancy. *Diabetes* (2000) 49(8):1325–33. doi: 10.2337/diabetes.49.8.1325
16. Lam CJ, Jacobson DR, Rankin MM, Cox AR, Kushner JA. β Cells Persist in T1D Pancreata Without Evidence of Ongoing β-Cell Turnover or Neogenesis. *J Clin Endocrinol Metab* (2017) 102(8):2647–59. doi: 10.1210/jc.2016-3806
17. Lam CJ, Cox AR, Jacobson DR, Rankin MM, Kushner JA. Highly Proliferative α-Cell-Related Islet Endocrine Cells in Human Pancreata. *Diabetes* (2018) 67(4):674–86. doi: 10.2337/db17-1114
18. Meier JJ, Butler AE, Saisho Y, Monchamp T, Galasso R, Bhushan A, et al. Beta-cell replication is the primary mechanism subserving the postnatal expansion of beta-cell mass in humans. *Diabetes* (2008) 57(6):1584–94. doi: 10.2337/db07-1369
19. Castaing M, Duvillié B, Quemeneur E, Basmaciogullari A, Scharfmann R. Ex vivo analysis of acinar and endocrine cell development in the human embryonic pancreas. *Dev Dyn* (2005) 234(2):339–45. doi: 10.1002/dvdy.20547
20. Gregg BE, Moore PC, Demozay D, Hall BA, Li M, Husain A, et al. Formation of a human β-cell population within pancreatic islets is set early in life. *J Clin Endocrinol Metab* (2012) 97(9):3197–206. doi: 10.1210/jc.2012-1206
21. Wang YJ, Golson ML, Schug J, Traum D, Liu C, Vivek K, et al. Single-Cell Mass Cytometry Analysis of the Human Endocrine Pancreas. *Cell Metab* (2016) 24(4):616–26. doi: 10.1016/j.cmet.2016.09.007
22. Jermendy A, Toschi E, Aye T, Koh A, Aguayo-Mazzucato C, Sharma A, et al. Rat neonatal beta cells lack the specialised metabolic phenotype of mature beta cells. *Diabetologia* (2011) 54(3):594–604. doi: 10.1007/s00125-010-2036-x
23. Jacovetti C, Matkovich SJ, Rodriguez-Trejo A, Guay C, Regazzi R. Postnatal β-cell maturation is associated with islet-specific microRNA changes induced by nutrient shifts at weaning. *Nat Commun* (2015) 6:8084. doi: 10.1038/ncomms9084
24. Dai C, Hang Y, Shostak A, Poffenberger G, Hart N, Prasad N, et al. Age-dependent human β cell proliferation induced by glucagon-like peptide 1 and calcineurin signaling. *J Clin Invest* (2017) 127(10):3835–44. doi: 10.1172/JCI91761
25. Willcox A, Richardson SJ, Bone AJ, Foulis AK, Morgan NG. Evidence of increased islet cell proliferation in patients with recent-onset type 1 diabetes. *Diabetologia* (2010) 53(9):2020–8. doi: 10.1007/s00125-010-1817-6
26. Hanley SC, Austin E, Assouline-Thomas B, Kapeluto J, Blaichman J, Moosavi M, et al. β-Cell mass dynamics and islet cell plasticity in human type 2 diabetes. *Endocrinology* (2010) 151(4):1462–72. doi: 10.1210/en.2009-1277
27. Dirice E, De Jesus DF, Kahraman S, Basile G, Ng RW, El Ouaamari A, et al. Human duct cells contribute to β cell compensation in insulin resistance. *JCI Insight* (2019) 4(8):e99576. doi: 10.1172/jci.insight.99576
28. Butler AE, Janson J, Bonner-Weir S, Ritzel R, Rizza RA, Butler PC. Beta-cell deficit and increased beta-cell apoptosis in humans with type 2 diabetes. *Diabetes* (2003) 52(1):102–10. doi: 10.2337/diabetes.52.1.102
29. Butler AE, Cao-Minh L, Galasso R, Rizza RA, Corradin A, Cobelli C, et al. Adaptive changes in pancreatic beta cell fractional area and beta cell turnover in human pregnancy. *Diabetologia* (2010) 53(10):2167–76. doi: 10.1007/s00125-010-1809-6
30. Parnaud G, Bosco D, Berney T, Pattou F, Kerr-Conte J, Donath MY, et al. Proliferation of sorted human and rat beta cells. *Diabetologia* (2008) 51(1):91–100. doi: 10.1007/s00125-007-0855-1
31. Aamodt KI, Aramandla R, Brown JJ, Fiaschi-Taesch N, Wang P, Stewart AF, et al. Development of a reliable automated screening system to identify small molecules and biologics that promote human β-cell regeneration. *Am J Physiol Endocrinol Metab* (2016) 311(5):E859–68. doi: 10.1152/ajpendo.00515.2015
32. Rosado-Olivieri EA, Aigha II, Kenty JH, Melton DA. Identification of a LIF-Responsive, Replication-Competent Subpopulation of Human β Cells. *Cell Metab* (2020) 31(2):327–38.e326. doi: 10.1016/j.cmet.2019.12.009
33. Maachi H, Ghislain J, Tremblay C, Poitout V. Pronounced proliferation of non-beta cells in response to beta-cell mitogens in isolated human islets of Langerhans. *Biorxiv* (2020). Vol Preprint. doi: 10.1101/2020.10.28.359158
34. Wang P, Karakose E, Liu H, Swartz E, Ackeifi C, Zlatanic V, et al. Combined Inhibition of DYRK1A, SMAD, and Trithorax Pathways Synergizes to Induce Robust Replication in Adult Human Beta Cells. *Cell Metab* (2019) 29(3):638–52.e635. doi: 10.1016/j.cmet.2018.12.005
35. Wang P, Alvarez-Perez JC, Felsenfeld DP, Liu H, Sivendran S, Bender A, et al. A high-throughput chemical screen reveals that harmine-mediated inhibition of DYRK1A increases human pancreatic beta cell replication. *Nat Med* (2015) 21(4):383–8. doi: 10.1038/nm.3820
36. Dirice E, Walpita D, Vetere A, Meier BC, Kahraman S, Hu J, et al. Inhibition of DYRK1A Stimulates Human β-Cell Proliferation. *Diabetes* (2016) 65(6):1660–71. doi: 10.2337/db15-1127
37. Yang K, Lee M, Jones PA, Liu SS, Zhou A, Xu J, et al. A 3D culture platform enables development of zinc-binding prodrugs for targeted proliferation of β cells. *Sci Adv* (2020) 6(47):eabc3207. doi: 10.1126/sciadv.abc3207
38. Tschen SI, Georgia S, Dhawan S, Bhushan A. Skp2 is required for incretin hormone-mediated β-cell proliferation. *Mol Endocrinol* (2011) 25(12):2134–43. doi: 10.1210/me.2011-1119
39. Krishnamurthy J, Ramsey MR, Ligon KL, Torrice C, Koh A, Bonner-Weir S, et al. p16INK4a induces an age-dependent decline in islet regenerative potential. *Nature* (2006) 443(7110):453–7. doi: 10.1038/nature05092
40. Zhong L, Georgia S, Tschen SI, Nakayama K, Bhushan A. Essential role of Skp2-mediated p27 degradation in growth and adaptive expansion of pancreatic beta cells. *J Clin Invest* (2007) 117(10):2869–76. doi: 10.1172/JCI32198
41. Hogrebe NJ, Augsornworawat P, Maxwell KG, Velazco-Cruz L, Millman JR. Targeting the cytoskeleton to direct pancreatic differentiation of human pluripotent stem cells. *Nat Biotechnol* (2020) 38(4):460–70. doi: 10.1038/s41587-020-0430-6
42. Korytnikov R, Nostro MC. Generation of polyhormonal and multipotent pancreatic progenitor lineages from human pluripotent stem cells. *Methods* (2016) 101:56–64. doi: 10.1016/j.ymeth.2015.10.017
43. McGaugh EC, Nostro MC. Efficient Differentiation of Pluripotent Stem Cells to NKX6-1+ Pancreatic Progenitors. *J Vis Exp* (2017) 121:e55265. doi: 10.3791/55265
44. Trott J, Tan EK, Ong S, Titmarsh DM, Denil SLIJ, Giam M, et al. Long-Term Culture of Self-renewing Pancreatic Progenitors Derived from Human Pluripotent Stem Cells. *Stem Cell Rep* (2017) 8(6):1675–88. doi: 10.1016/j.stemcr.2017.05.019
45. Wesolowska-Andersen A, Jensen RR, Alcántara MP, Beer NL, Duff C, Nylander V, et al. Analysis of Differentiation Protocols Defines a Common Pancreatic Progenitor Molecular Signature and Guides Refinement of Endocrine Differentiation. *Stem Cell Rep* (2020) 14(1):138–53. doi: 10.1016/j.stemcr.2019.11.010
46. Konagaya S, Iwata H. Chemically defined conditions for long-term maintenance of pancreatic progenitors derived from human induced pluripotent stem cells. *Sci Rep* (2019) 9(1):640. doi: 10.1038/s41598-018-36606-7
47. Ameri J, Borup R, Prawiro C, Ramond C, Schachter KA, Scharfmann R, et al. Efficient Generation of Glucose-Responsive Beta Cells from Isolated GP2. *Cell Rep* (2017) 19(1):36–49. doi: 10.1016/j.celrep.2017.03.032

48. Rosado-Olivieri EA, Anderson K, Kenty JH, Melton DA. YAP inhibition enhances the differentiation of functional stem cell-derived insulin-producing β cells. *Nat Commun* (2019) 10(1):1464. doi: 10.1038/s41467-019-09404-6
49. Sui L, Xin Y, Georgieva D, Diedenhofen G, Haataja L, et al. Reduced replication fork speed promotes pancreatic endocrine differentiation and controls graft size. *JCI Insight* (2020) In press preview. doi: 10.1172/jci.insight.141553
50. Russ HA, Parent AV, Ringler JJ, Hennings TG, Nair GG, Shveygert M, et al. Controlled induction of human pancreatic progenitors produces functional beta-like cells in vitro. *EMBO J* (2015) 34(13):1759–72. doi: 10.15252/embj.201591058
51. Romer AI, Singer RA, Sui L, Egli D, Sussel L. Murine Perinatal β-Cell Proliferation and the Differentiation of Human Stem Cell-Derived Insulin-Expressing Cells Require NEUROD1. *Diabetes* (2019) 68(12):2259–71. doi: 10.2337/db19-0117
52. Li X, Yang KY, Chan VW, Leung KT, Zhang XB, Wong AS, et al. Single-Cell RNA-Seq Reveals that CD9 Is a Negative Marker of Glucose-Responsive Pancreatic β-like Cells Derived from Human Pluripotent Stem Cells. *Stem Cell Rep* (2020) 15(5):1111–26. doi: 10.1016/j.stemcr.2020.09.009
53. Sharon N, Vanderhooft J, Straubhaar J, Mueller J, Chawla R, Zhou Q, et al. Wnt Signaling Separates the Progenitor and Endocrine Compartments during Pancreas Development. *Cell Rep* (2019) 27(8):2281–91.e2285. doi: 10.1016/j.celrep.2019.04.083
54. Basile G, Kulkarni RN, Morgan NG. How, When, and Where Do Human β-Cells Regenerate? *Curr Diabetes Rep* (2019) 19(8):48. doi: 10.1007/s11892-019-1176-8
55. Xiao X, Fischbach S, Song Z, Gaffar I, Zimmerman R, Wiersch J, et al. Transient Suppression of TGFβ Receptor Signaling Facilitates Human Islet Transplantation. *Endocrinology* (2016) 157(4):1348–56. doi: 10.1210/en.2015-1986
56. Dhawan S, Dirice E, Kulkarni RN, Bhushan A. Inhibition of TGF-β Signaling Promotes Human Pancreatic β-Cell Replication. *Diabetes* (2016) 65(5):1208–18. doi: 10.2337/db15-1331
57. Bartolome A, Zhu C, Sussel L, Pajvani UB. Notch signaling dynamically regulates adult β cell proliferation and maturity. *J Clin Invest* (2019) 129 (1):268–80. doi: 10.1172/JCI98098
58. Bar Y, Russ HA, Knoller S, Ouziel-Yahalom L, Efrat S. HES-1 is involved in adaptation of adult human beta-cells to proliferation in vitro. *Diabetes* (2008) 57(9):2413–20. doi: 10.2337/db07-1323
59. Jaafar R, Tran S, Shah AN, Sun G, Valdearcos M, Marchetti P, et al. mTORC1 to AMPK switching underlies β-cell metabolic plasticity during maturation and diabetes. *J Clin Invest* (2019) 129(10):4124–37. doi: 10.1172/JCI127021
60. Veres A, Faust AL, Bushnell HL, Engquist EN, Kenty JH, Harb G, et al. Charting cellular identity during human in vitro β-cell differentiation. *Nature* (2019) 569(7756):368–73. doi: 10.1038/s41586-019-1168-5
61. Mamidi A, Prawiro C, Seymour PA, de Lichtenberg KH, Jackson A, Serup P, et al. Mechanosignalling via integrins directs fate decisions of pancreatic progenitors. *Nature* (2018) 564(7734):114–8. doi: 10.1038/s41586-018-0762-2
62. Pepper AR, Pawlick R, Bruni A, Wink J, Rafiei Y, O'Gorman D, et al. Transplantation of Human Pancreatic Endoderm Cells Reverses Diabetes Post Transplantation in a Prevascularized Subcutaneous Site. *Stem Cell Rep* (2017) 8(6):1689–700. doi: 10.1016/j.stemcr.2017.05.004
63. Bruin JE, Rezania A, Xu J, Narayan K, Fox JK, O'Neil JJ, et al. Maturation and function of human embryonic stem cell-derived pancreatic progenitors in macroencapsulation devices following transplant into mice. *Diabetologia* (2013) 56(9):1987–98. doi: 10.1007/s00125-013-2955-4
64. Talavera-Adame D, Woolcott OO, Ignatius-Irudayam J, Arumugaswami V, Geller DH, Dafoe DC. Effective endothelial cell and human pluripotent stem cell interactions generate functional insulin-producing beta cells. *Diabetologia* (2016) 59(11):2378–86. doi: 10.1007/s00125-016-4078-1
65. Yung T, Poon F, Liang M, Coquenlorge S, McGaugh EC, Hui CC, et al. Sufu- and Spop-mediated downregulation of Hedgehog signaling promotes beta cell differentiation through organ-specific niche signals. *Nat Commun* (2019) 10 (1):4647. doi: 10.1038/s41467-019-12624-5
66. Cheng X, Ying L, Lu L, Galvão AM, Mills JA, Lin HC, et al. Self-renewing endodermal progenitor lines generated from human pluripotent stem cells. *Cell Stem Cell* (2012) 10(4):371–84. doi: 10.1016/j.stem.2012.02.024
67. Hannan NR, Fordham RP, Syed YA, Moignard V, Berry A, Bautista R, et al. Generation of multipotent foregut stem cells from human pluripotent stem cells. *Stem Cell Rep* (2013) 1(4):293–306. doi: 10.1016/j.stemcr.2013.09.003
68. Schulz TC, Young HY, Agulnick AD, Babin MJ, Baetge EE, Bang AG, et al. A scalable system for production of functional pancreatic progenitors from human embryonic stem cells. *PLoS One* (2012) 7(5):e37004. doi: 10.1371/journal.pone.0037004
69. Ludwig TE, Levenstein ME, Jones JM, Berggren WT, Mitchen ER, Frane JL, et al. Derivation of human embryonic stem cells in defined conditions. *Nat Biotechnol* (2006) 24(2):185–7. doi: 10.1038/nbt1177
70. Robitaille K, Rourke JL, McBane JE, Fu A, Baird S, Du Q, et al. High-throughput Functional Genomics Identifies Regulators of Primary Human Beta Cell Proliferation. *J Biol Chem* (2016) 291(9):4614–25. doi: 10.1074/jbc.M115.683912
71. Takane KK, Kleinberger JW, Salim FG, Fiaschi-Taesch NM, Stewart AF. Regulated and reversible induction of adult human β-cell replication. *Diabetes* (2012) 61(2):418–24. doi: 10.2337/db11-0580
72. Tanaka A, Watanabe A, Nakano Y, Matsumoto M, Okazaki Y, Miyajima A. Reversible expansion of pancreatic islet progenitors derived from human induced pluripotent stem cells. *Genes Cells* (2020) 25(5):302–11. doi: 10.1111/gtc.12759

14

Protecting Stem Cell Derived Pancreatic Beta-Like Cells from Diabetogenic T Cell Recognition

Roberto Castro-Gutierrez[1], *Aimon Alkanani*[1], *Clayton E. Mathews*[2], *Aaron Michels*[1] *and Holger A. Russ*[1*]

[1] Barbara-Davis Center for Diabetes, University of Colorado Anschutz, Aurora, CO, United States, [2] Department of Pathology, Immunology, and Laboratory Medicine, College of Medicine, University of Florida, Gainesville, FL, United States

***Correspondence:**
Holger A. Russ
holger.russ@cuanschutz.edu

Type 1 diabetes results from an autoimmune attack directed at pancreatic beta cells predominantly mediated by T cells. Transplantation of stem cell derived beta-like cells (sBC) have been shown to rescue diabetes in preclinical animal models. However, how sBC will respond to an inflammatory environment with diabetogenic T cells in a strict human setting has not been determined. This is due to the lack of model systems that closely recapitulates human T1D. Here, we present a reliable *in vitro* assay to measure autologous CD8 T cell stimulation against sBC in a human setting. Our data shows that upon pro-inflammatory cytokine exposure, sBC upregulate Human Leukocyte Antigen (HLA) class I molecules which allows for their recognition by diabetogenic CD8 T cells. To protect sBC from this immune recognition, we utilized genome engineering to delete surface expression of HLA class I molecules and to integrate an inducible overexpression system for the immune checkpoint inhibitor Programmed Death Ligand 1 (PD-L1). Genetically engineered sBC that lack HLA surface expression or overexpress PD-L1 showed reduced stimulation of diabetogenic CD8 T cells when compared to unmodified cells. Here, we present evidence that manipulation of HLA class I and PD-L1 receptors on sBC can provide protection from diabetes-specific immune recognition in a human setting.

Keywords: autoimmune type 1 diabetes, stem cell derived pancreatic insulin producing beta-like cells, beta-immune cell interface, PD-L1 and HLA class I, beta cell replacement therapy, genome engineering, human pluripotent stem cells, direct differentiation

INTRODUCTION

Type 1 diabetes mellitus (T1D) results from the autoimmune-mediated destruction of insulin producing beta cells in the pancreas (1). This specific beta cell depletion results in life threatening hyperglycemia if left untreated. Exogenous administration of insulin is necessary to treat the metabolic disturbances in T1D. However, management of the disease with insulin therapy remains challenging with risk for acute hypoglycemia and severe long-term complications. Despite the advancements in T1D technology including subcutaneous insulin pumps, continuous glucose monitors, and hybrid closed loop insulin delivery systems (2), a practical cure is still lacking.

Replacing lost beta cells by transplantation of cadaveric donor islets, which contain beta cells, along with simultaneous immune suppression has been proposed as a cure for T1D (3). Islet transplantation into T1D patients results in insulin independence for an average of 34 months (4), demonstrating a proof of principle that beta cell transplantation could indeed represent a practical cure for T1D patients. However, important caveats still remain: insufficient donor tissue availability, donor incompatibility, and remaining autoimmunity (5). To address the shortage of donor tissue for transplantation, research efforts have focused on using direct differentiation of human pluripotent stem cells (hPSC) into pancreatic and insulin producing, stem cell derived beta-like cells (sBC) to effectively create an abundant source of functional beta cells (6–10). At present it is still largely unknown how sBC respond to a human diabetogenic environment. This is critically important knowledge that is needed to protect sBC from immune rejection and autoimmune-related injury prior to successful transplantation into T1D patients.

Transplantation of macroencapsulated sBC progenitor cells into animal models is a viable approach to confer protection from the host immune system, but cell survival and function has been disappointing. Indeed, scar tissue formation, foreign body reactions, and poor vascularization are critical concerns for this technology (11). Microencapsulation of sBC using alginate derivatives allows for a better exchange of nutrient and gases which are important for sBC function, thus providing an attractive approach for cell replacement therapy (12). However, transplantation of naked sBC, without the need for cell encapsulation, would provide distinct benefits. Naked sBC grafts would be readily vascularized, improving survival and function, while mitigating foreign body reactions. On the other hand, naked sBC are more likely vulnerable to an immediate inflammatory environment and an immune attack without a protective layer provided by encapsulation approaches. Thus, alternative strategies need to be developed to address the challenges associated with cell replacement therapy using sBC for patients suffering from diabetes.

Human beta cells upregulate multiple receptors in response to an inflammatory environment in order to interact with the immune system. It is known that human beta cells express human leukocyte antigen (HLA) class I molecules which are implicated in the pathogenesis of T1D by presentation of diabetogenic antigens to CD8 T cells (13, 14). Self-reactive cytotoxic CD8 T cells with T cell receptors (TCRs) specific for beta cell antigens have been identified in T1D patients (15). However, not much is known regarding how sBC will react to these self-reactive, diabetogenic CD8 T cells that are already in circulation in T1D patients. This gap in knowledge is primary due to lack of a reliable assay to measure T cell responses towards sBC in a strictly human autologous setting.

Furthermore, cadaveric beta cells can upregulate the immune checkpoint inhibitor Programmed Death-Ligand 1 (PD-L1) receptor in response to pro-inflammatory cytokines (16–18) as a protecting mechanism. Interestingly, anti-PD-1 and anti-PD-L1 cancer therapy has resulted in spontaneous development of autoimmune diabetes in a subset of patients (19–21). Additionally, a recent study indicates that sBC overexpressing PD-L1 survive longer after transplantation into immune competent mice (22), implicating an important role for PD-L1 in providing peripheral tolerance to beta cells and preventing autoimmune diabetes. However, this study was limited by the use of a xenogeneic model.

Here we established a reproducible, well-defined assay system to measure T cell stimulation *in vitro* in order to study human sBC and T-cell interactions. We confirmed that sBC upregulate HLA class I molecules similar to human islet beta cells when exposed to pro-inflammatory cytokines. We further demonstrated that sBC are able to stimulate diabetogenic human T cells reactive to the self-antigen preproinsulin (PPI) in an HLA-peptide-TCR dependent manner. Then we sought to evaluate the hypothesis that sBC lacking HLA class I molecules and overexpressing PD-L1 would not be recognized by diabetogenic CD8 T cells when exposed to inflammatory cytokines. Using this novel system we provided direct evidence that sBC stimulate human diabetogenic CD8 T cells in an HLA-dependent manner and that PD-L1 overexpression can effectively reduce this stimulation.

MATERIALS AND METHODS

hPSC Culture and Differentiation of Stem Cell Derived Beta-Like Cells (sBC)

Undifferentiated human pluripotent stem (hPSC) Mel1$^{INS\text{-}GFP}$ reporter cells (23) were maintained on hES qualified Matrigel (Corning #354277) in mTeSR+ media (STEMCELL Technologies #05826). Differentiation to stem cell derived beta like cells (sBC) was carried out in suspension-based, bioreactor magnetic stirring system (Reprocell #ABBWVS03A-6, #ABBWVDW-1013, #ABBWBP03N0S-6) as follows. Confluent hPSC cultures were dissociated into single-cell suspension by incubation with TrypLE (Gibco #12-604-021) for 8 min at 37C. Detached cells were quenched with mTESR+ media. Cells were then counted using a MoxiGo II cell counter (Orflow), followed by seeding 0.5×10^6 cells/ml in mTeSR+ media supplemented with 10uM ROCK inhibitor (Y-27632, R&D Systems #1254-50). Bioreactors were placed on a magnetic stirring system set at 60rpm in a cell culture incubator with 5% CO_2 to induce sphere formation for 48 hours. To induce definitive endoderm differentiation, spheres were collected in a 50ml Falcon tube, allowed to settle by gravity, washed once with RPMI (Gibco #11-875-093) + 0.2% FBS, and re-suspended in d1 media [RPMI containing 0.2% FBS, 1:5,000 ITS (Gibco #41400-045), 100ng/ml Activin A (R&D Systems #338-AC-01M), and 3uM CHIR99021 (STEMCELL Technologies #72054)]. Differentiation media was changed daily by letting spheres settle by gravity for 3-10min. Most supernatant was removed by aspiration; fresh media was added, and bioreactors were placed back on stirrer system. sBC differentiation was based on published protocol (Russ et al, 2015) with modifications as outlined below. Differentiation medias are as: day 2-3: RPMI containing 0.2% FBS, 1:2,000 ITS, and 100ng/

ml Activin A; d4-5: RPMI containing 2% FBS, 1:1,000 ITS, and 50ng/ml KGF (Prepotech #100-19-1MG); d6: DMEM with 4.5g/L D-glucose (Gibco #11960-044) containing 1:100 SM1 (STEMCELL Technologies #5711), 1:100 NEAA (Gibco #11140-050), 1mM Sodium Pyruvate (Gibco #11360-070), 1:100 GlutaMAX (Gibco #35050-061), 3nM TTNPB, (R&D Systems #0761), 250nM Sant-1 (R&D Systems #1974), 250nM LDN (STEMCELL Technologies #72149), 30nM PMA (Sigma Aldrich #P1585-1MG), 50ug/ml 2-phospho-L-ascorbic acid trisodium salt (VitC) (Sigma #49752-10G); d7: DMEM containing 1:100 SM1, 1:100 NEAA, 1mM Sodium Pyruvate, 1:100 GlutaMAX, 3nM TTNPB, and 50ug/ml VitC; d8-9: DMEM containing 1:100 SM1, 1:100 NEAA, 1mM Sodium Pyruvate, 1:100 GlutaMAX, 100ng/ml EGF (R&D Systems #236-EG-01M), 50ng/ml KGF, and 50ug/ml VitC; d10-16: DMEM containing 2% fraction V BSA, 1:100 NEAA, 1mM Sodium Pyruvate, 1:100 GlutaMAX, 1:100 ITS, 10ug/ml Heparin (Sigma #H3149-250KU), 2mM N-Acetyl-L-cysteine (Cysteine) (Sigma #A9165-25G), 10uM Zinc sulfate heptahydrate (Zinc) (Sigma #Z0251-100g), 1x BME, 10uM Alk5i II RepSox (R&D Systems #3742/50), 1uM 3,3',5-Triiodo-L-thyronine sodium salt (T3) (Sigma #T6397), 0.5uM LDN, 1uM Gamma Secretase Inhibitor XX (XXi) (AsisChem #ASIS-0149) and 1:250 1M NaOH to adjust pH to ~7.4; d17-23: CMRL (Gibco #11530-037) containing 1% BSA, 1:100 NEAA, 1mM Sodium Pyruvate, 1:100 GlutaMAX, 10ug/ml Heparin, 2mM Cysteine, 10uM Zinc, 1x BME, 10uM Alk5i II RepSox, 1uM T3, 50ug/ml VitC, and 1:250 NaOH to adjust pH to ~7.4. All medias also contained 1x PenStrep.

Human Islet Culture

Human islets were cultured in CMRL media containing 10% FBS, 1X Pen/Strep, 1X BME.

Unos ID/Batch Type	Purity	Institute	BMI	Age
R290	80%	University of Alberta	35.8	74
R299	75%	University of Alberta	25.4	44
RRID : SAMN11000549	85%	The Scharp-Lacy Research Institute	32.9	47

Immunofluorescence

sBC and human islet clusters were dissociated for (i) cytospin studies or (ii) embedded for cryo-sectioning. (i) huma islets and sBC clusters were washed with PBS and incubated with 0.05% Trypsin with EDTA at 37 C for 12 min to create a single cell suspension. Cells were quenched with 2% FBS in PBS and filtered using a cell strainer. Cells were cytospun into glass slides and fixed for 15 min at room temperature with 4% paraformaldehyde. (ii) sBC clusters or human islets were fixed with 4% paraformaldehyde then washed twice with PBS. Fixed clusters for cryo-sectioning were incubated overnight in 30% sucrose (Sigma #S0389) before embedding in tissue-tek OCT (Sakura #4583) and storing at -80C overnight. OCT blocks containing fixed clusters were cryo-sectioned (10um thickness) and transferred to glass slides. Blocking and staining of cryo-sections proceeded as per whole mount staining protocol above. Antibody dilutions were prepared as indicated in **Table 1**. Images were acquired using confocal microscopy (Carl Zeiss LSM 800) using 10X, and 20X objectives. Where appropriate, cells were counted using Image J software.

Flow Cytometry

hPSC and differentiating clusters were collected and dissociated as outlined above. Single cells were filtered through cell strainer into FACS 5ml tubes and incubated for 30 min on ice for surface markers or overnight at 4C for intracellular markers. After incubation, the cells were washed and strained again through cell strainer and resuspended in FACS buffer for analyses on CYTEK Aurora. Analysis and graphs were made using FloJo software v10.6.2.

RT-qPCR

Total RNA was isolated using micro-RNeasy kit (Qiagen #74104) and reverse transcribed using the iSCRIPT cDNA kit (BioRad #1708891) as per manufacturer's instructions. qPCR analysis was performed on BioRad CFX96 Real Time System using TaqMan probes described in **Table 1**.

Western Blotting

Total protein was extracted from ~4x10^6 cell using RIPA lysis buffer. Lysis buffer was supplemented with EDTA-free protease inhibitor (4693159001) and phosphatase inhibitor (Roche, 04906837001) cocktails. Lysates were resolved on SDS–PAGE, transferred to a methanol-activated PVDF membrane (BioRad, 1620177), blocked in 5% milk for 30 min, followed by incubation with the indicated primary antibodies (**Table 1**) overnight at 4C. After washing, membranes were then incubated for 1hr at room temperature with secondary antibody conjugated to horseradish peroxidase (HRP) as indicated in **Table 1**. After incubation with Clarity Western ECL Substrate (BioRad, 1705061), bands were detected with Azure Biosystems Western Blotting imager.

CRIPSR-Cas9 and TALEN Mediated Genome Engineering

hPSC Mel1$^{INS\text{-}GFP}$ cells were dissociated into single cells using TrypLE incubation at 37C for 8 min. Cells were then quenched with mTeSR+ media and counted using MoxiGo II cell counter. 2x10^6 cells were transferred into microcentrifuge tubes and washed twice with PBS. Washed cells were then prepared for nucleofection of TALEN mediated knock-in (KI) of a Tet-On PD-L1 inducible system (i) or a CRIPSR-Cas9 mediated B2M gene knock-out (ii). Cells were nucleofected in P3 buffer following the Amaxa P3 Primary cell 4D-Nucleofector kit protocol (V4XP-3024) using the CB-150 program. (i) AAVS1-TALEN-L and AAVS1-TALEN-R (gift from Danwei Huangfu, Addgene plasmid # 59025; http://n2t.net/addgene:59025; RRID: Addgene 59025) as well as a *Tet-On* inducible PD-L1 overexpression plasmid (generated in house) were nucleofected into hPSC Mel1$^{INS\text{-}GFP}$ cells. Nucleofected cells were then plated in 10cm plates with 10uM ROCK inhibitor. After 48hr of plating, puromycin selection (0.5ug/ml) was performed for 48 hr then removed for 48hr followed by neomycin selection (50ug/ml) for 6

TABLE 1 | Reagents.

	Dilution	Company	Catalog #
1ry Antibodies			
rat anti-CPEP	1:400	Hybridoma/DSHB	GN-ID4
ms anti-PD-L1	1:100 IF/1:1000 WB	Cell signaling	41726S
rb anti-OCT4	1:50	Santa Cruz	sc-9081
ms anti-NKX6.1	1:200	Hybridoma/DSHB	F55A10
gt anti-PDX1	1:200	R&D	AF2419
rb anti-GAPDH	1:1000 WB	Abcam	ab9485
Conjugated Antibodies			
ms anti-TRA160 AF647	1:100	Biolegend	330606
ms anti-SOX2 AF594	1:100	Biolegend	656106
ms anti-FOXA2 PE	1:200	BD Biosciences	561589
ms anti-SOX17 AF488	1:200	BD Biosciences	562205
mIgG2a k Iso ctrl	1:100	Biolegend	400252 B217621
ms anti-HLA-ABC PerCP Cy5.5	1:100	Biolegend	311420 B227388
ms anti-PD-L1 PE-Cy7	1:100	Invitrogen	25-5983-41
ms anti-PDX1 PE	1:25	BD Biosciences	562161
ms anti-HLA-A*0201 BV421	1:100	BD Biosciences	740082
anti-PD1 PE	1:100	BD Biosciences	551892
2ry Antibodies			
anti-ms AF555	1:500	Thermo Fisher	A-31570
anti-rb AF555	1:500	Thermo Fisher	A-31572
anti-ms AF647	1:500	Thermo Fisher	A-31571
anti-rat AF488	1:500	Thermo Fisher	A-21208
anti-rb AF488	1:500	Thermo Fisher	R37118
anti-gt AF647	1:500	Thermo Fisher	A-21447
anti-rb HRP	1:1000	Invitrogen	31460
anti-ms HRP	1:1000	Invitrogen	31430
RNA Probes			
SOX2	1:20	BioRad	qHsaCEP0039595
NANOG	1:20	BioRad	qHsaCEP0050656
PDL1	1:20	BioRad	Hs00204257_m1
GAPDH	1:60	BioRad	qHsaCEP0041396
ACTINB	1:60	BioRad	Hs99999903_m1
HLA-A	1:20	BioRad	Hs01058806_g1
HLA-B	1:20	BioRad	Hs00818803_g1
HLA-C	1:20	BioRad	Hs00740298_g1
HLA-E	1:20	BioRad	Hs03045171_m1
HLA-F	1:20	BioRad	Hs01587840_m1
HLA-G	1:20	BioRad	Hs00365950_g1

days. Remaining colonies were picked and amplified for further characterization. (ii) guide RNAs (gRNA) targeting exon 1 and exon 2 of the B2M gene were generated in house and nucleofected along with CRISPR-Cas9. 24hr after plating, cells were selected for 48hr with puromycin (0.5ug/ml). Single colonies were picked and amplified for further characterization. Genomic DNA was extracted from targeted colonies and PCR analysis for TALEN KI and B2M KO was performed to identify potential modified clones. Further Sanger sequencing of PCR products confirm B2M mutations in the genome of hPSC Mel1$^{INS\text{-}GFP}$ cells.

T Cell Stimulation Assay

sBC clusters were washed with PBS and incubated with 0.05% Trypsin with EDTA at 37C for 12 min to create single cell suspension and quenched with 2% FBS in PBS. Cells were counted and plated into Matrigel coated 96 well plates at different densities for T cell stimulation assays or in 24 well plates for flow cytometry characterization. After 24hr sBC received an acute pro-inflammatory cytokine treatment (48hr IFN-γ, IL-1β, TNF-α) with or without 2μg/ml of Doxycycline hyclate (DOX; Sigma-Aldrich #D9891-1G). Cells were then washed twice with PBS and incubated with PPI: 15-24 (ALWGPDPAAA) at 10μg/ml or insulin B 13-23 (EALYLVCGERG) peptide at 100μg/ml for 4 hours. The insulin peptides used for these stimulation assays were obtained from Genemed Synthesis Inc. at > 95% purity and dissolved in PBS at a neutral pH. $1x10^5$ 5KC-1E6 T cells (TCR transductant responding to PPI: 15-24 presented by HLA-A*02:01) or 5KC-clone 5 (TCR transductant responding to insulin B 13-23 presented by HLA-DQ8) were co-culture with sBC overnight. 5KC T cells cultured without sBC or peptide were used as a negative control, and those treated with anti-CD3 monoclonal antibody (eBioscience, clone 2C11) at a concentration of 10μg/ml as positive control for each experiment. Murine IL-2 secreted by the 5KC cells was measured in the culture supernatant using a highly sensitive ELISA (V-PLEX IL-2 kit, Meso Scale Diagnostics, LLC) followed by detection on the MESO QuickPlex SQ120 instrument.

Statistical Analysis

All statistical analyses were performed in GraphPad Prism (Version 9.0.0). Student t test was used for samples with 2 groups. For multiple group analyses One- or Two-way ANOVA's was used with suggested multiple comparison tests. Paired test was employed with samples within the same bioreactor differentiation experiment.

Data and Resource Availability

No datasets were generated or analyzed during the current study.

RESULTS

Human Islets and sBC Differentially Express HLA and PD-L1 Molecules Upon Cytokine Exposure

During inflammation, beta cells are known to upregulate HLA class I and PD-L1 molecules in response to pro-inflammatory cytokines (13, 14, 16–18). These molecules are expressed on the cell surface of beta cells and help regulate immune cell responses. In order to determine if sBC exposed to inflammatory conditions exhibit a respond akin to bona fide beta cells, hPSC Mel1$^{INS\text{-}GFP}$ cells were differentiated into sBC using a directed differentiation protocol. Mel1$^{INS\text{-}GFP}$ cells contain a green fluorescence protein (GFP) reporter under the control of the endogenous insulin promoter that allows live monitoring of sBC generation (23). sBC and human islets in suspension culture were subjected to a 6-day cytokine treatment of IFN-γ 10ng/ml, IL-1β 0.5ng/ml and TNF-α 1ng/ml. Immunofluorescence staining of dissociated sBC or of sBC cluster sections indicates that sBC and human islets upregulate HLA-ABC molecules upon chronic cytokine treatment (**Figures 1A, B** and **Supplementary Figure 1A**). However, we do note that untreated sBCs exhibit considerable variability of HLA-ABC protein expression. Quantification shows that ~90% of treated sBC and human islets express HLA-ABC when exposed to pro-inflammatory cytokines (**Figures 1C, D**). Beta cell specific quantification demonstrates similar results for beta cells (**Figures 1E, F**). pan-HLA-ABC antibody used in immunofluorescence experiments cannot distinguish expression between different HLA class I isotypes. For this reason, quantitative PCR for specific isotypes was performed. This analysis shows a comparable upregulation of all HLA class I molecules tested in human islets (**Figure 1G**). Interestingly, sBC preferentially upregulated HLA-C upon pro-inflammatory cytokine treatment, while other HLA molecules were mostly unchanged compared to control conditions (**Figure 1H**). As preferential HLA-C expression has been previously observed in fetal tissues (24, 25), this could be an indicator that sBC are more closely aligned to fetal rather than adult beta cells.

Next, we evaluated the expression changes of PD-L1 upon cytokine treatment in human islets and sBC. PD-L1 is an important immune checkpoint inhibitor that can dampen immune cell activation in peripheral tissues (18). Thus, expression of PD-L1 in sBC could be important for regulating immune recognition and destruction by activated lymphocytes. Immunofluorescence staining and gene expression analysis confirmed a robust PD-L1 upregulation in human islets but not sBC after cytokine treatment (**Figures 1C, D, G, H**).

Taken together, this data indicates that induction of HLA class I and PD-L1 expression in respond to cytokine treatment differs between sBC and human islets *in vitro*. This data also shows that human islets and sBC react differently to an inflammatory environment.

Integration of a PD-L1 Inducible System and HLA Class I Knock-Out in hPSC

In order to engineer sBC that can potentially withstand an (auto) immune attack, a Tet-On doxycycline (DOX) inducible PD-L1 expression system (iP) was integrated site specifically into the AAVS1 locus of hPSC using Transcription activator-like effector nucleases (TALEN) technology (**Figures 2A, B**) (26). To create a clonal line, double selection for puromycin and neomycin was applied to Mel1$^{INS\text{-}GFP}$ cells that were nucleofected with two targeting and two TALEN plasmids followed by clonal selection after 11-13 days. Site specific integration of both targeting constructs was confirmed by genomic PCR analysis using specific primers for detecting neomycin and puromycin resistance genes in the AAVS1 loci (**Supplementary Figure 2A**). Functional testing of established iP line by DOX treatment for 48 hours revealed robust induction of PD-L1 mRNA expression while untreated controls displayed minimal leakiness of the inducible expression system (**Figure 2C**). Expression of key pluripotency markers SRY-Box Transcription Factor 2 (SOX2) and the Homeobox Transcription Factor Nanog (NANOG) were not changed in iP cells compared to unmodified wildtype (WT) cells (**Figure 2C**). Confirming the qPCR analysis, PD-L1 protein expression was readily detected by western blot at in DOX treated iP cells but not WT cells (**Figure 2D** and **Supplementary Figure 2B**). Immunofluorescence and surface flow cytometric analysis further demonstrated that PD-L1 protein was localized to the surface cell membrane in ~95% (+/- 4.4 STDV) of cells treated with DOX (**Figures 2E–G**). This data shows the generation of a clonal and functional Tet-On inducible PD-L1 expression system integrated into the AAVS1 loci of Mel1$^{INS\text{-}GFP}$ hPSC.

HLA class I molecules play an important role interfacing beta cells with cytotoxic CD8 T cells. Removing HLA class I surface expression from sBC could protect them from immune recognition and destruction by CD8 T cells. HLA class I molecules are composed of 3 polymorphic alpha helixes and 1 invariable beta-2 microglobulin (B2M) chain (**Figure 2H**). B2M is common in all HLA class I isotypes and is required for proper localization of the HLA class I complex to the cell membrane. Thus, we employed CRISPR/Cas9 technology to target exon 1 and 2 of the invariable B2M gene in the iP line using specific guide RNAs (**Figure 2I**). Sanger sequencing further confirmed a thymine insertion and an exon 1-2 deletion modifications, effectively generating a B2M knock-out (KO) line, termed iP-BKO (**Supplementary Figure 2C**). To confirm that the clonal iP-BKO line displayed impaired HLA class I surface expression, WT

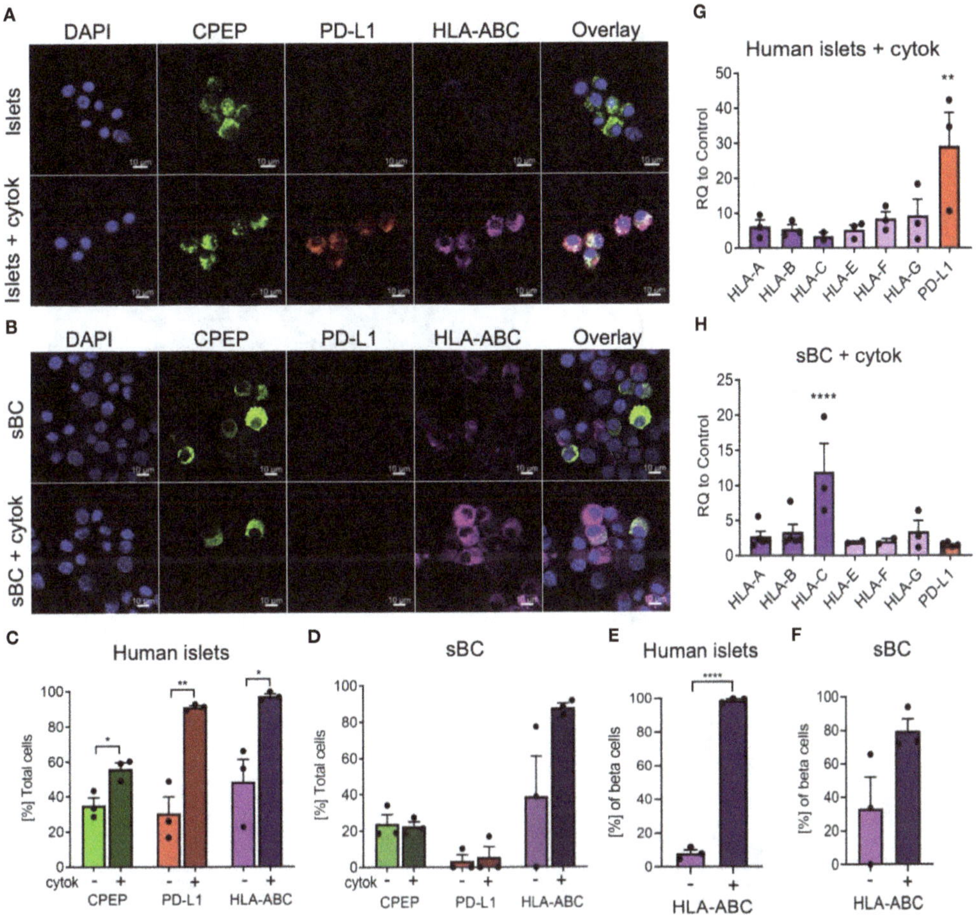

FIGURE 1 | Human islets and sBC differentially upregulate HLA class I and PD-L1 molecules upon inflammatory conditions. **(A, B)** Representative immunofluorescence images of control and cytokine treated human islets **(A)** or sBC **(B)** for 6 days. Samples were stained for DAPI (blue), CPEP (green), HLA-ABC (purple), and PD-L1 (red). Scale bar represents 10um. **(C–F)** Quantification of percentage CPEP, HLA-ABC, or PD-L1 positive cells over total cell count **(C, D)** or over total CPEP cell count **(E, F)** of control and cytokine treated human islets **(C, E)** or sBC (D, F). Results are shown as mean +/- SEM of n=3 independent donors. Unpaired student t test was performed with *p ≤ 0.05, **p ≤ 0.01, and ****p ≤ 0.0001. **(G, H)** qPCR gene expression analysis of multiple HLA class I isoforms and PD-L1 after 6 days of cytokine treatment in human islets **(G)** or sBC **(H)**. Values are normalized to ACTB and relative to control. Results are shown as mean +/- SEM of n=3 independent donors and differentiation experiments. Ordinary One-Way Anova with Dunnett's multiple test with **p ≤ 0.01, and ****p ≤ 0.0001.

and iP-BKO cells were subjected to IFN-γ(100ng/ml) stimulation for 48 hr. Flow cytometric and immunofluorescence staining confirmed the lack of HLA class I surface expression in iP-BKO cells (**Figures 2J, K**). Quantification of flow cytometric data indicates that around ~99% of all treated iP-BKO cells lack cell surface expression of HLA class I molecules (**Supplementary Figure 2D**). Taken together, these data confirm the generation of a B2M KO in iP cells resulting in a functional deletion of surface HLA class I expression.

iP and iP-BKO Cell Lines Can Differentiate Into sBC With Similar Efficiencies

In order to test whether genetic modifications impair the ability of iP and iP-BKO hPSC lines to differentiate into sBC, both lines were subjected to a 3D beta cell differentiation protocol based on previous published work by us that generates glucose responsive sBCs (**Figure 3A**) (9, 10). Samples were taken at subsequent differentiation stages for morphological and flow cytometric analysis. No significant morphological differences were observed between the two cell lines at any stage throughout the entire differentiation protocol (**Figure 3B**). Flow cytometric analysis for pluripotency markers SOX2 and TRA160 showed that ~97% of cells stained positive for both markers after hPSC 3D cluster formation (**Figure 3C**).

Differentiation into the definitive endoderm lineage, marked by double expression of transcription factors Forkhead Box A2 (FOXA2) and SRY-Box 17 (SOX17), was highly efficient in both cell lines with >80% of all cells expressing both markers (**Figure 3C**). Subsequent differentiation into pancreatic progenitors, marked by the transcription factor Pancreatic and Duodenal Homeobox 1 (PDX1), was similarly efficient in both cell lines with >70% of all cells expressing PDX1 at this stage.

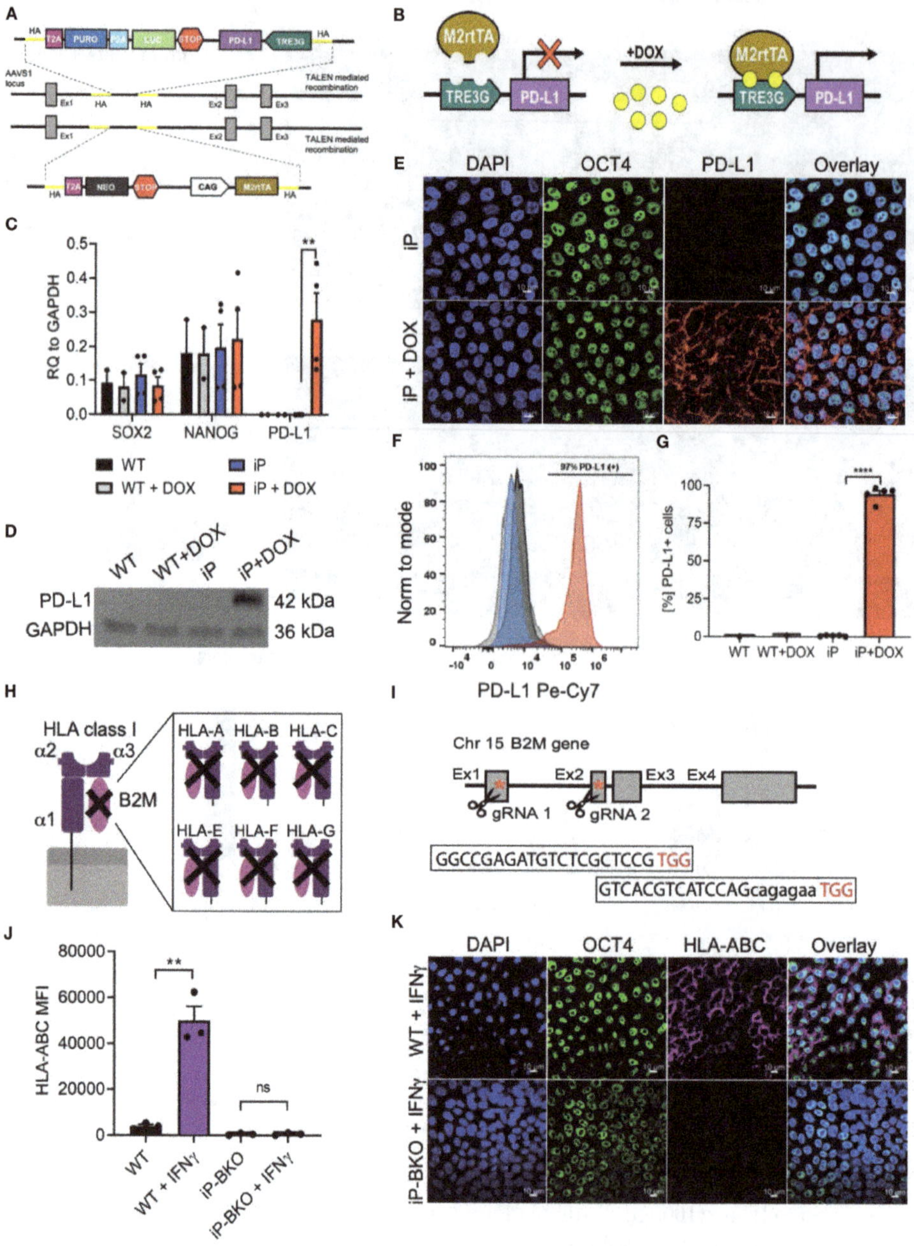

FIGURE 2 | Integration of immune protective gene editing in hPSC using genome engineering. **(A, B)** Schematic of TALEN-mediated recombination strategy for the integration **(A)** and mechanism of action **(B)** of a Tet-On DOX inducible PD-L1 expression system between exon1 and 2 of the AAVS1 locus in the hPSC Mel1$^{INS\text{-}GFP}$ line. **(C)** qPCR analysis of control hPSC (WT) and inducible PD-L1 hPSC (iP) cells with and without DOX treatment (2ug/ml for 48hr). Pluripotency markers SOX2 and NANOG were also analyzed. Values are normalized to GAPDH and shown as mean +/- SEM of n = 2 for WT and n = 4 for iP samples. 2way-Anova with Tukey's multiple comparison test was performed with $^{**}p \leq 0.01$. **(D)** Immunoblotting for PD-L1 and loading control GAPDH proteins in WT and iP cells with and without DOX treatment. Representative immunoblot of 3 independent repeats. Complete blot is presented in **Supplementary Figure 2B**. **(E)** Representative immunofluorescence images of iP cells with and without DOX treatment. Sample were stain for DAPI (blue), OCT4 (green), and PD-L1 (red). Representative image of 4 independent experiments. Scale bar represents 10um. **(F, G)** Representative flow cytometry histogram **(F)** and quantification **(G)** for the detection of surface PD-L1 protein on WT (black), WT + DOX (light gray), iP (blue), and iP + DOX (red) cells. Results from n = 5 for iP and n = 1 for WT are shown as mean +/- SEM. Unpaired student's t test was performed with $^{****}p \leq 0.0001$. **(H, I)** Structural schematic of HLA class I complex with membrane bound $\alpha1$ region, peptide binding grove composed of $\alpha2$ and $\alpha3$ regions and the invariable B2M region **(H)**. Inhibition of surface expression of all HLA class I isoforms is achieved by deletion of functional B2M using CRIPSR-Cas9 technology. Sequence for the two gRNAs specific for exons 1 and 2 with corresponding PAM sequences (in red) and location in relation to the B2M gene is shown with red asterisks **(I)**. **(J)** Flow cytometry quantification of surface HLA-ABC expression on WT and iP cells with B2M KO (iP-BKO) after IFN-γ treatment (100ng/ml for 48hr). MFI values are presented as mean +/- SEM for 3 independent experiments. Unpaired student's t test was performed with $^{**}p \leq 0.01$; ns, not significant. **(K)** Representative immunofluorescence images of iP-BKO cells after IFN-γ stimulation. Samples were stain for DAPI (blue), OCT4 (green), and HLA-ABC (purple). Scale bar represents 10um.

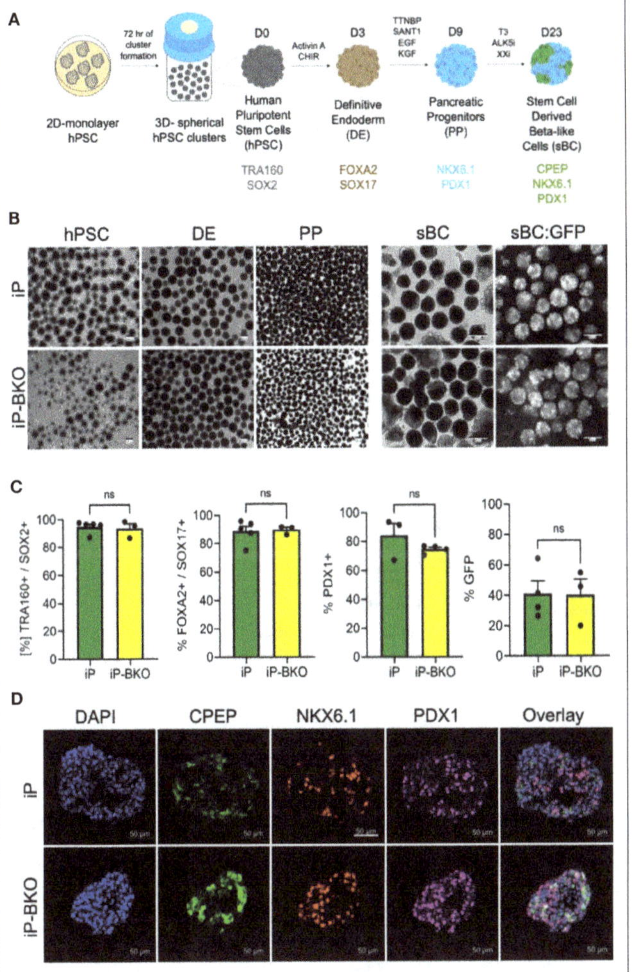

FIGURE 3 | iP and iP-BKO hPSC lines differentiate into sBC with similar efficiencies. **(A)** Schematic of stepwise, suspension culture-based, direct differentiation approach for the generation of sBC clusters from hPSC. Critical induction molecules and gene markers at key differentiation stages are shown. **(B)** Representative bright field images of differentiating hPSC clusters in bioreactors at key developmental stages (left panel). Live images of insulin promoter-driven GFP expression at sBC stage (right panel). Note the higher magnification for sBC compared to earlier hPSC, DE and PE stages. All scale bars represent 200um. **(C)** Quantification of stage specific marker expression for iP and iP-BKO differentiating clusters using flow cytometric analysis. Data is presented as SEM +/- percentage of positive cells for indicated markers. Unpaired student t test was performed with no significant changes denoted as ns. **(D)** Representative immunofluorescence images of sBC derived from iP and iP-BKO cell lines. Samples were stain for DAPI (blue), CPEP (green), NKX6.1 (red), and PDX1 (purple) are shown. Scale bar represents 50um.

Finally, sBC generation was monitored by GFP expression which is controlled by the endogenous insulin promoter. At the end of the differentiation protocol ~40% of all cells expressed GFP, indicating efficient generation of sBC. We did not find statistical differences between iP and iP-BKO cell lines at any stage of the sBC differentiation protocol (**Figure 3C**). Immunofluorescence analysis of sBC sections showed expression of key beta cell markers CPEP, NKX6.1 and PDX1 in both cell lines (**Figure 3D**). These data indicate that both iP and iP-BKO cells can differentiate into sBC at similar efficiencies, and that iP and iP-BKO genetic modifications do not alter differentiation potential into sBC.

PD-L1 Overexpression and HLA Class I KO Abrogate Diabetogenic CD8 T Cell Activation

In order to test whether sBC will stimulate human diabetogenic CD 8 T cells, a reliable *in vitro* assay to measure T cell stimulation was developed. iP and iP-BKO hPSC were differentiated into sBC, followed by dissociation and plating on Matrigel coated 96 well plates to generate a 2D sBC monolayer for subsequent experiments. First, we exposed 2D sBC to a 2-day acute (IFN-γ 100ng/ml, IL-1β 5ng/ml and TNF-α 10ng/ml) cytokine treatment alone to upregulate HLA class I molecules, or to a 2-day acute cytokine treatment plus DOX to additionally induce PD-L1 expression (**Figure 4A**). Flow cytometric analysis revealed that iP sBC upregulated HLA class I molecules at similar levels in both conditions, while HLA class I surface expression was not detected on iP-BKO sBC (**Figure 4B**). PD-L1 surface expression was detected on 2D sBC from both cell lines after cytokine treatment, differing from 3D sBC cultures exposed to chronic cytokine treatment (**Figure 1E**). Moreover, adding DOX to the cytokine treated sBC further increased PD-L1 expression significantly (**Figure 4C**). Nevertheless, this data confirms that iP sBC can be induced to express HLA class I and PD-L1 surface proteins, while iP-BKO sBC can only express PD-L1.

Diabetogenic CD8 T cells are known to play an important role in beta cell destruction and diabetes progression (15). Preproinsulin (PPI) is a major self-antigen in T1D. CD8 T cells responding to PPI: 15-24 presented by HLA-A*02:01 have been shown to lyse human beta cells (27, 28). In order to test whether sBC can be targeted by a preproinsulin specific CD8 T cell, iP sBC were co-cultured with a T cell receptor (TCR) transductant expressing a TCR that responds to preproinsulin PPI: 15-24 peptide presented by HLA-A*02:01 (termed 5KC-1E6). The TCR α/β genes from the 1E6 T cell clone were retrovirally expressed in an immortalized T cell devoid of endogenous TCR expression, thus creating a TCR transductant that secretes interleukin-2 (IL-2) when stimulated with PP1: 15-24 presented by HLA-A*02:01 (29). Of note, the parental hPSC line Mel1, from which iP and iP-BKO lines were derived, contains the HLA-A*02:01 allele (**Supplementary Figure 3A**). T cell activation was measured by quantifying IL-2 secreted into the media after co-culture with sBC. Co-cultures of 5KC-1E6 T cells and iP sBC without cytokine treatment did not result in significant IL-2 secretion (**Figure 4D**). This is likely due to the low abundance of the PPI: 15-24 produced by sBC in culture and the low expression of HLA class I molecules at steady state. However, co-culture experiments with prior exposure of 2D sBC to pro-inflammatory cytokines resulted in a low but reproducible detection of IL-2 indicating activation of a PPI specific T cell with an inflammatory stimulus (**Figure 4D**). These results indicate that sBC can activate diabetogenic CD8 T cells in an inflammatory environment.

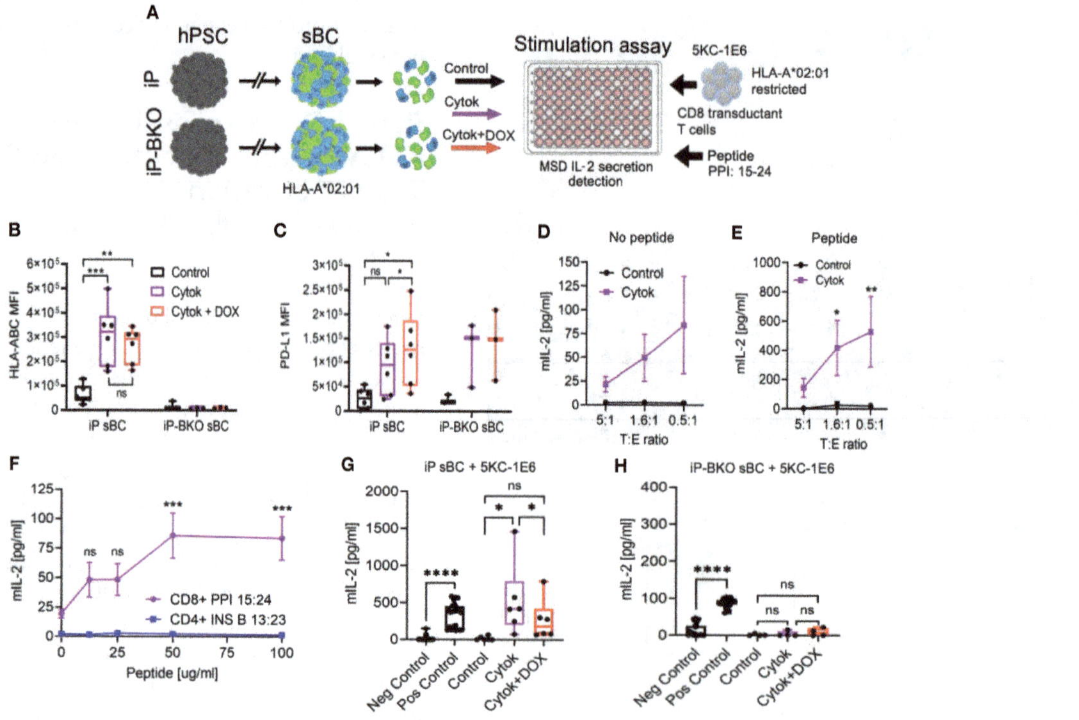

FIGURE 4 | sBC activate a preproinsulin specific CD8 T cell receptor transductant in an HLA and PD-L1 dependent manner. **(A)** Schematic depicting the sBC and CD8 T cell receptor transductant (5KC-1E6) co-culture experimental assay set up. iP and iP-BKO cells were differentiated into sBC as clusters, followed by dissociation into single cells. sBC were treated with control (black), 2-day pro-inflammatory cytokines (purple) or with pro-inflammatory cytokines and DOX (red) for 48hr. sBC were then incubated with preproinsulin (PPI) 15-24 peptide for 4 hours. This was followed by 5KC-1E6 co-culture overnight with measurement of mIL-2 from the cell culture supernatants. **(B, C)** Flow cytometry quantification of surface HLA-ABC **(B)** or PD-L1 **(C)** expression on iP and iP-BKO sBC after treatments. Data is shown as MFI in box and whisker plots. Each dot represents a single differentiation experiment. Ordinary One-Way Anova was performed with Tukey's multiple comparison test with $*p \leq 0.05$, $**p \leq 0.01$, $***p \leq 0.001$ and ns denoting no significant changes. **(D, E)** Stimulation assay of sBC treated with (purple) or without cytokines (black) followed by co-culture with 5KC-1E6 T cells at denoted target to effector rations (T:E). Note that sBC were cultured with **(E)** or without **(D)** exogenous PPI: 15-24 peptide. Data is presented as mean secreted mIL-2 concentration +/- SEM of 5 independent differentiation experiments. 2-way Anova with Sidak's multiple comparison test was performed with $*p \leq 0.05$, and $**p \leq 0.01$. **(F)** Peptide titration experiment of different peptide-TCR combinations. sBC were treated with cytokines and incubated with different concentrations of PPI: 15-24 (purple line) or INS B 13:23 (blue line) peptides. sBC were co-culture with corresponding CD8 T cell receptor transductants. Data is presented as mean secreted mIL-2 concentration +/- SEM of 3 independent differentiation experiments. 2-way Anova with Sidak's multiple comparison test was performed with $***p \leq 0.001$, and ns denoting no significant changes. **(G, H)** Stimulation assay of co-culture of 5KC-1E6 cells with iP **(G)** or iP-BKO **(H)** sBC after denoted treatments and incubated with exogenous PPI: 15-24. 5KC-1E6 alone or treated with anti-CD3 antibody served as negative and positive controls, respectively. Note the significant reduction in T cell stimulation upon PD-L1 expression (red) compared to only cytokine treatment (purple). Data is shown as box and whisker plots of secreted mIL-2. Each dot represents an independent stimulation assay with sBC from independent differentiation experiments. A paired t-test was performed with ns as no significant and $*p \leq 0.05$, $****p \leq 0.0001$.

Next, similar experiments were repeated with the addition of exogenous PPI: 15-24 peptide to sBC cultures prior to co-culturing with 5KC-1E6 T cells. This resulted in a more robust, approximately 8-fold increased, secretion of IL-2 in cytokine treated sBC cultures (**Figure 4E**, **Supplementary Figures 3B, C**). As expected, sBC cultures not treated with cytokines did not trigger T cell stimulation despite the addition of exogenous PPI: 15-24 peptide (**Figure 4E**), indicating the importance and specificity of upregulating HLA class I molecules for CD8 T cell recognition. To determine if the IL-2 release in cytokine treated cultures was a specifically due to PPI: 15-24 bound to HLA-A*02:01 on sBC, a peptide titration was performed. There is a peptide binding curve for increased IL-2 with increasing PPI peptide concentration with a saturation level around 50ug/ml of PPI: 15-24 (**Figure 4F**). As a T cell specificity control, an insulin-specific CD4 TCR transductant (5KC-clone 5) that responds to insulin B chain amino acids 13-23 presented by the T1D risk HLA class II molecule, HLA-DQ8, was used in these experiments (30). 5KC-clone 5 was not activated to secrete IL-2 despite the presence of cytokines and exogeneous addition of cognate insulin peptide suggesting that activation of T cell is dependent on peptide presentation by sBC (**Figure 4F**). Taken together, these experiments indicate sBC can specifically activate a PPI reactive CD8 T cell in the presence of pro-inflammatory cytokines and the response is increased with the addition of PPI: 15-24 presented by HLA-A*02:01. Thus, sBC are susceptible to diabetogenic CD8 T cell mediated targeting in an HLA-peptide-TCR dependent manner.

Using this human sBC - CD8 T cell interaction model, whether PD-L1 overexpression and/or HLA KO reduces PPI

CD8 T cell activation was investigated. iP and iP-BKO 2D sBC were treated with 2-days cytokines alone or 2-days cytokines plus DOX, followed by PPI: 15:24 peptide addition and co-culture with 5KC-1E6 T cells. As anticipated, iP sBC that were treated with cytokines stimulated T cells by releasing IL-2 (**Figure 4G**). However, T cell stimulation was significantly reduced when PD-L1 overexpression was induced with DOX (**Figure 4G**). Of note, 5KC-1E6 T cells do express mouse PD-1 protein on their surface (**Supplementary Figure 3D**). Experiments conducted using iP-BKO sBC (e.g. those lacking HLA class I molecules) showed no stimulation of 5KC-1E6 T cells even when exogenous PPI: 15-24 peptide was added (**Figure 4H**).

Collectively, this study and presents a novel model system assay that allows for detailed studies into human beta cell - T cell interaction. This system is also amenable to test genetic engineering models of immune protection. Finally, the protective role of PD-L1 in sBC was confirmed using this model system.

DISCUSSION

The use of hPSC for the generation of sBC could provide an unlimited source of functional beta-like cells for cell replacement therapy. However, the response of sBC to an inflammatory environment has not been well studied. Here we show that sBC upregulate HLA class I molecules upon pro-inflammatory cytokine exposure, similar to the response of cadaveric human islets to an inflammatory insult. Furthermore, HLA class I isotype specific gene expression analysis revealed that HLA-C is the predominant HLA class I molecule upregulated in sBC. This result differs from human islets where all HLA isotypes are upregulated at similar levels. It has been shown that HLA-C is the main isotype expressed in fetal trophoblasts (24, 25). It is intriguing to speculate that sBC, which have been proposed to closer resemble fetal rather than adult beta cells (31), might display an immature immune profile when challenged. However, further studies are warranted to clarify this possibility. Cadaveric beta cells have been shown to upregulate PD-L1 in response to pro-inflammatory cytokines (16–18) and our data confirms this result. However, we find that sBCs differ from cadaveric beta cells and do not express PD-L1 under the proinflammatory conditions employed in our study. This is in contrast to other reports that show PD-L1 expression on sBCs after treatment with high levels of proinflammatory cytokines (22). While further studies are required to clarify this point, we anticipate that regulation of gene expression of immune regulatory proteins in sBCs is highly context dependent.

Although sBC have been demonstrated to reverse diabetes in multiple mouse models, we studied their interaction with diabetogenic immune cells in a strictly human context. Co-culture of sBC and peripheral blood mononuclear cells derived from the same type 1 diabetic patient could represent a true autologous human system to study autoimmune diabetes *in vitro* (32). However, the low frequency of diabetogenic T cells found within the peripheral blood of T1D patients complicates the interpretation of these studies. Here, we provide an innovative platform to model HLA matched T cell – beta cell interactions for the study of autoimmune diabetes. Unlike peripheral blood immune cells, T cell receptor transductants represent an abundant and reusable source of antigen specific T cells (29, 33).

Here, we demonstrate that PPI reactive CD8 T cells transductants are stimulated by sBC in a TCR-peptide-HLA dependent manner when an inflammatory stimulus is present. This model system can be used for future studies to examine T cells of other specificities such as those responding to glutamic acid decarboxylase (GAD), insulinoma antigen-2 (IA-2), zinc transporter 8, hybrid insulin peptides and neoepitopes, and other immune cell subsets involved in T1D pathogenesis. This data supports previous studies in which immunogenicity of sBC to PPI: 15-24 specific cytotoxic T lymphocytes (CTLs) was described (34). Moreover, this study further demonstrates that genetic engineering of sBC could represent a viable strategy to avoid diabetogenic CD8 T cell activation. Future studies using diabetogenic CTLs co-culture with immune engineered sBCs could further substantiate this notion.

Manipulating HLA class I molecules using genetic engineering of hPSC that are differentiated into sBC could potentially provide an unlimited source of beta cells that are protected from a diabetogenic immune attack. Here, we knocked-out B2M which resulted in a functional KO for all HLA class I isotypes. hPSC that lack expression of HLA class I molecules successfully differentiated into sBC and did not stimulate PPI reactive T cells. We found that WT sBC exposed to an inflammatory environment upregulated HLA class I molecules and induce higher levels of activation of T cell transductants compared to sBC cultures at steady state with low levels of HLA class I expression. These results highlight the important role of cytokine-induced HLA class I molecules in immune cell activation and peptide presentation from stem cell derived cells. However, complete ablation of all HLA class I molecules has been shown to render cells vulnerable to a natural killer (NK) cell lysis (35, 36). Different alternative approaches to overcome NK cell targeting have been investigated. Targeted disruption of classical HLA-A, -B and -C while retaining expression of non-classical HLA-E alone (37) or by the combinatorial expression of HLA-G, CD47 and PD-L1 (38, 39) have been shown to protect hPSC from NK cell recognition and lysis. For these reasons, we explored the possibility of overexpressing the PD-L1 receptor and keeping expression of all HLA class I molecules to avoid NK cell lysis.

We successfully generated a knock-in of a Tet-On PD-L1 inducible expression system in the safe harbor locus AAVS1 in hPSC. This resulted in the induction of PD-L1 in ~95% of the hPSC. Our results further show that by overexpressing the immune checkpoint receptor PD-L1 in sBC, T cell stimulation is abrogated. A recent study showed that PD-L1 overexpression resulted in increased survival of sBC after 4 weeks of transplantation into the kidney capsule of immune competent and humanized mouse models (22). These studies showed increased survival and decreased immune infiltration of sBC transplants in a xenogeneic model system. Moreover, the specificity of immune – beta cell interface was not analyzed in this study. In contrast, our present study shows the direct interaction and activation of diabetogenic CD8 T cells by sBC and that PD-L1 expression can reduce T cell stimulation in an *in vitro* model of human T1D devoid of xenogeneic inputs.

In conclusion, we provide a reliable assay for the study of human sBC and diabetogenic T cell interactions *in vitro*. Using this assay, we further confirmed that sBC are vulnerable to diabetogenic T cell activation and that PD-L1 expression on sBC is sufficient to reduce T cell stimulation. Finally we provide a proof of principle that genetic engineering of human pluripotent stem cells can be used to generate immune protected stem cell derived beta-like cells that do not activate diabetogenic T cells.

AUTHOR CONTRIBUTIONS

Conception and study design: RC-G, AM, and HR. Execution of experiments: RC-G, AA, and HR. Data analysis and interpretation: RC-G, AA, CM, AM, and HR. Manuscript writing: RC-G, AM, and HR. Final approval of manuscript: RC-G, AA, CM, AM, and HR. HR is the guarantor of this work and takes full responsibility for the work as a whole. All authors contributed to the article and approved the submitted version.

FUNDING

This work was supported by NIH Grants (DK108868, DK032083, DK099317 and DK116073 to AM, DK104194, DK127497, DK122638 and AI042288 to CEM, and DK120444 and AI140044 to HAR, P30-DK116073 to University of Colorado Diabetes Research Center), a new investigator award from the NIDDK- supported Human Islets Research Network (HIRN, RRID : SCR_014393; UC24 DK1041162) (HR), a Culshaw Junior Investigator Award in Diabetes (HR), the Juvenile Diabetes Research Foundation (JDRF 2-SRA-2019-781-S-B) (HR), the Children's Diabetes Foundation, and the Colorado Clinical and Translational Science Institute (TR002535 to AM). RC-G was supported by an RNA Bioscience Initiative Scholar award.

SUPPLEMENTARY MATERIAL

Supplementary Figure 1 | (A) Representative immunofluorescence images of cluster sections from control and 6-day cytokine treated sBC stained for DAPI (blue), CPEP (green), HLA-ABC (purple), and PD-L1 (red). Scale bar represents 50um. **(B, C)** qPCR analysis of HLA class I isoforms and PD-L1 gene expression in control and 6-day cytokine treated human islets **(B)** and sBC **(C)**. Values are normalized to ACTB and results are shown as mean +/- SEM of n = 3 independent donors and experiments.

Supplementary Figure 2 | (A) Schematic illustrating AAVS1 genomic region with Tet-On inducible PD-L1 system integration. Arrows delineate primers generated to test for correct integration of the inducible PD-L1 expression system (top panel). Agarose DNA gel showing genomic PCR amplification results from the AAVS1 region of WT and iP cells using indicated primer combinations (bottom panel). **(B)** Complete immunoblotting image of WT and iP cells with and without DOX treatment (2ug/ml for 48hr). PD-L1 band was detected at ~42kDa, while loading control GAPDH band was detected at ~37kDa as marked by protein standards. **(C)** Sanger sequencing results from bands obtained after genomic DNA amplification of the targeted B2M region. Green, yellow and blue boxes represent different DNA amplification bands that were extracted for Sanger sequencing analysis. Compared to WT genomic B2M gene sequencing (green box), iP-BKO B2M gene sequencing showed a thymine insertion (yellow box) on one allele and a deletion of ~3906 base pairs resulting in exon 1 and exon 2 fusion in the second allele (blue box) demonstrating two different genomic alterations in each of two clonal cell lines analyzed. **(D)** Flow cytometry histogram of surface HLA-ABC protein expression on WT and iP-BKO cells treated with IFN-γ(100ng/ml for 48hr). WT (pink) WT+IFN-γ(purple), iP-BKO (light green), iP-BKO+IFN-γ(dark green), unstained control (black) and isotype control (gray). A representative histogram normalized to mode from 3 independent experiments is presented.

Supplementary Figure 3 | (A) Flow cytometry histogram of surface HLA-A2 on iP sBC after pro-inflammatory cytokine treatment (purple) compared to unstained sample (black). Data is presented in a representative histogram normalized to mode. **(B, C)** Stimulation assay of 5KC-1E6 PPI: 15-24 specific T cells after overnight co-culture at different T:E ratios with iP **(A)** or iP-BKO **(B)** sBC pre-treated with or without pro-inflammatory cytokines and with different concentrations of exogeneous peptide. Data is shown as mean of secreted mIL-2 +/- SEM of 5 independent experiments. 2-way Anova with Sidak's multiple comparison test was performed in which $^{*}p \leq 0.05$, and $^{****}p \leq 0.0001$. **(D)** Flow cytometry histogram of surface murine PD1 on 5KC-1E6 (PPI 15-24 specific) T cells (red) compared to unstained cells (gray). Data is presented in a representative histogram normalized with overall cell count.

REFERENCES

1. Atkinson MA, Eisenbarth GS, Michels AW. Type 1 Diabetes. *Lancet* (2014) 383:69–82. doi: 10.1016/S0140-6736(13)60591-7
2. Beck RW, Bergenstal RM, Laffel LM, Pickup JC. Advances in Technology for Management of Type 1 Diabetes. *Lancet* (2019) 394:1265–73. doi: 10.1016/S0140-6736(19)31142-0
3. Shapiro AJ, Lakey JR, Ryan EA, Korbutt GS, Toth E, Warnock GL, et al. Islet Transplantation in Seven Patients With Type 1 Diabetes Mellitus Using a Glucocorticoid-Free Immunosuppressive Regimen. *New Engl J Med* (2000) 343:230–8. doi: 10.1056/NEJM200007273430401
4. Ryan EA, Lakey JR, Paty BW, Imes S, Korbutt GS, Kneteman NM, et al. Successful Islet Transplantation: Continued Insulin Reserve Provides Long-Term Glycemic Control. *Diabetes* (2002) 51:2148–57. doi: 10.2337/diabetes.51.7.2148
5. Castro-Gutierrez R, Michels AW, Russ HA. β Cell Replacement: Improving on the Design. *Curr Opin Endocrinol Diabetes Obes* (2018) 25:251–7. doi: 10.1097/MED.0000000000000418
6. Pagliuca FW, Millman JR, Gürtler M, Segel M, Van Dervort A, Ryu JH, et al. Generation of Functional Human Pancreatic β Cells *In Vitro*. *Cell* (2014) 159:428–39. doi: 10.1016/j.cell.2014.09.040
7. Rezania A, Bruin JE, Arora P, Rubin A, Batushansky I, Asadi A, et al. Reversal of Diabetes With Insulin-Producing Cells Derived *In Vitro* From

Human Pluripotent Stem Cells. *Nat Biotechnol* (2014) 32:1121. doi: 10.1038/nbt.3033

8. Velazco-Cruz L, Song J, Maxwell KG, Goedegebuure MM, Augsornworawat P, Hogrebe NJ, et al. Acquisition of Dynamic Function in Human Stem Cell-Derived β Cells. *Stem Cell Rep* (2019) 12:351–65. doi: 10.1016/j.stemcr.2018.12.012
9. Russ HA, Parent AV, Ringler JJ, Hennings TG, Nair GG, Shveygert M, et al. Controlled Induction of Human Pancreatic Progenitors Produces Functional Beta-Like Cells *In Vitro*. *EMBO J* (2015) 34:1759–72. doi: 10.15252/embj.201591058
10. Nair GG, Liu JS, Russ HA, Tran S, Saxton MS, Chen R, et al. Giacometti S: Recapitulating Endocrine Cell Clustering in Culture Promotes Maturation of Human Stem-Cell-Derived β Cells. *Nat Cell Biol* (2019) 21:263–74. doi: 10.1038/s41556-018-0271-4
11. Desai T, Shea LD. Advances in Islet Encapsulation Technologies. *Nat Rev Drug Discovery* (2017) 16:338–50. doi: 10.1038/nrd.2016.232
12. Vegas AJ, Veiseh O, Gürtler M, Millman JR, Pagliuca FW, Bader AR, et al. Long-Term Glycemic Control Using Polymer-Encapsulated Human Stem Cell–Derived Beta Cells in Immune-Competent Mice. *Nat Med* (2016) 22:306. doi: 10.1038/nm.4030
13. Richardson SJ, Rodriguez-Calvo T, Gerling IC, Mathews CE, Kaddis JS, Russell MA, et al. Islet Cell Hyperexpression of HLA Class I Antigens: A Defining Feature in Type 1 Diabetes. *Diabetologia* (2016) 59:2448–58. doi: 10.1007/s00125-016-4067-4
14. Gonzalez-Duque S, Azoury ME, Colli ML, Afonso G, Turatsinze J-V, Nigi L, et al. Conventional and Neo-Antigenic Peptides Presented by β Cells are Targeted by Circulating Naïve CD8+ T Cells in Type 1 Diabetic and Healthy Donors. *Cell Metab* (2018) 28:946–60.e946. doi: 10.1016/j.cmet.2018.07.007
15. Pinkse GG, Tysma OH, Bergen CA, Kester MG, Ossendorp F, van Veelen PA, et al. Autoreactive CD8 T Cells Associated With β Cell Destruction in Type 1 Diabetes. *Proc Natl Acad Sci* (2005) 102:18425–30. doi: 10.1073/pnas.0508621102
16. Colli ML, Hill JL, Marroquí L, Chaffey J, Dos Santos RS, Leete P, et al. PDL1 is Expressed in the Islets of People With Type 1 Diabetes and is Up-Regulated by Interferons-α and-γ *via* IRF1 Induction. *EBioMedicine* (2018) 36:367–75. doi: 10.1016/j.ebiom.2018.09.040
17. Demine S, Schiavo AA, Marín-Cañas S, Marchetti P, Cnop M, Eizirik DL. Pro-Inflammatory Cytokines Induce Cell Death, Inflammatory Responses, and Endoplasmic Reticulum Stress in Human iPSC-Derived Beta Cells. *Stem Cell Res Ther* (2020) 11:7. doi: 10.1186/s13287-019-1523-3
18. Osum KC, Burrack AL, Martinov T, Sahli NL, Mitchell JS, Tucker CG, et al. Interferon-Gamma Drives Programmed Death-Ligand 1 Expression on Islet β Cells to Limit T Cell Function During Autoimmune Diabetes. *Sci Rep* (2018) 8:1–12. doi: 10.1038/s41598-018-26471-9
19. Stamatouli AM, Quandt Z, Perdigoto AL, Clark PL, Kluger H, Weiss SA, et al. Collateral Damage: Insulin-Dependent Diabetes Induced With Checkpoint Inhibitors. *Diabetes* (2018) 67:1471–80. doi: 10.2337/dbi18-0002
20. Akturk HK, Alkanani A, Zhao Z, Yu L, Michels AW. PD-1 Inhibitor Immune-Related Adverse Events in Patients With Preexisting Endocrine Autoimmunity. *J Clin Endocrinol Metab* (2018) 103:3589–92. doi: 10.1210/jc.2018-01430
21. Akturk H, Kahramangil D, Sarwal A, Hoffecker L, Murad M, Michels A. Immune Checkpoint Inhibitor-Induced Type 1 Diabetes: A Systematic Review and Meta-Analysis. *Diabetic Med* (2019) 36:1075–81. doi: 10.1111/dme.14050
22. Yoshihara E, O'Connor C, Gasser E, Wei Z, Oh TG, Tseng TW, et al. Immune-Evasive Human Islet-Like Organoids Ameliorate Diabetes. *Nature* (2020) 586:606–11. doi: 10.1038/s41586-020-2631-z
23. Micallef SJ, Li X, Schiesser JV, Hirst CE, Yu QC, Lim SM, et al. INS GFP/w Human Embryonic Stem Cells Facilitate Isolation of *In Vitro* Derived Insulin-Producing Cells. *Diabetologia* (2012) 55:694–706. doi: 10.1007/s00125-011-2379-y
24. King A, Burrows T, Hiby S, Bowen J, Joseph S, Verma S, et al. Surface Expression of HLA-C Antigen by Human Extravillous Trophoblast. *Placenta* (2000) 21:376–87. doi: 10.1053/plac.1999.0496
25. Hiby SE, Walker JJ, O'Shaughnessy KM, Redman CW, Carrington M, Trowsdale J, et al. Combinations of Maternal KIR and Fetal HLA-C Genes Influence the Risk of Preeclampsia and Reproductive Success. *J Exp Med* (2004) 200:957–65. doi: 10.1084/jem.20041214
26. González F, Zhu Z, Shi Z-D, Lelli K, Verma N, Li QV, et al. An iCRISPR Platform for Rapid, Multiplexable, and Inducible Genome Editing in Human Pluripotent Stem Cells. *Cell Stem Cell* (2014) 15:215–26. doi: 10.1016/j.stem.2014.05.018
27. Skowera A, Ellis RJ, Varela-Calviño R, Arif S, Huang GC, Van-Krinks C, et al. CTLs are Targeted to Kill β Cells in Patients With Type 1 Diabetes Through Recognition of a Glucose-Regulated Preproinsulin Epitope. *J Clin Invest* (2008) 118:3390–402. doi: 10.1172/JCI35449
28. Knight RR, Kronenberg D, Zhao M, Huang GC, Eichmann M, Bulek A, et al. Human β-Cell Killing by Autoreactive Preproinsulin-Specific CD8 T Cells is Predominantly Granule-Mediated With the Potency Dependent Upon T-Cell Receptor Avidity. *Diabetes* (2013) 62:205–13. doi: 10.2337/db12-0315
29. Mann SE, Zhou Z, Landry LG, Anderson AM, Alkanani AK, Fischer J, et al. Multiplex T Cell Stimulation Assay Utilizing a T Cell Activation Reporter-Based Detection System. *Front Immunol* (2020) 11:633. doi: 10.3389/fimmu.2020.00633
30. Michels AW, Landry LG, McDaniel KA, Yu L, Campbell-Thompson M, Kwok WW, et al. Islet-Derived CD4 T Cells Targeting Proinsulin in Human Autoimmune Diabetes. *Diabetes* (2017) 66:722–34. doi: 10.2337/db16-1025
31. Hrvatin S, O'Donnell CW, Deng F, Millman JR, Pagliuca FW, DiIorio P, et al. Differentiated Human Stem Cells Resemble Fetal, Not Adult, β Cells. *Proc Natl Acad Sci* (2014) 111:3038–43. doi: 10.1073/pnas.1400709111
32. Leite NC, Sintov E, Meissner TB, Brehm MA, Greiner DL, Harlan DM, et al. Modeling Type 1 Diabetes *In Vitro* Using Human Pluripotent Stem Cells. *Cell Rep* (2020) 32:107894. doi: 10.1016/j.celrep.2020.107894
33. Williams T, Krovi HS, Landry LG, Crawford F, Jin N, Hohenstein A, et al. Development of T Cell Lines Sensitive to Antigen Stimulation. *J Immunol Methods* (2018) 462:65–73. doi: 10.1016/j.jim.2018.08.011
34. van der Torren CR, Zaldumbide A, Duinkerken G, Brand-Schaaf SH, Peakman M, Stangé G, et al. Immunogenicity of Human Embryonic Stem Cell-Derived Beta Cells. *Diabetologia* (2017) 60:126–33. doi: 10.1007/s00125-016-4125-y
35. Xu H, Wang B, Ono M, Kagita A, Fujii K, Sasakawa N, et al. Targeted Disruption of HLA Genes *via* CRISPR-Cas9 Generates iPSCs With Enhanced Immune Compatibility. *Cell Stem Cell* (2019) 24. 566–78.e567. doi: 10.1016/j.stem.2019.02.005
36. Wang D, Quan Y, Yan Q, Morales JE, Wetsel RA. Targeted Disruption of the β2-Microglobulin Gene Minimizes the Immunogenicity of Human Embryonic Stem Cells. *Stem Cells Trans Med* (2015) 4:1234–45. doi: 10.5966/sctm.2015-0049
37. Gornalusse GG, Hirata RK, Funk SE, Riolobos L, Lopes VS, Manske G, et al. HLA-E-Expressing Pluripotent Stem Cells Escape Allogeneic Responses and Lysis by NK Cells. *Nat Biotechnol* (2017) 35:765. doi: 10.1038/nbt.3860
38. Han X, Wang M, Duan S, Franco PJ, Kenty JH-R, Hedrick P, et al. Generation of Hypoimmunogenic Human Pluripotent Stem Cells. *Proc Natl Acad Sci* (2019) 116:10441–6. doi: 10.1073/pnas.1902566116
39. Shi L, Li W, Liu Y, Chen Z, Hui Y, Hao P, et al. Generation of Hypoimmunogenic Human Pluripotent Stem Cells *via* Expression of Membrane-Bound and Secreted β2m-HLA-G Fusion Proteins. *Stem Cells* (2020) 38:1423–37. doi: 10.1002/stem.3269

15

Multipotent Mesenchymal Stromal Cells Interact and Support Islet of Langerhans Viability and Function

*Naomi Koehler[1], Leo Buhler[2], Bernhard Egger[1,2] and Carmen Gonelle-Gispert[1]**

[1] *Surgical Research Unit, Faculty of Science and Medicine, University of Fribourg, Fribourg, Switzerland,* [2] *Department of Surgery, Cantonal Hospital Fribourg, Fribourg, Switzerland*

Correspondence:
Carmen Gonelle-Gispert
carmen.gonelle@unifr.ch

Type 1 diabetes (T1D) is a widespread disease, affecting approximately 41.5 million people worldwide. It is generally treated with exogenous insulin, maintaining physiological blood glucose levels but also leading to long-term therapeutic complications. Pancreatic islet cell transplantation offers a potential alternative treatment to insulin injections. Shortage of human organ donors has raised the interest for porcine islet xenotransplantation. Neonatal porcine islets are highly available, can proliferate and mature *in vitro* as well as after transplantation *in vivo*. Despite promising preclinical results, delayed insulin secretion caused by immaturity and immunogenicity of the neonatal porcine islets remains a challenge for their clinical application. Multipotent mesenchymal stromal cells (MSCs) are known to have pro-angiogenic, anti-inflammatory and immunomodulatory effects. The current state of research emphasizes the great potential of co-culture and co-transplantation of islet cells with MSCs. Studies have shown enhanced islet proliferation and maturation, insulin secretion and graft survival, resulting in an improved graft outcome. This review summarizes the immunomodulatory and anti-inflammatory properties of MSC in the context of islet transplantation.

Keywords: MSC, neonatal porcine islets, diabetes, cell encapsulation, xenotransplantion

INTRODUCTION

Human pancreatic islet transplantation through portal vein infusion, is a current clinical beta-cell replacement therapy to treat patients with advanced Type I Diabetes (T1D). However, live-long immunosuppression, difficulties to achieve long-term islet graft function and insulin independence as well as the shortage of suitable pancreata from heart-beating brain-dead donors for islet isolation, are still important limitations for ongoing allo-transplantation programs.

Pig islet xenotransplantation is a promising alternative to overcome the bottleneck of islet availability for the treatment of T1D. However, clinical application of pig to human islet transplantation will depend on genetic engineering of pigs to overcome immune barriers and to reduce risks of pathogen infection of porcine viruses (1). Recently, significant progress has been achieved with the transplantation of pig organs presenting several genomic modifications to prevent hyperacute rejection (2, 3) and cellular immune responses (4, 5). In immunosuppressed nonhuman primates, long-term control of diabetes by the transplantation of adult porcine islets had been successfully achieved (6, 7). Other strategies to protect porcine islets from the host immune system include islet encapsulation in semi-permeable hydrogel, such as alginate (8) functionalized by bioactive ligands or by poly(ethylene glycol) (PEG) derivatives (9, 10). Mesenchymal stem cells (MSCs) are multipotent cells and play an important role in tissue repair, angiogenesis and their immunomodulatory action on immune cells have been widely studied (11, 12). In the field of islet transplantation MSC are investigated for the improvement of islet function and graft survival after transplantation. Numerous studies of co-culture and co-transplantation with MSCs indicate a functional support. However, due to variable transplantation settings and origins of MSCs the immunomodulatory role, as well as their ability to reduce inflammatory processes *in vivo* remains controversial. This review summarizes the immunomodulatory and anti-inflammatory properties of MSC in the context of islet transplantation and evokes some of the current challenges of islet xenotransplantation.

MULTIPOTENT MESENCHYMAL STROMAL CELLS (MSCs), ALSO CALLED MESENCHYMAL STEM CELLS

Multipotent mesenchymal stromal cells (MSCs) are self-renewing multipotential progenitor cells, differentiating along the osteogenic, chondrogenic and adipogenic lineages (13). MSCs have first been isolated from the bone marrow over 50 years ago and bone marrow-derived MSCs (BM-MSCs) still represent the most conventional source. A variety of other tissues also contain MSCs, including adipose tissue, umbilical cord blood, Wharton's jelly, amniotic fluid, endometrium, skin and skeletal muscle (14–22). It remains unknown which source is most suitable for the clinical use in the context of islet cell transplantation and further research is needed concerning this matter.

To facilitate isolation and expansion as well as to standardize characterization, the International Society for Cellular Therapy (ISCT) proposed three minimal criteria for defining mesenchymal stem cells. First, MSCs must be plastic-adherent when maintained in standard culture conditions. Second, MSCs must express CD105, CD73 and CD90, and lack expression of CD45, CD34, CD14 or CD11b, CD79alpha or CD19 and HLA-DR surface molecules. Third, MSCs must differentiate to osteoblasts, adipocytes and chondroblasts *in vitro* (23). MSCs have been shown to perform various beneficial functions, making them highly interesting for application in cell-based therapy, especially also for islet transplantation.

MSCs SUSTAIN ANGIOGENESIS

One major limitation of islet graft survival is a delayed revascularization after transplantation. After isolation, islet cells are cut off from their oxygenation *via* micro-vascularization and are temporarily dependent after transplantation on diffusion of nutrients and oxygen in order to ensure survival. Neovascularization is finalized after approximately two weeks, however with a lower capillary density and a significantly reduced perfusion compared to islets prior to transplantation (24). Further remodeling takes up to another three months (16).

Several studies have shown a pro-angiogenic potential of MSCs. MSCs promote angiogenesis through expression and release of different pro-angiogenic cytokines, including vascular endothelial growth factor (VEGF), fibroblast growth factor (FGF), transforming growth factor beta (TGF-β), as well as annexin 1 (ANXA1), matrix metalloproteinase (MMP) and Angiopoietin-1 (Ang-1) (16). The impact of VEGF is contentious, showing not only beneficial proangiogenic but also damaging proinflammatory effects (16). MSCs seem to keep a balance through its valuable anti-inflammatory property, discussed later. Kinnaird et al. demonstrated further that co-culturing islets with MSCs enhanced neovascularization of islets through promotion of proliferation and migration of endothelial and smooth muscle cells (25).

Also, other studies showed that co-transplantation of islets with MSCs improves graft survival and function by increased neovascularization, shortening the post-transplantation ischemia period (26–28).

IMMUNOMODULATORY PROPERTIES OF MSCs

MSCs do not express co-stimulatory molecules that activate the immune system, such as CD40, CD80 or CD86 (29). Originally, it was thought that MSCs express only low or no human leukocyte antigen (HLA) class I and II molecules. It has since then been demonstrated that MSCs, like all somatic tissues, express MHC class I molecules constitutively and have the ability to express MHC class II when exposed to inflammatory cues such as interferon-γ (30). *In vitro* studies showed that, attracted by a number of complement proteins, growth factors, proinflammatory cytokines and chemokines, MSCs migrate towards sites of inflammation supporting the hypothesis that MSCs possess anti-inflammatory properties (17, 31, 32). MSCs can express potent inhibitory molecules of both, innate and adaptive immune effectors (33), however, after transplantation, this may not allow to circumvent acquired alloimmunization, as observed in human trials (30). Nevertheless, immediate events such as acute toxicity associated with the administration of MSCs have not been described (34, 35).

MSCs have also shown to exert an immunomodulatory effect through phenotype-alteration of different immune cells, including dendritic cells (DC), T- and B-cells, as well as natural killer cells (NK cells) (36). Several authors have described an inhibitory effect of MSCs on immune cell proliferation, generating an immunosuppressive local milieu (16, 37–40). Research has further shown that MSCs induce modifications of the adaptive immune system, notably T-cells, entailing T-cell anergy. MSCs act on T-cells through physically hindering contact with antigen-presenting cells (APCs) (41) or by an indirect suppression of T-cell activation *via* MSCs by hindering the maturation of DCs through cell-to-cell contact. These semi-mature DCs possess a tolerogenic phenotype, thus restraining T-cell activation (42). Also, MSCs are able to inhibit T-cell reactivity through the downregulation of proinflammatory cytokines (37, 43) and to escape cytotoxic T-cell-mediated apoptosis (44, 45). Importantly, MSCs inhibit T-lymphocyte proliferation through soluble factors, such as TGF-β1 and HGF (41, 46) or nitric oxide (47). TGF-β1 plays a well-documented role in MSCs immunomodulation, including a role in regulatory T cell (Treg) induction and/or expansion (48–50). MSCs promote the expression of regulatory T-cells (Treg) (43, 51). Early studies showed that stable islet allograft function in cynomolgus monkey was associated with increased numbers of regulatory T-cells in peripheral blood (43). Further, when co-transplanted with allogeneic islets in diabetic cynomolgus monkeys, MSCs derived from islet recipient were more efficient to prolong islet survival, when compared with 3rd party MSCs or islet derived MSCs from the donor. Using recipient-derived MSCs, they observed decreased number of memory T cells, reduced anti-donor T cell proliferation and higher Treg:T cell ratios (52).

CO-TRANSPLANTATION OF ISLETS WITH MSCs FROM DIFFERENT SOURCES

Various possible sources of MSCs have been tested for co-transplantation with islet cells so far.

In murine models, several studies showed improved and prolonged graft survival, function, morphology and revascularization, as well as induction of beta cell proliferation following transplantation of murine islets with autologous (53), syngeneic (27, 54, 55), allogeneic (27, 56–59) or xenogeneic MSCs (60). In mice, co-transplantation of autologous MSCs delayed islet allograft rejection and generated a local immune-privileged site in mice (53). Rackham et al. studied the effects of co-transplantation of syngeneic murine MSCs and islet cells, and observed an improved graft outcome (54). In a subsequent study they examined the underlying factors, suggesting Annexin A1 to be a key contributor to the improved graft function through direct and indirect mechanisms (61). The exact mechanisms remain unclear.

Co-transplantation of islets with MSCs in syngeneic rodent models showed better outcomes of islet survival and function than islets transplanted alone (26, 27, 62). Karaoz et al. described an improved islet function after co-culturing allogenic rat MSCs and islet cells, suggesting paracrine actions through IL-6, TGF-β1, osteopontin and fibronectin (59). Further, allogeneic MSCs resulted in improved islet xenograft survival and function in immune-competent diabetic mice (63). In cynomolgus monkeys, intraportal co-infusion of allogenic MSCs and islets, increased islet engraftment and function, shown by a reduced number of islets necessary to reach normoglycemia (43).

Co-encapsulation studies using islets versus islets and MSC also showed beneficial effects on islet function (64–66). Intraperitoneally syngeneic transplantation of co-encapsulated islets and MSCs showed significantly lower glycaemia compared to islets encapsulated alone. By week 6, 71% of mice transplanted with islets and MSCs were cured, whereas only 16% of the islets-alone group was cured at that point. Interestingly, islet area in recovered capsules was significantly higher when co-encapsulated with MSCs suggesting that MSCs promote survival of islet cells independently from its effects on revascularization. In this study co-encapsulation of islets with MSC did not inhibit pericapsular fibrotic overgrowth (PFO), suggesting that MSCs have no influence on the inflammatory process that causes fibrotic overgrowth (64). PFO is an inflammatory host reaction, induced through the leakage of antigens from semi-permeable microcapsules, that severely impairs islet viability and graft function. However, in a mouse model of islet allotransplantation, co-encapsulation of MSCs (stimulated or not with a cocktail of pro-inflammatory cytokines) with islets in alginate microcapsules, prevented pericapsular fibrotic overgrowth (PFO) compared to islets encapsulated alone (66). Further mice receiving islets co-encapsulated with stimulated and unstimulated MSC achieved higher percentages of normoglycemic mice (100% versus 71.4%, respectively) compared to mice transplanted with islets encapsulated alone (9.1%). Similarly, *in vitro* rat MSCs and rat islet cells when co-encapsulated in a ligand-functionalized polyethylene glycol (PEG) hydrogel (67) led to a doubling of the stimulation index compared to islets encapsulated alone. Co-encapsulation of islet cells and MSCs in addition with cell adhesion peptides led to a significant sevenfold increase of the stimulation index compared to islets encapsulated alone (67).

Human islets co-cultured in direct cell contact with human MSCs compared to islets co-cultured with MSCs but without cell-to-cell contact, displayed significantly enhanced insulin secretion in the presence of cell-to-cell contact. This effect was identified to be dependent on N-cadherin interaction, since impeding N-cadherin interaction with antibodies led to a reversal of the enhanced insulin secretion. Additionally, mice transplanted intraperitoneally with human islets co-encapsulated with MSCs in hydrogel microspheres, composed of calcium alginate and covalently crosslinked to polyethylene glycol showed significantly lower blood glucose levels and prolonged islet graft survival (57). Others have shown that improved graft function correlates with enhanced revascularization of islets transplanted under the kidney capsule (68–70). Accordingly, research findings revealed significantly higher apoptosis rates in islet cells cultured without MSCs (16).

Taken altogether, these findings support the hypothesis that co-transplantation of MSCs and islet cells is beneficial and that MSCs are useful for future therapeutic applications.

IMPROVED NEONATAL PORCINE ISLET FUNCTION, SURVIVAL AND GRAFT OUTCOME

The main disadvantage of neonatal or juvenile porcine islets, also called porcine pancreatic islet cell clusters (ICCs), is their lack of integrity and maturity. ICCs are obtained by *in vitro* digestion of neonatal or juvenile pig pancreas with subsequent short time culture in a specific maturation media to increase the beta cell mass for transplantation (**Figure 1**) (71). Also, porcine pancreatic ICC co-transplanted with human MSC into immune deficient diabetic mice reached normoglycemia significantly earlier than mice transplanted with ICC alone (60).

He et al. demonstrated an improved and accelerated development of ICCs in diabetic rhesus monkeys after co-transplantation with allogeneic simian MSCs into diabetic rhesus monkeys (72). Additionally, the group described an enhanced expression of genes implicated in the development of endocrine cells and insulin and further demonstrated enhanced expression and activation of PDGFR-α in neonatal islets through MSCs confirming earlier studies demonstrating the capability of PDGFR-α to stimulate beta-cell proliferation (73). Further, He et al. suggest an inhibition of the Notch1 signaling provoked by PDGFR-α, leading to an improved islet development and maturation. It is known that Notch1 downregulates the expression of several genes and transcription factors implicated in the development of endocrine cells and insulin (72, 74). Juvenile porcine exocrine pancreas-derived MSCs (pMSCs) co-cultured with direct cell to cell contact of juvenile porcine ICCs significantly enhanced beta-cell function, suggesting that cell signaling *via* adhesion molecules are important (57, 65). However, co-encapsulation of such ICCs with pMSCs do not effectively prevent PFO and graft survival was rapidly impaired after transplantation of capsules in immunocompetent mice. Therefore, further research is required to enable efficient long-term survival of encapsulated juvenile porcine islets. Possible approaches being evaluation of modified alginate chemical composition (75) or the use of different anti-fibrotic polymers (65).

IMMUNOMODULATION STRATEGIES TO INCREASE XENOGRAFT SURVIVAL

To overcome the immunological barrier between pig and humans, genetic modifications have been performed in pig strains to reduce immunogenicity of organs and tissue. The first genetically modified pig, i.e. with a single human transgene for a complement regulating gene (hDAF), allowed survival of pig organs in immunosuppressed non-human primates for several months. Since then, genetic engineering, using CRISPR-CAS9, allowed cloning of animals with additional genetic modifications. Today, pigs with over 10 genetic modifications, both, deletions of pig antigens and inclusions of human transgenes are under investigation for transplantation purposes (76). Immunosuppressive regimens are still necessary but recently heart transplantation from genetically modified pigs [α1,3-galactosyltransferase-knockout and knock in human CD46 (77, 78) and thrombomodulin (79)] to a non-human primate (baboon) reached long term survival of 195 days (80).

Furthermore, immunoregulatory therapies (tolerance induction) using Treg-based therapeutic approaches are under investigation. Regulatory T cells (Tregs) are immune-suppressive T cells that are critical for the maintenance of tolerance *in vivo* (81). Chimeric antigen receptors (CARs) are synthetic fusion proteins that have been developed to genetically modify T cells in order to create a specificity toward designated antigens. The application of the CAR technology to Tregs, may allow to reduce immune responses for solid organ and cell transplantation. CAR Treg therapies are currently developed using genetic modifications for xenogenic pig antigens with the aim to improve graft acceptance of xenotransplanted tissue i.e. porcine islets. This might be achieved through infusion of *ex vivo* expansion of donor-specific Tregs (55, 82).

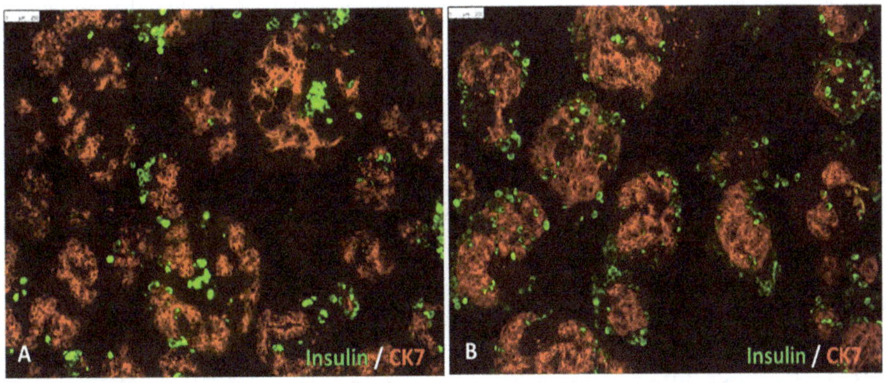

FIGURE 1 | *In vitro* differentiation of isolated porcine pancreatic islet cell clusters: Panel **(A)** shows cell clusters containing insulin-positive beta cells (green) and CK7-positive pancreatic exocrine tissue (red) at 3 days after isolation. Panel **(B)** shows pancreatic islet cell clusters 7 days after culture in neonatal pig islet differentiation media (scale bar represents 250 μm).

CAR-Tregs technology started with a study of MacDonald and colleagues which successfully transduced human Tregs with a CAR targeting the human leukocyte antigen (HLA) class I-A2 (A2-CAR) (83). In a human skin xenograft transplant model, HLA class I-A2 specific CAR-Tregs alleviated rejection of skin transplants (84). Since co-transplantation of autologous MSC delayed islet allograft rejection, it is possible that genetically modified, MSC, could be exploited as a target cell in porcine ICC xenografts to foster islet function and to increase trafficking and activation of adoptive transferred CAR-Treg cells to increase tolerance toward pig ICC xenografts.

An additional challenge for islet transplantation is the precise quantification of beta cell mass (BCM) or endocrine cell mass (ECM) *in vivo*. Imaging the progressive loss of beta cells following islet transplantation should allow the development of individualized therapies for the management of patients post-transplant (85). Recently, a suitable biomarker for beta cell quantification, the dipeptidyl aminopeptidase-like protein 6 (DPP6) has been identified as a promising target for human BCM imaging in healthy individuals as well as diabetic patients (86, 87). First imaging and biodistribution studies using SPECT/CT and radiolabeled high-affinity camelid single-domain antibody (nanobody) directed specifically against human DPP6, allowed to visualize transplanted DPP6-expressing Kelly neuroblastoma cells or insulin-producing human EndoC-βH1 cells in immunodeficient mice. Importantly, neonatal pig islets expressing near-infrared fluorescent protein (iRFP) were non-invasively monitored through multispectral optoacoustic tomography (MSOT). MSOT signals, obtained after islet transplantation under the kidney capsule in mice, and obtained after subcutaneous and intramuscular islet transplantation in pigs, allowed to distinguish graft mass changes (88). Such reporter gene-expressing islets are also promising tools to evaluate the efficacy of newly developed biomaterials for encapsulation and transplantation of porcine islets.

CONCLUSION

Diabetes is a worldwide disease, affecting over 40 million people and putting an important burden on the healthcare system. Exogenous insulin represents the predominant treatment modality for type 1 diabetes, but is associated with long-term complications. Islet cell transplantation is a highly promising approach for treating type 1 diabetes aiming at reestablishing a physiological insulin secretion through replacement of the endocrine tissue. Despite improving preclinical and clinical results over the past decades, the need for immunosuppression and donor shortage limits the clinical application of this procedure. The implementation of porcine pancreatic ICCs with porcine MSC might represent a promising alternative to help to overcome the problem of donor shortage; especially neonatal or juvenile pigs providing high islet yields. Encapsulation techniques could resolve the need for immunosuppression, shielding the islets from immune attacks while still enabling the exchange of oxygen, insulin and nutrients. Yet, delayed and impaired graft outcome due to immature islet cells and the formation of pericapsular fibrosis continue to severely limit the clinical application of encapsulated islet transplantation.

AUTHOR CONTRIBUTIONS

NK: designed and wrote the manuscript; LB contributed to manuscript writing; LB and BE: critically revised the manuscript; CGG: designed, contributed to manuscript writing; supervised the writing of the manuscript. All authors contributed to the article and approved the submitted version.

ACKNOWLEDGMENTS

This review article has been supported by the Insuleman Foundation, the Child Foundation and the Foundation La Colombe.

REFERENCES

1. Denner J. Porcine Lymphotropic Herpesviruses (Plhvs) and Xenotranplantation. *Viruses* (2021) 13(6):1072. doi: 10.3390/v13061072
2. Lai L, Kolber-Simonds D, Park KW, Cheong HT, Greenstein JL, Im GS, et al. Production of Alpha-1,3-Galactosyltransferase Knockout Pigs by Nuclear Transfer Cloning. *Science* (2002) 295(5557):1089–92. doi: 10.1126/science.1068228
3. Phelps CJ, Koike C, Vaught TD, Boone J, Wells KD, Chen SH, et al. Production of Alpha 1,3-Galactosyltransferase-Deficient Pigs. *Science* (2003) 299(5605):411–4. doi: 10.1126/science.1078942
4. Liu Z, Hu W, He T, Dai Y, Hara H, Bottino R, et al. Pig-to-Primate Islet Xenotransplantation: Past, Present, and Future. *Cell Transplant* (2017) 26 (6):925–47. doi: 10.3727/096368917X694859
5. Hryhorowicz M, Lipinski D, Hryhorowicz S, Nowak-Terpilowska A, Ryczek N, Zeyland J. Application of Genetically Engineered Pigs in Biomedical Research. *Genes (Basel)* (2020) 11(6):670. doi: 10.3390/genes11060670
6. Shin JS, Kim JM, Kim JS, Min BH, Kim YH, Kim HJ, et al. Long-Term Control of Diabetes in Immunosuppressed Nonhuman Primates (NHP) by the Transplantation of Adult Porcine Islets. *Am J Transplant* (2015) 15 (11):2837–50. doi: 10.1111/ajt.13345
7. Shin JS, Kim JM, Min BH, Yoon IH, Kim HJ, Kim JS, et al. Pre-Clinical Results in Pig-to-Non-Human Primate Islet Xenotransplantation Using Anti-CD40 Antibody (2C10R4)-Based Immunosuppression. *Xenotransplantation* (2018) 25(1):e12356. doi: 10.1111/xen.12356
8. Ghasemi A, Akbari E, Imani R. An Overview of Engineered Hydrogel-Based Biomaterials for Improved Beta-Cell Survival and Insulin Secretion. *Front Bioeng Biotechnol* (2021) 9:662084. doi: 10.3389/fbioe.2021.662084
9. Laporte C, Tubbs E, Pierron M, Gallego A, Moisan A, Lamarche F, et al. Improved Human Islets' Viability and Functionality With Mesenchymal Stem Cells and Arg-Gly-Asp Tripeptides Supplementation of Alginate Micro-Encapsulated Islets In Vitro. *Biochem Biophys Res Commun* (2020) 528 (4):650–7. doi: 10.1016/j.bbrc.2020.05.107
10. Passemard S, Szabo L, Noverraz F, Montanari E, Gonelle-Gispert C, Buhler LH, et al. Synthesis Strategies to Extend the Variety of Alginate-Based Hybrid Hydrogels for Cell Microencapsulation. *Biomacromolecules* (2017) 18 (9):2747–55. doi: 10.1021/acs.biomac.7b00665
11. Shi M, Liu ZW, Wang FS. Immunomodulatory Properties and Therapeutic Application of Mesenchymal Stem Cells. *Clin Exp Immunol* (2011) 164(1):1–8. doi: 10.1111/j.1365-2249.2011.04327.x
12. Castro-Manrreza ME, Montesinos JJ. Immunoregulation by Mesenchymal Stem Cells: Biological Aspects and Clinical Applications. *J Immunol Res* (2015) 2015:394917. doi: 10.1155/2015/394917

13. Sudo K, Kanno M, Miharada K, Ogawa S, Hiroyama T, Saijo K, et al. Mesenchymal Progenitors Able to Differentiate Into Osteogenic, Chondrogenic, and/or Adipogenic Cells In Vitro Are Present in Most Primary Fibroblast-Like Cell Populations. *Stem Cells* (2007) 25(7):1610–7. doi: 10.1634/stemcells.2006-0504
14. Kern S, Eichler H, Stoeve J, Klüter H, Bieback K. Comparative Analysis of Mesenchymal Stem Cells From Bone Marrow, Umbilical Cord Blood, or Adipose Tissue. *Stem Cells* (2006) 24(5):1294–301. doi: 10.1634/stemcells.2005-0342
15. da Silva Meirelles L, Chagastelles PC, Nardi NB. Mesenchymal Stem Cells Reside in Virtually All Post-Natal Organs and Tissues. *J Cell Sci* (2006) 119(Pt 11):2204–13. doi: 10.1242/jcs.02932
16. Gamble A, Pawlick R, Pepper AR, Bruni A, Adesida A, Senior PA, et al. Improved Islet Recovery and Efficacy Through Co-Culture and Co-Transplantation of Islets With Human Adipose-Derived Mesenchymal Stem Cells. *PLoS One* (2018) 13(11):e0206449. doi: 10.1371/journal.pone.0206449
17. Kang EJ, Lee YH, Kim MJ, Lee YM, Kumar BM, Jeon BG, et al. Transplantation of Porcine Umbilical Cord Matrix Mesenchymal Stem Cells in a Mouse Model of Parkinson's Disease. *J Tissue Eng Regener Med* (2013) 7(3):169–82. doi: 10.1002/term.504
18. Zhang W, Liu XC, Yang L, Zhu DL, Zhang YD, Chen Y, et al. Wharton's Jelly-Derived Mesenchymal Stem Cells Promote Myocardial Regeneration and Cardiac Repair After Miniswine Acute Myocardial Infarction. *Coron Artery Dis* (2013) 24(7):549–58. doi: 10.1097/MCA.0b013e3283640f00
19. Miernik K, Karasinski J. Porcine Uterus Contains a Population of Mesenchymal Stem Cells. *Reproduction* (2012) 143(2):203–9. doi: 10.1530/REP-11-0202
20. Subbarao RB, Ullah I, Kim EJ, Jang SJ, Lee WJ, Jeon RH, et al. Characterization and Evaluation of Neuronal Trans-Differentiation With Electrophysiological Properties of Mesenchymal Stem Cells Isolated From Porcine Endometrium. *Int J Mol Sci* (2015) 16(5):10934–51. doi: 10.3390/ijms160510934
21. Park BW, Kang DH, Kang EJ, Byun JH, Lee JS. Peripheral Nerve Regeneration Using Autologous Porcine Skin-Derived Mesenchymal Stem Cells. *J Tissue Eng Regener Med* (2012) 6(2):113–24. doi: 10.1002/term.404
22. Kang EJ, Byun JH, Choi YJ, Maeng GH, Lee SL, Kang DH, et al. In Vitro and In Vivo Osteogenesis of Porcine Skin-Derived Mesenchymal Stem Cell-Like Cells With a Demineralized Bone and Fibrin Glue Scaffold. *Tissue Eng Part A* (2010) 16(3):815–27. doi: 10.1089/ten.tea.2009.0439
23. Dominici M, Le Blanc K, Mueller I, Slaper-Cortenbach I, Marini F, Krause D, et al. Minimal Criteria for Defining Multipotent Mesenchymal Stromal Cells. The International Society for Cellular Therapy Position Statement. *Cytotherapy* (2006) 8(4):315–7. doi: 10.1080/14653240600855905
24. Carlsson P-O, Palm F, Mattsson G. Low Revascularization of Experimentally Transplanted Human Pancreatic Islets. *J Clin Endocrinol Metab* (2003) 87:5418–23. doi: 10.1210/jc.2002-020728
25. Kinnaird T, Stabile E, Burnett MS, Lee CW, Barr S, Fuchs S, et al. Marrow-Derived Stromal Cells Express Genes Encoding a Broad Spectrum of Arteriogenic Cytokines and Promote In Vitro and In Vivo Arteriogenesis Through Paracrine Mechanisms. *Circ Res* (2004) 94(5):678–85. doi: 10.1161/01.RES.0000118601.37875.AC
26. Figliuzzi M, Cornolti R, Perico N, Rota C, Morigi M, Remuzzi G, et al. Bone Marrow-Derived Mesenchymal Stem Cells Improve Islet Graft Function in Diabetic Rats. *Transplant Proc* (2009) 41(5):1797–800. doi: 10.1016/j.transproceed.2008.11.015
27. Ito T, Itakura S, Todorov I, Rawson J, Asari S, Shintaku J, et al. Mesenchymal Stem Cell and Islet Co-Transplantation Promotes Graft Revascularization and Function. *Transplantation* (2010) 89(12):1438–45. doi: 10.1097/TP.0b013e3181db09c4
28. Päth G, Perakakis N, Mantzoros CS, Seufert J. Stem Cells in the Treatment of Diabetes Mellitus - Focus on Mesenchymal Stem Cells. *Metabolism* (2019) 90:1–15. doi: 10.1016/j.metabol.2018.10.005
29. Wang Q, Ding G, Xu X. Immunomodulatory Functions of Mesenchymal Stem Cells and Possible Mechanisms. *Histol Histopathol* (2016) 31(9):949–59. doi: 10.14670/HH-11-750
30. Ankrum JA, Ong JF, Karp JM. Mesenchymal Stem Cells: Immune Evasive, Not Immune Privileged. *Nat Biotechnol* (2014) 32(3):252–60. doi: 10.1038/nbt.2816
31. English K. Mechanisms of Mesenchymal Stromal Cell Immunomodulation. *Immunol Cell Biol* (2013) 91(1):19–26. doi: 10.1038/icb.2012.56
32. Azhar M, Yin M, Bommireddy R, Duffy JJ, Yang J, Pawlowski SA, et al. Generation of Mice With a Conditional Allele for Transforming Growth Factor Beta 1 Gene. *Genesis* (2009) 47(6):423–31. doi: 10.1002/dvg.20516
33. Uccelli A, Moretta L, Pistoia V. Mesenchymal Stem Cells in Health and Disease. *Nat Rev Immunol* (2008) 8(9):726–36. doi: 10.1038/nri2395
34. Lalu MM, McIntyre L, Pugliese C, Fergusson D, Winston BW, Marshall JC, et al. Safety of Cell Therapy With Mesenchymal Stromal Cells (Safecell): A Systematic Review and Meta-Analysis of Clinical Trials. *PLoS One* (2012) 7(10):e47559. doi: 10.1371/journal.pone.0047559
35. Koç ON, Gerson SL, Cooper BW, Dyhouse SM, Haynesworth SE, Caplan AI, et al. Rapid Hematopoietic Recovery After Coinfusion of Autologous-Blood Stem Cells and Culture-Expanded Marrow Mesenchymal Stem Cells in Advanced Breast Cancer Patients Receiving High-Dose Chemotherapy. *J Clin Oncol* (2000) 18(2):307–16. doi: 10.1200/JCO.2000.18.2.307
36. Aggarwal S, Pittenger MF. Human Mesenchymal Stem Cells Modulate Allogeneic Immune Cell Responses. *Blood* (2005) 105(4):1815–22. doi: 10.1182/blood-2004-04-1559
37. Glennie S, Soeiro I, Dyson PJ, Lam EW-F, Dazzi F. Bone Marrow Mesenchymal Stem Cells Induce Division Arrest Anergy of Activated T Cells. *Blood* (2005) 105(7):2821–7. doi: 10.1182/blood-2004-09-3696
38. Corcione A, Benvenuto F, Ferretti E, Giunti D, Cappiello V, Cazzanti F, et al. Human Mesenchymal Stem Cells Modulate B-Cell Functions. *Blood* (2006) 107(1):367–72. doi: 10.1182/blood-2005-07-2657
39. Spaggiari GM, Capobianco A, Becchetti S, Mingari MC, Moretta L. Mesenchymal Stem Cell-Natural Killer Cell Interactions: Evidence That Activated NK Cells Are Capable of Killing Mscs, Whereas Mscs can Inhibit IL-2-Induced NK-Cell Proliferation. *Blood* (2006) 107(4):1484–90. doi: 10.1182/blood-2005-07-2775
40. Bartholomew A, Sturgeon C, Siatskas M, Ferrer K, McIntosh K, Patil S. Mesenchymal Stem Cells Suppress Lymphocyte Proliferation In Vitro and Prolong Skin Graft Survival In Vivo. *Exp Hematol* (2002) 30(1):42–8. doi: 10.1016/S0301-472X(01)00769-X
41. Krampera M, Glennie S, Dyson J, Scott D, Laylor R, Simpson E, et al. Bone Marrow Mesenchymal Stem Cells Inhibit the Response of Naive and Memory Antigen-Specific T Cells to Their Cognate Peptide. *Blood* (2003) 101(9):3722–9. doi: 10.1182/blood-2002-07-2104
42. Beyth S, Borovsky Z, Mevorach D, Liebergall M, Gazit Z, Aslan H, et al. Human Mesenchymal Stem Cells Alter Antigen-Presenting Cell Maturation and Induce T-Cell Unresponsiveness. *Blood* (2005) 105(5):2214–9. doi: 10.1182/blood-2004-07-2921
43. Berman DM, Willman MA, Han D, Kleiner G, Kenyon NM, Cabrera O, et al. Mesenchymal Stem Cells Enhance Allogeneic Islet Engraftment in Nonhuman Primates. *Diabetes* (2010) 59(10):2558–68. doi: 10.2337/db10-0136
44. Ryan JM, Barry FP, Murphy JM, Mahon BP. Mesenchymal Stem Cells Avoid Allogeneic Rejection. *J Inflamm (London England)* (2005) 2:8–8. doi: 10.1186/1476-9255-2-8
45. Rasmusson I, Uhlin M, Le Blanc K, Levitsky V. Mesenchymal Stem Cells Fail to Trigger Effector Functions of Cytotoxic T Lymphocytes. *J Leukocyte Biol* (2007) 82(4):887–93. doi: 10.1189/jlb.0307140
46. Di Nicola M, Carlo-Stella C, Magni M, Milanesi M, Longoni PD, Matteucci P, et al. Human Bone Marrow Stromal Cells Suppress T-Lymphocyte Proliferation Induced by Cellular or Nonspecific Mitogenic Stimuli. *Blood* (2002) 99(10):3838–43. doi: 10.1182/blood.V99.10.3838
47. Sato K, Ozaki K, Oh I, Meguro A, Hatanaka K, Nagai T, et al. Nitric Oxide Plays a Critical Role in Suppression of T-Cell Proliferation by Mesenchymal Stem Cells. *Blood* (2007) 109(1):228–34. doi: 10.1182/blood-2006-02-002246
48. Carrillo-Galvez AB, Cobo M, Cuevas-Ocana S, Gutierrez-Guerrero A, Sanchez-Gilabert A, Bongarzone P, et al. Mesenchymal Stromal Cells Express GARP/LRRC32 on Their Surface: Effects on Their Biology and Immunomodulatory Capacity. *Stem Cells* (2015) 33(1):183–95. doi: 10.1002/stem.1821
49. Patel SA, Meyer JR, Greco SJ, Corcoran KE, Bryan M, Rameshwar P. Mesenchymal Stem Cells Protect Breast Cancer Cells Through Regulatory T Cells: Role of Mesenchymal Stem Cell-Derived TGF-Beta. *J Immunol* (2010) 184(10):5885–94. doi: 10.4049/jimmunol.0903143

50. Salazar KD, Lankford SM, Brody AR. Mesenchymal Stem Cells Produce Wnt Isoforms and TGF-Beta1 That Mediate Proliferation and Procollagen Expression by Lung Fibroblasts. *Am J Physiol Lung Cell Mol Physiol* (2009) 297(5):L1002–11. doi: 10.1152/ajplung.90347.2008
51. De Miguel MP, Fuentes-Julián S, Blázquez-Martínez A, Pascual CY, Aller MA, Arias J, et al. Immunosuppressive Properties of Mesenchymal Stem Cells: Advances and Applications. *Curr Mol Med* (2012) 12(5):574–91. doi: 10.2174/156652412800619950
52. Kenyon NS, Willman MA, Han D, Leeman RS, Rabassa A, Diaz WL, et al. Extended Survival vs Accelerated Rejection of Nonhuman Primate Islet Allografts: Effect of Mesenchymal Stem Cell Source and Timing. *Am J Transplant* (2021) 21(11):3524–37. doi: 10.1111/ajt.16693
53. Ben Nasr M, Vergani A, Avruch J, Liu L, Kefaloyianni E, D'Addio F, et al. Co-Transplantation of Autologous Mscs Delays Islet Allograft Rejection and Generates a Local Immunoprivileged Site. *Acta Diabetol* (2015) 52(5):917–27. doi: 10.1007/s00592-015-0735-y
54. Rackham CL, Chagastelles PC, Nardi NB, Hauge-Evans AC, Jones PM, King AJF. Co-Transplantation of Mesenchymal Stem Cells Maintains Islet Organisation and Morphology in Mice. *Diabetologia* (2011) 54(5):1127–35. doi: 10.1007/s00125-011-2053-4
55. Huang D, Wang Y, Hawthorne WJ, Hu M, Hawkes J, Burns H, et al. Ex Vivo-Expanded Baboon CD39 + Regulatory T Cells Prevent Rejection of Porcine Islet Xenografts in NOD-SCID IL-2rgamma(-/-) Mice Reconstituted With Baboon Peripheral Blood Mononuclear Cells. *Xenotransplantation* (2017) 24 (6):e12344. doi: 10.1111/xen.12344
56. Longoni B, Szilagyi E, Quaranta P, Paoli GT, Tripodi S, Urbani S, et al. Mesenchymal Stem Cells Prevent Acute Rejection and Prolong Graft Function in Pancreatic Islet Transplantation. *Diabetes Technol Ther* (2010) 12(6):435–46. doi: 10.1089/dia.2009.0154
57. Montanari E, Meier RPH, Mahou R, Seebach JD, Wandrey C, Gerber-Lemaire S, et al. Multipotent Mesenchymal Stromal Cells Enhance Insulin Secretion From Human Islets via N-Cadherin Interaction and Prolong Function of Transplanted Encapsulated Islets in Mice. *Stem Cell Res Ther* (2017) 8(1):199. doi: 10.1186/s13287-017-0646-7
58. Gao X, Song L, Shen K, Wang H, Qian M, Niu W, et al. Bone Marrow Mesenchymal Stem Cells Promote the Repair of Islets From Diabetic Mice Through Paracrine Actions. *Mol Cell Endocrinol* (2014) 388(1):41–50. doi: 10.1016/j.mce.2014.03.004
59. Karaoz E, Genç ZS, Demircan PÇ, Aksoy A, Duruksu G. Protection of Rat Pancreatic Islet Function and Viability by Coculture With Rat Bone Marrow-Derived Mesenchymal Stem Cells. *Cell Death Dis* (2010) 1(4):e36–6. doi: 10.1038/cddis.2010.14
60. Hayward JA, Ellis CE, Seeberger K, Lee T, Salama B, Mulet-Sierra A, et al. Cotransplantation of Mesenchymal Stem Cells With Neonatal Porcine Islets Improve Graft Function in Diabetic Mice. *Diabetes* (2017) 66(5):1312–21. doi: 10.2337/db16-1068
61. Rackham CL, Vargas AE, Hawkes RG, Amisten S, Persaud SJ, Austin ALF, et al. Annexin A1 Is a Key Modulator of Mesenchymal Stromal Cell–Mediated Improvements in Islet Function. *Diabetes* (2016) 65(1):129–39. doi: 10.2337/db15-0990
62. Borg DJ, Weigelt M, Wilhelm C, Gerlach M, Bickle M, Speier S, et al. Mesenchymal Stromal Cells Improve Transplanted Islet Survival and Islet Function in a Syngeneic Mouse Model. *Diabetologia* (2014) 57(3):522–31. doi: 10.1007/s00125-013-3109-4
63. Perez-Basterrechea M, Obaya AJ, Meana A, Otero J, Esteban MM. Cooperation by Fibroblasts and Bone Marrow-Mesenchymal Stem Cells to Improve Pancreatic Rat-to-Mouse Islet Xenotransplantation. *PloS One* (2013) 8(8):e73526. doi: 10.1371/journal.pone.0073526
64. Kerby A, Jones ES, Jones PM, King AJ. Co-Transplantation of Islets With Mesenchymal Stem Cells in Microcapsules Demonstrates Graft Outcome can be Improved in an Isolated-Graft Model of Islet Transplantation in Mice. *Cytotherapy* (2013) 15(2):192–200. doi: 10.1016/j.jcyt.2012.10.018
65. Montanari E, Szabo L, Balaphas A, Meyer J, Perriraz-Mayer N, Pimenta J, et al. Multipotent Mesenchymal Stromal Cells Derived From Porcine Exocrine Pancreas Improve Insulin Secretion From Juvenile Porcine Islet Cell Clusters. *Xenotransplantation* (2021) 28(3):e12666. doi: 10.1111/xen.12666
66. Vaithilingam V, Evans MDM, Lewy DM, Bean PA, Bal S, Tuch BE. Co-Encapsulation and Co-Transplantation of Mesenchymal Stem Cells Reduces Pericapsular Fibrosis and Improves Encapsulated Islet Survival and Function When Allografted. *Sci Rep* (2017) 7(1):10059. doi: 10.1038/s41598-017-10359-1
67. Bal T, Nazli C, Okcu A, Duruksu G, Karaöz E, Kizilel S. Mesenchymal Stem Cells and Ligand Incorporation in Biomimetic Poly(Ethylene Glycol) Hydrogels Significantly Improve Insulin Secretion From Pancreatic Islets. *J Tissue Eng Regenerative Med* (2017) 11(3):694–703. doi: 10.1002/term.1965
68. Bertera S, Knoll MF, Knoll C, Hara H, Kimbrel EA, Kouris NA, et al. Human Hemangioblast-Derived Mesenchymal Stem Cells Promote Islet Engraftment in a Minimal Islet Mass Transplantation Model in Mice. *Front Med (Lausanne)* (2021) 8:660877. doi: 10.3389/fmed.2021.660877
69. Cunha JP, Leuckx G, Sterkendries P, Korf H, Bomfim-Ferreira G, Overbergh L, et al. Human Multipotent Adult Progenitor Cells Enhance Islet Function and Revascularisation When Co-Transplanted as a Composite Pellet in a Mouse Model of Diabetes. *Diabetologia* (2017) 60(1):134–42. doi: 10.1007/s00125-016-4120-3
70. Selmani Z, Naji A, Zidi I, Favier B, Gaiffe E, Obert L, et al. Human Leukocyte Antigen-G5 Secretion by Human Mesenchymal Stem Cells Is Required to Suppress T Lymphocyte and Natural Killer Function and to Induce CD4 +CD25highFOXP3+ Regulatory T Cells. *Stem Cells* (2008) 26(1):212–22. doi: 10.1634/stemcells.2007-0554
71. Korbutt GS, Elliott JF, Ao Z, Smith DK, Warnock GL, Rajotte RV. Large Scale Isolation, Growth, and Function of Porcine Neonatal Islet Cells. *J Clin Invest* (1996) 97(9):2119–29. doi: 10.1172/JCI118649
72. He S, Wang C, Du X, Chen Y, Zhao J, Tian B, et al. Mscs Promote the Development and Improve the Function of Neonatal Porcine Islet Grafts. *FASEB J* (2018) 32(6):3242–53. doi: 10.1096/fj.201700991R
73. Chen H, Gu X, Liu Y, Wang J, Wirt SE, Bottino R, et al. PDGF Signalling Controls Age-Dependent Proliferation in Pancreatic β-Cells. *Nature* (2011) 478(7369):349–55. doi: 10.1038/nature10502
74. Hald J, Hjorth JP, German MS, Madsen OD, Serup P, Jensen J. Activated Notch1 Prevents Differentiation of Pancreatic Acinar Cells and Attenuate Endocrine Development. *Dev Biol* (2003) 260(2):426–37. doi: 10.1016/S0012-1606(03)00326-9
75. Vaithilingam V, Bal S, Tuch BE. Encapsulated Islet Transplantation: Where do We Stand? *Rev Diabetes Stud* (2017) 14(1):51–78. doi: 10.1900/RDS.2017.14.51
76. Meier RPH, Muller YD, Balaphas A, Morel P, Pascual M, Seebach JD, et al. Xenotransplantation: Back to the Future? *Transpl Int* (2018) 31(5):465–77. doi: 10.1111/tri.13104
77. Cooper DKC, Hara H, Iwase H, Yamamoto T, Jagdale A, Kumar V, et al. Clinical Pig Kidney Xenotransplantation: How Close Are We? *J Am Soc Nephrol* (2020) 31(1):12–21. doi: 10.1681/ASN.2019070651
78. Jagdale A, Nguyen H, Li J, Burnette K, Ayares D, Cooper DKC, et al. Does Expression of a Human Complement-Regulatory Protein on Xenograft Cells Protect Them From Systemic Complement Activation? *Int J Surg* (2020) 83:184–8. doi: 10.1016/j.ijsu.2020.09.034
79. Hara H, Iwase H, Nguyen H, Miyagawa Y, Kuravi K, Foote JB, et al. Stable Expression of the Human Thrombomodulin Transgene in Pig Endothelial Cells Is Associated With a Reduction in the Inflammatory Response. *Cytokine* (2021) 148:155580. doi: 10.1016/j.cyto.2021.155580
80. Langin M, Mayr T, Reichart B, Michel S, Buchholz S, Guethoff S, et al. Consistent Success in Life-Supporting Porcine Cardiac Xenotransplantation. *Nature* (2018) 564(7736):430–3. doi: 10.1038/s41586-018-0765-z
81. Pilat N, Sprent J. Treg Therapies Revisited: Tolerance Beyond Deletion. *Front Immunol* (2020) 11:622810. doi: 10.3389/fimmu.2020.622810
82. Ezzelarab MB. Regulatory T Cells From Allo- to Xenotransplantation: Opportunities and Challenges. *Xenotransplantation* (2018) 25(3):e12415. doi: 10.1111/xen.12415
83. MacDonald KG, Hoeppli RE, Huang Q, Gillies J, Luciani DS, Orban PC, et al. Alloantigen-Specific Regulatory T Cells Generated With a Chimeric Antigen Receptor. *J Clin Invest* (2016) 126(4):1413–24. doi: 10.1172/JCI82771
84. Boardman DA, Philippeos C, Fruhwirth GO, Ibrahim MA, Hannen RF, Cooper D, et al. Expression of a Chimeric Antigen Receptor Specific for Donor HLA Class I Enhances the Potency of Human Regulatory T Cells in

Preventing Human Skin Transplant Rejection. *Am J Transplant* (2017) 17 (4):931–43. doi: 10.1111/ajt.14185

85. Demine S, Schulte ML, Territo PR, Eizirik DL. Beta Cell Imaging-From Pre-Clinical Validation to First in Man Testing. *Int J Mol Sci* (2020) 21(19):7274. doi: 10.3390/ijms21197274
86. Balhuizen A, Massa S, Mathijs I, Turatsinze JV, De Vos J, Demine S, et al. A Nanobody-Based Tracer Targeting DPP6 for Non-Invasive Imaging of Human Pancreatic Endocrine Cells. *Sci Rep* (2017) 7(1):15130. doi: 10.1038/s41598-017-15417-2
87. Demine S, Garcia Ribeiro R, Thevenet J, Marselli L, Marchetti P, Pattou F, et al. A Nanobody-Based Nuclear Imaging Tracer Targeting Dipeptidyl Peptidase 6 to Determine the Mass of Human Beta Cell Grafts in Mice. *Diabetologia* (2020) 63(4):825–36. doi: 10.1007/s00125-019-05068-5
88. Kemter E, Citro A, Wolf-van Buerck L, Qiu Y, Bottcher A, Policardi M, et al. Transgenic Pigs Expressing Near Infrared Fluorescent Protein-a Novel Tool for Noninvasive Imaging of Islet Xenotransplants. *Xenotransplantation* (2021) 21:e12719. doi: 10.1111/xen.12719

16

Microvascular Complications in Type 1 Diabetes: A Comparative Analysis of Patients Treated with Autologous Nonmyeloablative Hematopoietic Stem-Cell Transplantation and Conventional Medical Therapy

Jaquellyne G. Penaforte-Saboia[1], Renan M. Montenegro Jr.[1,2], Carlos E. Couri[3,4], Livia A. Batista[1], Ana Paula D. R. Montenegro[2], Virginia O. Fernandes[2], Hussain Akhtar[2,5], Carlos A. Negrato[6], Kelen Cristina Ribeiro Malmegrim[3], Daniela Aparecida Moraes[3,4], Juliana B. E. Dias[3,4], Belinda P. Simões[3,4], Marilia Brito Gomes[7] and Maria Carolina Oliveira[3,4]*

[1] Post Graduate Program in Medical Sciences, Federal University of Ceará, Ceará, Brazil, [2] Federal University of Ceará, Ceará, Brazil, [3] Center for Cell-Based Therapy, Regional Blood Center of Ribeirão Preto, Ribeirão Preto Medical School, University of São Paulo, Ribeirão Preto, Brazil, [4] Department of Internal Medicine, Ribeirão Preto Medical School, University of São Paulo, Ribeirão Preto, Brazil, [5] University of Oslo, UIO, Oslo, Noruega, [6] Brazilian Society of Diabetes, São Paulo, Brazil, [7] State University of Rio de Janeiro, Rio de Janeiro, Brazil

***Correspondence:**
Renan M. Montenegro Jr.
renanmmjr@gmail.com*

Objective: To explore the impact on microvascular complications, long-term preservation of residual B-cell function and glycemic control of patients with type 1 diabetes treated with autologous nonmyeloablative hematopoietic stem-cell transplantation (AHST) compared with conventional medical therapy (CT).

Research design and methods: Cross-sectional data of patients treated with AHST were compared with patients who received conventional therapy from the Brazilian Type 1 Diabetes Study Group, the largest multicenter observational study in type 1 diabetes mellitus in Brazil. Both groups of patients had diabetes for 8 years on average. An assessment comparison was made on the presence of microvascular complications, residual function of B cell, A1c, and insulin dose of the patients.

Results: After a median of 8 years of diagnosis, none of the AHST-treated patients (n = 24) developed microvascular complications, while 21.5% (31/144) had at least one ($p < 0.005$) complication in the CT group (n = 144). Furthermore, no case of nephropathy was reported in the AHST group, while 13.8% of CT group ($p < 0.005$) developed nephropathy during the same period. With regard of residual B-cell function, the percentage of individuals with predicted higher C-peptide levels (IDAA1C $\leq$ 9) was about 10-fold higher in the AHST group compared with CT (75 vs. 8.3%) ($p < 0.001$) group. Among AHST patients, 54.1% (13/24) had the HbA1c < 7.0 compared with 13.1% in the CT ($p < 0.001$) group.

Conclusion: Patients with newly diagnosed type 1 diabetes treated with AHST presented lower prevalence of microvascular complications, higher residual B-cell function, and better glycemic control compared with the CT group.

Keywords: type 1 diabetes, autologous nonmyeloablative hematopoietic stem-cell transplantation, residual B-cell function, glycemic control, microvascular complications

INTRODUCTION

Type 1 diabetes mellitus (T1DM) is an autoimmune disease that results from autoreactive $CD4^+$ and $CD8^+$ lymphocytes attack against pancreatic beta cells (1), leading to symptomatic diabetes and lifelong insulin dependence (2). Islet autoimmunity is the first stage of the disease (3) and, at the time of clinical diagnosis, about 60–90% of the beta cells are destroyed or dysfunctional (1). Immunotherapy aims to preserve the remaining mass of B cells by attenuation of the activated, autoreactive T cells (4).

Bone marrow transplantation (BMT) has been studied as a therapeutic approach for T1DM since the 1980s (5). Initially, studies suggested that allogeneic BMT primarily for hematological disease could potentially transfer T1DM and in some instances, BMT could reverse it (6). Additionally, it was demonstrated that immunological intervention in newly diagnosed type 1 diabetic patients resulted in a slower decline of beta-cell function (7).

Between 2004 and 2010, Voltarelli et al. performed autologous nonmyeloablative hematopoietic stem-cell transplantation (AHST) in individuals with newly diagnosed T1DM (8–10). Twenty-one out of 25 patients became insulin-free during hospitalization for AHST. In June 2015, after a mean follow-up of 87 months, three were continuously insulin-free and 18 patients resumed exogenous insulin use after mean period of 33 months (ranging from 6 to 100 months). Mean area under the curve of C-peptide levels significantly increased from baseline compared with 6, 12, 24, 36, and 48 months post-AHST (10). At 60 and 72 months, most patients resumed insulin therapy and their C-peptide declined. Residual B-cell function was similar to that before transplant (10).

Several studies, including the Diabetes Control and Complications Trial (DCCT), demonstrated the association between high C-peptide levels and lower frequency of microvascular complications such as retinopathy, neuropathy, and diabetic nephropathy (11–13). For this reason, it was hypothesized that therapy with AHST could preserve B-cell function, improve metabolic control, and reduce microvascular complications.

Thus, the aim of this study was to compare the patients treated with AHST and those treated with the conventional medical therapy (CT) from the Brazilian Type 1 Diabetes Study Group (BrazDiab1). The following parameters were used in the comparison: the occurrence of microvascular complications, long-term preservation of residual B-cell function, and the patient's glycemic control. This is the first analysis of the AHST through a comparator group for T1DM.

MATERIALS AND METHODS

Study Design and Subjects

Hematopoietic Stem-Cell Transplantation (AHST) Group

Twenty-five patients with <6 weeks of diagnosis of t1dm aged between 13 and 31 years old were included in this group from January 2004 until May 2010 at the Center for Cell-based Therapy at the Ribeirão Preto Medical School—University of São Paulo, Brazil.

The rationale of AHST is based on the theory of "immunologic reset." A complete study protocol was presented elsewhere (8–10). In brief, peripheral hematopoietic stem cells were mobilized with cyclophosphamide plus a granulocyte-colony stimulating factor. Stem cells were then harvested from the peripheral blood through apheresis, which was continued daily until the number of harvested progenitor cells reached a minimum of 3.0×10^6 $CD34^+$ cells/kg body weight. After approximately 15 days from beginning of mobilization, patients were conditioned with rabbit anti-thymoglobulin (ATG) for 5 days plus cyclophosphamide in the first 4 days. Intravenous infusion of autologous hematopoietic stem cells was performed on the sixth day with the aim of restoring immune balance. The AHST database was obtained from patient chart review in 2015. One patient lost the follow-up and was not included in this analysis.

The Conventional Therapy from Brazilian Type 1 Diabetes Study Group (CT)

The BrazDiab1 study group performed a multicenter retrospective, cross-sectional study on individuals with T1DM between December 2008 and December 2010, in 28 public clinics from the secondary and tertiary level of health care, from 20 cities (population > 100,000) from all the five Brazilian geographical regions (North, Northeast, Southeast, South and Midwest). All patients received health care from the National Brazilian Health Care System (NBHCS), and were regularly treated by an endocrinologist. All data were obtained from a single examination performed by trained medical professionals.

Details of BrazDiab1 have been published elsewhere (14). Briefly, participants were interviewed and demographic data were collected through a questionnaire. Furthermore, the following variables were assessed through an interview during a clinical visit and the patient's medical records review.

Eligibility Criteria and Patient Selection

Patients enrolled in the BrazDiab1 (n = 3,591) were primarily included. In the second step, the patients who were clinical screened for retinopathy, neuropathy, diabetic nephropathy, and

HbA1c [National Glycohemoglobin Standardization Program method, including high-performance liquid chromatography (HPLC) and immunoturbidimetry] were selected. It was considered a 1-year timeline from the data collection (n = 1,613). Then, at the third stage of selection, the patients were matched for the age at diagnosis of diabetes (n = 538) and duration of diabetes (n = 251) to the AHST group. For the final stage, a matching of gender was made; BrazDiab1 women underwent a random selection through the SPSS program to maintain the approximate proportion of 30% of women present in the transplant group. A total of 144 patients were included in the final analysis (**Figure 1**).

Microvascular Complications

In both groups, diabetic retinopathy was screened by fundoscopy by a local ophthalmologist; nephropathy by microalbuminuria and glomerular filtration rate, defined by the Cockcroft–Gault equation (15); distal symmetric polyneuropathy was assessed by questioning the patients about symptoms of neuropathy, including paresthesia, dulled sensation, and pain in the feet and a neurological exam on the sensation of perception of vibration and 10-g monofilament pressure at the distal plantar of both great toes and metatarsal joints. The presence of microvascular complications was defined according to the American Diabetes Association Recommendations (16).

Residual Beta-Cell Function

The C-peptide values were measured only on the AHST group. Consequently, the residual B-cell function was evaluated indirectly in both groups using a model of insulin-dose adjusted HbA1c (IDAA1C), calculated as [A1C (%) + 4 × insulin dose (units per kilogram per 24 h)]. Values of IDAA1C ≤ 9 were associated with a predicted stimulated C-peptide (SCP) > 300 pmol/L and proposed as a way to evaluate clinically meaningful changes in intervention therapies that are aimed to preserve/regenerate B-cell function in new onset of T1DM (17). We followed the same conversion ratio to assess residual B-cell reserve.

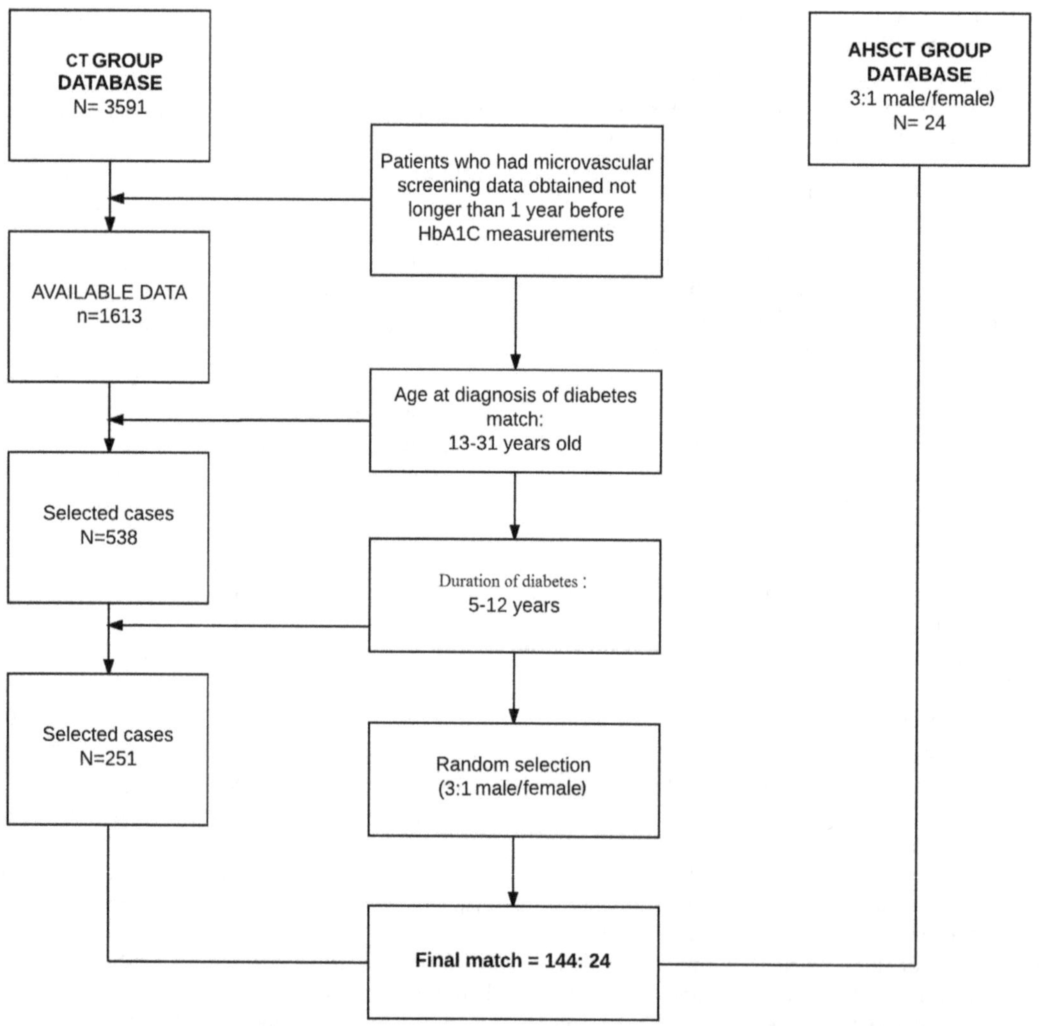

FIGURE 1 | Case selection from both compared groups for data analysis.

Glycemic Control and Insulin Dose

Insulin-dose adjusted HbA1c < 7.0 (53 mmol/mol) was used as the marker for the long-term good metabolic control, according to the ADA recommendation (16). HbA1c was assessed using National Glycohemoglobin Standardization Program methods, including HPLC and immunoturbidimetry. Additionally, we also evaluated the percentage of patients from each treatment group with daily insulin doses <0.5 U/kg and HbA1C of <6.5% (48 mmol/mol) (4, 18).

Statistical Analysis

Continuous variables were presented as median and SD, and the categorical variables as absolute frequency and percentage. The comparative analysis between the two groups was estimated using the unpaired Student *t*-test or Mann–Whitney *U* test for continuous variables and the Pearson's chi-square test or Fisher's exact test for the categorical variables. Fisher's exact test was used to compare the microvascular and glycemic control parameters of the analysis. The chi-square test was performed for the residual beta-cell function analysis and for insulin-dose comparison, Mann–Whitney test. Results were considered significant at $p < 0.05$.

RESULTS

Patient Characteristics

Data of 24 subjects from AHST were contrasted against the data of 144 subjects from the CT (70% males in both, $p = 0.945$). The groups were well matched; in both, the age at diabetes diagnosis varied from 13 to 31 years, and the median was 16.0 years (p25–75: 15.5–20.5) for AHST group and 17.0 (p25–75: 14–22) for CT group ($p = 0.964$). Diabetes duration varied from 5 to 12 years for both groups, and the median was 9.0 years (p25–75: 7.5–10.0) for the AHST group and 8.0 years (p25–75: 6.0–10.0) for the CT group ($p = 0.325$).

Demographic characteristics of the study population are summarized in **Table 1**.

Microvascular Complications

The AHST treatment group had lower prevalence of at least one microvascular complication. While none of the patients in the AHST group developed any microvascular complication, 21.5% (31/144) from CT group had at least one ($p = 0.005$). When microvascular complications were evaluated separately, the prevalence of nephropathy was significantly higher in the CT group (20 vs. no cases) ($p = 0.037$). Regarding neuropathy and retinopathy, there were nine (6.25%) and eight (5.5%) cases in the CT group vs. no cases in the AHST group ($p = 0.28$ and $p = 0.24$, respectively) (**Table 2**).

TABLE 1 | Gender, t1dm duration and age range distribution for the AHST and CT groups.

Variable	AHST group	CT group	*p*-Value
	N = 24	*N* = 144	
Gender, M (%)	70.8	70.1	0.945
Duration of diabetes, years Median (p25–75%)	9.0 (7.5–10.0)	8.0 (6.0–10.0)	0.325[a]
Age at diagnosis, years Median (p25–75%)	16.0 (15.5–20.5)	17 (14.0–22.0)	0.964[a]
13–16 years (%)	54.2	44.4	
17–21 years (%)	29.2	29.2	
22–33 years (%)	16.7	25.0	

[a]*Mann–Whitney test.*

AHST, autologous nonmyeloablative hematopoietic stem-cell transplantation; CT, conventional therapy group from Brazilian Type 1 Diabetes Study Group.

Residual B-Cell Function

The AHST group showed a better residual B-cell function through the IDAA1C method. The percentage of patients with IDAA1C ≤ 9 was about 10-fold higher in the AHST group as compared with the CT group (75 vs. 8.3%; $p < 0.001$). Among patients <18 years old, 66.6% from AHST group vs. 5.4% from CT group were presented with IDAA1C ≤ 9 ($p < 0.001$); among patients ≥18 years old, the difference was found to be 88.8 vs. 11.4%, respectively ($p < 0.001$).

Glycemic Control and Insulin Dose

The median dose of insulin was 65% lower among AHST group as compared with CT group (0.30 UI/kg/day vs. 0.85 UI/kg/day; $p < 0.001$). The median HbA1c values were significantly lower in AHST group as compared with CT group (6.99 vs. 9.32%; $p < 0.001$) despite the use of lower insulin doses. Among AHST patients, 54.1% (13/24) had HbA1c < 7.0 (53 mmol/mol) as compared with 13.1% in CT group ($p < 0.001$). In addition, at any given HbA1C intervals, patients in the AHST group achieved the intervals at lower insulin-dose cutoffs than in the conventional group (**Table 2**). The proportion of patients who presented with HbA1C of <6.5% (48 mmol/mol) and insulin dose <0.5 UI/kg/day was higher in AHST group than in conventional group—41.6 vs. 0.67% ($p < 0.001$). Only 33.3% of patients in the AHST group used rapid-acting and intermediate/long-acting insulins compared with 87.5% in CT group ($p < 0.001$). None and two patients, respectively, were on insulin pump therapy in the AHST and CT groups. Additionally, 12.5%

TABLE 2 | Comparison of occurrence of diabetes microvascular complications and residual B-cell function in the AHST and CT groups.

Variable	AHST group	CT group	*p*-Value
	No./total	No./total	
Microvascular outcomes			
Any microvascular complication	0/24	37/144	0.005[a]
Diabetic nephropathy	0/24	20/144	0.037[a]
Diabetic retinopathy	0/24	8/144	0.283[a]
Diabetic neuropathy	0/24	9/144	0.241[a]
Residual B-cell function (IDAA1C ≤ 9.0)	%	%	
Total	75.0	8.3	$p < 0.001$[a]
<18 years old	66.6	5.4	$p < 0.001$[a]
≥18 years old	88.8	11.4	$p < 0.001$[a]

[a]*Fisher test.*

IDAA1C, insulin-dose adjusted HbA1c; AHST, autologous nonmyeloablative hematopoietic stem-cell transplantation; CT, conventional therapy group from Brazilian Type 1 Diabetes Study Group.

(3/24) of patients from AHST group were not taking insulin compared with no cases in the CT group ($p < 0.001$). All these three patients had HbA1C of <7.0% (53 mmol/mol). In the year 2009, with a mean follow-up of 2.4years, 52.1% patients from AHST were insulin-free. Since 2008, every patient who stopped insulin use started to use sitagliptin 100 mg/day. A summary of the data about glycemic control, insulin use, and partial remission T1DM for both groups is described in **Table 3**.

DISCUSSION

Trials analyzing the effects of AHST in new-onset t1dm have shown promise in significantly attenuating the loss of insulin secretion (8, 9, 19, 20). The Voltarelli et al.'s publication was the first study to analyze the safety and efficacy of autologous stem-cell transplantation in humans. However, one of the recognized limitations of this study was the absence of a control group.

To the best of our knowledge, this is the first paper that compared late outcomes from patients submitted to AHST and subjects treated with conventional therapy, matched by gender, age at diagnosis of T1DM, and duration of disease.

According to literature, the prevalence of microvascular complications in T1DM varied considerably. However, it has been established that the strict control of glucose levels, especially in patients with a recent diagnosis of diabetes, and the maintenance of higher levels of C-peptide secretion result in a lower frequency of microvascular complications regardless of the insulin regimen (11–13). A recent Cochrane Library review showed that even in intensive glucose control group, the microvascular complication incidence was 6.2, 16.3, and 4.9% for retinopathy (follow-up: 5–6.5 years), nephropathy (follow-up: 3.5–6.5 years), and neuropathy (follow-up: 5–6.5 years), respectively (13), which is similar to our data in the group treated with conventional therapy. Another recent study conducted at the University of Poznan in subjects with new-onset type 1 diabetes demonstrated an association between the absence of partial remission of diabetes at any time in the follow-up (defined as Hba1C of <6.5% (48 mmol/mol) associated with insulin dose ≤0.3 UI/kg/day and stimulated C-peptide levels > 500 pmol/mL) and the incidence of microvascular complications. In the remitter group, at the end of 7 years, 7.6% of the patients had at least one microvascular complication, while in the non-remitter group the incidence was 46.4% (21). In contrast, none of the AHST patients after an average follow-up of 8.7 years presented any microvascular complications.

Based on the fact that the majority of patients resumed insulin use and C-peptide levels started do decline after 60 months after AHST (10), these data suggest that the effectiveness transplantation in reducing or delaying the development of microvascular complications is multifactorial. It is probably also dependent on mechanisms that go beyond the preservation of beta cells and improvement of glycemic control, and it should be considered in the remodeling of the diabetic microenvironment as proposed by Ahmed El-Badawy et al. (22).

The direct measurement of C-peptide has been recommended to provide the most appropriate primary outcome in trials evaluating the efficacy of therapies to preserve B-cell function (23). However, IDAA1C has been shown to be a good predictor of SCP levels, with a strong inverse correlation between SCP and IDAA1C, despite the fact that its substance did not fully predict C-peptide responses (24). It was evaluated initially in children and adolescents with T1DM (17). According to this model, an IDAA1C ≤ 9 corresponds to a predicted level of SCP > 300 pmol/L (17); however, an IDDAC ≤ 9 can significantly decrease the number of subjects with SCP ≥ 200 pmol/L (25). Additionally, Buckingham et al. in evaluating the patients at aged between 7 and 45 years old with a recent diagnosis of t1dm found that virtually all participants with IDAA1C < 9 retained substantial insulin secretion, with a peak of C-peptide levels > 200 pmol/L (26).

The same group mentioned above, who initially described the IDAA1c (17), also carried out a study in 129 Danish children and adolescents. They have confirmed the findings of the first trial, which established the relationship between IDAA1C ≤ 9 and partial remission of T1DM (SCP > 300 pmol/L). Although IDAA1C ≤ 9 has a good sensitivity in predicting an SCP > 300 pmol/L, the specificity was reduced, with 34 patients presented a level of SCP > 300 pmol/L and IDAA1C > 9 at 12 months. Most of them were >10 years old and presented poor glycemic control. Thus, the higher C-peptide levels in this group may not reflect a lower severity of T1DM, but, in part, may be a consequence of reduced insulin sensitivity during puberty and use of mixed meal with more carbohydrates in the Danish cohort than in the initial cohort. Ultimately, the authors were able to prove the prognostic and diagnostic power of the IDAA1c measure that presents a good estimate of final insulin production, and thus it can be used in interventional studies aimed

TABLE 3 | Glycemic control and insulin use for the AHST and CT groups.

Patients characteristics	AHST group		CT group		p-Value
HbA1C (%) Median (p25–75)	6.9 (6.0–7.6)		8.7 (7.7–10.4)		<0.001[b]
HbA1C (%) (mmol/mol) Median (p25–75)	52 (42–60)		72 (61–90)		
Insulin dose (UI/kg) Median (p25–75)	0.3 (0.23–0.35)		0.8 (0.7–0.9)		<0.001[b]
Insulin use, *n* (%)	21[a] (87.5)		144 (100.0)		<0.001[b]
Only intermediate/long acting, *n* (%)	13 (54.1)		16 (11.1)		<0.001[b]
Intermediate/long acting plus short acting, *n* (%)	8 (33.3)		128 (88.8)		<0.001[b]
HbA1C < 6.5% (48 mmol/mol) with insulin dose <0.5 UI/kg, *n* (%)	10 (41.6)		1 (0.7)		<0.001[b]
HbA1C intervals (% and mmol/mol)	**n (%)**	**Ins/kg/day**	**n (%)**	**Ins/kg/day**	
≤6.99 (53)	13 (54.1)	0.29	19 (3.1)	0.68	<0.001[b]
7.0–7.99 (53–64)	8 (33.3)	0.31	25 (17.3)	1.01	
8.0–8.99 (64–75)	1 (4.1)	0.40	33 (22.9)	0.82	
>9.0 (75)	2 (8.3)	0.25	67 (46.5)	0.91	

Glycemic control goal: Hba1c < 7.0%.

[a] *The three patients who were not taking insulin presented with HbA1c < 7.0% and were using sitagliptin 100 mg/day.*

[b] *Mann–Whitney test.*

AHST, autologous nonmyeloablative hematopoietic stem-cell transplantation; CT, conventional therapy group from Brazilian Type 1 Diabetes Study Group.

at reducing the progression of t1dm and preserving beta-cell function (27).

Analysis of natural history of residual insulin secretion in T1DM showed that the level of SCP continued to diminish over time, and at the end of 4 years, regardless of age, only a small percentage of subjects maintain stable B-cell function. In this period, only 2.7 and 20.4% of the patients aged between 12 and 17 and ≥18 years, respectively, maintained IDAA1C ≤ 9 (25). These results correspond well with our conventional therapy group after an average follow-up of 8.2 years. In contrast, in AHST group the prevalence of IDAA1C ≤ 9 was 66.6 and 88.8% for respective age groups. Another trial evaluating 3,657 patients with T1DM (below 18 years old at diagnosis) found that in only 5% of the IDAA1c was ≤9 after 6 years of diagnosis (24). Thus, despite the limitations of a cross-sectional design derived from a retrospective cohort, our data demonstrated that the therapy with AHST has the potential for improving long-term beta-cell function.

Specific mechanisms of how AHST improves B-cell function are yet not fully understood. This is due to attaining the immunological tolerance, hematopoietic stem-cell infusion, or both (28). The analysis of the immunological status after AHST showed a newly regenerated immune system that is more tolerant to auto-antigens, with an increase in thymic production of naïve T-cell populations, a normalization of the T-cell receptor repertoire, and an increase in the population of T-regulatory cells (Tregs) (10, 29). Tregs promote self-tolerance through the modulation of activated effector cells during antigen reexperiencing after transplantation (30) and they may also inhibit spontaneous autoreactive proliferation in lymphopenic environments (31). They increased after AHST was associated with prolonged remission of disease. In addition, lower autoreactive islet-specific CD8^{+} T-cell (aCTL) frequencies were also associated with longer insulin-free periods. However, AHSC was not able to affect specific islet autoreactivity, i.e., with no significant changes in a CTL frequencies following transplantation (10).

The main therapeutic mechanism of AHST seems to be the immunological resetting. However, stem cells may have therapeutic potential associated with their intrinsic regenerative capacity as well as the immunomodulatory potential. The immunomodulatory properties could arrest beta-cell destruction, preserve residual beta-cell function, facilitate endogenous beta-cell regeneration, and prevent the recurrence of autoimmunity. The regenerative capacity could promote a self-replenishing supply of glucose-responsive insulin-secretin cells for transplantation (32–34).

However, the ability of transplanted hematopoietic cells to directly differentiate into pancreatic beta cells has not yet been clearly demonstrated. While human stem-cell embryos are capable of differentiating and expressing insulin-producing genes, rat stem-cell embryos are able to generate cells of the pancreatic islets through genetic modification (35, 36).

The general goal of insulin therapy in t1dm is to reduce hemoglobin A1C (A1C) to ≤7.0% (53 mmol/mol) (16). However, this better glycemic control may be accompanied by undesirable effects, such as increased episodes of hypoglycemia (13) need for larger and more frequent doses of insulin. In clinical practice, it has been observed that only a minority of diabetic patients, irrespective of types of insulin or dosing schedules, can maintain HbA1c goal. In a recent publication, Baxter et al. in evaluating a representative cohort of adults with t1dm in the United Kingdom health system found that, despite the use of human insulin, insulin analogs, basal-bolus insulin, and an increasing use of insulin pumps <30% patients with t1dm have HbA1c levels <7.5% (59 mmol/mol) (37).

Similarly, a Brazilian population-based study observed that only 11.6% of the patients presented with HbA1c < 7.0% (53 mmol/mol) (14), similar to that observed in our population in the standard treated group. Garg et al. in a very recent randomized, multicenter trial evaluating the use of sotagliflozin—a new oral inhibitor of sodium-glucose cotransporters 1 and 2—in patients with T1DM showed that the association of this drug with existing insulin regimens was associated with a significant reduction in glycated hemoglobin values and in daily insulin dose. However, at the end of 24 weeks, even among patients in the sotagliflozin group, only 29.6% had HbA1C < 7.0%, and the placebo-corrected reductions from baseline in the mean daily total insulin dose was 9.7% (38). In contrast, the majority of patients in the AHST group (54.1%) presented with HbA1C < 7.0% (53 mmol/mol) and the median dose of insulin was <65% in AHST group as compared with CT group.

Patients in the AHST group were presented with lesser use of rapid insulin and with lower values of HbA1C ($p < 0.001$) as compared with CT group. In addition, 12.5% (3/24) of patients in the AHST group did not need to take insulin; none of such patients were observed in the group treated with standard therapy. Furthermore, a study evaluating teplizumab in recently diagnosed individuals, with mean age of 19 years, just 5% (19/415) of the patients were insulin-free after a shorter period of 1 year (4). However, it should be mentioned that the percentage of insulin-free patients presented decreased with a larger follow-up in AHST group. Previous published data involving median follow-up of 2.4 years, relative to 23 of the 24 patients from AHST group, showed that 52.1% were free (8), as compared with 12.5% found here. This effect occurred in parallel with the reduction of beta-cell function over time. Nevertheless, mounting evidence is observed that AHST therapy appears to be able to achieve better glycemic control with a remarkable decrease needed for insulin administration.

In conclusion, AHST seems to induce a reduction in the prevalence of microvascular complications/diabetic nephropathy as compared with the conventional therapy group. Additionally, our data suggest enhanced preservation of the residual beta-cell function and an improvement in the glycemic control with lesser or no use of insulin.

However, important limitations of this study include a retrospective design, small number of patients, and indirect measurement of C-peptide levels. Another limiting factor is a relatively short duration of diabetes for the analysis of microvascular complications. Joint efforts should be maintained for the discovery of new therapeutic strategies for this disease including its long-term effect. The data presented here have important implications for t1dm treatment in Brazil and potentially elsewhere in the world.

AUTHOR CONTRIBUTIONS

JS contributed to the analysis of data and writing of the manuscript. CC, KM, DM, JD, BS, and MG contributed to the analysis of data. RM, LB, AM, VF, AH, and CN contributed to the writing, reviewing, and editing of the manuscript.

ACKNOWLEDGMENTS

The authors thank Andrezza Ohana, Ant Brazil Viana Jr., Lia Karbage, Izabella Tamira, and Milena Sousa for their technical support.

REFERENCES

1. Van Belle TL, Coppieters KT, von Herrath MG. Type 1 diabetes: etiology, immunology, and therapeutic strategies. *Physiol Rev* (2011) 91:79–118. doi:10.1152/physrev.00003.2010
2. Atkinson MA, Eisenbarth GS, Michels AW. Type 1 diabetes. *Lancet* (2014) 383:69–82. doi:10.1016/S0140-6736(13)60591-7
3. Insel RA, Dunne JL, Atkinson MA, Chiang JL, Dabelea D, Gottlieb PA, et al. Staging presymptomatic type 1 diabetes: a scientific statement of JDRF, the Endocrine Society, and the American Diabetes Association. *Diabetes Care* (2015) 38:1964–74. doi:10.2337/dc15-1419
4. Sherry N, Hagopian W, Ludvigsson J, Jain SM, Wahlen J, Ferry RJ Jr, et al. Teplizumab for treatment of type 1 diabetes (Protégé study): 1-year results from a randomised, placebo-controlled trial. *Lancet* (2011) 378:487–97. doi:10.1016/S0140-6736(11)60931-8
5. Ikehara S, Ohtsuki H, Good RA, Asamoto H, Nakamura T, Sekita K, et al. Prevention of type I diabetes in nonobese diabetic mice by allogenic bone marrow transplantation. *Proc Natl Acad Sci U S A* (1985) 82:7743–7. doi:10.1073/pnas.82.22.7743
6. Lampeter EF, Homberg M, Quabeck K, Schaefer UW, Wernet P, Bertrams J, et al. Transfer of insulin-dependent diabetes between HLA-identical siblings by bone marrow transplantation. *Lancet* (1993) 341:1243–4. doi:10.1016/0140-6736(93)91148-F
7. Skyler J. Immunotherapy for interdicting the type 1 diabetes disease process. 3rd ed. In: Pickup J, Williams G, editors. *Textbook of Diabetes*. Oxford, England: Blackwell Publishing (2003). p. 74.1–74.12.
8. Couri CE, Oliveira MC, Stracieri AB, Moraes DA, Pieroni F, Barros GM, et al. C-peptide levels and insulin independence following autologous nonmyeloablative hematopoietic stem cell transplantation in newly diagnosed type 1 diabetes mellitus. *JAMA* (2009) 301:1573–9. doi:10.1001/jama.2009.470
9. Voltarelli JC, Couri CE, Stracieri AB, Oliveira MC, Moraes DA, Pieroni F, et al. Autologous nonmyeloablative hematopoietic stem cell transplantation in newly diagnosed type 1 diabetes mellitus. *JAMA* (2007) 297:1568–76. doi:10.1001/jama.297.14.1568
10. Malmegrim KC, Azevedo JT, Arruda LC, Abreu JR, Couri CE, de Oliveira GL, et al. Immunological balance is associated with clinical outcome after autologous hematopoietic stem cell transplantation in type 1 diabetes. *Front Immunol* (2017) 8:167. doi:10.3389/fimmu.2017.00167
11. Steffes MW, Sibley S, Jackson M, Thomas W. Beta-cell function and the development of diabetes-related complications in the diabetes control and complications trial. *Diabetes Care* (2003) 26:832–6. doi:10.2337/diacare.26.3.832
12. Kuhtreiber WM, Washer SL, Hsu E, Zhao M, Reinhold P III, Burger D, et al. Low levels of C-peptide have clinical significance for established type 1 diabetes. *Diabet Med* (2015) 32:1346–53. doi:10.1111/dme.12850
13. Fullerton B, Jeitler K, Seitz M, Horvath K, Berghold A, Siebenhofer A. Intensive glucose control versus conventional glucose control for type 1 diabetes mellitus. *Cochrane Database Syst Rev* (2014) 2:CD009122. doi:10.1002/14651858.CD009122.pub2
14. Gomes MB, Cobas RA, Matheus AS, Tannus LR, Negrato CA, Rodacki M, et al. Regional differences in clinical care among patients with type 1 diabetes in Brazil: Brazilian Type 1 Diabetes Study Group. *Diabetol Metab Syndr* (2012) 4:44. doi:10.1186/1758-5996-4-44
15. Pierrat A, Gravier E, Saunders C, Caira MV, Aït-Djafer Z, Legras B, et al. Predicting GFR in children and adults: a comparison of the Cockcroft-Gault, Schwartz, and modification of diet in renal disease formulas. *Kidney Int* (2003) 64:1425–36. doi:10.1046/j.1523-1755.2003.00208.x
16. American Diabetes Association. Standards of medical care in diabetes – 2013. *Diabetes Care* (2013) 36:S11–66. doi:10.2337/dc13-S011
17. Mortensen HB, Hougaard P, Swift P, Hansen L, Holl RW, Hoey H, et al. New definition for the partial remission period in children and adolescents with type 1 diabetes. *Diabetes Care* (2009) 32:1384–90. doi:10.2337/dc08-1987
18. Herold KC, Hagopian W, Auger JA, Poumian-Ruiz E, Taylor L, Donaldson D, et al. Anti-CD3 monoclonal antibody in new-onset type 1 diabetes mellitus. *N Engl J Med* (2002) 346:1692–8. doi:10.1056/NEJMoa012864
19. Li L, Shen S, Ouyang J, Hu Y, Hu L, Cui W, et al. Autologous hematopoietic stem cell transplantation modulates immunocompetent cells and improves β-cell function in Chinese patients with new onset of type 1 diabetes. *J Clin Endocrinol Metab* (2012) 97:1729–36. doi:10.1210/jc.2011-2188
20. Snarski E, Milczarczyk A, Torosian T, Paluszewska M, Urbanowska E, Król M, et al. Independence of exogenous insulin following immunoablation and stem cell reconstitution in newly diagnosed diabetes type I. *Bone Marrow Transplant* (2011) 46:562–6. doi:10.1038/bmt.2010.147
21. Niedzwiecki P, Pilacinski S, Uruska A, Adamska A, Naskret D, Zozulinska-Ziolkiewicz D. Influence of remission and its duration on development of early microvascular complications in young adults with type 1 diabetes. *J Diabetes Complications* (2015) 29:1105–11. doi:10.1016/j.jdiacomp.2015.09.002
22. El-Badawy A, El-Badri N. Clinical efficacy of stem cell therapy for diabetes mellitus: a meta-analysis. *PLoS One* (2016) 11:e0151938. doi:10.1371/journal.pone.0151938
23. Palmer JP, Fleming GA, Greenbaum CJ, Herold KC, Jansa LD, Kolb H, et al. C-peptide is the appropriate outcome measure for type 1 diabetes clinical trials to preserve beta-cell function: report of an ADA workshop, 21-22 October 2001. *Diabetes* (2004) 53:250–64. doi:10.2337/diabetes.53.1.250
24. Nagl K, Hermann JM, Plamper M, Schröder C, Dost A, Kordonouri O, et al. Factors contributing to partial remission in type 1 diabetes: analysis based on the insulin dose-adjusted HbA1c in 3657 children and adolescents from Germany and Austria. *Pediatr Diabetes* (2017) 18:428–34. doi:10.1111/pedi.12413
25. Hao W, Gitelman S, DiMeglio LA, Boulware D, Greenbaum CJ; Type 1 Diabetes TrialNet Study Group. Fall in C-peptide during first 4 years from diagnosis of type 1 diabetes: variable relation to age, HBA1C, and insulin dose. *Diabetes Care* (2016) 39:1664–70. doi:10.2337/dc16-0360
26. Buckingham B, Cheng P, Beck RW, Kollman C, Ruedy KJ, Weinzimer SA, et al. CGM-measured glucose values have a strong correlation with C-peptide, HbA1c and IDAAC, but do poorly in predicting C-peptide levels in the two years following onset of diabetes. *Diabetologia* (2015) 58:1167–74. doi:10.1007/s00125-015-3559-y
27. Max Andersen MLC, Hougaard P, Pörksen S, Nielsen LB, Fredheim S, Svensson J, et al. Partial remission definition: validation based on the insulin dose-adjusted HbA1c (IDAA1C) in 129 Danish children with new-onset type 1 diabetes. *Pediatr Diabetes* (2014) 15:469–76. doi:10.1111/pedi.12208
28. Couri CE, de Oliveira MC, Simões BP. Risks, benefits, and therapeutic potential of hematopoietic stem cell transplantation for autoimmune diabetes. *Curr Diab Rep* (2012) 12:604–11. doi:10.1007/s11892-012-0309-0
29. McFarland RD, Douek DC, Koup RA, Picker LJ. Identification of a human recent thymic emigrant phenotype. *Proc Natl Acad Sci U S A* (2000) 97:4215–20. doi:10.1073/pnas.070061597
30. O'Gorman WE, Dooms H, Thorne SH, Kuswanto WF, Simonds EF, Krutzik PO, et al. The initial phase of an immune response functions to activate regulatory T cells. *J Immunol* (2009) 183:332–9. doi:10.4049/jimmunol.0900691

31. Winstead CJ, Fraser JM, Khoruts A. Regulatory CD4+CD25+Foxp3+ T cells selectively inhibit the spontaneous form of lymphopenia-induced proliferation of naive T cells. *J Immunol* (2008) 180:7305–17. doi:10.4049/jimmunol.180.11.7305
32. Fiorina P, Voltarelli J, Zavazava N. Immunological applications of stem cells in type 1 diabetes. *Endocr Rev* (2011) 32:725–54. doi:10.1210/er.2011-0008
33. BarcalaTabarrozzi AE, Castro CN, Dewey RA, Sogayar MC, Labriola L, Perone MJ. Cell-based interventions to halt autoimmunity in type 1 diabetes mellitus. *Clin Exp Immunol* (2013) 171:135–46. doi:10.1111/cei.12019
34. Fändrich F, Ungefroren H. Customized cell-based treatment options to combat autoimmunity and restore beta-cell function in type 1 diabetes mellitus: current protocols and future perspectives. *Adv Exp Med Biol* (2010) 654:641–65. doi:10.1007/978-90-481-3271-3_28
35. Soria B, Roche E, Berná G, León-Quinto T, Reig JA, Martín F. Insulin-secreting cells derived from embryonic stem cells normalize glycemia in streptozotocin-induced diabetic mice. *Diabetes* (2000) 49:157–62. doi:10.2337/diabetes.49.2.157
36. Lumelsky N, Blondel O, Laeng P, Velasco I, Ravin R, McKay R. Differentiation of embryonic stem cells to insulin-secreting structures similar to pancreatic islets. *Science* (2001) 292:1389–94. doi:10.1126/science.1058866
37. Baxter M, Hudson R, Mahon J, Bartlett C, Samyshkin Y, Alexiou D, et al. Estimating the impact of better management of glycaemic control in adults with Type 1 and Type 2 diabetes on the number of clinical complications and the associated financial benefit. *Diabet Med* (2016) 33:1575–81. doi:10.1111/dme.13062
38. Garg SK, Henry RR, Banks P, Buse JB, Davies MJ, Fulcher GR, et al. Effects of sotagliflozin added to insulin in patients with type 1 diabetes. *N Engl J Med* (2017). doi:10.1056/NEJMoa1708337

17

Lower Insulin-Dose Adjusted A1c (IDAA1c) is Associated with Less Complications in Individuals with Type 1 Diabetes Treated with Hematopoetic Stem-Cell Transplantation and Conventional Therapy

*Jaquellyne Gurgel Penaforte-Saboia [1,2], Carlos Eduardo Barra Couri [3,4], Virginia Oliveira Fernandes [1,2,5], Ana Paula Dias Rangel Montenegro [1,2,5], Lívia Aline De Araújo Batista [1,5], Lenita Zajdenverg [6], Carlos Antonio Negrato [7], Kelen Cristina Ribeiro Malmegrim [3], Daniela Aparecida Moraes [3,4], Juliana Bernardes Elias Dias [3,4], Maria Carolina Oliveira [3,4], Akhtar Hussain [1,8,9], Marilia Brito Gomes [10] and Renan Magalhães Montenegro Jr. [1,2,5]**

[1] Postgraduate Program in Medical Sciences, Department of Clinical Medicine, Federal University of Ceará, Fortaleza, Brazil, [2] Clinical Research Unit, Walter Cantidio University Hospital, Federal University of Ceará/EBSERH, Fortaleza, Brazil, [3] Center for Cell-Based Therapy, Regional Blood Center of Ribeirão Preto, Ribeirão Preto Medical School, University of São Paulo, Ribeirão Preto, Brazil, [4] Department of Internal Medicine, Ribeirão Preto Medical School, University of São Paulo, Ribeirão Preto, Brazil, [5] Postgraduate Program in Public Health, Department of Community Health, Federal University of Ceará, Fortaleza, Brazil, [6] University Hospital, Federal University of Rio de Janeiro, Rio de Janeiro, Brazil, [7] Brazilian Society of Diabetes, São Paulo, Brazil, [8] Faculty of Health Sciences, Nord University, Bodø, Norway, [9] Centre for Global Health Research, Diabetic Association of Bangladesh, Dhaka, Bangladesh, [10] Diabetes Unit, Department of Internal Medicine, State University of Rio de Janeiro, Rio de Janeiro, Brazil

***Correspondence:**
Renan Magalhães Montenegro Jr.
renanmmjr@gmail.com

Objective: To evaluate the association between insulin-dose adjusted A1C (IDAA1c) and microvascular complications (MC) and hypoglycemia in a representative Brazilian population of Type 1 diabetes mellitus (T1DM) patients.

Research Design and Methods: This was a cross-sectional study based on a previous study, "Microvascular Complications in Type 1 Diabetes: a comparative analysis of patients treated with autologous nonmyeloablative hematopoietic stem-cell transplantation (AHST) and conventional medical therapy (CT)". The 168 patients in that study (144 from CT plus 24 from AHST) were re-subdivided into two groups, according to their IDAA1c values (30 patients had IDAA1c ≤ 9; 138 had IDAA1c >9). Then, the prevalence of MC (diabetic renal disease, neuropathy, and retinopathy), hypoglycemia (blood glucose <60 mg/dL), and severe hypoglycemic (episode of hypoglycemia that required the assistance of another person to treat) events were compared between the groups. The groups were well-matched on these factors: duration of disease, sex, and age at the time of diagnosis of T1DM.

Results: After an average of 8 years after diagnosis, only 6.6% (2/30) of the patients from IDAA1c $\leq$ 9 group developed any MC, whereas 21.0% (29/138) from the IDAA1c >9 group had at least one complication (p = 0.044). Regarding hypoglycemic events, the proportion of individuals who reported at least 1 episode of hypoglycemia in the last month was 43.3 and 64.7% from the IDAA1c $\leq$9 and IDAA1c >9 groups, respectively (p = 0.030). Regarding severe hypoglycemia, the proportion of patients presenting at least one episode in the last month and the rate of episode/patient/month were similar between groups (6.7 vs. 13.2%; p = 0.535; and 0.1/patient/month vs. 0.25/patient/month; p = 0.321).

Conclusion: In a representative Brazilian population of T1DM patients, those with IDAA1c $\leq$9 presented a lower frequency of MC, as well as fewer episodes of hypoglycemia, in the month prior to the analysis.

Keywords: type 1 diabetes (T1D), IDAA1c, residual B-cell function, glycemic control, microvascular complications

INTRODUCTION

There is a highly variable rate of decline in β-cell function after diagnosis of type 1 diabetes mellitus (T1DM); many patients retain detectable insulin secretion for years or decades (1). The assessment of β-cell function can help physicians to evaluate and manage T1DM patients. Several studies, including the Diabetes Control and Complications Trial (DCCT), have demonstrated that preservation of β-cell mass, in parallel with good glucose control, reduces the risk of serious hypoglycemia in the long term. Moreover, the higher the C-peptide levels, the lower the incidence of microvascular complications (MC), such as diabetic retinopathy, neuropathy, and diabetic renal disease (1–14).

Direct measurement of endogenous insulin secretion would be the most accurate method of evaluating β-cell function; however, it presents some limitations such as, low accuracy for very low levels of insulin, inability to differ between insulin and its intermediates, such as proinsulin, inability to differ between endogenous and exogenous insulin, and low reproducibility of the insulin dosage in peripheral blood by virtue of its first-pass hepatic extraction (15).

Thus, C-peptide measurements during the mixed-meal tolerance test (MMTT) have been recommended as gold standard to evaluate the degree of β-cell function. C-peptide is a byproduct of the enzymatic cleavage of proinsulin to insulin; it is co-secreted by pancreatic β-cells in equimolar concentrations with insulin (16–27). However, the time demand necessary to perform stimulation tests, as well as their usual unavailability, limits their use in clinical practice (28).

In the face of this caveat, insulin-dose adjusted A1c (IDAA1c) is an easy and fast alternative to evaluate pancreatic β-cell function. IDAA1c is a model of insulin-dose adjusted glycated hemoglobin A1c, calculated as "A1c (%) + 4x insulin dose (units per kilogram per 24 h) (28)".

IDAA1c was first described in 2009 by the Hvidoere Study Group (28). The rationale behind the use of both total insulin dose and glycated hemoglobin in the same formula is to reduce the influence of the treatment regimen, since patients with different β-cell functions could present similar glycated hemoglobin values only by intensifying the treatment, which would limit the use of any of the two parameters separately. In that study, stimulated C-peptide, insulin dose (IU/kg), and A1c levels were periodically evaluated in 275 T1DM patients during the first year after diagnosis. A negative correlation among stimulated C-peptide, A1c, and insulin dose was demonstrated. Statistical analysis of regression coefficients showed a factor of ~4 between the coefficients of these parameters (A1c, 0.21; insulin dosage, 0.94), leading to the IDAA1c formulation as A1c (%) + 4x insulin dose. A linear correlation between IDAA1c and C-peptide over a continuum of stimulated C-peptide values, 6 and 12 months after diagnosis was observed. In addition, it was demonstrated that the IDAA1c threshold, nine, corresponds to a predicted level of 300 pmol/l for the corresponding peak-stimulated C-peptide (28).

In this study we aim to analyze the association between IDAA1c and MC in patients with T1DM as well as with hypoglycemia, in a representative Brazilian population of T1DM patients.

MATERIALS AND METHODS

Study Design and Subjects

This study is based on analyses from a previous study, "microvascular complications in T1DM—a comparative analysis of patients treated with autologous non-myeloablative hematopoietic stem-cell transplantation (AHST) and conventional medical therapy." The study design and data collection methods have been described elsewhere (29). Briefly, cross-sectional data of 24 T1DM patients, treated with AHST, were compared with those of 144 patients who received conventional therapy (CT) from the Brazilian Type 1 Diabetes Study (BrazDiab1).

The BrazDiab1 is the largest observational multicenter study in T1DM in Brazil. Assessed patients from 20 cities (population> 100,000) from all five Brazilian geographic regions (North, Northeast, Southeast, South, and Midwest). A total of 3,591

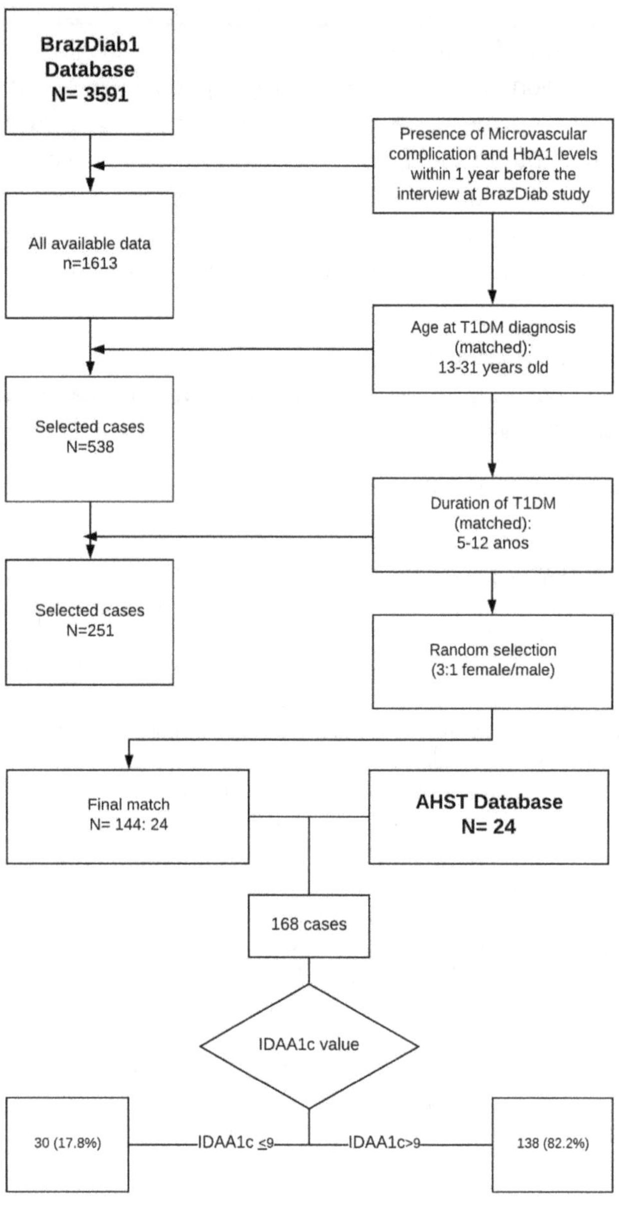

FIGURE 1 | Case selection from AHST and CT group for data analysis.

patients were enrolled in BrazDiab1, but only 1,613 had microvascular screening data available for this study. After AHST matching data (age at diagnosis of diabetes, duration of diabetes and gender) a total of 144 patients were included in the final analysis (**Figure 1**). All the 144 patients of CT were regularly followed by endocrinologists and were on insulin therapy as treatment for T1DM, with 87.5% of them using rapid acting and intermediate/long-acting insulin, 11.1% only intermediate/long-acting insulin and 1.4% were on insulin pump therapy (29, 30).

Complete study protocol of AHST was presented elsewhere (31). In brief, after peripheral hematopoietic stem cells mobilization with cyclophosphamide plus a granulocyte-colony stimulating factor, stem cells were harvested from the peripheral blood through apheresis. After mobilization, patients were conditioned with rabbit anti-thymoglobulin (ATG) plus cyclophosphamide. Posteriorly, intravenous infusion of autologous hematopoietic stem cells was performed with the aim of restoring immune balance.

The primary endpoint was to analyze the presence of MC, residual function of β-cell (through IDAA1c), insulin dose, and A1c in the well-matched groups.

In the present study, 168 patients of the initial study (144 from CT plus 24 from AHST) were re-subdivided into two groups, according to their IDAA1c values (IDAA1c ≤9 or IDAA1c >9). Then, the prevalence of MC, hypoglycemia, and severe hypoglycemic events were compared between groups. Data on hypoglycemia were not presented in the original study (29).

Microvascular Complications

Diabetic retinopathy was screened by fundoscopy by a local ophthalmologist; diabetic renal disease (DRD) was screened by measurements of albuminuria and glomerular filtration rate, estimated by the Cockcroft–Gault equation (32). Distal symmetric polyneuropathy was assessed by asking the patients about symptoms of neuropathy, including paresthesia, dulled sensation, and pain in the feet and using a neurological examination on the sensation of perception of vibration and 10-g monofilament pressure at the distal plantar of both great toes and metatarsal joints. The presence of MC was defined according to the American Diabetes Association Recommendations (33).

Hypoglycemia

Hypoglycemia was assessed by asking the patient if he/she had any episode of hypoglycemia and/or severe hypoglycemia (SH) in the month prior to the interview, as well as the number of hypoglycemic episodes in the same period. Hypoglycemia was defined as capillary blood glucose level <60 mg/dL; severe hypoglycemia (SH), as an episode of hypoglycemia that required the assistance of another person to treat with oral carbohydrate or intravenous glucose due to altered consciousness or seizure (34).

Statistical Analysis

Continuous variables were presented as mean and SD, and the categorical variables as absolute frequency and percentage. The comparative analysis between the two groups was estimated using the unpaired Student *t*-test, or Mann–Whitney *U*-test for continuous variables and the Pearson's chi-square test, Likelihood ratio test or Fisher's exact test for the categorical variables. Likelihood ratio test and Fisher's exact test was used to compare the MC of the analysis. The Pearson's chi-square, Mann–Whitney and Fisher test were performed for comparison of hypoglycemic events. Results were considered significant at $p < 0.05$.

RESULTS

Patient Characteristics

The patient data selection was showed on **Figure 1**. Of the 168 patients, 30 (17.8%) and 138 (82.2%) had IDAA1c ≤9 and

TABLE 1 | Gender, duration of T1DM and age distribution for the IDAA1c ≤9 and IDAA1c>9 groups.

Patient characteristics	IDAA1c≤9 group (*n* = 30)	IDAA1c>9 group (*n* = 138)	*p*-value
Male gender, *n* (%)	17 (56.7)	101 (73.2)	0.073[a]
Duration of diabetes, Years#	8.8 ± 1.88	8.23 ± 2.34	0.214[b]
Age at diagnosis, Years#	19.67 ± 4.94	18.29 ± 4,81	0.159[b]

[a]Q² Fisher Test; [b] T Test; #mean ± SD.

TABLE 2 | Comparison of occurrence of diabetes microvascular complications in the IDAA1c ≤9 and IDAA1c >9 groups.

Patient characteristics	IDAA1c≤9 group (*n* = 30) !	IDAA1c>9 group (*n* = 138)	*p*-value
Any microvascular complication	6.6% (2/30)	21.0% (29/138)	0.044[c]
Diabetic renal disease	6.6% (2/30)	13.0% (18/138)	0.067[d]
Microalbuminuria	1	15	
Macroalbuminuria	1	3	
Diabetic neuropathy	0	6.5% (9/138)	0.365[d]
Diabetic retinopathy	0	5.7% (8/138)	0.353[d]

! IDAA1c ≤9 group N=30 (18 AHST +12 BrazDiab1). [c]Likelihood ratio test; [d]Q² Fisher Test.

TABLE 3 | Comparison of Hypoglycemic events in the IDAA1C ≤9 and IDAA1c>9 groups.

Patient characteristics	IDAA1c≤9 group (*n* = 30)!	IDAA1c>9 group (*n* = 138)	*p*-value
Hypoglycemia event, at least 1 episode in the last month	43.3% (13/30)	64.7% (88/136)[e]	0.030[g]
Hypoglycemia per patient, any event in the last month	3.1 (93/30)	2.9 (407/136)[e]	0.088[h]
Severe hypoglycemia, at least 1 episode in the last month	6.7% (2/30)	13.2% (18/136)[f]	0.535[i]
Severe hypoglycemia per patient, any event in the last month	0.1/patient (3/30)	0.25/patient (33/130)[f]	0.321[h]

! IDAA1c ≤9 group N = 30 (18 AHST +12 BrazDiab1); [e]2 patients did not report hypoglycemic events; [f]8 patients did not report hypoglycemic events; [g]Pearson [h]Mann-Whitney; [i]Fisher.

IDAA1c >9, respectively. The proportion of males was 56.7 and 73.2% in the IDAA1c≤9 and IDAA1c>9 groups, respectively ($p = 0.073$). During the data collection, the mean duration of diabetes was 8.8 years in the IDAA1c ≤9 group and 8.2 years in the IDAA1c >9 group ($p = 0.214$). The age at diabetes diagnosis varied from 13 to 31 years; the mean was 19.6 years in the IDAA1c ≤9 group and 18.2 years in the IDAA1c >9 group ($p = 0.159$). Demographic characteristics of the study population are summarized in **Table 1**.

Microvascular Complications

The prevalence of MC (diabetic renal disease, neuropathy, or retinopathy) was lower in the IDAA1c ≤9 group. Whereas, only 6.6% (2/30) of patients in this group developed any form of MC, 21.0% (29/138) of the IDAA1c >9 group had at least one of them ($p = 0.044$). In IDAA1c≤9 both patients with MC had diabetic renal disease, one had microalbuminuria and the other one had macroalbuminuria, either treated with CT. When the individual components of MC were assessed separately, the number of cases was larger in IDAA1c >9 group, but the difference was not statistically significant (**Table 2**).

Hypoglycemic Events

The proportion of patients who reported at least one episode of hypoglycemia in the last month was lesser in the IDAA1c ≤9 group (43.3%) than in the IDAA1c > 9 group (64.7%) ($p = 0.030$); however, the rate of hypoglycemic episodes per patient per month was similar between groups (3.1/patient/month vs. 2.9/patient/month; $p = 0.088$). Regarding severe hypoglycemia, the proportion of patients presenting at least one episode in the last month and the rate of episode/patient/month were similar between groups (6.7 vs. 13.2%; $p = 0.535$; and 0.1/patient/month vs. 0.25/patient/month; $p = 0.321$). In the IDAA1c > 9 group, there were no data on the presence of hypoglycemia in two patients and no data on severe hypoglycemia in eight patients (**Table 3**).

DISCUSSION

In this study, we analyzed the association between IDAA1c levels and hypoglycemia and MC. We found that in the group of patients with IDAA1c ≤9, there was lower prevalence of MC, as well as fewer episodes of hypoglycemia, in the month prior to the analysis.

Since its first description in 2009, the relationship between IDAA1c and C-peptide levels have been studied severally, increasing the credibility of this parameter as a marker of β-cell function (35–38). In 2014, the same group replicated its early findings in a different population. Assessing 129 Danish children and adolescents with new-onset T1DM, they have confirmed the relationship between IDAA1c ≤9 and peak-stimulated C-peptide levels >300 pmol/L; this shows a linear relation between these two variables at 6 and 12 months after diagnosis (35).

Another trial that included an adult population (67 individuals aged 7–45 years) with 24 months of follow-up, confirmed the good specificity of IDAA1c in predicting C-peptide levels. In 99% of cases, when IDAA1c was <9, peak C-peptide levels were >200 pmol/L (34). However, it is important to emphasize that in both studies, the sensitivity of IDAA1c was shown to be lower than specificity, with some individuals presenting peak-stimulated C-peptide levels >200–300 pmol/L, even when IDAA1c was >9 (35, 36).

Recently, a new formula to estimate residual beta-cell function in children was published. In addition to HbA1c, and insulin dose, age, gender and BMI were included. The predictive value for 90-min stimulated C-peptide was significantly higher than IDAA1c. However, this clinical model was evaluated only in a small number of children with recent-onset type 1 diabetes (37).

In addition to its correlation with C-peptide levels, other studies have indirectly evaluated the relationship between IDAA1c and β-cell function. A periodic measurement of IDAA1c during 6 years in a large longitudinal observational study of 3,657 children and adolescents with newly diagnosed type 1 diabetes showed that older age at onset of diabetes, absence of Diabetic ketoacidosis (DKA) and initial autoantibody negativity were associated with a higher possibility of patients presenting values of IDAA1c ≤9 at any moment during the 6 years of study (38). On the other hand, positive autoantibodies at T1DM onset are predictors for a shorter period of IDAA1c values ≤9 during the observational study period (38). Previously, we have applied IDAA1c to evaluate and compare β-cell function in patients with newly diagnosed T1DM treated with AHST and patients who received conventional therapy from the Brazilian Type 1 Diabetes Study Group (BrazDiab1). We demonstrated that after a median of 8 years of diagnosis, the proportion of individuals with IDAA1c ≤ 9 was about 10-fold higher in the first group (29). Similarly, Type 1 NEw ONset Study (NeOn) that evaluated 1,048 patients <19 years old, showed that the number of DKA episodes increased in parallel with IDAA1c levels. While the proportion of participants with IDAA1c ≤9 decreased from 23% at 12 months to 7% at 36 months, the rate of patients developing diabetic ketoacidosis (DKA) increased from 1% in the first year after diagnosis to 6% in years 2 and 3 (39). It is known that both advanced age at the onset of diabetes and the absence of DKA correlate with better residual β-cell function (40–46), whereas the presence of serum autoantibodies, as well as DKA at T1DM onset, may be related to an aggressive autoimmune reaction and accelerated destruction of beta cells (46, 47).

On the other hand, numerous studies show that the preservation of β-cell function results in fewer MC (1–15), where the DCCT provided the strongest evidence. In DCCT, C-peptide responders (C-peptide response above 0.2 nmol/l) had significantly lower rates of albumin excretion ($P = 0.027$) and nearly significantly lower prevalence of retinopathy ($P = 0.057$) than non-responders, after adjusting for age and sex (12, 15). Thus, since IDAA1c ≤9 has been associated with better pancreatic β-cell reserve, and this condition reflects in lower MC, this may justify our finding of less prevalence of MC and hypoglycemia in the IDAA1c ≤9 group.

Data from literature show a reduced risk of hypoglycemia in patients with better residual β-cell function (48, 49). In our study, 43.3% of patients in the IDAA1c ≤9 group presented with at least 1 hypoglycemic episode in the month prior to the analysis and 63.7% in the IDAA1c >9 group ($p = 0.03$). This is in accordance with the DCCT trial that showed reduced incidence of hypoglycemia in individuals with better β-cell function (48).

A recent study that included patients undergoing immunomodulatory therapy, showed that a 4-h area under the curve (AUC) of C-peptide levels is a robust predictor of number of hypoglycemic events and demonstrated an inverse correlation between a continuous scale of IDAA1c values and the 4-h AUC of C-peptide. However, there was no difference in hypoglycemic rates between IDAA1c ≤9 and IDAA1c >9 groups in our study. Probably, the arbitrary categorization of two groups simply based on the threshold, nine, would not be appropriated to this population. Recent reports indicate that reduced rates of hypoglycemia are observed across a range of C-peptide values with no threshold or breakpoint (50).

By evaluating episodes of severe hypoglycemia in our study, the proportion of patients with at least one episode, and the rate of episodes of severe hypoglycemia per patient were similar between groups. In line with our findings, the NEON study showed no significant difference in IDAA1c values between those participants with or without severe hypoglycemia. Whereas, the percentage of participants with IDAA1c ≤9 decreased from 23% at 12 months to 7% at 36 months, the rate of patients reporting severe hypoglycemic episodes only increased from 2 to 4% in the same period, but without statistical significance (51). Another study that included data from 3,320 Danish patients aged 0–18 years with mean diabetes duration of 4 years showed similar results. Although initially IDAA1c ≤9 had been correlated with fewer episodes of severe hypoglycemia, the difference was no longer significant in the final adjusted model (52).

In this study, we showed a relationship between IDAA1c and MC in T1DM patients. However, its retrospective nature, cross-sectional design, population size, and the use of self-reported assessment of hypoglycemic episodes (which could lead to memory bias), are recognized limitations. Considering that continuous glucose monitoring levels were not evaluated, silent episodes would not be accounted. On the other hand, strengths were the evaluation of IDAA1c in a population with long term disease and the inclusion of a large number of patients undergoing immunomodulatory therapy, which allowed us to evaluate a significant proportion of patients presenting IDAA1c ≤9, even after a long disease duration. Therefore, this simple parameter of estimation of β-cell function may be useful in defining patients with T1DM who deserve greater attention in the sense of prevention-associated complications.

In conclusion, patients with T1DM with IDAA1c ≤9 presented a lower frequency of MC, as well as hypoglycemic episodes. Thus, the use of IDAA1c may be a very simple method to help physicians in the prediction of patients with higher risk of these complications and in the management of patients with T1DM. However, since most patients in the IDAA1C ≤9 group (18 of 30) were treated with AHST, these results may not be widespread for general T1DM population with long duration of diabetes. Thus, further studies with a greater number of patients, who are followed for a longer period of time, are necessary for confirmation of these findings and determine the clinical application of IDAA1c.

AUTHOR CONTRIBUTIONS

JP-S contributed to the analysis of data and writing of the manuscript. CC, KM, DM, JD, MO, and MG contributed with the analysis of data. RM, LZ, LB, AM, VF, AH, and CN contributed to the writing, reviewing, and editing of the manuscript.

REFERENCES

1. Kuhtreiber WM, Washer SL, Hsu E, Zhao M, Reinhold P III, Burger D, Zheng H, et al. Low levels of C-peptide have clinical significance for established Type 1 diabetes. *Diabet Med.* (2015) 32:1346–53. doi: 10.1111/dme.12850
2. Nakanishi K, Kobayashi T, Miyashita H, Ohkubo M, Sugimoto T, Murase T, et al. Relationships among islet cell antibodies, residual β-cell function, and metabolic control in patients with insulin-dependent diabetes mellitus of long duration: use of a sensitive C-peptide radioimmunoassay. *Metabolism.* (1990) 39:925–30.
3. Nakanishi K, Kobayashi T, Inoko H, Tsuji K, Murase T, Kosaka K. Residual B-cell function and HLA-A24 in IDDM, markers of glycemic control and subsequent development of diabetic retinopathy. *Diabetes.* (1995) 44:1334–9.
4. Sjöberg S, Gunnarsson R, Gjötterberg M, Lefvert AK, Persson A, Ostman J. Residual insulin production, glycaemic control and prevalence of microvascular lesions and polyneuropathy in long-term type 1 (insulin dependent) diabetes mellitus. *Diabetologia.* (1987) 30:208–13.
5. Ludvigsson J, Heding LG, Larsson Y, Leander E. C-peptide in juvenile diabetics beyond the postinitial remission period. *Acta Paediatr Scand.* (1977) 66:177–84.
6. Madsbad S, McNair P, Faber OK, Binder C, Christiansen C, Transbøl I. B-Cell function and metabolic control in insulin treated diabetics. *Acta Endocrinol.* (1980) 93:196–200.
7. Sjöberg S, Gjötterberg M, Berglund L, Möller E, Ostman J. Residual C-peptide excretion is associated with a better long-term glycemic control and slower progress of retinopathy in type 1 (insulin-dependent) diabetes mellitus. *J Diabet Complications.* (1991) 5:18–22.
8. Grajwer LA, Pildes RS, Horwitz DL, Rubenstein AH. Control of juvenile diabetes mellitus and its relationship to endogenous insulin secretion as measured by C-peptide immunoreactivity. *J Pediatr.* (1977) 90:42–8.
9. Gonen B, Goldman J, Baldwin D, Goldberg RB, Ryan WG, Blix PM, et al. Metabolic control in diabetic patients: effect of insulin-secretory reserve (measured by plasma C-peptide levels) and circulating insulin antibodies. *Diabetes.* (1979) 28:749–53.
10. Fukuda M, Tanaka A, Tahara Y, Ikegami H, Yamamoto Y, Kumahara Y, et al. Correlation between minimal secretory capacity of pancreatic B-cells and stability of diabetic control. *Diabetes.* (1988) 37:81–8.
11. Johansson BL, Borg K, Fernqvist-Forbes E, Kernell A, Odergren T, Wahren J. Beneficial effects of C-peptide on incipient nephropathy and neuropathy in patients with type 1 diabetes mellitus. *Diabet Med.* (2000) 17:181–9. doi: 10.1046/j.1464-5491.2000.00274.x
12. Steffes MW, Sibley S, Jackson M, Thomas W. Beta-cell function and the development of diabetes-related complications in the diabetes control and complications trial. *Diabetes Care.* (2003) 26:832–6. doi: 10.2337/diacare.26.3.832
13. Fullerton B, Jeitler K, Seitz M, Horvath K, Berghold A, Siebenhofer A. Intensive glucose control versus conventional glucose control for type 1 diabetes mellitus. *Cochrane Database Syst Rev.* (2014) 2:CD009122. doi: 10.1002/14651858.CD009122.pub2
14. Cooperberg BA, Cryer PE. Beta-cell-mediated signaling predominates over direct alpha-cell signaling in the regulation of glucagon secretion in humans. *Diabetes Care.* (2009) 32:2275–80. doi: 10.2337/dc09-0798
15. Palmer JP, Fleming GA, Greenbaum CJ, Herold KC, Jansa LD, Kolb H, et al. C-peptide is the appropriate outcome measure for type 1 diabetes clinical trials to preserve β-cell function. *Diabetes.* (2004) 53:250–64. doi: 10.2337/diabetes.53.1.250
16. Jones AG, Hattersley AT. The clinical utility of C-peptide measurement in the care of patients with diabetes. *Diabet Med.* (2013) 30:803–17. doi: 10.1111/dme.12159
17. Greenbaum CJ, Mandrup-Poulsen T, McGee PF, Battelino T, Haastert B, Ludvigsson J, et al. Mixed-meal tolerance test versus glucagon stimulation test for the assessment of beta-cell function in therapeutic trials in type 1 diabetes. *Diabetes Care.* (2008) 31:1966–71. doi: 10.2337/dc07-2451
18. Polonsky K, Jaspan J, Pugh W, Cohen D, Schneider M, Schwartz T, et al. Metabolism of C-peptide in the dog: *in vivo* demonstration of the absence of hepatic extraction. *J Clin Invest.* (1983) 72:1114–23.
19. Polonsky KS, Pugh W, Jaspan JB, Cohen DM, Karrison T, Tager HS, et al. C-peptide and insulin secretion: relationship between peripheral concentrations of C-peptide and insulin and their secretion rates in the dog. *J Clin Invest.* (1984) 74:1821–9.
20. Bratusch-Marrain PR, Waldhäusl WK, Gasić S, Hofer A. Hepatic disposal of biosynthetic human insulin and porcine proinsulin in humans. *Metabolism.* (1984) 33:151–7.
21. Licinio-Paixao J, Polonsky KS, Given BD, Pugh W, Ostrega D, Frank BF, et al. Ingestion of a mixed meal does not affect the metabolic clearance rate of biosynthetic human C-peptide. *J Clin Endocrinol Metab.* (1986) 63:401–3.
22. Gumbiner B, Polonsky KS, Beltz WF, Griver K, Wallace P, Brechtel G, et al. Effects of weight loss and reduced hyperglycemia on the kinetics of insulin secretion in obese non-insulin dependent diabetes mellitus. *J Clin Endocrinol Metab.* (1990) 70:1594–602.
23. Faber OK, Hagen C, Binder C, Markussen J, Naithani VK, Blix PM, et al. Kinetics of human connecting peptide in normal and diabetic subjects. *J Clin Invest.* (1978) 62:197–203. doi: 10.1172/JCI109106
24. Polonsky KS, Licinio-Paixao J, Given BD, Pugh W, Rue P, Galloway J, et al. Use of biosynthetic human C-peptide in the measurement of insulin secretion rates in normal volunteers and type I diabetic patients. *J Clin Invest.* (1986) 77:98–105.
25. Shapiro ET, Tillil H, Rubenstein AH, Polonsky KS. Peripheral insulin parallels changes in insulin secretion more closely than C-peptide after bolus intravenous glucose administration. *J Clin Endocrinol Metab.* (1988) 67:1094–9.
26. Eaton RP, Allen RC, Schade DS, Erickson KM, Standefer J. Prehepatic insulin production in man: kinetic analysis using peripheral connecting peptide behavior. *J Clin Endocrinol Metab.* (1980) 5:520–8.
27. Van Cauter E, Mestrez F, Sturis J, Polonsky KS. Estimation of insulin secretion rates from C-peptide levels: comparison of individual and standard kinetic parameters for C-peptide clearance. *Diabetes.* (1992) 41:368–77.
28. Mortensen HB, Hougaard P, Swift P, Hansen L, Holl RW, Hoey H, et al. Hvidoere study group on childhood diabetes. new definition for the partial remission period in children and adolescents with type 1 diabetes. *Diabetes Care.* (2009) 32:1384–90. doi: 10.2337/dc08-1987
29. Penaforte-Saboia JG, Montenegro RM Jr, Couri CE, Batista LA, Montenegro APDR, Fernandes VO, et al. Microvascular complications in Type 1 diabetes: a comparative analysis of patients treated with autologous nonmyeloablative hematopoietic stem-cell transplantation and conventional medical therapy. *Front Endocrinol.* (2017) 8:331. doi: 10.3389/fendo.2017.00331
30. Gomes MB, Cobas RA, Matheus AS, Tannus LR, Negrato CA, Rodacki M, et al. Regional differences in clinical care among patients with type 1 diabetes in Brazil: Brazilian Type 1 Diabetes Study Group. *Diabetol Metab Syndr.* (2012) 4:44. doi: 10.1186/1758-5996-4-44
31. Couri CE, Oliveira MC, Stracieri AB, Moraes DA, Pieroni F, Barros GM, et al. C-peptide levels and insulin independence following autologous nonmyeloablative hematopoietic stem cell transplantation

in newly diagnosed type 1 diabetes mellitus. *JAMA.* (2009) 301:1573–9. doi: 10.1001/jama.2009.470

32. Pierrat A, Gravier E, Saunders C, Caira MV, Aït-Djafer Z, Legras B, et al. Predicting GFR in children and adults: a comparison of the Cockcroft-Gault, Schwartz, and modification of diet in renal disease formulas. *Kidney Int.* (2003) 64:1425–36.
33. American Diabetes Association. Standards of medical care in diabetes - 2013. *Diabetes Care.* (2013) 36:S11–66. doi: 10.2337/dc13-S011
34. Seaquist ER, Anderson J, Childs B, Cryer P, Dagogo-Jack S, Fish L, et al. Hypoglycemia and diabetes: a report of a workgroup of the American Diabetes Association and the Endocrine Society. *Diabetes Care.* (2013) 36:1384–95. doi: 10.2337/dc12-2480
35. Max Andersen ML, Hougaard P, Pörksen S, Nielsen LB, Fredheim S, Svensson J, et al. Partial remission definition: validation based on the insulin dose-adjusted HbA1c (IDAA1C) in 129 Danish children with new-onset type 1 diabetes. *Pediatr Diabetes.* (2014) 15:469–76. doi: 10.1111/pedi.12208
36. Buckingham B, Cheng P, Beck RW, Kollman C, Ruedy KJ, Weinzimer SA, et al. CGM-measured glucose values have a strong correlation with C-peptide, HbA1c and IDAAC, but do poorly in predicting C-peptide levels in the two years following onset of diabetes. *Diabetologia.* (2015) 58:1167–74. doi: 10.1007/s00125-015-3559-y
37. Buchanan K, Mehdi AM, Hughes I, Cotterill A, Le Cao KA, Thomas R, et al. An improved clinical model to predict stimulated C-peptide in children with recent-onset type 1 diabetes. *Pediatr Diabetes.* (2019) 20:166–171. doi: 10.1111/pedi.12808
38. Nagl K, Hermann JM, Plamper M, Schröder C, Dost A, Kordonouri O, et al. Factors contributing to partial remission in type 1 diabetes: analysis based on the insulin dose-adjusted HbA1c in 3657 children and adolescents from Germany and Austria. *Pediatr Diabetes.* (2017) 18:428–34. doi: 10.1111/pedi.12413
39. Davis AK, DuBose SN, Haller MJ, Miller KM, DiMeglio LA, Bethin KE, et al. Prevalence of detectable C-peptide according to age at diagnosis and duration of type 1 diabetes. *Diabetes Care.* (2015) 38:476–81. doi: 10.2337/dc 14-1952
40. Ludvigsson J, Heding LG. C-peptide in children with juvenile diabetes. A preliminary report. *Diabetologia.* (1976) 12:627–30.
41. Ostman J, Landin-Olsson M, Törn C, Palmer J, Lernmark A, Arnqvist H, et al.Ketoacidosis in young adults is not related to the islet antibodies at the diagnosis of Type 1 diabetes mellitus–a nationwide study. *Diabet Med.* (2000) 17:269–74. doi: 10.1046/j.1464-5491.2000.00265.x
42. Madsbad S, Alberti KG, Binder C, Burrin JM, Faber OK, Krarup T, et al. Role of residual insulin secretion in protecting against ketoacidosis in insulin-dependent diabetes. *Br Med J.* (1979) 2:1257–9.
43. Komulainen J, Lounamaa R, Knip M, Kaprio EA, Akerblom HK. Ketoacidosis at the diagnosis of type 1 (insulin dependent) diabetes mellitus is related to poor residual beta cell function. *Arch Dis Child.* (1996) 75:410–5.
44. Zamaklar M, Jotic A, Lalic N, Lalic K, Rajkovic N, Milicic T. Relation between course of disease in type 1 diabetes and islet cell antibodies. *Ann N Y Acad Sci.* (2002) 958:251–3. doi: 10.1111/j.1749-6632.2002.tb02980.x
45. Schölin A, Törn C, Nyström L, Berne C, Arnqvist H, Blohmé G, et al. Normal weight promotes remission and low number of islet antibodies prolong the duration of remission in Type 1 diabetes. *Diabet Med.* (2004) 21:447–55. doi: 10.1111/j.1464-5491.2004.01175.x
46. Pörksen S, Laborie LB, Nielsen L, Louise Max Andersen M, Sandal T, de Wet H, et al. Disease progression and search for monogenic diabetes among children with new onset type 1 diabetes negative for ICA, GAD- and IA-2 Antibodies. *BMC Endocr Disord.* (2010) 10:16. doi: 10.1186/1472-6823-10-16
47. Bowden SA, Duck MM, Hoffman RP. Young children (<5 yr) and adolescents (>12 yr) with type 1 diabetes mellitus have low rate of partial remission: diabetic ketoacidosis is an important risk factor. *Pediatr Diabetes.* (2008) 9:197–201. doi: 10.1111/j.1399-5448.2008.00376.x
48. Diabetes Control and Complications Trial Research Group. Effect of intensive therapy on residual beta-cell function in patients with type1diabetes in the diabetes control and complications trial. A randomized, controlled trial. *Ann Intern Med.* (1998) 128:517–23.
49. Pinckney A, Rigby MR, Keyes-Elstein L, Soppe CL, Nepom GT, Ehlers MR. Correlation among hypoglycemia, glycemic variability, and c-peptide preservation after alefacept therapy in patients with type 1 diabetes mellitus: analysis of data from the immune tolerance network T1DAL trial. *Clin Ther.* (2016) 38:1327–1339. doi: 10.1016/j.clinthera.2016.04.032
50. Lachin JM, McGee P, Palmer JP, DCCT/EDIC Research Group. Impact of C-peptide preservation on metabolic and clinical outcoms in the Diabetes Control and Complications Trial. *Diabetes.* (2014) 63:739–48. doi: 10.2337/db13-0881
51. Cengiz E, Cheng P, Ruedy KJ, Kollman C, Tamborlane WV, Klingensmith GJ, et al. Clinical outcomes in youth beyond the first year of type 1 diabetes: results of the Pediatric Diabetes Consortium (PDC) type 1 diabetes new onset (NeOn) study. *Pediatr Diabetes.* (2017) 18:566–573. doi: 10.1111/pedi.12459
52. Johansen A, Kanijo B, Fredheim S, Olsen B, Hertz B, Lauridsen MH, et al. Prevalence and predictors of severe hypoglycemia in Danish children and adolescents with diabetes. *Pediatr Diabetes.* (2015) 16:354–60. doi: 10.1111/pedi.12171

18

Stem Cell Transplantation in the Treatment of Type 1 Diabetes Mellitus: From Insulin Replacement to Beta-Cell Replacement

Xin-Xing Wan[1], Dan-Yi Zhang[2], Md. Asaduzzaman Khan[3], Sheng-Yuan Zheng[2], Xi-Min Hu[4], Qi Zhang[4], Rong-Hua Yang[5] and Kun Xiong[4,6,7*]*

[1] Department of Endocrinology, Third Xiangya Hospital, Central South University, Changsha, China, [2] Clinical Medicine Eight-Year Program, Xiangya School of Medicine, Central South University, Changsha, China, [3] The Research Centre for Preclinical Medicine, Southwest Medical University, Luzhou, China, [4] Department of Anatomy and Neurobiology, School of Basic Medical Science, Central South University, Changsha, China, [5] Department of Burn and Plastic Surgery, Guangzhou First People's Hospital, School of Medicine, South China University of Technology, Guangzhou, China, [6] Hunan Key Laboratory of Ophthalmology, Changsha, China, [7] Key Laboratory of Emergency and Trauma, Ministry of Education, College of Emergency and Trauma, Hainan Medical University, Haikou, China

***Correspondence:**
Kun Xiong
xiongkun2001@163.com
Rong-Hua Yang
21720091@qq.com

Type 1 diabetes mellitus (T1DM) is an autoimmune disease that attacks pancreatic β-cells, leading to the destruction of insulitis-related islet β-cells. Islet β-cell transplantation has been proven as a curative measure in T1DM. However, a logarithmic increase in the global population with diabetes, limited donor supply, and the need for lifelong immunosuppression restrict the widespread use of β-cell transplantation. Numerous therapeutic approaches have been taken to search for substitutes of β-cells, among which stem cell transplantation is one of the most promising alternatives. Stem cells have demonstrated the potential efficacy to treat T1DM by reconstitution of immunotolerance and preservation of islet β-cell function in recent research. cGMP-grade stem cell products have been used in human clinical trials, showing that stem cell transplantation has beneficial effects on T1DM, with no obvious adverse reactions. To better achieve remission of T1DM by stem cell transplantation, in this work, we explain the progression of stem cell transplantation such as mesenchymal stem cells (MSCs), human embryonic stem cells (hESCs), and bone marrow hematopoietic stem cells (BM-HSCs) to restore the immunotolerance and preserve the islet β-cell function of T1DM in recent years. This review article provides evidence of the clinical applications of stem cell therapy in the treatment of T1DM.

Keywords: type 1 diabetes mellitus, stem cell, β-cell, immunotolerance, transplantation

1 INTRODUCTION

Diabetes mellitus (DM) characterized by hyperglycemia, caused by insufficient insulin secretion or insulin resistance, is a group of chronic metabolic diseases. According to the International Diabetes Federation (IDF), the global adult diabetes population will exceed 537 million by 2021, and more than three-fourths of people with diabetes live in low- and middle-income countries, indicating that diabetes disproportionately affects the poor (http://www.diabetesatlas.org/). Diabetes is classified into four types: type 1 diabetes mellitus (T1DM), type 2 diabetes mellitus (T2DM), gestational diabetes mellitus (GDM), and monogenic diabetes mellitus (1–5). T1DM is an autoimmune disease, where autoreactive T cells attack pancreatic β-cells, leading to insulitis-related islet β-cell destruction, which results in an absolute lack of insulin secretion causing hyperglycemia, abnormal glucose metabolism, and lifelong dependence on exogenous insulin. The majority of T1DM patients have poor blood glucose control and large blood sugar fluctuation. Chronic hyperglycemia results in the development of serious complications associated with diabetes, such as microvascular and macrovascular complications, reducing the quality of life and causing a considerable economic burden on T1DM patients and the society (6). The incidence rate of T1DM is increasing every year around the world (7, 8). Although there is evidence that a combination of genetic susceptibility and environmental factors can increase the risk of immune disorder in T1DM patients, the exact etiology of the impaired immune system in T1DM is still unclear. More scientific efforts are needed to prevent β-cell loss and improve the quality of life in T1DM.

2 THE DIFFICULTIES OF INSULIN REPLACEMENT AND β-CELL REPLACEMENT IN T1DM

At present, the treatment and preventive options for T1DM are limited, mainly through insulin replacement therapy. T1DM cannot be cured; patients must rely on exogenous insulin injections for the rest of their lives to maintain glycemic control. Lente and NPH insulin were the only effective methods for the treatment of T1DM in the past (9, 10). In recent years, novel approaches to insulin treatment, such as the introduction of glycosylated hemoglobin assays (HbA1c) and continuous glucose monitoring (CGM), have been used, and the effectiveness of basal/bolus therapy using portable continuous subcutaneous insulin infusion (CSII) pumps and closed-loop artificial pancreas system has been demonstrated. Artificial pancreas combining CGM with CSII pumps could automatically administer an appropriate insulin dose *via* a dosing algorithm. Some randomized controlled trials proved that the artificial pancreas system could efficiently adjust the glycemic index by automatically delivering exogenous insulin with dosing algorithms based on sensor glucose levels (11). However, the lag time of glycemia detected by CGM and the risk of hypoglycemia and infections limit the application of artificial pancreas, and some of the patients with unawareness of hypoglycemic events such as brittle type T1DM are not qualified to use the artificial pancreas (12, 13). Also, insulin replacement therapy can only supplement the missing insulin and cannot fundamentally restore the function of the pancreas. Although these achievements can better manage blood glucose and large blood sugar fluctuation in T1DM, they can hardly prevent the occurrence of a series of complications, including microvascular, macrovascular, and neuropathy complications (14, 15). As a result, many adjunctive therapies, such as dietary and weight management, nutrition therapy, physical activity and exercise, and some drugs used to treat T2DM, have been proposed to treat T1DM, which alleviate blood glucose fluctuation and reduce the lifetime risk of complications to some extent, but their effectiveness is limited. Therefore, it is very important to develop better technology and equipment for diagnosis and treatment options to prevent T1DM (16–18).

β-Cell replacement has also been proven as a curative measure in T1DM, which may be achieved through pancreas or islet transplantation in selective candidates (19). Pancreatic transplantation has the potential of re-establishing physiologic-regulated insulin production, obviously decreasing the risk of hypoglycemic unawareness and finally decreasing the longtime risk of mortality from severe hypoglycemic complications (20). Since 2000, β-cell replacement through intrahepatic isolated islet transplantation has proven efficacious, indicating that islet transplantation is also an important option in the treatment of T1DM (21). Compared with the artificial pancreas system, islet transplantation and pancreatic transplantation were the better options to relieve the symptom of T1DM patients with unawareness of hypoglycemic events such as brittle type T1DM for a long time (22). T1DM patients can be clinically alleviated through improved control of the levels of blood glucose and restored awareness of hypoglycemia, resulting in the prevention of several life-threatening complications associated with diabetes, such as diabetic foot, microvascular and macrovascular diseases, kidney failure, nerve damage, and blindness (23). During the process of pancreatic or islet transplantation, both the autoimmune and alloimmune systems are still major threats to increase the transplantation risk. Patients treated with cell replacement therapies require immunosuppressive drugs as life-long treatment, and in many cases, these drugs lead to toxicities and side effects that made the adoption of this treatment strategy limited to only the most severe disease cases, inhibiting the widespread adoption of pancreatic or islet transplantation therapies in T1DM (24).

Besides the immune problem, the logarithmic increase in the global population of people with diabetes, the limited donor supply, and the need for lifelong immunosuppression restrict its widespread use (25). Numerous therapeutic approaches have been reported to solve this problem, including the search for β-cell substitutes, porcine islet xenotransplantation, and stem cell transplantations, which present solutions to the donor shortage and may be the most likely alternatives (26, 27).

Although the artificial pancreas system and pancreatic transplantation in T1DM can normalize and improve glycemic control in T1DM, the application of artificial pancreas systems and pancreatic transplantation is still limited due to their shortage. To solve the problem, stem cell transplant is a promising new strategy for patients with T1DM. There are many advantages of stem cells in the treatment of T1DM: first of all, stem cells such as bone marrow-derived stem cells (MSCs) can easily be obtained from bone marrow, umbilical cord blood, adipose tissue, etc. compared with islet and pancreas; secondly, the pluripotent stem cells could differentiate into β-cells and increase the secretion of insulin; thirdly, stem cells can moderate the immunome effect by inhibiting T-cell proliferation and reduce the inflammatory response, which can protect β-cells from autoimmune attack; and finally, stem cells can secrete cytokines by paracrine effects to enhance the antioxidant and proliferation ability of cells, which can help improve the survival of β-cells. To better understand the constitution of immunotolerance and preservation of islet β-cell function, we reviewed the progression of stem cells in recent years and tried to provide support for the clinical applications of stem cell therapy in the treatment of T1DM, especially in the brittle type T1DM.

3 STEM CELL TRANSPLANTATION THERAPY FOR T1DM

Stem cells are undifferentiated cells capable of self-renewal, giving rise to virtually any tissue or organ (28–33). Stem cells can be grouped into four broad categories based on their origin: adult stem cells (ASCs), fetal stem cells (FSCs), embryonic stem cells (ESCs), and induced pluripotent stem cells (iPSCs). iPSCs and ESCs are pluripotent stem cells (PSCs), whereas ASCs are unipotent or oligopotent (34–36). PSCs, such as human-induced PSCs (iPSCs) and human embryonic stem cells (ESCs), offer a reproducible source of human cells at a very early developmental stage with the potential to form any cell type in the adult body (37–39). iPSCs, human cord blood-derived multipotent stem cells (CB-SCs), hematopoietic stem cells (HSCs), and MSCs were used for the preservation of β-cells by islet protection and regeneration, and another potent function of stem cells is the ability to re-establish peripheral tolerance toward β-cells through remodeling of the immune response as well as through inhibition of autoreactive T-cell function (40, 41). In general, stem cells can increase the mass of islets by the ability of differentiation to β-cells-like organoids, and reconstitute immunotolerance by inhibiting the immune response of T cell and Th1 cells through TGF-β and inflammatory pathways (**Figure 1**). As T1DM is featured as an autoimmune disease by activating immune cells to attack and destroy pancreatic β-cells, the immunomodulatory properties of stems cells and its potential ability of differentiation into insulin-producing cells should be considered when using stem cell therapy for T1DM treatment.

3.1 MSC Transplantation in T1DM

MSCs are one of the best candidate cells used as cell therapy for T1DM. MSCs are fibroblast-like, multipotent stromal, non-hematopoietic cells that could easily be sourced from various tissues, including adipose tissue, bone marrow, and umbilical cord blood (42). MSCs rapidly undergo mesodermal lineage differentiation, such as adipocytes, myoblasts, cardiomyocytes, chondrocytes, and β-cell-like cells (43–45). The bone marrow and umbilical cord blood could be separated over a gradient of Percoll by density gradient centrifugation to collect the MNCs, and the MNCs were washed with PBS and transferred to a 100-mm culture dish to induce MSCs. The redundant tissues such as arteries and veins were removed from the adipose tissue, human umbilical cord, etc. The region of the remaining adipose or human umbilical cord tissues was diced into small fragments and seeded into a 100-mm culture dish to collect the MSCs. The induced MSCs were stored in liquid nitrogen and cultured for up to five passages for transplantation *via* intravenous injection. The characteristics of MSCs were defined by the International Society for Cell Therapy (ISCT) as follows: adherence to plastic; expression of the surface molecules CD73, CD90, and CD105 in the absence of CD34, CD45, HLA-DR, and CD14 or CD11b and CD79a or CD19; and the capacity for differentiation to adipocytes, osteoblasts, and chondroblasts *in vitro* (46). The potential of MSCs as a cell-based therapy in the treatment of immunologic disorders has been well established (47). MSCs can alter the microenvironment in tissues and promote existing β-cell survival and regeneration, resulting in increased β-cell mass and normal blood glucose recovery (48–51). Injection of bone marrow MSCs into diabetic mice can increase insulin levels and downregulate hyperglycemia; the exosomes derived from human umbilical cord stem cells (hUCMSCs) can enhance insulin sensitivity (52). Similarly, monotherapy with human umbilical cord MSCs reverses autoimmunity, promotes islet cell regeneration, and improves blood glucose control (53–55).

The allogeneic MSCs have been attempted in clinical trials, which can improve the level of insulin and C-peptide and reduce blood glucose. Although MSC xenotransplantation was not used in the clinics, several pieces of evidence showed that human-derived MSCs could alleviate the diabetic symptom through β-cell-like organoid differentiation and immunomodulation in NOD mice, rats, and monkeys, while far-red light, gene editing, and other modifications could enhance the function of MSCs in a T1DM animal model, indicating that the intervening and xenograft MSCs are the potential option for T1DM treatment.

3.1.1 Immunomodulatory Ability of MSCs

MSCs can protect β-cell, increase the secretion of insulin, and reduce glycemia in patients with T1DM by regulating the immune system. The application of MSCs in eliminating autoimmune diseases has been fully proven in an animal model, and MSCs have a wide range of regulatory effects on immune cells. Domouky et al. showed that MSCs could reduce hyperglycemia in diabetic rats on day 15 (56). The inhibition of T-cell proliferation in islets and the presence of increased Treg in T1DM were features of MSCs' autoimmune properties (57). Shigemoto-Kuroda et al. developed T1DM mouse models for autoimmune diseases and discovered that MSCs could suppress type 1 helper T cell (Th1) development and delay the onset of T1DM in mice. $CD4^+$ cells were found in significant numbers in the islets of mice treated with PBS, while fewer $CD4^+$ cells were found in the islets of MSC-treated mice. The level of insulin in plasma was increased by MSC treatment, and there was a significant reduction in the production of IL-12, IFN-γ, p70, and tumor necrosis factor (TNF) (58). Bassi et al. isolated MSCs from epididymal fat tissue from 8-week-old male Balb/c mice and characterized through immunophenotyping its capacity to prevent the proliferation of $CD4^+$ T cells (59, 60). Treatment of NOD mice with MSCs attenuated hyperglycemia of early-onset

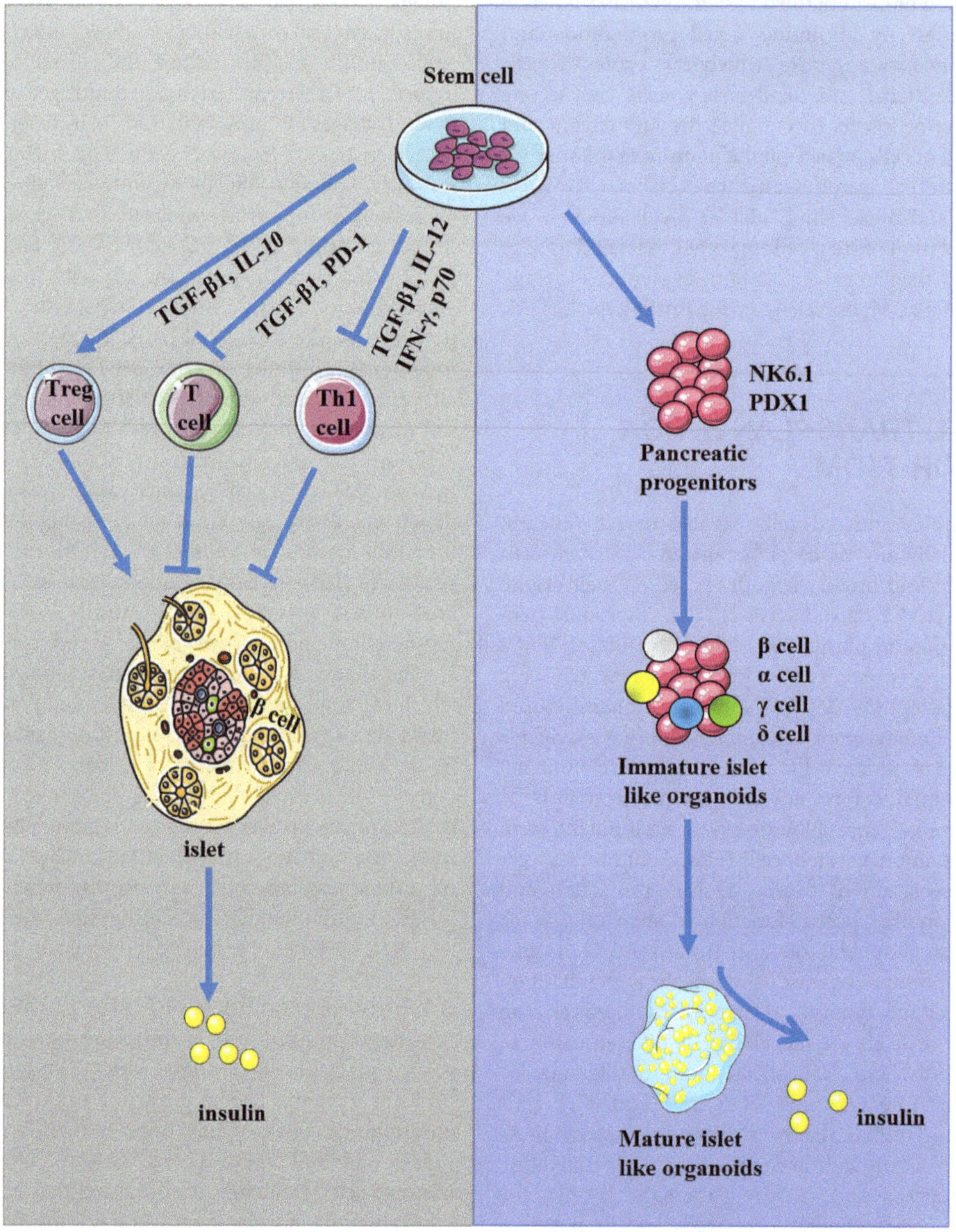

FIGURE 1 | The possible mechanism of stem cells in the treatment of T1DM. Stem cells were used for the reconstitution of immunotolerance through activating T regulatory cells (Treg) and inhibiting T and Th1 cells, and they could also be used for the preservation and regeneration of β-cells.

autoimmune diabetes and increased amylin levels, reflective of autoimmune diabetes improvement; reduced the amount of inflammatory cell infiltration, maintaining insulin expression in pancreatic islets by suppressing the Th1 immune response in the pancreas; and promoted the high expression of active TGF-β1 (60). Meanwhile, syngeneic MSCs were detected for a significantly longer period, albeit with diminishing persistence in immune-deficient mice model (61). In another study, van Megen et al. found that activation of MSCs can take up and process antigen and increase HLA-DR expression and immune inhibitory markers, while their metabolic profile was maintained without enhancing T-cell proliferation. MSCs can also enhance immunosuppressive capacity without stimulating alloreactive T cells (62). In an *in-vitro* study, Montanucci et al. provided preliminary evidence that immunoisolatory microcapsule-hHUCMS (CpS-hUCMS) may represent a functional biohybrid artificial system, where molecular products can induce effective immunomodulatory effects *in vitro* and in T1DM patients, making it possible to further clarify their therapeutic potential in humans (63). Montanucci et al. isolated and microencapsulated human umbilical cord Wharton jelly-derived mesenchymal stem cells (hUCMS) for xenograft (TX) in a spontaneous T1DM mouse model (NOD mice). At 10 days of TX, Treg cells did not increase, while at 216 days of TX, $CD4^+$ CD25 high cells increased in terms of both percentage and number. Further research found that at 216 days of TX, only the mild T1DM NOD mice presented sustained and full alleviation of hyperglycemia, while no alleviation of hyperglycemia was observed in severe T1DM NODs. These findings suggested that the successful hUCMS therapy approach for the treatment of T1DM in NOD mice depended on the stage of the T1DM disease process, with severe T1DM NODs exhibiting a continuous decrease in residual β-cell mass (64). All these results provide encouraging first steps in the clinical translation of the use of preactivated MSCs as a cellular immune intervention therapy, which helps to treat inflammatory and autoimmune disorders, including T1DM.

3.1.2 Islet Protection and Regeneration Ability of MSCs

MSCs can increase the mass of islets and the survival of β-cells by differentiation to β-cell-like cells. MSCs rapidly undergo mesodermal lineage differentiation to β-cell-like cells, and the transdifferentiation of MSCs into insulin-producing cells was successfully attempted *in vitro*, with the pancreatic and duodenal homeobox 1 (PDX-1), the key marker that was present in the transdifferentiation of MSCs to insulin-producing cells (65). Chen et al. successfully differentiated MSCs into pancreatic islet β-cell-like cells by pre-inducing in L-DMEM with 10 mmol/L nicotinamide and 1 mmol/L β-mercaptoethanol for 24 h and re-inducing in serum-free H-DMEM with 10 mmol/L nicotinamide and 1 mmol/L mercaptoethanol for another 10 h. These induced islet β-cell-like cells with similar morphological characters to pancreatic islet cells and promoted the transcription, translation, and excretion of insulin, which could effectively control the level of blood glucose in diabetic rats (66). Similar results were reported by several other groups that islet-like clusters can be formatted *in vitro* by cultured MSCs given the appropriate procedure (67, 68). Human umbilical cord Wharton jelly cells (hUCWJCs) are a subtype of MSCs, which were transplanted into a T1DM mouse model with renal damage, and the therapeutic effect of transplantation was evaluated. It was found that hUCWJCs can promote the level of C-peptide and insulin in mice, which certified the potential of intraperitoneal injection of hUCWJCs and the ability of hUCWJCs to migrate to damaged tissues to enhance the secretion of insulin from non-pancreatic local cells (69).

3.1.3 MSC Transplantation in Clinical Trials

MSCs have been used in human clinical trials, showing that stem cell transplantation has beneficial effects on T1DM. In an open-label, non-randomized, parallel-armed prospective study, Lu et al. enrolled 53 participants including 33 adult-onset (≥18 years) and 20 juvenile-onset T1DM (ChiCTR2100045434). The results revealed that an intravenous dose of allogeneic UC-MSCs was safe in people with newly diagnosed T1DM at 12 months of follow-up, which probably led to better islet β-cell protection compared with standard treatment alone during the first year after diagnosis (70). Cai et al. proved that transplantation of UC-MSCs was safe and associated with moderate improvement of metabolic measures in patients with established T1DM too (NCT01374854) (50). Another clinical trial revealed that MSC injection through liver puncture could successfully decrease the levels of insulin, islet cells, and glutamic acid decarboxylase (GAD) antibody in two patients within 1 year, with a decreased concentration of blood glucose and HbA1c and increased concentration of C-peptide, indicating immune regulatory cell tolerance (71).

3.2 HSC Transplantation in T1DM

The conception of HSCs was generated in the 1950s with the discovery that intravenously injected bone marrow cells could rescue irradiated mice from lethality through re-establishing blood cell production (72). The ability to manage expansion and the characteristics of self-renewal of the hematopoietic compartment while maintaining the capacity for differentiation into HSCs were demonstrated (72, 73). Peripheral hematopoietic stem cells are mobilized with cyclophosphamide and granulocyte colony-stimulating factors. Leukapheresis using a continuous-flow blood cell separator was initiated when the rebounding $CD34^+$ cells reached 10 cells/μl. Apheresis was continued daily until the number of harvested progenitor cells reached a minimum of 3.0×10^6 $CD34^+$ cells/kg body weight. Unmanipulated peripheral blood stem cells were frozen in 10% dimethyl sulfoxide in a rate-controlled freezer and stored in the vapor phase of liquid nitrogen (74). Then, the collected cells were injected intravenously. HSCs have proven to be safe in human subjects and have been widely utilized as an effective treatment for hematological malignancies (75). Recently, HSCs have been used in T1DM for the suppressed function of the immune system response in both *in-vitro* and *in-vivo* studies.

3.2.1 Immunomodulatory Ability of HSCs

Immunomodulatory activity is the most important ability of HSCs in patients with T1DM. HSCs can inhibit the occurrence of T1DM (76–78). Patients with recent-onset T1DM have been triumphantly reverted to euglycemia by autologous hematopoietic stem and progenitor cell transplantation (AHSCT), and modulation of autologous hematopoietic stem and progenitor cells (HSPCs) with prostaglandins (PGs) *in vitro* enhances their immunoregulatory properties through increasing the expression of the immune checkpoint-signaling molecule PD-L1 (79). Wang et al. demonstrated a lower proportion of proliferating T conventional cells (Tcon) and a higher absolute number and percentage of Treg cells in pancreatic lymph nodes from resistant mice among the younger recipients compared to the rapid progressors among the older recipients, and older NOD mice progressed more rapidly to the end stage of diabetes (80). Although mixed chimerism with MHC-matched non-autoimmune donor bone marrow (BM) transplants did not prevent T1DM in NOD mice models, induction of either mixed or complete chimerism with MHC-mismatched BM transplants inhibited T1DM in the same mice (81). This limited the translational applications of HSCs to reshape the autoimmune response by myeloablative agents/approaches. The genetically modified HSCs were used to overcome the disadvantage. *Ex-vivo* genetic manipulation of NOD HSCs to encode proinsulin and transgenically target MHC class II could successfully prevent T1DM onset (78, 82). The increased CXCL12 (SDF-1) level in bone marrow-derived HSCs of NOD mice is considered to change the transport of HSCs and peripheral dendritic cells, which is conducive to the occurrence of T1DM (78, 83).

3.2.2 HSC Transplantation in Clinical Trials

D'Addio et al. enrolled 65 individuals with newly diagnosed T1DM in three independent clinical trials, where the patients transplanted with HSCs showed enhanced C-peptide levels at 6 months after treatment compared with baseline, and the immune system showed an overall stabilization in the remaining follow-up period (84). Gu et al. performed a parallel-assignment, phase-II prospective, non-randomized trial, in which 20 patients were treated only with insulin injections and 20 received autologous hematopoietic stem cell therapy (AHSCT). The results demonstrated the beneficial effects of AHSCT in patients with recent-onset T1DM by increasing the concentration of C-peptide and inducing insulin independence, and the safety and good tolerability of AHSCT compared with conventional intensive insulin therapy was also certified (77). Another clinical trial in Ning's research also proved that AHSCT was safe without a reduction in the diversity of T-cell receptor (TCR) repertoires, and TCR repertoires tended to be more stable after AHSCT (85). The clinical trial data also showed significant direct correlations between HSPC levels and the coefficient of variation of glucose levels or time in hypoglycemia, which were weaker in patients with long-standing diabetes than in those with short-term diabetes (86). HSC transplantation improves glycated hemoglobin levels in a time-dependent manner (87).

3.3 ESC and iPSC Transplantation in T1DM

ESCs and iPSCs are PSCs that can regenerate the islet β-cells and immune cells through differentiating, which helps to increase the mass of β-cells. D'Amour et al. firstly proved that ESC-derived β-cells could be successfully generated through the *in-vitro* recapitulation of pancreatic islets and β-cell physiological development by stepwise application of specific factors (88, 89). iPSCs, reprogrammed from somatic cells, have a similar ability to differentiate and proliferate like ESCs. iPSCs collected from the umbilical cord at birth have the potential for self-renewable multipotency and can differentiate into various lineages such as islets (31, 59, 90). Hence, iPSCs provide a promising platform to produce insulin-secreting cells *in vitro*. However, the utilization of ESCs and iPSCs is less due to law restrictions in many countries, so there is little clinical research on ESCs and iPSCs.

3.3.1 Islet Protection and Regeneration Ability of ESCs and iPSCs

Transplantation of ESCs or iPSCs in T1DM can regenerate the islet β-cells and increase β-cell mass through differentiating to insulin-producing cells (IPCs), pancreatic progenitors, islet organoids, and interspecific pancreatic chimeras, which benefited the treatment of T1DM. Rezania et al. cultured iPSCs expressing key markers of mature β-cells such as insulin *in vitro* and obtained cells which have functional similarities to human islets; the iPSCs rapidly reversed hyperglycemia in streptozotocin (STZ)-induced diabetic mice through increasing the level of insulin and C-peptide when transplanted *in vivo* (91). iPSCs can be generated from the skin fibroblasts of T1DM patients. These iPSCs can differentiate into pancreatic cell lineages and generate T1DM SC β-cells, making autologous stem cell-derived pancreatic progeny transplantation for T1DM possible (92). Korytnikov and Nostro isolated hPSCs successfully in a lab and then transplanted them to mice models to monitor their developmental potential *in vivo*, and they found that mice transplanted with multipotent pancreatic progenitor cells can form all pancreatic lineages *in vivo* (93). Nadav et al. used single-cell RNA sequencing of differentiating β-cells and revealed that ESC differentiation toward the mature β-cell phenotype can be tracked at each stage through monitoring the expression of markers identifying each intermediate progenitor, such as the β-cell marker insulin, endocrine precursor marker neurogenin 3, NK6 homeobox (NKX6.1), and PDX-1 (89, 94). They also proved that WNT inhibition and bone morphogenetic protein (BMP) activation could modulate the ratio of progenitors and endocrine cells, which shed light on a possible gene editing target for ESC differentiation toward the mature β-cells (89).

3.3.2 Immunomodulatory Ability of ESCs and iPSCs

Haque et al. characterized autoantigen-specific naturally occurring Treg-like iPSC-Tregs and proved that adoptive transfer of ovalbumin (OVA)-specific iPSC-Tregs greatly suppressed autoimmunity in the mouse model preventing the β-cells from destruction. These tissue-associated Tregs can effectively inhibit the migration and activity of the pathogenic immune cells and accumulate in the diabetic pancreas causing T1DM by downregulating the production of proinflammatory cytokine IFN-γ and suppressing the expression of ICAM-1 (95). Another report declared that pancreatic endoderm derived from hESCs can generate functional insulin-producing cells *in vivo* regardless of the presence of innate lymphoid cell elements, and the combination of CTLA4Ig and anti-CD40L mAbs can block hESC-PE graft rejection in immunocompetent mice, while regulatory T cells were not needed for the tolerance during hESC-PE transplantation (96).

To increase the function of stem cells, it is very important to sustain the regeneration and differentiation ability of stem cells and prevent the programmed death of stem cells *in vivo* (30, 97). Several approaches and conditions, including far-red light, genetic engineering, biological material scaffolds, nanofiber tubular, combination treatment with insulin or other drugs, microcapsules, and co-transplantation with more than one type of stem cells, were utilized to promote the survival, differentiation, and immunomodulatory ability of stem cells *in vivo* and *in vitro*. These preclinical attempts tried to derive pancreas islet cells, increase the number and function of Tregs, ameliorate the function of islets, and prevent β-cells more effectively (**Table 1**). Also, stem cells have been used in human clinical trials, which showed that stem cell transplantation had beneficial effects on T1DM, with no obvious adverse reactions (**Table 2**). Recently, an allogeneic, gene-edited, immune-evasive, stem cell-derived therapy for the treatment of T1DM was approved in Canada for clinical trial application (CTA)

TABLE 1 | Intervention strategies to improve the treatment effect of stem cells.

Type	Donor	Recipient	Pretreatment condition	Doses	Effect	Pathway/target/mechanism	Year/ reference
hESCs	Human	Mice	Aggregates of hESC-derived PP with rat-derived MV in hydrogels	N/A	Accelerate the normalization of glycemia and persist long-term and promote beta-cell maturation	Improve the survival of PP cells by reducing hypoxia and apoptosis	2021 (98)
hMSCs	Human	Mice	Far-red light-activated human islet-like designer	N/A	Improve glucose tolerance, sustain blood glucose control, and attenuate both oxidative stress and development of multiple diabetes-related complications in the kidneys	Sustain fine-tuned secretion of insulin	2021 (99)
hPSCs	Human	*In vitro*	PD-L1 overexpression and HLA class I knockout by genome engineering	N/A	Provide protection from diabetes-specific immune recognition	Abrogate human diabetogenic CD8 T cells in activation	2021 (100)
hPSCs	Human	*In vitro*	Superporous agarose scaffolds	N/A	Enhance the function of the bioartificial pancreas	Sustain insulin production	2021 (101)
hESCs	Human	Mice/dog	Nanofibrous encapsulation	2,500 cluster cells/ mouse	Corrected glucose levels immediately (within a week)	Sustain the survival of human SC-β cells in immunodeficient mice and immunocompetent mice and dogs	2021 (102)
hBM-cMSCs	Human	Rhesus monkey	Combined therapy with liraglutide	1.5×10^6/kg	Reduce FBG and HbA1c	Immunomodulation by increased Tregs, IL-4, IL-10, and TGF-β1 and decreased IL-6 and IL-1β	2021 (103)
ADMSCs	Human	Mice	Co-culture with MIN6 cells	0.5×10^6/mouse	Provide protection to MIN6 cells from streptozotocin	AKT and ERK pathway	2021 (43)
MSCs	Human	Mice	Generated human alpha-1 antitrypsin-engineered MSCs	N/A	Increased self-renewal, better migration, and multilineage differentiation abilities	Upregulate the expression of WNT3A, KDR, ICAM 1, VICAM1, MMP2, and IGF1	2021 (104)
hUCMS and hIDCs	Human	Mice	Co-microencapsulation of hUCMS and hIDC in AG	1×10^6 hIDC + 1×10^6 hUCMS/1.3 ml AG	Reverse the recent onset hyperglycemia in mice	Couple the immunoregulatory activities of hUCMS and insulin production by hIDC	2020 (105)
MSCs	Human	Rats	Combined therapy with insulin	2×10^6 MSCs/kg	Decrease blood glucose level	Regulate the expression of leptin receptor, neuropeptide Y, and melanocortin-4 receptor	2020 (106)
ADMSCs	Human	Mice	Co-transplant with neonatal porcine islets	1×10^6 MSCs/mouse	Improve glucose tolerance and stimulate serum porcine insulin	Secrete anti-inflammatory and proangiogenic factors	2019 (107)
MSCs	Mice	Mice Allograft	Intraperitoneal injection MSCs and MSC-conditioned medium together	1×10^6 MSCs/mouse	More effective in induction of immunosuppressive effects	Reduce the inflammatory cytokines and increase inflammatory cytokine	2020 (108)

(Continued)

TABLE 1 | Continued

Type	Donor	Recipient	Pretreatment condition	Doses	Effect	Pathway/target/mechanism	Year/ reference
ESCs	Mice	Mice Allograft	MHC-mismatched	N/A	Prevent insulitis and T1D development	Increase the number and function of Tregs	2019 (109)
HSPCs	Human	Murine	Overexpression PD-L1 by genome engineering	1×10^6/mouse	Inhibit the autoimmune response and revert diabetes	PD-L1/PD-1 pathway	2017 (110)
hESCs	Human	Mice	Alginate microencapsulation with CXCL12	1×10^6/mouse	Enhance insulin secretion and enhance immunoisolation	CXCL12 pathway	2019 (111)
HSPCs	Human	Human Autograft	Modulation with prostaglandins	1×10^5/mouse	Abrogate the autoreactive T-cell response	PD-L1 pathway	2018 (79)
MSCs	Mice	Mice Allograft	Co-transplantation with immature dendritic cells	2×10^5/mouse	Decrease blood glucose and glycosylated hemoglobin levels	Inhibit the proliferation of T lymphocytes to induce immune tolerance	2017 (112)
iPSCs	Human	Mice	Transient demethylation treatment	N/A	Significantly improve the yield of PDX-1$^+$ and NKX6.1$^+$ cells	Rescue and generate the islet-like compact cell clusters	2017 (113)
MSCs	Mice	Mice Allograft	Overexpress TGF-β by genome engineering	5×10^5/mouse	Improve insulin levels and suppress adverse immune responses	IFN-γ/IL-4 pathway	2017 (114)
hESCs	Human	*In vitro*	Barium alginate capsules	N/A	Increase cell proliferation and pancreatic differentiation.	TGF-β signaling	2016 (115)
MSCs	Rats	*In vitro*	Co-encapsulation within GLP-1 ligand-functionalized polyethylene glycol hydrogel	N/A	Improve islet function and stimulate cell survival	GLP-1 promotes the stimulation of insulin gene transcription, islet growth, and neogenesis	2017 (116)
ADMSCs	Mice	Mice Allograft	Intrasplenic injection or intrapancreatic injection	1×10^6/mouse	Intrasplenic administration improves β-cell mass and insulin production	Increase pancreatic TGF-β levels	2015 (117)

hESCs, human embryonic stem cells; PP, pancreatic progenitors; hMSCs, human mesenchymal stem cells; hPSCs, human pluripotent stems; hBM-cMSCs, human clonal mesenchymal stem cells; FBG, fasting blood glucose; ADMSCs, adipose tissue-derived MSCs; N/A, not applicable; MSCs, mesenchymal stem cells; WNT3A, Wnt family member 3A; KDR, kinase insert domain receptor; ICAM-1, intercellular adhesion molecule 1; VICAM-1, vascular cell adhesion protein 1; MMP2, matrix metalloproteinase-2; IGF-1, insulin-like growth factor; iPSCs, induced pluripotent stem cells; hUCMS, human umbilical cord-derived mesenchymal stem cells; hIDCs, pancreatic islet-derived insulin-producing cells; AG, sodium alginate; HSPCs, hematopoietic stem and progenitor cells; PD-L1, programmed death ligand 1.

TABLE 2 | Stem cell treatment of T1DM in human clinical trials.

Type	Case	Transplantation method	Observation	Effect	Side effect	Number of clinical trials	Year/ reference
MSCs	27 (MSC-treated); 26 (control)	The MSC-treated group received repeat transplantation with an interval of 3 months, and each time for 1.0×10^6 cells/kg was given.	Participants were followed up at 3, 6, and 12 months and yearly afterward	HbA1c levels decreased and C-peptide was significantly increased in the MSC-treated group	Had mild fever after MSC infusion	ChiCTR2100045434	2021 (70)
ASCs	8 (ASC treated); 5 (control)	1×10^6 cells/kg ASCs and cholecalciferol (Vit D) 2,000 IU/day for 3 months	Participants were followed up at 3, 6, and 12 months and yearly afterward	Stability of C-peptide, better glucose control, and lower insulin requirement	Transient headache ($n = 8$), mild local infusion reactions ($n = 7$), tachycardia ($n = 4$), and abdominal cramps ($n = 1$)	NCT03920397	2020 (118)
AHSCT	20 (AHSCT); 20 (control)	HSCs were mobilized and collected from peripheral blood by leukapheresis and cryopreserved. Cells were injected intravenously after conditioning with CTX (200 mg/kg) and rabbit ATG (4.5 mg/kg)	Participants were followed up at 3, 6, 12, 18, 24, 36, and 48 months	Reduction of insulin dosage, decreased HbA1c and increased Cmax and area under the curve for C-peptide (AUCC)	Febrile neutropenia ($n = 9$), nausea and vomiting ($n = 11$), alopecia ($n = 19$), and blood component transfusions due to bone marrow suppression ($n = 5$)	NCT00807651	2018 (77)
CB-SCs	15	CB-SC-treated mononuclear cells (interaction for 2~3 h) were returned to the patient's blood circulation *via* a dorsal vein in the hand with physiological saline; after 3 months, subjects received a similar second treatment	Follow-up visits were scheduled 2, 8, 12, 18, 26, 40, and 56 weeks after treatment for clinical assessments and laboratory tests	Restored the regeneration of naive CD^{4+} T cells and the function of β-cell in patients with residual β-cell function was rescued and without a significant linear decline	N/A	NCT01350219	2015 (119)
UC-MSCs	21 (UC-MSCs); 21 (control)	The dorsal pancreatic artery or its substitute was identified, and 60~80 ml BM-MNCs (106.8×10^6/kg) plus 30~50 ml UC-MSCs (1.1×10^6/kg) were sequentially infused within 30 min	N/A	HbA1c levels decreased, FBG decreased, and insulin dose requirements reduced	N/A	NCT01374854	2016 (50)
CB-SCs	12 (with CB-SCs educated); 3 (without CB-SCs educated)	The collected lymphocytes were transferred into the device for exposure to allogeneic CB-SCs for 2~3 h, then were returned to the patient's circulation *via* a dorsal vein in the hand under gravity flow control (2 to 3 ml/min) with physiological saline	Follow-up visits were scheduled 4, 12, 24, and 40 weeks after treatment for clinical assessments and laboratory tests	Improved fasting C-peptide levels, reduced daily dose of insulin, increased Tregs, and reduced HbA1c	N/A	NCT01350219	2012 (120)
ASCs	7 (ASCs + VIT D); 4 (VIT D); 6 (control)	Allogenic ASC (1×10^6 cells/kg) and cholecalciferol 2,000 UI/day for 6 months	Participants were followed up at T0 and after 1, 3, and 6 months	Improved fasting C-peptide levels and reduced HbA1c level	Four patients developed local thrombophlebitis within the first week and two had transient mild eye floaters during infusion, with no subsequent visual abnormalities. One patient developed central retinal vein occlusion at T3, with complete resolution at T6	NCT03920397	2021 (121)

T1DM, type 1 diabetes mellitus; BMI, body mass index; SCs, stem cells; MSCs, mesenchymal stem cells; hESCs, human embryonic stem cells; ASCs, adipose stem cells; HSCs, hematopoietic stem cells; BM-MSCs, bone marrow-derived mesenchymal stem cells; HbA1c, glycosylated hemoglobin assays; FBG, fasting blood glucose; Treg, T regulatory; hUCMS, human umbilical cord matrix stem cells; sBCs, stem cell-derived pancreatic beta-like cells; ADMSCs, adipose tissue-derived MSCs; CB-SCs, human cord blood-derived multipotent stem cells; UC-MSCs, umbilical cord mesenchymal stromal cells; VIT D, vitamin D; N/A, not applicable.

(CRISPR Therapeutics and ViaCyte, Inc. to start clinical trial of the first gene-edited cell replacement therapy for the treatment of T1DM, retrieved on November 16, 2021). This CRISPR therapeutics offered novel β-cell replacement therapies to address unmet T1DM needs. All these efforts are aimed at better promoting the effectiveness of stem cells, which proved to be a more viable option for the treatment of T1DM, lessening the suffering of the patients.

4 CONCLUDING REMARKS

In recent research, stem cell therapy has demonstrated itself as a rapidly expanding and potentially limitless source of β-cells to arrive at a cure for T1DM by reconstitution of immunotolerance and differentiation into islet β-cell clusters. As the immunosuppression affected the effect of transplantation of stem cells, stem cell intervention before transplantation could help preserve β-cells and remodel the immune response. However, several challenges, such as the ethical problem of autologous and allogeneic stem cells used to preserve the function of β-cells, still need resolution. Although research into β-cell replacement derived from stem cells is increasing every year, we must make more efforts in the future on the intervention with stem cell transplantation, which can help achieve remission of T1DM by β-cell replacement.

AUTHOR CONTRIBUTIONS

X-XW reviewed the literature, wrote the manuscript, and created the descriptive figures. X-MH and QZ edited the tables and figures. D-YZ and S-YZ assisted in the literature review. MK edited the manuscript. KX and R-HY revised the manuscript. All authors read and approved the final manuscript.

REFERENCES

1. Campbell MR. Review of Current Status of Molecular Diagnosis and Characterization of Monogenic Diabetes Mellitus: A Focus on Next-Generation Sequencing. *Expert Rev Mol Diagn* (2020) 20(4):413–20. doi: 10.1080/14737159.2020.1730179
2. Miller KM, Hermann J, Foster N, Hofer SE, Rickels MR, Danne T, et al. Longitudinal Changes in Continuous Glucose Monitoring Use Among Individuals With Type 1 Diabetes: International Comparison in the German and Austrian DPV and U.S. T1D Exchange Registries. *Diabetes Care* (2020) 43(1):e1–2. doi: 10.2337/dc19-1214
3. Quansah DY, Gross J, Gilbert L, Pauchet A, Horsch A, Benhalima K, et al. Cardiometabolic and Mental Health in Women With Early Gestational Diabetes Mellitus: A Prospective Cohort Study. *J Clin Endocrinol Metab* (2022) 107(3):e996–e1008. doi: 10.1210/clinem/dgab791
4. Vinuela A, Varshney A, van de Bunt M, Prasad RB, Asplund O, Bennett A, et al. Genetic Variant Effects on Gene Expression in Human Pancreatic Islets and Their Implications for T2D. *Nat Commun* (2020) 11(1):4912.
5. Wang J, Lv B, Chen X, Pan Y, Chen K, Zhang Y, et al. An Early Model to Predict the Risk of Gestational Diabetes Mellitus in the Absence of Blood Examination Indexes: Application in Primary Health Care Centres. *BMC Pregnancy Childbirth* (2021) 21(1):814. doi: 10.1186/s12884-021-04295-2
6. Pang H, Luo S, Xiao Y, Xia Y, Li X, Huang G, et al. Emerging Roles of Exosomes in T1DM. *Front Immunol* (2020) 11:593348. doi: 10.3389/fimmu.2020.593348
7. Rosell Rask S, Krarup Hansen T, Bjerre M. FGF21 and Glycemic Control in Patients With T1D. *Endocrine* (2019) 65(3):550–7. doi: 10.1007/s12020-019-02027-3
8. Terrazzano G, Bruzzaniti S, Rubino V, Santopaolo M, Palatucci AT, Giovazzino A, et al. T1D Progression Is Associated With Loss of CD3(+) CD56(+) Regulatory T Cells That Control CD8(+) T Cell Effector Functions. *Nat Metab* (2020) 2(2):142–52. doi: 10.1038/s42255-020-0173-1
9. Katsarou A, Gudbjornsdottir S, Rawshani A, Dabelea D, Bonifacio E, Anderson BJ, et al. Type 1 Diabetes Mellitus. *Nat Rev Dis Primers* (2017) 3:17016. doi: 10.1038/nrdp.2017.16
10. Owens DR. Insulin Preparations With Prolonged Effect. *Diabetes Technol Ther* (2011) 13 Suppl 1:S5–14. doi: 10.1089/dia.2011.0068
11. Haidar A, Legault L, Matteau-Pelletier L, Messier V, Dallaire M, Ladouceur M, et al. Outpatient Overnight Glucose Control With Dual-Hormone Artificial Pancreas, Single-Hormone Artificial Pancreas, or Conventional Insulin Pump Therapy in Children and Adolescents With Type 1 Diabetes: An Open-Label, Randomised Controlled Trial. *Lancet Diabetes Endocrinol* (2015) 3(8):595–604. doi: 10.1016/S2213-8587(15)00141-2
12. Garg SK, Weinzimer SA, Tamborlane WV, Buckingham BA, Bode BW, Bailey TS, et al. Glucose Outcomes With the In-Home Use of a Hybrid Closed-Loop Insulin Delivery System in Adolescents and Adults With Type 1 Diabetes. *Diabetes Technol Ther* (2017) 19(3):155–63. doi: 10.1089/dia.2016.0421
13. Nijhoff MF, de Koning EJP. Artificial Pancreas or Novel Beta-Cell Replacement Therapies: A Race for Optimal Glycemic Control? *Curr Diabetes Rep* (2018) 18(11):110. doi: 10.1007/s11892-018-1073-6
14. Nally LM, Sherr JL, Van Name MA, Patel AD, Tamborlane WV. Pharmacologic Treatment Options for Type 1 Diabetes: What's New? *Expert Rev Clin Pharmacol* (2019) 12(5):471–9. doi: 10.1080/17512433.2019.1597705
15. Yan WT YY, Hu XM, Ning WY, Liao LS, Lu S, Zhao W, et al. Do Pyroptosis, Apoptosis, and Necroptosis (PANoptosis) Exist in Cerebral Ischemia? Evidence From Cell and Rodent Studies. *Neural Regener Res* (2022) 17 (8):1761–8. doi: 10.4103/1673-5374.331539
16. Atkinson MA, Eisenbarth GS, Michels AW. Type 1 Diabetes. *Lancet* (2014) 383(9911):69–82. doi: 10.1016/S0140-6736(13)60591-7
17. Gregg EW, Cheng YJ, Srinivasan M, Lin J, Geiss LS, Albright AL, et al. Trends in Cause-Specific Mortality Among Adults With and Without Diagnosed Diabetes in the USA: An Epidemiological Analysis of Linked National Survey and Vital Statistics Data. *Lancet* (2018) 391(10138):2430–40. doi: 10.1016/S0140-6736(18)30314-3
18. Rawshani A, Sattar N, Franzen S, Rawshani A, Hattersley AT, Svensson AM, et al. Excess Mortality and Cardiovascular Disease in Young Adults With Type 1 Diabetes in Relation to Age at Onset: A Nationwide, Register-Based Cohort Study. *Lancet* (2018) 392(10146):477–86. doi: 10.1016/S0140-6736(18)31506-X
19. Chiang JL, Kirkman MS, Laffel LM, Peters AL. Type 1 Diabetes Sourcebook A: Type 1 Diabetes Through the Life Span: A Position Statement of the American Diabetes Association. *Diabetes Care* (2014) 37(7):2034–54. doi: 10.2337/dc14-1140

20. Ryan AJ, O'Neill HS, Duffy GP, O'Brien FJ. Advances in Polymeric Islet Cell Encapsulation Technologies to Limit the Foreign Body Response and Provide Immunoisolation. *Curr Opin Pharmacol* (2017) 36:66–71. doi: 10.1016/j.coph.2017.07.013
21. Vantyghem MC, de Koning EJP, Pattou F, Rickels MR. Advances in Beta-Cell Replacement Therapy for the Treatment of Type 1 Diabetes. *Lancet* (2019) 394(10205):1274–85. doi: 10.1016/S0140-6736(19)31334-0
22. Gruessner RW, Sutherland DE, Kandaswamy R, Gruessner AC. Over 500 Solitary Pancreas Transplants in Nonuremic Patients With Brittle Diabetes Mellitus. *Transplantation* (2008) 85(1):42–7. doi: 10.1097/01.tp.0000296820.46978.3f
23. Abdulreda MH, Rodriguez-Diaz R, Cabrera O, Caicedo A, Berggren PO. The Different Faces of the Pancreatic Islet. *Adv Exp Med Biol* (2016) 938:11–24. doi: 10.1007/978-3-319-39824-2_2
24. Sneddon JB, Tang Q, Stock P, Bluestone JA, Roy S, Desai T, et al. Stem Cell Therapies for Treating Diabetes: Progress and Remaining Challenges. *Cell Stem Cell* (2018) 22(6):810–23. doi: 10.1016/j.stem.2018.05.016
25. Coe TM, Markmann JF, Rickert CG. Current Status of Porcine Islet Xenotransplantation. *Curr Opin Organ Transplant* (2020) 25(5):449–56. doi: 10.1097/MOT.0000000000000794
26. Liang CC, Shaw SW, Huang YH, Lee TH. Human Amniotic Fluid Stem Cells can Improve Cerebral Vascular Remodelling and Neurological Function After Focal Cerebral Ischaemia in Diabetic Rats. *J Cell Mol Med* (2021) 25 (21):10185–96. doi: 10.1111/jcmm.16956
27. Zhang Q WX, Hu XM, Zhao WJ, Ban XX, Huang YX, Yan WT, et al. Targeting Programmed Cell Death to Improve Stem Cell Therapy: Implications for Treating Diabetes and Diabetes-Related Diseases. *Front Cell Dev Biol* (2021) 9:809656. doi: 10.3389/fcell.2021.809656
28. Wen SJ, Zheng XM, Liu LF, Li NN, Mao HA, Huang L, et al. Effects of Primary Microglia and Astrocytes on Neural Stem Cells in *In Vitro* and *In Vivo* models of ischemic stroke. *Neural Regener Res* (2021) 16(9):1677–85.
29. Sultan N, Amin LE, Zaher AR, Grawish ME, Scheven BA. Dental Pulp Stem Cells Stimulate Neuronal Differentiation of PC12 Cells. *Neural Regener Res* (2021) 16(9):1821–8. doi: 10.4103/1673-5374.306089
30. Hu XM, Li ZX, Zhang DY, Yang YC, Fu SA, Zhang ZQ, et al. A Systematic Summary of Survival and Death Signalling During the Life of Hair Follicle Stem Cells. *Stem Cell Res Ther* (2021) 12(1):453. doi: 10.1186/s13287-021-02527-y
31. Francese R, Fiorina P. Immunological and Regenerative Properties of Cord Blood Stem Cells. *Clin Immunol* (2010) 136(3):309–22. doi: 10.1016/j.clim.2010.04.010
32. Hu XM, Zhang Q, Zhou RX, Wu YL, Li ZX, Zhang DY, et al. Programmed Cell Death in Stem Cell-Based Therapy: Mechanisms and Clinical Applications. *World J Stem Cells* (2021) 13(5):386–415. doi: 10.4252/wjsc.v13.i5.386
33. Yang R, Liu F, Wang J, Chen X, Xie J, Xiong K. Epidermal Stem Cells in Wound Healing and Their Clinical Applications. *Stem Cell Res Ther* (2019) 10(1):229. doi: 10.1186/s13287-019-1312-z
34. Ilic D, Polak JM. Stem Cells in Regenerative Medicine: Introduction. *Br Med Bull* (2011) 98:117–26. doi: 10.1093/bmb/ldr012
35. Bongso A, Richards M. History and Perspective of Stem Cell Research. *Best Pract Res Clin Obstet Gynaecol* (2004) 18(6):827–42. doi: 10.1016/j.bpobgyn.2004.09.002
36. Yang R, Yang S, Zhao J, Hu X, Chen X, Wang J, et al. Progress in Studies of Epidermal Stem Cells and Their Application in Skin Tissue Engineering. *Stem Cell Res Ther* (2020) 11(1):303. doi: 10.1186/s13287-020-01796-3
37. Parmar M, Grealish S, Henchcliffe C. The Future of Stem Cell Therapies for Parkinson Disease. *Nat Rev Neurosci* (2020) 21(2):103–15. doi: 10.1038/s41583-019-0257-7
38. Thomson JA, Itskovitz-Eldor J, Shapiro SS, Waknitz MA, Swiergiel JJ, Marshall VS, et al. Embryonic Stem Cell Lines Derived From Human Blastocysts. *Science* (1998) 282(5391):1145–7. doi: 10.1126/science.282.5391.1145
39. Takahashi K, Tanabe K, Ohnuki M, Narita M, Ichisaka T, Tomoda K, et al. Induction of Pluripotent Stem Cells From Adult Human Fibroblasts by Defined Factors. *Cell* (2007) 131(5):861–72. doi: 10.1016/j.cell.2007.11.019
40. Tahbaz M, Yoshihara E. Immune Protection of Stem Cell-Derived Islet Cell Therapy for Treating Diabetes. *Front Endocrinol (Lausanne)* (2021) 12:716625. doi: 10.3389/fendo.2021.716625
41. Buron F, Reffet S, Badet L, Morelon E, Thaunat O. Immunological Monitoring in Beta Cell Replacement: Towards a Pathophysiology-Guided Implementation of Biomarkers. *Curr Diabetes Rep* (2021) 21(6):19. doi: 10.1007/s11892-021-01386-4
42. Han C, Wang YJ, Wang YC, Guan X, Wang L, Shen LM, et al. Caveolin-1 Downregulation Promotes the Dopaminergic Neuron-Like Differentiation of Human Adipose-Derived Mesenchymal Stem Cells. *Neural Regener Res* (2021) 16(4):714–20.
43. Khatri R, Petry SF, Linn T. Intrapancreatic MSC Transplantation Facilitates Pancreatic Islet Regeneration. *Stem Cell Res Ther* (2021) 12(1):121. doi: 10.1186/s13287-021-02173-4
44. Wan X, Zhu L, Zhao L, Peng L, Xiong J, Yang W, et al. Hper3 Promotes Adipogenesis *via* Hhsp90aa1-Mediated Inhibition of Notch1 Pathway. *Cell Death Dis* (2021) 12(4):301. doi: 10.1038/s41419-021-03584-0
45. Wang ZL, He RZ, Tu B, He JS, Cao X, Xia HS, et al. Drilling Combined With Adipose-Derived Stem Cells and Bone Morphogenetic Protein-2 to Treat Femoral Head Epiphyseal Necrosis in Juvenile Rabbits. *Curr Med Sci* (2018) 38(2):277–88. doi: 10.1007/s11596-018-1876-3
46. Dominici M, Le Blanc K, Mueller I, Slaper-Cortenbach I, Marini F, Krause D, et al. Minimal Criteria for Defining Multipotent Mesenchymal Stromal Cells. The International Society for Cellular Therapy Position Statement. *Cytotherapy* (2006) 8(4):315–7.
47. Davies LC, Alm JJ, Heldring N, Moll G, Gavin C, Batsis I, et al. Type 1 Diabetes Mellitus Donor Mesenchymal Stromal Cells Exhibit Comparable Potency to Healthy Controls *In Vitro*. *Stem Cells Transl Med* (2016) 5 (11):1485–95. doi: 10.5966/sctm.2015-0272
48. Gerace D, Martiniello-Wilks R, Nassif NT, Lal S, Steptoe R, Simpson AM. CRISPR-Targeted Genome Editing of Mesenchymal Stem Cell-Derived Therapies for Type 1 Diabetes: A Path to Clinical Success? *Stem Cell Res Ther* (2017) 8(1):62. doi: 10.1186/s13287-017-0511-8
49. Carlsson PO, Schwarcz E, Korsgren O, Le Blanc K. Preserved Beta-Cell Function in Type 1 Diabetes by Mesenchymal Stromal Cells. *Diabetes* (2015) 64(2):587–92. doi: 10.2337/db14-0656
50. Cai J, Wu Z, Xu X, Liao L, Chen J, Huang L, et al. Umbilical Cord Mesenchymal Stromal Cell With Autologous Bone Marrow Cell Transplantation in Established Type 1 Diabetes: A Pilot Randomized Controlled Open-Label Clinical Study to Assess Safety and Impact on Insulin Secretion. *Diabetes Care* (2016) 39(1):149–57. doi: 10.2337/dc15-0171
51. Gimble JM, Katz AJ, Bunnell BA. Adipose-Derived Stem Cells for Regenerative Medicine. *Circ Res* (2007) 100(9):1249–60. doi: 10.1161/01.RES.0000265074.83288.09
52. Chen MT, Zhao YT, Zhou LY, Li M, Zhang Q, Han Q, et al. Exosomes Derived From Human Umbilical Cord Mesenchymal Stem Cells Enhance Insulin Sensitivity in Insulin Resistant Human Adipocytes. *Curr Med Sci* (2021) 41(1):87–93. doi: 10.1007/s11596-021-2323-4
53. Zhao Y. Stem Cell Educator Therapy and Induction of Immune Balance. *Curr Diabetes Rep* (2012) 12(5):517–23. doi: 10.1007/s11892-012-0308-1
54. Nauta AJ, Fibbe WE. Immunomodulatory Properties of Mesenchymal Stromal Cells. *Blood* (2007) 110(10):3499–506. doi: 10.1182/blood-2007-02-069716
55. Ren G, Zhang L, Zhao X, Xu G, Zhang Y, Roberts AI, et al. Mesenchymal Stem Cell-Mediated Immunosuppression Occurs *via* Concerted Action of Chemokines and Nitric Oxide. *Cell Stem Cell* (2008) 2(2):141–50. doi: 10.1016/j.stem.2007.11.014
56. Domouky AM, Hegab AS, Al-Shahat A, Raafat N. Mesenchymal Stem Cells and Differentiated Insulin Producing Cells Are New Horizons for Pancreatic Regeneration in Type I Diabetes Mellitus. *Int J Biochem Cell Biol* (2017) 87:77–85. doi: 10.1016/j.biocel.2017.03.018
57. Boumaza I, Srinivasan S, Witt WT, Feghali-Bostwick C, Dai Y, Garcia-Ocana A, et al. Autologous Bone Marrow-Derived Rat Mesenchymal Stem Cells Promote PDX-1 and Insulin Expression in the Islets, Alter T Cell Cytokine Pattern and Preserve Regulatory T Cells in the Periphery and Induce Sustained Normoglycemia. *J Autoimmun* (2009) 32(1):33–42. doi: 10.1016/j.jaut.2008.10.004
58. Shigemoto-Kuroda T, Oh JY, Kim DK, Jeong HJ, Park SY, Lee HJ, et al. MSC-Derived Extracellular Vesicles Attenuate Immune Responses in Two Autoimmune Murine Models: Type 1 Diabetes and Uveoretinitis. *Stem Cell Rep* (2017) 8(5):1214–25. doi: 10.1016/j.stemcr.2017.04.008

59. Abdi R, Fiorina P, Adra CN, Atkinson M, Sayegh MH. Immunomodulation by Mesenchymal Stem Cells: A Potential Therapeutic Strategy for Type 1 Diabetes. *Diabetes* (2008) 57(7):1759–67. doi: 10.2337/db08-0180
60. Bassi EJ, Moraes-Vieira PM, Moreira-Sa CS, Almeida DC, Vieira LM, Cunha CS, et al. Immune Regulatory Properties of Allogeneic Adipose-Derived Mesenchymal Stem Cells in the Treatment of Experimental Autoimmune Diabetes. *Diabetes* (2012) 61(10):2534–45. doi: 10.2337/db11-0844
61. Gerace D, Martiniello-Wilks R, Habib R, Ren B, Nassif NT, O'Brien BA, et al. *Ex Vivo* Expansion of Murine MSC Impairs Transcription Factor-Induced Differentiation Into Pancreatic Beta-Cells. *Stem Cells Int* (2019) 2019:1395301. doi: 10.1155/2019/1395301
62. van Megen KM, van 't Wout ET, Lages Motta J, Dekker B, Nikolic T, Roep BO. Activated Mesenchymal Stromal Cells Process and Present Antigens Regulating Adaptive Immunity. *Front Immunol* (2019) 10:694. doi: 10.3389/fimmu.2019.00694
63. Montanucci P, Alunno A, Basta G, Bistoni O, Pescara T, Caterbi S, et al. Restoration of T Cell Substes of Patients With Type 1 Diabetes Mellitus by Microencapsulated Human Umbilical Cord Wharton Jelly-Derived Mesenchymal Stem Cells: An *In Vitro* Study. *Clin Immunol* (2016) 163:34–41. doi: 10.1016/j.clim.2015.12.002
64. Montanucci P, Pescara T, Alunno A, Bistoni O, Basta G, Calafiore R. Remission of Hyperglycemia in Spontaneously Diabetic NOD Mice Upon Transplant of Microencapsulated Human Umbilical Cord Wharton Jelly-Derived Mesenchymal Stem Cells (hUCMS). *Xenotransplantation* (2019) 26 (2):e12476. doi: 10.1111/xen.12476
65. Karnieli O, Izhar-Prato Y, Bulvik S, Efrat S. Generation of Insulin-Producing Cells From Human Bone Marrow Mesenchymal Stem Cells by Genetic Manipulation. *Stem Cells* (2007) 25(11):2837–44. doi: 10.1634/stemcells.2007-0164
66. Chen LB, Jiang XB, Yang L. Differentiation of Rat Marrow Mesenchymal Stem Cells Into Pancreatic Islet Beta-Cells. *World J Gastroenterol* (2004) 10 (20):3016–20. doi: 10.3748/wjg.v10.i20.3016
67. Gao F, Wu DQ, Hu YH, Jin GX, Li GD, Sun TW, et al. *In Vitro* Cultivation of Islet-Like Cell Clusters From Human Umbilical Cord Blood-Derived Mesenchymal Stem Cells. *Transl Res* (2008) 151(6):293–302. doi: 10.1016/j.trsl.2008.03.003
68. Chao KC, Chao KF, Fu YS, Liu SH. Islet-Like Clusters Derived From Mesenchymal Stem Cells in Wharton's Jelly of the Human Umbilical Cord for Transplantation to Control Type 1 Diabetes. *PloS One* (2008) 3(1):e1451. doi: 10.1371/journal.pone.0001451
69. Maldonado M, Huang T, Yang L, Xu L, Ma L. Human Umbilical Cord Wharton Jelly Cells Promote Extra-Pancreatic Insulin Formation and Repair of Renal Damage in STZ-Induced Diabetic Mice. *Cell Commun Signal* (2017) 15(1):43. doi: 10.1186/s12964-017-0199-5
70. Lu J, Shen SM, Ling Q, Wang B, Li LR, Zhang W, et al. One Repeated Transplantation of Allogeneic Umbilical Cord Mesenchymal Stromal Cells in Type 1 Diabetes: An Open Parallel Controlled Clinical Study. *Stem Cell Res Ther* (2021) 12(1):340. doi: 10.1186/s13287-021-02417-3
71. Mesples A, Majeed N, Zhang Y, Hu X. Early Immunotherapy Using Autologous Adult Stem Cells Reversed the Effect of Anti-Pancreatic Islets in Recently Diagnosed Type 1 Diabetes Mellitus: Preliminary Results. *Med Sci Monit* (2013) 19:852–7. doi: 10.12659/MSM.889525
72. Eaves CJ. Hematopoietic Stem Cells: Concepts, Definitions, and the New Reality. *Blood* (2015) 125(17):2605–13. doi: 10.1182/blood-2014-12-570200
73. Carroll D, St Clair DK. Hematopoietic Stem Cells: Normal Versus Malignant. *Antioxid Redox Signal* (2018) 29(16):1612–32. doi: 10.1089/ars.2017.7326
74. Voltarelli JC, Couri CE, Stracieri AB, Oliveira MC, Moraes DA, Pieroni F, et al. Autologous Nonmyeloablative Hematopoietic Stem Cell Transplantation in Newly Diagnosed Type 1 Diabetes Mellitus. *JAMA* (2007) 297(14):1568–76. doi: 10.1001/jama.297.14.1568
75. Copelan EA. Hematopoietic Stem-Cell Transplantation. *N Engl J Med* (2006) 354(17):1813–26. doi: 10.1056/NEJMra052638
76. Tian C, Ansari MJ, Paez-Cortez J, Bagley J, Godwin J, Donnarumma M, et al. Induction of Robust Diabetes Resistance and Prevention of Recurrent Type 1 Diabetes Following Islet Transplantation by Gene Therapy. *J Immunol* (2007) 179(10):6762–9. doi: 10.4049/jimmunol.179.10.6762
77. Gu B, Miao H, Zhang J, Hu J, Zhou W, Gu W, et al. Clinical Benefits of Autologous Haematopoietic Stem Cell Transplantation in Type 1 Diabetes Patients. *Diabetes Metab* (2018) 44(4):341–5. doi: 10.1016/j.diabet.2017.12.006
78. Pastore I, Assi E, Ben Nasr M, Bolla AM, Maestroni A, Usuelli V, et al. Hematopoietic Stem Cells in Type 1 Diabetes. *Front Immunol* (2021) 12:694118. doi: 10.3389/fimmu.2021.694118
79. Ben Nasr M, D'Addio F, Malvandi AM, Faravelli S, Castillo-Leon E, Usuelli V, et al. Prostaglandin E2 Stimulates the Expansion of Regulatory Hematopoietic Stem and Progenitor Cells in Type 1 Diabetes. *Front Immunol* (2018) 9:1387. doi: 10.3389/fimmu.2018.01387
80. Wang N, Rajasekaran N, Hou T, Macaubas C, Mellins ED. Immunological Basis for Rapid Progression of Diabetes in Older NOD Mouse Recipients Post BM-HSC Transplantation. *PloS One* (2015) 10(5):e0128494. doi: 10.1371/journal.pone.0128494
81. Racine J, Wang M, Zhang C, Lin CL, Liu H, Todorov I, et al. Induction of Mixed Chimerism With MHC-Mismatched But Not Matched Bone Marrow Transplants Results in Thymic Deletion of Host-Type Autoreactive T-Cells in NOD Mice. *Diabetes* (2011) 60(2):555–64. doi: 10.2337/db10-0827
82. Chan J, Clements W, Field J, Nasa Z, Lock P, Yap F, et al. Transplantation of Bone Marrow Genetically Engineered to Express Proinsulin II Protects Against Autoimmune Insulitis in NOD Mice. *J Gene Med* (2006) 8 (11):1281–90. doi: 10.1002/jgm.968
83. Leng Q, Nie Y, Zou Y, Chen J. Elevated CXCL12 Expression in the Bone Marrow of NOD Mice Is Associated With Altered T Cell and Stem Cell Trafficking and Diabetes Development. *BMC Immunol* (2008) 9:51. doi: 10.1186/1471-2172-9-51
84. D'Addio F, Valderrama Vasquez A, Ben Nasr M, Franek E, Zhu D, Li L, et al. Autologous Nonmyeloablative Hematopoietic Stem Cell Transplantation in New-Onset Type 1 Diabetes: A Multicenter Analysis. *Diabetes* (2014) 63 (9):3041–6. doi: 10.2337/db14-0295
85. Zhang J, Hu M, Wang B, Gao J, Wang L, Li L, et al. Comprehensive Assessment of T-Cell Repertoire Following Autologous Hematopoietic Stem Cell Transplantation for Treatment of Type 1 Diabetes Using High-Throughput Sequencing. *Pediatr Diabetes* (2018) 19(7):1229–37. doi: 10.1111/pedi.12728
86. Boscari F, D'Anna M, Bonora BM, Tresso S, Cappellari R, Avogaro A, et al. Effects of Glucose Variability on Hematopoietic Stem/Progenitor Cells in Patients With Type 1 Diabetes. *J Endocrinol Invest* (2021) 44(1):119–26. doi: 10.1007/s40618-020-01278-6
87. Snarski E, Szmurlo D, Halaburda K, Krol M, Urbanowska E, Milczarczyk A, et al. An Economic Analysis of Autologous Hematopoietic Stem Cell Transplantation (AHSCT) in the Treatment of New Onset Type 1 Diabetes. *Acta Diabetol* (2015) 52(5):881–8. doi: 10.1007/s00592-015-0724-1
88. D'Amour KA, Agulnick AD, Eliazer S, Kelly OG, Kroon E, Baetge EE. Efficient Differentiation of Human Embryonic Stem Cells to Definitive Endoderm. *Nat Biotechnol* (2005) 23(12):1534–41. doi: 10.1038/nbt1163
89. Loretelli C, Assi E, Seelam AJ, Ben Nasr M, Fiorina P. Cell Therapy for Type 1 Diabetes. *Expert Opin Biol Ther* (2020) 20(8):887–97. doi: 10.1080/14712598.2020.1748596
90. Tyndall A, Walker UA, Cope A, Dazzi F, De Bari C, Fibbe W, et al. Immunomodulatory Properties of Mesenchymal Stem Cells: A Review Based on an Interdisciplinary Meeting Held at the Kennedy Institute of Rheumatology Division, London, UK, 31 October 2005. *Arthritis Res Ther* (2007) 9(1):301. doi: 10.4103/1673-5374.331539
91. Rezania A, Bruin JE, Arora P, Rubin A, Batushansky I, Asadi A, et al. Reversal of Diabetes With Insulin-Producing Cells Derived *In Vitro* From Human Pluripotent Stem Cells. *Nat Biotechnol* (2014) 32(11):1121–33. doi: 10.1038/nbt.3033
92. Maehr R, Chen S, Snitow M, Ludwig T, Yagasaki L, Goland R, et al. Generation of Pluripotent Stem Cells From Patients With Type 1 Diabetes. *Proc Natl Acad Sci USA* (2009) 106(37):15768–73. doi: 10.1073/pnas.0906894106
93. Korytnikov R, Nostro MC. Generation of Polyhormonal and Multipotent Pancreatic Progenitor Lineages From Human Pluripotent Stem Cells. *Methods* (2016) 101:56–64. doi: 10.1016/j.ymeth.2015.10.017
94. Zhu C, Ishikami S, Wang P, Zhao H, Li H. Optimal Design and Fabrication of Multichannel Helical Long-Period Fiber Gratings Based on Phase-Only Sampling Method. *Opt Express* (2019) 27(3):2281–91. doi: 10.1364/OE.27.002281

95. Haque M, Lei F, Xiong X, Das JK, Ren X, Fang D, et al. Stem Cell-Derived Tissue-Associated Regulatory T Cells Suppress the Activity of Pathogenic Cells in Autoimmune Diabetes. *JCI Insight* (2019) 4(7):e126471. doi: 10.1172/jci.insight.126471
96. Szot GL, Yadav M, Lang J, Kroon E, Kerr J, Kadoya K, et al. Tolerance Induction and Reversal of Diabetes in Mice Transplanted With Human Embryonic Stem Cell-Derived Pancreatic Endoderm. *Cell Stem Cell* (2015) 16(2):148–57. doi: 10.1016/j.stem.2014.12.001
97. Yan WT, Lu S, Yang YD, Ning WY, Cai Y, Hu XM, et al. Research Trends, Hot Spots and Prospects for Necroptosis in the Field of Neuroscience. *Neural Regener Res* (2021) 16(8):1628–37.
98. Aghazadeh Y, Poon F, Sarangi F, Wong FTM, Khan ST, Sun X, et al. Microvessels Support Engraftment and Functionality of Human Islets and hESC-Derived Pancreatic Progenitors in Diabetes Models. *Cell Stem Cell* (2021) 28(11):1936–1949.e1938. doi: 10.1016/j.stem.2021.08.001
99. Yu G, Zhang M, Gao L, Zhou Y, Qiao L, Yin J, et al. Far-Red Light-Activated Human Islet-Like Designer Cells Enable Sustained Fine-Tuned Secretion of Insulin for Glucose Control. *Mol Ther J Am Soc Gene Ther* (2022) 30(1):341–54. doi: 10.1016/j.ymthe.2021.09.004
100. Castro-Gutierrez R, Alkanani A, Mathews CE, Michels A, Russ HA. Protecting Stem Cell Derived Pancreatic Beta-Like Cells From Diabetogenic T Cell Recognition. *Front Endocrinol* (2021) 12:707881. doi: 10.3389/fendo.2021.707881
101. Shaheen R, Gurlin RE, Gologorsky R, Blaha C, Munnangi P, Santandreu A, et al. Superporous Agarose Scaffolds for Encapsulation of Adult Human Islets and Human Stem-Cell-Derived β Cells for Intravascular Bioartificial Pancreas Applications. *J Biomed materials Res Part A* (2021) 109(12):2438–48. doi: 10.1002/jbm.a.37236
102. Wang X, Maxwell KG, Wang K, Bowers DT, Flanders JA, Liu W, et al. A Nanofibrous Encapsulation Device for Safe Delivery of Insulin-Producing Cells to Treat Type 1 Diabetes. *Sci Trans Med* (2021) 13(596):eabb4601. doi: 10.1126/scitranslmed.abb4601
103. Navabi R, Negahdari B, Hajizadeh-Saffar E, Hajinasrollah M, Jenab Y, Rabbani S, et al. Combined Therapy of Mesenchymal Stem Cells With a GLP-1 Receptor Agonist, Liraglutide, on an Inflammatory-Mediated Diabetic Non-Human Primate Model. *Life Sci* (2021) 276:119374. doi: 10.1016/j.lfs.2021.119374
104. Song L, Gou W, Wang J, Wei H, Lee J, Strange C, et al. Overexpression of Alpha-1 Antitrypsin in Mesenchymal Stromal Cells Improves Their Intrinsic Biological Properties and Therapeutic Effects in Nonobese Diabetic Mice. *Stem Cells Trans Med* (2021) 10(2):320–31. doi: 10.1002/sctm.20-0122
105. Montanucci P, Pescara T, Greco A, Leonardi G, Marini L, Basta G, et al. Co-Microencapsulation of Human Umbilical Cord-Derived Mesenchymal Stem and Pancreatic Islet-Derived Insulin Producing Cells in Experimental Type 1 Diabetes. *Diabetes-Metabolism Res Rev* (2021) 37(2):e3372. doi: 10.1002/dmrr.3372
106. Sarvestani FS, Zare MA, Saki F, Koohpeyma F, Al-Abdullah IH, Azarpira N. The Effect of Human Wharton's Jelly-Derived Mesenchymal Stem Cells on MC4R, NPY, and LEPR Gene Expression Levels in Rats With Streptozotocin-Induced Diabetes. *Iranian J Basic Med Sci* (2020) 23 (2):214–23.
107. Kuppan P, Seeberger K, Kelly S, Rosko M, Adesida A, Pepper AR, et al. Co-Transplantation of Human Adipose-Derived Mesenchymal Stem Cells With Neonatal Porcine Islets Within a Prevascularized Subcutaneous Space Augments the Xenograft Function. *Xenotransplantation* (2020) 27(4): e12581. doi: 10.1111/xen.12581
108. Hashemi SM, Hassan ZM, Hossein-Khannazer N, Pourfathollah AA, Soudi S. Investigating the Route of Administration and Efficacy of Adipose Tissue-Derived Mesenchymal Stem Cells and Conditioned Medium in Type 1 Diabetic Mice. *Inflammopharmacology* (2020) 28(2):585–601. doi: 10.1007/s10787-019-00661-x
109. Su M, Lin Y, He Z, Lai L. Transplantation of MHC-Mismatched Mouse Embryonic Stem Cell-Derived Thymic Epithelial Progenitors and MHC-Matched Bone Marrow Prevents Autoimmune Diabetes. *Stem Cell Res Ther* (2019) 10(1):239. doi: 10.1186/s13287-019-1347-1
110. Ben Nasr M, Tezza S, D'Addio F, Mameli C, Usuelli V, Maestroni A, et al. PD-L1 Genetic Overexpression or Pharmacological Restoration in Hematopoietic Stem and Progenitor Cells Reverses Autoimmune Diabetes. *Sci Trans Med* (2017) 9(416):eaam7543. doi: 10.1126/scitranslmed.aam7543
111. Alagpulinsa DA, Cao JJL, Driscoll RK, Sîrbulescu RF, Penson MFE, Sremac M, et al. Alginate-Microencapsulation of Human Stem Cell-Derived β Cells With CXCL12 Prolongs Their Survival and Function in Immunocompetent Mice Without Systemic Immunosuppression. *Am J Transplant Off J Am Soc Transplant Am Soc Transplant Surgeons* (2019) 19(7):1930–40.
112. Long G, Zhang G, Zhang F, Li M, Ye D, Yang D, et al. Cotransplantation of Mesenchymal Stem Cells and Immature Dendritic Cells Potentiates the Blood Glucose Control of Islet Allografts. *BioMed Res Int* (2017) 2017:4107943. doi: 10.1155/2017/4107943
113. Manzar GS, Kim EM, Zavazava N. Demethylation of Induced Pluripotent Stem Cells From Type 1 Diabetic Patients Enhances Differentiation Into Functional Pancreatic β Cells. *J Biol Chem* (2017) 292(34):14066–79. doi: 10.1074/jbc.M117.784280
114. Daneshmandi S, Karimi MH, Pourfathollah AA. TGF-β Engineered Mesenchymal Stem Cells (TGF-β/MSCs) for Treatment of Type 1 Diabetes (T1D) Mice Model. *Int Immunopharmacol* (2017) 44:191–6. doi: 10.1016/j.intimp.2017.01.019
115. Richardson T, Barner S, Candiello J, Kumta PN, Banerjee I. Capsule Stiffness Regulates the Efficiency of Pancreatic Differentiation of Human Embryonic Stem Cells. *Acta biomaterialia* (2016) 35:153–65. doi: 10.1016/j.actbio.2016.02.025
116. Bal T, Nazli C, Okcu A, Duruksu G, Karaöz E, Kizilel S. Mesenchymal Stem Cells and Ligand Incorporation in Biomimetic Poly(Ethylene Glycol) Hydrogels Significantly Improve Insulin Secretion From Pancreatic Islets. *J Tissue Eng Regenerative Med* (2017) 11(3):694–703. doi: 10.1002/term.1965
117. Yaochite JN, Caliari-Oliveira C, de Souza LE, Neto LS, Palma PV, Covas DT, et al. Therapeutic Efficacy and Biodistribution of Allogeneic Mesenchymal Stem Cells Delivered by Intrasplenic and Intrapancreatic Routes in Streptozotocin-Induced Diabetic Mice. *Stem Cell Res Ther* (2015) 6(1):31. doi: 10.1186/s13287-015-0017-1
118. Araujo DB, Dantas JR, Silva KR, Souto DL, Pereira MFC, Moreira JP, et al. Allogenic Adipose Tissue-Derived Stromal/Stem Cells and Vitamin D Supplementation in Patients With Recent-Onset Type 1 Diabetes Mellitus: A 3-Month Follow-Up Pilot Study. *Front Immunol* (2020) 11:993. doi: 10.3389/fimmu.2020.00993
119. Delgado E, Perez-Basterrechea M, Suarez-Alvarez B, Zhou H, Revuelta EM, Garcia-Gala JM, et al. Modulation of Autoimmune T-Cell Memory by Stem Cell Educator Therapy: Phase 1/2 Clinical Trial. *EBioMedicine* (2015) 2 (12):2024–36. doi: 10.1016/j.ebiom.2015.11.003
120. Zhao Y, Jiang Z, Zhao T, Ye M, Hu C, Yin Z, et al. Reversal of Type 1 Diabetes *via* Islet Beta Cell Regeneration Following Immune Modulation by Cord Blood-Derived Multipotent Stem Cells. *BMC Med* (2012) 10:3. doi: 10.1186/1741-7015-10-3
121. Dantas JR, Araujo DB, Silva KR, Souto DL, de Fatima Carvalho Pereira M, Luiz RR, et al. Adipose Tissue-Derived Stromal/Stem Cells + Cholecalciferol: A Pilot Study in Recent-Onset Type 1 Diabetes Patients. *Arch Endocrinol Metab* (2021) 65(3):342–51. doi: 10.20945/2359-3997000000368

19

Mesenchymal Stem Cell Exosomes as a New Strategy for the Treatment of Diabetes Complications

Jiachao Xiong†, Hao Hu†, Rong Guo, Hui Wang and Hua Jiang**

Department of Plastic Surgery, Shanghai East Hospital, Tongji University School of Medicine, Shanghai, China

***Correspondence:**
Hui Wang
pathology2000@163.com
Hua Jiang
dosjh@126.com

†These authors have contributed equally to this work

Diabetes mellitus (DM) is a metabolic disease, now prevalent worldwide, which is characterized by a relative or absolute lack of insulin secretion leading to chronically increased blood glucose levels. Diabetic patients are often accompanied by multiple macrovascular complications, such as coronary heart disease, hypertension, macrovascular arteriosclerosis, and microvascular complications. Microvascular complications include diabetic kidney injury, diabetic encephalopathy, and diabetic foot, which reduce the quality of life and survival status of patients. Mesenchymal stem cell exosomes (MSC-Exos) possess repair functions similar to MSCs, low immunogenicity, and ease of storage and transport. MSC-Exos have been proven to possess excellent repair effects in repairing various organ damages. This study reviews the application of MSC-Exos in the treatment of DM and its common complications. MSC-Exos may be used as an effective treatment for DM and its complications.

Keywords: diabetes mellitus, complication, microvascular damage, mesenchymal stem cells, exosomes

INTRODUCTION

Diabetes mellitus (DM) is a metabolic disease, now prevalent worldwide, characterized by chronic hyperglycemia, accompanied by disorders of glucose, adipose tissue, and protein metabolism, which are caused by defects in insulin secretion or action (1). The global prevalence of diabetes is continuously rising, perhaps due to social and economic development, changes in lifestyles, and population aging (2). DM is clinically common with type 1 and type 2 diabetes. Currently, type 1 diabetes is common in adolescents and requires lifelong insulin therapy because of the elimination of pancreatic islet β-cells by the immune response, resulting in an absolute lack of insulin secretion. Type 2 diabetes, commonly arising in elderly and obese patients, occurs due to the decreased insulin sensitivity of peripheral tissues, accompanied by defective insulin secretion in pancreatic islet β-cells (3, 4). Patients with type 2 diabetes mainly control blood glucose by taking drugs that increase the sensitivity of insulin target tissues or increase pancreatic islet β-cells, combined with individualized therapy, such as regulating blood lipids, controlling their body weight, and improving their lifestyle (5). Unfortunately, most patients with diabetes have unstable blood glucose levels. Long-term high levels of blood glucose can cause systemic macrovascular and microvascular damage, potentially leading to chronic complications in multiple tissues and organs, such as the eyes, kidneys, nerves, and heart, which is an important cause of cardiorenal failure, blindness, amputation, and even death (6). Therefore, the management of DM and its related chronic complications is particularly important.

Chronic diabetic patients often have multiple complications, such as diabetic kidney injury, diabetic encephalopathy, and diabetic foot, which reduce the quality of life and survival status of patients. Currently, there is no way to eliminate diabetes, and changes in patients' lifestyles combined with drug therapy are the main ways to control blood glucose. However, the therapeutic methods for diabetes complications are insufficient and new methods are urgently required to improve the quality of life and survival rate of patients.

Mesenchymal stem cells (MSCs) are pluripotent progenitor cells that can differentiate into adipocytes, osteoblasts, chondrocytes, and other cell types of mesodermal origin (7). MSCs are characterized by their high self-renewal ability, low immunogenicity, and immune regulation ability, and play an important role in clinical cell therapy. MSCs originate from a wide range of sources and were first isolated from bone marrow (8). Subsequent studies have found that MSCs can be isolated from various human tissues, such as adipose tissue, umbilical cord, synovium, gingiva, menstrual blood, and urine (9, 10). The low immunogenicity of MSCs makes them a good material for transplantation. After transplantation, MSCs can chemoattract to the vicinity of damaged tissues and secrete a variety of growth factors and anti-inflammatory factors to promote the repair of damaged tissues (11). However, MSCs and mesenchymal tumor cells have numerous identical stem gene phenotypes, which strongly suggests that some early tumor cells are derived from MSCs (12). In addition, previous studies have found that MSCs promote tumor formation through vascularization, immune regulation, and the promotion of tumor interstitial remodeling (13, 14). These factors have greatly restricted the clinical applications of MSCs. Previous studies have found that exosomes isolated from MSC culture medium possess a repair function similar to MSCs and no risk of tumor formation (15, 16).

BIOLOGICAL FUNCTIONS OF MESENCHYMAL STEM CELL EXOSOMES

MSC exosomes (MSC-Exos) are extracellular vesicles between 30–150 nm in diameter that have the same lipid bilayer structure as the cell membrane (17). Additionally, MSC-Exos possess more advantages than MSCs, such as lower immunogenicity, high stability, and easy storage (18). MSC-Exos contain multiple biologically active substances, such as lipids, proteins, and RNAs that can regulate the biological activities of target cells *via* membrane fusion or endocytosis (19, 20). Guo et al. (21) injected MSC-Exos into rats with spinal cord injury by intranasal administration and found that MSC-Exos greatly enhanced axon growth and angiogenesis, reduced the proliferation of microglia and astrocytes, and significantly promoted the repair of spinal cord injury. Moreover, MSC-Exos are rich in the C-C motif chemokine receptor-2 that promotes ischemia-reperfusion kidney injury healing by inhibiting macrophage function (22). MSC-Exos have shown excellent repair effects in various tissue injuries, such as liver, cardiovascular, and skin wounds that involve mechanisms of angiogenesis, regulation of cell proliferation, and immune regulation (17, 23, 24). The use of MSC-Exos, as an alternative to MSCs, has become a new strategy for tissue regeneration (**Figure 1**).

ROLE OF MSC-Exos IN DM

Both type 1 diabetes and type 2 diabetes are accompanied by a certain degree of pancreatic islet damage. The progression of DM can be delayed by promoting islet regeneration and improving the sensitivity of insulin target tissues, which is a potential new treatment for diabetic patients (25) (**Table 1**).

Insulin deficiency in patients with type 1 diabetes occurs due to the autoimmune destruction of islet tissue, and MSC-Exos have the unique ability of immune regulation and can repair pancreatic islet tissue to alleviate DM. MSC-Exos can chemoattract pancreatic tissue and promote the regeneration of pancreatic β-cells and insulin secretion through the pancreatic and duodenal homeobox 1 pathway (26). Accumulating evidence has reported that MSC-Exos have a great therapeutic and regenerative effect on islet injury in type 1 diabetes by upregulating the levels of regulatory T cells, interleukin (IL)-4, IL-10, and transforming growth factor β (TGF-β), while downregulating IL-17 and interferon γ, ultimately improving the autoimmune response of diabetic mice and the regeneration of islets (27, 28).

Glucose transporter 4 transfer from intracellular vesicles to the plasma membrane is the final step of insulin signaling, in which the phosphorylation of insulin receptor substrate 1 and protein kinase B is an essential step, while the phosphorylation in type 2 diabetes patients is often blocked, leading to insulin resistance (33). Meanwhile, adenosine 5'-monophosphate (AMP)-activated protein kinase (AMPK) plays an important role in the regulation of glucose and lipid metabolism in skeletal muscle and liver, and is regarded as an important target to reverse type 2 diabetes-related metabolic abnormalities (34, 35). MSC-Exos can activate autophagy through the AMPK pathway or restore the phosphorylation of insulin receptor substrate 1 and protein kinase B, which contribute to muscle glucose transporter 4 expression to reverse peripheral insulin resistance; it also relieves the apoptosis of islet β-cells and restores the insulin secretion function of type 2 diabetes (30). Patients with type 2 diabetes are commonly associated with obesity (36, 37), which induces the M2 phenotype of macrophages through the transactivation of arginase-1 that promotes hepatic glucose and lipid metabolic balance to reduce obesity (29, 31, 32).

ROLE OF MSC-Exos IN DM COMPLICATION

DM is the ninth most common cause of death globally, and most diabetic patients have at least one complication (38). A large

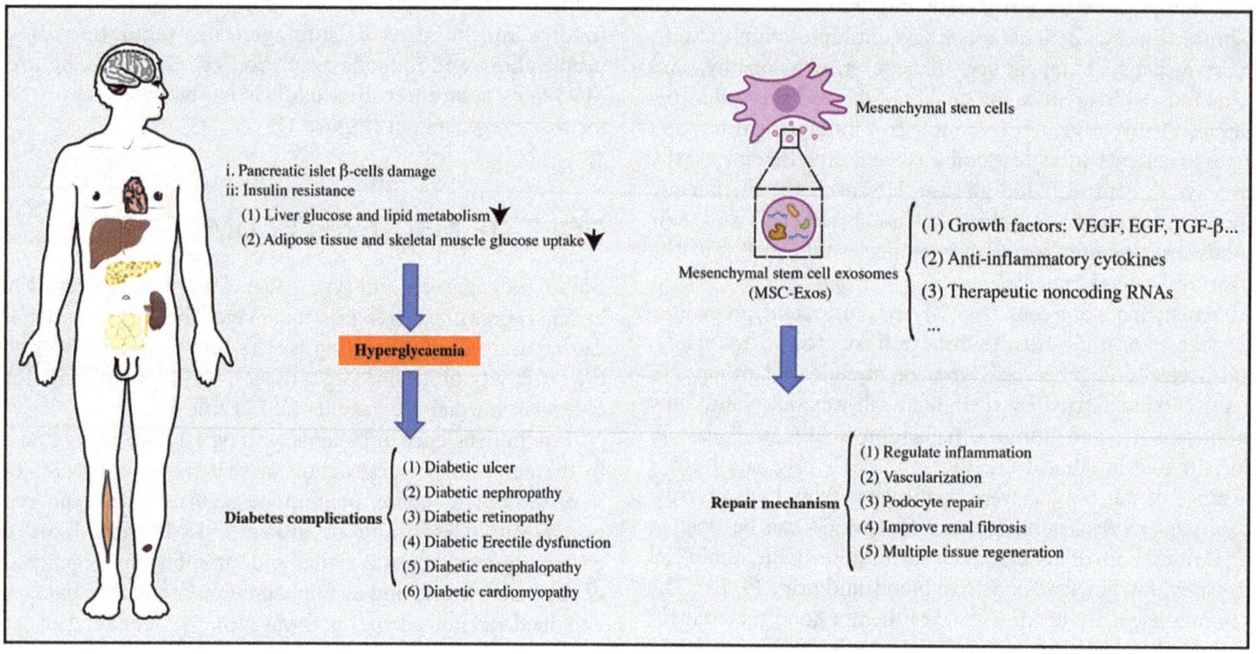

FIGURE 1 | Diabetes/diabetes complications and mesenchymal stem cell exosomes repair.

observational study showed that 50% of patients with type 2 diabetes had microvascular complications, and 27% were associated with macrovascular complications (39). The development of common DM complications, including diabetic kidney injury, diabetic encephalopathy, and diabetic chronic ulcers, determines the quality of life of patients. In recent years, MSC-Exos have played a substantial role in the treatment of diabetic complications, providing a new approach for its treatment (**Table 2**).

Diabetes Ulcer

Wound healing consists of four overlapping and dynamic processes of hemostasis, inflammation, proliferation, and remodeling, and the obstruction of one of these steps leads to delayed healing (78). The healing of diabetic ulcers (DUs) is often blocked by factors, such as infection, peripheral neurovascular disease, hypoxia, and neuropeptide conduction, which increase the patient's medical expenditure and prolong the hospitalization period. DUs, one of the most serious complications of DM, often occur on the feet of patients; therefore, they are also called diabetic foot. The incidence of DUs in patients with type 1 diabetes is approximately 20%, which is similar to that in patients with type 2 diabetes (79). In addition, DUs are an important cause of disability, and more than 15% of patients will undergo amputation (80). Despite the progress made in wound care, the United States Centers for Disease Control and Prevention reported that DUs are still the main cause of non-traumatic amputation (81). Therefore, it is necessary to improve the care and treatment of DUs. Similar to MSCs, MSC-Exos promote the healing of DUs by regulating the inflammatory microenvironment of the wound, promoting vascularization and anti-oxidant apoptosis (40).

Regulate Inflammation

Previous studies have shown that M1 macrophages can produce pro-inflammatory cytokines, such as IL-1β and tumor necrosis factor-alpha (TNF-α), while M2 macrophages can produce anti-inflammatory cytokines, such as TGF-β and VEGF (82, 83). Therefore, macrophage polarization plays a vital role in the regulation of the inflammatory response (84). Patients with

TABLE 1 | Mesenchymal stem cells (MSC) used in the treatment of diabetes mellitus (DM).

DM type	MSC types	Effect	Ref.
Type 1 diabetes	Menstrual blood MSC	Controlling blood glucose, increasing insulin level and promoting islet regeneration	(26)
	Adipose derived MSC		(27)
	Bone marrow MSC		(28)
Type 2 diabetes	Adipose derived MSC	Improve insulin resistance, improve liver sugar storage capacity, and promote islet regeneration	(29)
	Umbilical cord MSC		(30)
	Bone marrow MSC		(31)
	Umbilical cord MSC		(32)

TABLE 2 | MSC used in the treatment of DM complication.

DM complication	MSC types	Effect/involved noncoding RNA	Ref.
Diabetes wound	Bone marrow MSC	Vascularization	(40)
	Adipose derived MSC	Vascularization	(41)
	Adipose derived MSC	Vascularization/miR-21-5p	(42)
	Bone marrow MSC	Vascularization	(43)
	Bone marrow MSC	Regulate inflammation/lncRNA H19	(44)
	Adipose derived MSC	Vascularization	(45)
	Synovium MSC	Vascularization/miR-126-3p	(46)
	Urine MSC	Vascularization	(47)
	Adipose derived MSC	Vascularization and regulate inflammation	(48)
	Induced pluripotent stem cell	Vascularization	(49)
	menstrual blood-derived MSC	Vascularization and regulate inflammation	(50)
	Adipose derived MSC	Vascularization	(51)
	Adipose derived MSC	Vascularization	(52)
	Adipose derived MSC	Vascularization/mmu_circ_0000250	(53)
	Bone marrow MSC	Vascularization and regulate inflammation	(54)
	Bone marrow MSC	Vascularization/miR-221-3p	(55)
	Umbilical cord MSC	Vascularization	(56)
Diabetic nephropathy	Adipose derived MSC	Podocyte repair/miRNA-215-5p	(57)
	Urine MSC	Podocyte repair	(58)
	Bone marrow MSC	Anti-fibrosis and promote renal function recovery	(59)
	Bone marrow MSC	Anti-fibrosis and promote renal function recovery	(60)
	Adipose derived MSC	Podocyte repair/miR-486	(61)
	Urine MSC	Podocyte repair/miR-16-5	(62)
	Umbilical cord MSC	Reduce kidney inflammation and improve kidney function	(63)
	Umbilical cord MSC	Reduce kidney inflammation and improve kidney function	(64)
Diabetic retinopathy	Adipose derived MSC	Retinal repair/miR-222	(65)
	Umbilical cord MSC	Retinal repair and regulation of inflammation/miR-126	(66)
Erectile dysfunction	Adipose derived MSC	Vascularization and anti-apoptosis	(67)
	Adipose derived MSC	Promote angiogenesis and anti-fibrosis/miR-126, miR-130a, miR-132, miR-let7b, miR-let7c	(68)
	Adipose derived MSC	Vascularization and anti-inflammatory	(69)
	Bone marrow MSC	Vascularization and anti-inflammatory/miR-21-5p	(70)
Cognitive dysfunction	Bone marrow MSC	Nerve repair	(71)
	Bone marrow MSC	Nerve repair and anti-inflammatory/miR-146a	(72)
	Bone marrow MSC	Nerve repair	(73)
Diabetic stroke	Bone marrow MSC	Nerve repair/miR-9	(74)
	Bone marrow MSC	Nerve repair/miR-145	(75)
Submandibular gland dysfunction	Bone marrow MSC	Salivary gland function repair	(76)
Diabetic cardiomyopathy	MSC	Reduce myocardial injury and fibrosis	(77)

diabetes are in a state of chronic inflammation, and the number of pro-inflammatory M1 macrophages in the damaged wound is significantly higher than that in anti-inflammatory M2 macrophages (85). Excessive polarization of M1 macrophages will inevitably increase the secretion of IL-1β, TNF-α, and other pro-inflammatory cytokines that prolong the inflammatory period of wound repair, which further leads to wound collagen regeneration and scar hyperplasia (86–88).

MSC-Exos increase the M2/M1 polarization ratio, which reduces the inflammation of DUs wounds and promotes healing (50). MSC-Exos inhibit the activation of the phosphatidylinositol 3-kinase/protein kinase B (PI3K/AKT) pathway and weaken the phosphorylation of AKT by promoting the expression of pentaerythritol tetranitrate (PETN) to regulate the M1/M2 polarization ratio. This inhibits the inflammatory response of wounds in diabetic rats and accelerates the rapid transition from the inflammation phase to the tissue regeneration phase (44, 54). It can be seen that the PETN/PI3K/AKT axis is closely related to macrophage polarization, which regulates the local inflammatory response and enhances the proliferation and migration ability of fibroblasts at the injury site that promotes the healing of DU wounds.

Vascularization

The degree of wound vascularization determines the healing rate and remodeling of the wound. Neovascularization occurs through the wound repair process. New capillaries are formed in granulation tissue after three days of injury; they grow rapidly and are interwoven into a network to provide oxygen and nutrients to the cells in the damaged area (89). In the process of angiogenesis, pro-angiogenic factors play a role in maintaining vascular growth. However, DM significantly reduces the level of pro-angiogenic factors, leading to blocked angiogenesis and delayed wound healing (90).

MSC-Exos are rich in multiple growth factors and therapeutic noncoding RNAs that can effectively promote the vascularization of skin wounds and are safer and more stable compared to cell therapy (91, 92). Studies have demonstrated that MSC-Exos are rich in circular RNAs, long noncoding RNAs, and microRNAs (miRNAs) that can regulate the expression of related repair genes to promote the vascularization and healing of DU wounds. Exosomes derived from mmu_circ_0000250-modified adipose-derived mesenchymal stem cells were found to promote the activation of autophagy to inhibit cell apoptosis in a high-glucose environment, promote the expression of SIRT1 through miR-128-3p adsorption, promote wound vascularization, and ultimately accelerate the healing of DM wounds (53). Yu et al. (55) found that MSC-Exos can promote the angiogenesis ability of endothelial cells; the expression of VEGF upregulates the expression of miR-221-3p and activates the AKT/endothelial nitric oxide synthase (eNOS) pathway to promote the vascularization of DM wounds. Moreover, MSC-Exos can be used as a good carrier of therapeutic noncoding RNA for the healing of DU wounds. Studies have applied miR-126-3p or miR-21-5p engineered MSC-Exos to diabetic wounds and found that they can activate the PI3K/AKT, mitogen-activated protein kinase (MAPK)/extracellular signal-regulated kinase (ERK), and Wnt/β-catenin pathways, which are closely related to blood vessel formation, to promote the vascularization and re-epithelialization of DM wounds and enhance the efficiency of healing (42, 46). In addition, compared with MSC-Exos alone, combining MSC-Exos with hydrogel materials to improve the survival rate of MSC-Exos applied to DU wounds has shown better vascularization and wound closure rates (41, 45, 51). Thus, MSC-Exos are expected to become a new biological agent for the treatment of DUs.

Diabetic Nephropathy

Diabetic nephropathy (DN) is a severe type of microvascular kidney damage caused by DM that eventually develops into end-stage renal disease (ESRD), which is mostly characterized by a persistent protein or persistent glomerular filtration rate reduction. Once the course of DM exceeds 20 years, the incidence of DN can be as high as 35%, and approximately 8% of patients will progress to ESRD. Epidemiological statistics have shown that DM causes more than 40% of ESRD cases in the United States, and diabetes-related chronic kidney disease has also become the main cause of ESRD in hospitalized patients in China (38). Early DN manifests as glomerular hyperfiltration and microalbuminuria. As the disease progresses, a series of pathological changes appear in the kidneys, such as glomerular basement membrane thickening, mesangial expansion, glomerular sclerosis, podocyte loss, and renal interstitial fibrosis (93). The glomerular filtration rate gradually decreases and eventually develops into uremia (94). The various growth factors and therapeutic noncoding RNAs contained in MSC-Exos also have significant effects on improving renal function, delaying renal fibrosis, and repairing podocyte function, and are expected to become a new tool for the treatment of DN.

Podocyte Repair

Podocytes are an important part of the glomerular filtration barrier and are known to maintain barrier function together with vascular endothelial cells (95, 96). Studies have shown that hyperglycemia can induce podocyte apoptosis, which reduces the number of podocytes, resulting in proteinuria (97). Hence, preventing podocyte damage plays a vital role in the treatment of DN.

Studies have revealed that MSC-Exos have a significant protective effect against acute and chronic kidney injury (98, 99). MSC-Exos can effectively reduce podocyte damage induced by high glucose levels by delivering therapeutic miRNAs. MSC-Exos deliver therapeutic miRNAs, such as miR-215-5p, miR-486, miR-150, miR-134, and miR-16-5p to podocytes. This directly targets small mothers against decapentaplegic (Smad)-1 to weaken mammalian target of rapamycin (mTOR)-mediated autophagy or cooperates with vascular endothelial growth factor A (VEGFA) to protect podocytes from the effects of hyperglycemia, and improve the proliferation and migration of podocytes to protect renal function (57, 61, 62). In addition, exosomes from urine-derived stem cells were injected into a diabetic rat model through the tail vein, and it was observed that the overexpression of caspase-3 was inhibited, podocyte apoptosis was reduced, the proliferation of renal tubular endothelial cells was promoted, and the urine output and urinary microprotein excretion in DN rats were effectively reduced (58). These results indicate that MSC-Exos can alleviate podocyte injury and improve renal function recovery.

Improvement of Renal Fibrosis

Renal fibrosis is the central link in DN and ultimately leads to irreversible kidney damage. Renal fibrosis is closely related to inflammatory cell infiltration, epithelial-endothelial mesenchymal transition, and myofibroblast transdifferentiation. Inflammatory cells secrete multiple inflammatory cytokines, such as IL-1β, IL-6, TNF-α, and TGF-β1 (100, 101). In the early stage of DN, TGF-β1, an important inflammatory cytokine for renal fibrosis, interferes with the cell cycle and causes renal hypertrophy (102). Subsequently, TGF-β1 can activate the downstream Smad2/3, MAPKs, PI3K/AKT, RhoA, and Wnt/β-catenin signaling pathways to trigger the synthesis of the extracellular matrix and myofibroblast transdifferentiation, thereby accelerating the process of renal fibrosis (63).

Studies have reported that repeated administration of MSC-Exos to diabetic animal models can ameliorate glomerular hypertrophy, basement membrane thickening, and fibrosis, to reduce the progression of DN (59, 63, 64). MSC-Exos inhibit the secretion of TGF-β1 to reduce epithelial-endothelial mesenchymal transition and block the proliferation of mesangial cells induced by the MAPKs and PI3K/AKT/mTOR pathways, thus alleviating renal fibrosis (60). MSC-Exos contain growth cytokines, such as epidermal growth factor, fibroblast growth factor, hepatocyte growth factor, and VEGF, which have anti-inflammatory and anti-fibrotic effects. It can downregulate the expression of fibroblast markers, such as alpha-smooth muscle actin (α-SMA) and collagen IV in renal tubules, and improve renal fibrosis in DN rats.

Diabetic Retinopathy

Diabetic retinopathy (DR) is an important cause of vision loss in the elderly. Hyperglycemia can cause multiple pathological changes in the retinal neurovascular unit, including optic nerve inflammation, glial hyperplasia, abnormal vascular permeability, and blood-retinal barrier decomposition, eventually leading to retinal fibrosis, vision loss, and blindness in severe cases (103–105). Epidemiology demonstrates that the prevalence of DR is as high as 28% in the United States and 25% in Asian countries (38). Thus, it is important to develop effective treatments for DR.

Previous studies have observed the activation of the NOD-like receptor family pyrin domain containing 3 (NLRP3) inflammasome, which leads to the maturation of proinflammatory cytokines, such as IL-1β, IL-18, and caspase-1 in the retinas of DM rats to mediate the apoptosis of retinal cells (106, 107). Previous studies (66) have found that MSC-Exos significantly downregulated the expression of high-mobility group box 1 (HMGB1), NLRP3, and NF-kappaB/P65 protein in DR rats, inhibiting the production of various inflammatory cytokines and reducing retinal vascular endothelial injury. The use of MSC-Exos rich in therapeutic noncoding RNAs may become a new method for the treatment and prevention of DR. MSC-Exos with overexpressed miR-126 was used to more effectively inhibit the activation of the HMGB1 signaling pathway and improve the inflammatory response in DR rats. In addition, Safwat et al. (65) found that MSC-Exos can deliver miR-222 to retinal cells and regulate signal transducer and activator of transcription 5A (STAT5) protein expression to inhibit neovascularization in advanced DR, which promotes retinal regeneration.

Diabetic Erectile Dysfunction

Erectile dysfunction (ED) is a common chronic complication of DM and is defined as the inability to achieve or maintain an adequate erection during sexual intercourse (108). Epidemiology demonstrates that the prevalence of ED in men with diabetes is as high as 70% and is three times that of nondiabetic men (109). Previous studies reported that hyperglycemia caused VEGF signaling transduction, the synthesis of neuronal nitric oxide synthase and endothelial nitric oxide synthase to be blocked, and the level of oxygen free radicals to increase, which resulted in increased apoptosis of sponge endothelial cells and smooth muscle cells in ED patients (110, 111). However, the efficacy of oral phosphodiesterase type 5 inhibitors was unsatisfactory, and more effective methods need to be developed (112).

MSC-Exos injection therapy can significantly increase the ratio of intracavernosal pressure to mean arterial pressure and upregulate the expression of atrial natriuretic peptide, brain natriuretic peptide, and neuronal nitric oxide synthase to promote the recovery of erectile function in DM rats (69). MSC-Exos contain pro-angiogenic miRNAs (miR-126, miR-130a, and miR-132) and anti-fibrotic miRNAs (miR-let7b and miR-let7c), which may increase the proliferation of vascular endothelial cells and smooth muscle cells in the cavernous body by increasing the proliferation of vascular endothelial cells and the expression of smooth muscle markers (α-SMA) and anti-apoptotic proteins (Bcl-2) to alleviate ED (67, 68). In addition, MSC-Exos reduced apoptosis and promoted the proliferation of cavernous smooth muscle cells by delivering miR-21-5p to target programmed cell death 4, and significantly improved erectile function and smooth muscle density in DM rats (70).

Diabetic Cardiomyopathy

Diabetic cardiomyopathy (DC) is the systolic and diastolic dysfunction caused by DM, which eventually leads to heart failure. Coronary artery disease and ischemic cardiomyopathy are the main contributors to cardiac death in diabetic patients (113). Early DC manifests as impaired diastolic function, but no significant changes in systolic function (normal ejection fraction). As the disease progresses, cardiac systolic function is affected by reduced ejection fraction, and the pathological manifestations include left ventricular hypertrophy and interstitial fibrosis (114). The heart is a terminally differentiated organ, and it is difficult to regenerate cardiomyocytes after damage (115). Exosomes derived from MSC therapy may be a new approach for DC repair.

MSC-Exos have shown good therapeutic effects against cardiac ischemic diseases. Exosomes derived from bone marrow MSCs were directly injected into rats with myocardial infarction, which increased the expression of the miR-19a/AKT/ERK axis by inhibiting PTEN; thus, myocardial cell apoptosis was reduced with the significant recovery of myocardial contractile function and the reduction of infarct size (116). Notably, MSC-Exos may have a protective effect against myocardial injury. Lai et al. (117) used human embryonic stem cell-derived MSC-Exos perfusion buffer in a mouse model of myocardial ischemia-reperfusion and observed that the activation of the AKT/ERK pathway and the inhibition of the c-Jun NH2-terminal kinase pro-apoptotic pathway in the myocardial tissue was accompanied by a significant improvement in cardiac function at 1 h, 48 h, and 28 days after intervention. MSC-Exos were injected into the tail vein of a rat model of diabetic myocardial injury and it was observed that MSC-Exos inhibited the TGF-β1/Smad2 signaling pathway to improve myocardial injury and fibrosis induced by DM (77). At present, there are few studies on MSC-Exos used in DC, but MSC-Exos has a powerful regulating and repairing effect on myocardial injury; therefore, the repair of DCs has great application prospects.

Other Rare Complications

Other rare complications of DM include cognitive impairment, stroke, and submandibular gland dysfunction. Abnormal blood glucose metabolism results in central nervous system neuron damage, decreased hippocampal synaptic plasticity, astrocyte foot swelling, which leads to cognitive dysfunction, various vascular diseases and increased vascular permeability, which leads to ischemic stroke, and salivary gland function damage, which leads to salivary quality reduction and gland function disorder (71, 76). There are few reports on the application of MSC-Exos to the above-mentioned complications, but some curative effects have been achieved.

Studies have reported that MSC-Exos can act on damaged neurons and astrocytes to promote their repair and reverse cognitive dysfunction (72, 73). MSC-Exos can significantly reduce the expression of ATP-binding cassette A1 and type 1 insulin-like growth factor receptor by increasing the expression of miR-145 or reducing the expression of miR-9, and increasing the neurorepair and cognitive function improvement of stroke DM rats (74, 75). AbuBakr et al. (76) found that MSC-Exos inhibited the TGFβ signaling pathway through Smad2 and Smad3 to inhibit the damage of salivary glands caused by DM, which was manifested by the reduction of serum amylase and salivary IgA, which led to the restoration of salivary gland function.

CONCLUSION

DM is a metabolic disorder prevalent worldwide, and its incidence is increasing annually. Long-term hyperglycemia causes many chronic complications with regards to microvascular disease, and there is a lack of effective treatment methods. MSC-Exos have a repair function similar to MSCs, but do not have the shortcomings of MSCs in terms of promoting tumor formation and the difficulties associated with storage. MSC-Exos are rich in a variety of growth cytokines, repair proteins, and therapeutic noncoding RNAs, which can promote the repair of organs damaged by DM and its complications by regulating inflammation, vascularization, and anti-apoptotic mechanisms. The use of MSC-Exos may be an effective treatment strategy for DM and its complications.

AUTHOR CONTRIBUTIONS

HJ and HW designed the study. JX and HH performed and drafted the manuscript. RG revised manuscript. All authors contributed to the article and approved the submitted version.

REFERENCES

1. Cowie CC, Rust KF, Byrd-Holt DD, Eberhardt MS, Flegal KM, Engelgau MM, et al. Prevalence of Diabetes and Impaired Fasting Glucose in Adults in the U.S. Population: National Health And Nutrition Examination Survey 1999-2002. *Diabetes Care* (2006) 29:1263–8. doi: 10.2337/dc06-0062
2. Scott A, Chambers D, Goyder E, O'Cathain A. Socioeconomic Inequalities in Mortality, Morbidity and Diabetes Management for Adults With Type 1 Diabetes: A Systematic Review. *PloS One* (2017) 12:e0177210. doi: 10.1371/journal.pone.0177210
3. Atkinson MA, Eisenbarth GS, Michels AW. Type 1 Diabetes. *Lancet (London England)* (2014) 383:69–82. doi: 10.1016/S0140-6736(13)60591-7
4. Kahn SE, Cooper ME, Del Prato S. Pathophysiology and Treatment of Type 2 Diabetes: Perspectives on the Past, Present, and Future. *Lancet (London England)* (2014) 383:1068–83. doi: 10.1016/S0140-6736(13)62154-6
5. Regazzi R. MicroRNAs as Therapeutic Targets for the Treatment of Diabetes Mellitus and its Complications. *Expert Opin Ther Targets* (2018) 22:153–60. doi: 10.1080/14728222.2018.1420168
6. Forbes JM, Cooper ME. Mechanisms of Diabetic Complications. *Physiol Rev* (2013) 93:137–88. doi: 10.1152/physrev.00045.2011
7. Kim N, Cho SG. Clinical Applications of Mesenchymal Stem Cells. *Korean J Internal Med* (2013) 28:387–402. doi: 10.3904/kjim.2013.28.4.387
8. Vining KH, Mooney DJ. Mechanical Forces Direct Stem Cell Behaviour in Development and Regeneration. *Nat Rev Mol Cell Biol* (2017) 18:728–42. doi: 10.1038/nrm.2017.108
9. Kasoju N, Wang H, Zhang B, George JH, Gao S, Triffitt JT, et al. Transcriptomics of Human Multipotent Mesenchymal Stromal Cells: Retrospective Analysis and Future Prospects. *Biotechnol Adv* (2017) 35:407–18. doi: 10.1016/j.biotechadv.2017.04.005
10. Packer M. The Alchemist's Nightmare: Might Mesenchymal Stem Cells That Are Recruited to Repair the Injured Heart Be Transformed Into Fibroblasts Rather Than Cardiomyocytes? *Circulation* (2018) 137:2068–73. doi: 10.1161/CIRCULATIONAHA.117.032190
11. Jadalannagari S, Aljitawi OS. Ectodermal Differentiation of Wharton's Jelly Mesenchymal Stem Cells for Tissue Engineering and Regenerative Medicine Applications. *Tissue Eng Part B Rev* (2015) 21:314–22. doi: 10.1089/ten.teb.2014.0404
12. Galiè M, Konstantinidou G, Peroni D, Scambi I, Marchini C, Lisi V, et al. Mesenchymal Stem Cells Share Molecular Signature With Mesenchymal Tumor Cells and Favor Early Tumor Growth in Syngeneic Mice. *Oncogene* (2008) 27:2542–51. doi: 10.1038/sj.onc.1210920
13. Yu JM, Jun ES, Bae YC, Jung JS. Mesenchymal Stem Cells Derived From Human Adipose Tissues Favor Tumor Cell Growth In Vivo. *Stem Cells Dev* (2008) 17:463–73. doi: 10.1089/scd.2007.0181
14. Bagley RG, Weber W, Rouleau C, Yao M, Honma N, Kataoka S, et al. Human Mesenchymal Stem Cells From Bone Marrow Express Tumor Endothelial and Stromal Markers. *Int J Oncol* (2009) 34:619–27. doi: 10.3892/ijo_00000187
15. Cosenza S, Toupet K, Maumus M, Luz-Crawford P, Blanc-Brude O, Jorgensen C, et al. Mesenchymal Stem Cells-Derived Exosomes are More Immunosuppressive Than Microparticles in Inflammatory Arthritis. *Theranostics* (2018) 8:1399–410. doi: 10.7150/thno.21072
16. Timmers L, Lim SK, Arslan F, Armstrong JS, Hoefer IE, Doevendans PA, et al. Reduction of Myocardial Infarct Size by Human Mesenchymal Stem Cell Conditioned Medium. *Stem Cell Res* (2007) 1:129–37. doi: 10.1016/j.scr.2008.02.002
17. Zhang W, Bai X, Zhao B, Li Y, Zhang Y, Li Z, et al. Cell-Free Therapy Based on Adipose Tissue Stem Cell-Derived Exosomes Promotes Wound Healing *Via* the PI3K/Akt Signaling Pathway. *Exp Cell Res* (2018) 370:333–42. doi: 10.1016/j.yexcr.2018.06.035
18. Zhang ZG, Buller B, Chopp M. Exosomes - Beyond Stem Cells for Restorative Therapy in Stroke and Neurological Injury. *Nat Rev Neurol* (2019) 15:193–203. doi: 10.1038/s41582-018-0126-4
19. Raik S, Kumar A, Bhattacharyya S. Insights Into Cell-Free Therapeutic Approach: Role of Stem Cell "Soup-Ernatant". *Biotechnol Appl Biochem* (2018) 65:104–18. doi: 10.1002/bab.1561
20. Baglio SR, Rooijers K, Koppers-Lalic D, Verweij FJ, Pérez Lanzón M, Zini N, et al. Human Bone Marrow- and Adipose-Mesenchymal Stem Cells Secrete Exosomes Enriched in Distinctive miRNA and tRNA Species. *Stem Cell Res Ther* (2015) 6:127. doi: 10.1186/s13287-015-0116-z
21. Guo S, Perets N, Betzer O, Ben-Shaul S, Sheinin A, Michaelevski I, et al. Intranasal Delivery of Mesenchymal Stem Cell Derived Exosomes Loaded With Phosphatase and Tensin Homolog Sirna Repairs Complete Spinal Cord Injury. *ACS Nano* (2019) 13:10015–28. doi: 10.1021/acsnano.9b01892
22. Shen B, Liu J, Zhang F, Wang Y, Qin Y, Zhou Z, et al. Ccr2 Positive Exosome Released by Mesenchymal Stem Cells Suppresses Macrophage Functions and Alleviates Ischemia/Reperfusion-Induced Renal Injury. *Stem Cells Int* (2016) 2016:1240301. doi: 10.1155/2016/1240301
23. Lopez-Verrilli MA, Caviedes A, Cabrera A, Sandoval S, Wyneken U, Khoury M. Mesenchymal Stem Cell-Derived Exosomes From Different Sources Selectively Promote Neuritic Outgrowth. *Neuroscience* (2016) 320:129–39. doi: 10.1016/j.neuroscience.2016.01.061
24. Liang X, Ding Y, Zhang Y, Tse HF, Lian Q. Paracrine Mechanisms of Mesenchymal Stem Cell-Based Therapy: Current Status and Perspectives. *Cell Transplant* (2014) 23:1045–59. doi: 10.3727/096368913X667709
25. Guay C, Regazzi R. Exosomes as New Players in Metabolic Organ Cross-Talk. *Diabetes Obes Metab* (2017) 19 Suppl 1:137–46. doi: 10.1111/dom.13027
26. Mahdipour E, Salmasi Z, Sabeti N. Potential of Stem Cell-Derived Exosomes to Regenerate β Islets Through Pdx-1 Dependent Mechanism in a Rat Model of Type 1 Diabetes. *J Cell Physiol* (2019) 234:20310–21. doi: 10.1002/jcp.28631
27. Nojehdehi S, Soudi S, Hesampour A, Rasouli S, Soleimani M, Hashemi SM. Immunomodulatory Effects of Mesenchymal Stem Cell-Derived Exosomes on Experimental Type-1 Autoimmune Diabetes. *J Cell Biochem* (2018) 119:9433–43. doi: 10.1002/jcb.27260

28. Sabry D, Marzouk S, Zakaria R, Ibrahim HA, Samir M. The Effect of Exosomes Derived From Mesenchymal Stem Cells in the Treatment of Induced Type 1 Diabetes Mellitus in Rats. *Biotechnol Lett* (2020) 42:1597–610. doi: 10.1007/s10529-020-02908-y
29. Zhao H, Shang Q, Pan Z, Bai Y, Li Z, Zhang H, et al. Exosomes From Adipose-Derived Stem Cells Attenuate Adipose Inflammation and Obesity Through Polarizing M2 Macrophages and Beiging in White Adipose Tissue. *Diabetes* (2018) 67:235–47. doi: 10.2337/db17-0356
30. Sun Y, Shi H, Yin S, Ji C, Zhang X, Zhang B, et al. Human Mesenchymal Stem Cell Derived Exosomes Alleviate Type 2 Diabetes Mellitus by Reversing Peripheral Insulin Resistance and Relieving β-Cell Destruction. *ACS Nano* (2018) 12:7613–28. doi: 10.1021/acsnano.7b07643
31. Su T, Xiao Y, Xiao Y, Guo Q, Li C, Huang Y, et al. Bone Marrow Mesenchymal Stem Cells-Derived Exosomal Mir-29b-3p Regulates Aging-Associated Insulin Resistance. *ACS Nano* (2019) 13:2450–62. doi: 10.1021/acsnano.8b09375
32. He Q, Wang L, Zhao R, Yan F, Sha S, Cui C, et al. Mesenchymal Stem Cell-Derived Exosomes Exert Ameliorative Effects in Type 2 Diabetes by Improving Hepatic Glucose and Lipid Metabolism *Via* Enhancing Autophagy. *Stem Cell Res Ther* (2020) 11:223. doi: 10.1186/s13287-020-01731-6
33. Lin HV, Ren H, Samuel VT, Lee HY, Lu TY, Shulman GI, et al. Diabetes in Mice With Selective Impairment of Insulin Action in Glut4-expressing Tissues. *Diabetes* (2011) 60:700–9. doi: 10.2337/db10-1056
34. Navik U, Sheth VG, Kabeer SW, Tikoo K. Dietary Supplementation of Methyl Donor L-Methionine Alters Epigenetic Modification in Type 2 Diabetes. *Mol Nutr Food Res* (2019) 63:e1801401. doi: 10.1002/mnfr.201801401
35. Zhang BB, Zhou G, Li C. AMPK: An Emerging Drug Target for Diabetes and the Metabolic Syndrome. *Cell Metab* (2009) 9:407–16. doi: 10.1016/j.cmet.2009.03.012
36. Gilleron J, Bouget G, Ivanov S, Meziat C, Ceppo F, Vergoni B, et al. Rab4b Deficiency in T Cells Promotes Adipose Treg/Th17 Imbalance, Adipose Tissue Dysfunction, and Insulin Resistance. *Cell Rep* (2018) 25:3329–3341.e3325. doi: 10.1016/j.celrep.2018.11.083
37. Sud N, Zhang H, Pan K, Cheng X, Cui J, Su Q. Aberrant Expression of microRNA Induced by High-Fructose Diet: Implications in the Pathogenesis of Hyperlipidemia and Hepatic Insulin Resistance. *J Nutr Biochem* (2017) 43:125–31. doi: 10.1016/j.jnutbio.2017.02.003
38. Zheng Y, Ley SH, Hu FB. Global Aetiology and Epidemiology of Type 2 Diabetes Mellitus and its Complications. *Nat Rev Endocrinol* (2018) 14:88–98. doi: 10.1038/nrendo.2017.151
39. Litwak L, Goh SY, Hussein Z, Malek R, Prusty V, Khamseh ME. Prevalence of Diabetes Complications in People With Type 2 Diabetes Mellitus and its Association With Baseline Characteristics in the Multinational A1chieve Study. *Diabetol Metab Syndr* (2013) 5:57. doi: 10.1186/1758-5996-5-57
40. Ding J, Wang X, Chen B, Zhang J, Xu J. Exosomes Derived From Human Bone Marrow Mesenchymal Stem Cells Stimulated by Deferoxamine Accelerate Cutaneous Wound Healing by Promoting Angiogenesis. *BioMed Res Int* (2019) 2019:9742765. doi: 10.1155/2019/9742765
41. Shafei S, Khanmohammadi M, Heidari R, Ghanbari H, Taghdiri Nooshabadi V, Farzamfar S, et al. Exosome Loaded Alginate Hydrogel Promotes Tissue Regeneration in Full-Thickness Skin Wounds: An *In Vivo* Study. *J Biomed Mat Res Part A* (2020) 108:545–56. doi: 10.1002/jbm.a.36835
42. Lv Q, Deng J, Chen Y, Wang Y, Liu B, Liu J. Engineered Human Adipose Stem-Cell-Derived Exosomes Loaded With miR-21-5p to Promote Diabetic Cutaneous Wound Healing. *Mol Pharma* (2020) 17:1723–33. doi: 10.1021/acs.molpharmaceut.0c00177
43. Shabbir A, Cox A, Rodriguez-Menocal L, Salgado M, Van Badiavas E. Mesenchymal Stem Cell Exosomes Induce Proliferation and Migration of Normal and Chronic Wound Fibroblasts, and Enhance Angiogenesis In Vitro. *Stem Cells Dev* (2015) 24:1635–47. doi: 10.1089/scd.2014.0316
44. Li B, Luan S, Chen J, Zhou Y, Wang T, Li Z, et al. The MSC-Derived Exosomal Lncrna H19 Promotes Wound Healing in Diabetic Foot Ulcers by Upregulating PTEN *Via* Microrna-152-3p. *Mol Ther Nucleic Acids* (2020) 19:814–26. doi: 10.1016/j.omtn.2019.11.034
45. Shiekh PA, Singh A, Kumar A. Exosome Laden Oxygen Releasing Antioxidant and Antibacterial Cryogel Wound Dressing OxOBand Alleviate Diabetic and Infectious Wound Healing. *Biomaterials* (2020) 249:120020. doi: 10.1016/j.biomaterials.2020.120020
46. Tao SC, Guo SC, Li M, Ke QF, Guo YP, Zhang CQ. Chitosan Wound Dressings Incorporating Exosomes Derived From MicroRNA-126-Overexpressing Synovium Mesenchymal Stem Cells Provide Sustained Release of Exosomes and Heal Full-Thickness Skin Defects in a Diabetic Rat Model. *Stem Cells Trans Med* (2017) 6:736–47. doi: 10.5966/sctm.2016-0275
47. Chen CY, Rao SS, Ren L, Hu XK, Tan YJ, Hu Y, et al. Exosomal DMBT1 From Human Urine-Derived Stem Cells Facilitates Diabetic Wound repair by Promoting Angiogenesis. *Theranostics* (2018) 8:1607–23. doi: 10.7150/thno.22958
48. Li X, Xie X, Lian W, Shi R, Han S, Zhang H, et al. Exosomes from adipose-derived stem cells overexpressing Nrf2 Accelerate Cutaneous Wound Healing by Promoting Vascularization in a Diabetic Foot Ulcer Rat Model. *Exp Mol Med* (2018) 50:29. doi: 10.1038/s12276-018-0058-5
49. Kobayashi H, Ebisawa K, Kambe M, Kasai T, Suga H, Nakamura K, et al. Effects of Exosomes Derived From the Induced Pluripotent Stem Cells on Skin Wound Healing. *Nagoya J Med Sci* (2018) 80:141–53. doi: 10.18999/nagjms.80.2.141
50. Dalirfardouei R, Jamialahmadi K, Jafarian AH, Mahdipour E. Promising Effects of Exosomes Isolated From Menstrual Blood-Derived Mesenchymal Stem Cell on Wound-Healing Process in Diabetic Mouse Model. *J Tissue Eng Regen Med* (2019) 13:555–68. doi: 10.1002/term.2799
51. Wang C, Wang M, Xu T, Zhang X, Lin C, Gao W, et al. Engineering Bioactive Self-Healing Antibacterial Exosomes Hydrogel for Promoting Chronic Diabetic Wound Healing and Complete Skin Regeneration. *Theranostics* (2019) 9:65–76. doi: 10.7150/thno.29766
52. Wang J, Yi Y, Zhu Y, Wang Z, Wu S, Zhang J, et al. Effects of Adipose-Derived Stem Cell Released Exosomes on Wound Healing in Diabetic Mice. *Zhongguo Xiu Fu Chong Jian Wai Ke Za Zhi* (2020) 34:124–31. doi: 10.7507/1002-1892.201903058
53. Shi R, Jin Y, Hu W, Lian W, Cao C, Han S, et al. Exosomes Derived From Mmu_Circ_0000250-Modified Adipose-Derived Mesenchymal Stem Cells Promote Wound Healing in Diabetic Mice by Inducing miR-128-3p/SIRT1-mediated Autophagy. *Am J Physiol Cell Physiol* (2020) 318:C848–c856. doi: 10.1152/ajpcell.00041.2020
54. Liu W, Yu M, Xie D, Wang L, Ye C, Zhu Q, et al. Melatonin-Stimulated MSC-derived Exosomes Improve Diabetic Wound Healing Through Regulating Macrophage M1 and M2 Polarization by Targeting the PTEN/AKT Pathway. *Stem Cell Res Ther* (2020) 11:259. doi: 10.1186/s13287-020-01756-x
55. Yu M, Liu W, Li J, Lu J, Lu H, Jia W, et al. Exosomes Derived From Atorvastatin-Pretreated MSC Accelerate Diabetic Wound Repair by Enhancing Angiogenesis *Via* AKT/eNOS Pathway. *Stem Cell Res Ther* (2020) 11:350. doi: 10.1186/s13287-020-01824-2
56. Yang J, Chen Z, Pan D, Li H, Shen J. Umbilical Cord-Derived Mesenchymal Stem Cell-Derived Exosomes Combined Pluronic F127 Hydrogel Promote Chronic Diabetic Wound Healing and Complete Skin Regeneration. *Int J Nanomedicine* (2020) 15:5911–26. doi: 10.2147/IJN.S249129
57. Jin J, Wang Y, Zhao L, Zou W, Tan M, He Q. Exosomal Mirna-215-5p Derived From Adipose-Derived Stem Cells Attenuates Epithelial-Mesenchymal Transition of Podocytes by Inhibiting Zeb2. *BioMed Res Int* (2020) 2020:2685305. doi: 10.1155/2020/2685305
58. Jiang ZZ, Liu YM, Niu X, Yin JY, Hu B, Guo SC, et al. Exosomes Secreted by Human Urine-Derived Stem Cells Could Prevent Kidney Complications From Type I Diabetes in Rats. *Stem Cell Res Ther* (2016) 7:24. doi: 10.1186/s13287-016-0287-2
59. Nagaishi K, Mizue Y, Chikenji T, Otani M, Nakano M, Konari N, et al. Mesenchymal Stem Cell Therapy Ameliorates Diabetic Nephropathy *Via* the Paracrine Effect of Renal Trophic Factors Including Exosomes. *Sci Rep* (2016) 6:34842. doi: 10.1038/srep34842
60. Ebrahim N, Ahmed IA, Hussien NI, Dessouky AA, Farid AS, Elshazly AM, et al. Mesenchymal Stem Cell-Derived Exosomes Ameliorated Diabetic Nephropathy by Autophagy Induction Through the Mtor Signaling Pathway. *Cells* (2018) 7(12):226. doi: 10.20944/preprints201809.0153.v1
61. Jin J, Shi Y, Gong J, Zhao L, Li Y, He Q, et al. Exosome Secreted From Adipose-Derived Stem Cells Attenuates Diabetic Nephropathy by

Promoting Autophagy Flux and Inhibiting Apoptosis in Podocyte. *Stem Cell Res Ther* (2019) 10:95. doi: 10.1186/s13287-019-1177-1
62. Duan YR, Chen BP, Chen F, Yang SX, Zhu CY, Ma YL, et al. Exosomal microRNA-16-5p From Human Urine-Derived Stem Cells Ameliorates Diabetic Nephropathy Through Protection of Podocyte. *J Cell Mol Med* (2019) 1–16. doi: 10.1111/jcmm.14558
63. Li H, Rong P, Ma X, Nie W, Chen Y, Zhang J, et al. Mouse Umbilical Cord Mesenchymal Stem Cell Paracrine Alleviates Renal Fibrosis in Diabetic Nephropathy by Reducing Myofibroblast Transdifferentiation and Cell Proliferation and Upregulating Mmps in Mesangial Cells. *J Diabetes Res* (2020) 2020:3847171. doi: 10.1155/2020/3847171
64. Xiang E, Han B, Zhang Q, Rao W, Wang Z, Chang C, et al. Human Umbilical Cord-Derived Mesenchymal Stem Cells Prevent the Progression of Early Diabetic Nephropathy Through Inhibiting Inflammation and Fibrosis. *Stem Cell Res Ther* (2020) 11:336. doi: 10.1186/s13287-020-01852-y
65. Safwat A, Sabry D, Ragiae A, Amer E, Mahmoud RH, Shamardan RM. Adipose Mesenchymal Stem Cells-Derived Exosomes Attenuate Retina Degeneration of Streptozotocin-Induced Diabetes in Rabbits. *J Circulating Biomarkers* (2018) 7:1849454418807827. doi: 10.1177/1849454418807827
66. Zhang W, Wang Y, Kong Y. Exosomes Derived From Mesenchymal Stem Cells Modulate miR-126 to Ameliorate Hyperglycemia-Induced Retinal Inflammation Via Targeting Hmgb1. *Invest Ophthalmol Visual Sci* (2019) 60:294–303. doi: 10.1167/iovs.18-25617
67. Chen F, Zhang H, Wang Z, Ding W, Zeng Q, Liu W, et al. Adipose-Derived Stem Cell-Derived Exosomes Ameliorate Erectile Dysfunction in a Rat Model of Type 2 Diabetes. *J Sex Med* (2017) 14:1084–94. doi: 10.1016/j.jsxm.2017.07.005
68. Zhu LL, Huang X, Yu W, Chen H, Chen Y, Dai YT. Transplantation of Adipose Tissue-Derived Stem Cell-Derived Exosomes Ameliorates Erectile Function in Diabetic Rats. *Andrologia* (2018) 50(2). doi: 10.1111/and.12871
69. Wang J, Mi Y, Wu S, You X, Huang Y, Zhu J, et al. Exosomes From Adipose-Derived Stem Cells Protect Against High Glucose-Induced Erectile Dysfunction by Delivery of Corin in a Streptozotocin-Induced Diabetic Rat Model. *Regen Ther* (2020) 14:227–33. doi: 10.1016/j.reth.2020.03.002
70. Huo W, Li Y, Zhang Y, Li H. Mesenchymal Stem Cells-Derived Exosomal microRNA-21-5p Downregulates PDCD4 and Ameliorates Erectile Dysfunction in a Rat Model of Diabetes Mellitus. *FASEB J: Off Publ Fed Am Soc Exp Biol* (2020) 34(10):13345–60. doi: 10.1096/fj.202000102RR
71. Zhao W, Zhang H, Yan J, Ma X. An Experimental Study on the Treatment of Diabetes-Induced Cognitive Disorder Mice Model With Exosomes Deriving From Mesenchymal Stem Cells (Mscs). *Pakistan J Pharm Sci* (2019) 32:1965–70.
72. Kubota K, Nakano M, Kobayashi E, Mizue Y, Chikenji T, Otani M, et al. An Enriched Environment Prevents Diabetes-Induced Cognitive Impairment in Rats by Enhancing Exosomal miR-146a Secretion From Endogenous Bone Marrow-Derived Mesenchymal Stem Cells. *PloS One* (2018) 13:e0204252. doi: 10.1371/journal.pone.0204252
73. Nakano M, Nagaishi K, Konari N, Saito Y, Chikenji T, Mizue Y, et al. Bone Marrow-Derived Mesenchymal Stem Cells Improve Diabetes-Induced Cognitive Impairment by Exosome Transfer Into Damaged Neurons and Astrocytes. *Sci Rep* (2016) 6:24805. doi: 10.1038/srep24805
74. Venkat P, Zacharek A, Landschoot-Ward J, Wang F, Culmone L, Chen Z, et al. Exosomes Derived From Bone Marrow Mesenchymal Stem Cells Harvested From Type Two Diabetes Rats Promotes Neurorestorative Effects After Stroke in Type Two Diabetes Rats. *Exp Neurol* (2020) 334:113456. doi: 10.1016/j.expneurol.2020.113456
75. Cui C, Ye X, Chopp M, Venkat P, Zacharek A, Yan T, et al. Mir-145 Regulates Diabetes-Bone Marrow Stromal Cell-Induced Neurorestorative Effects in Diabetes Stroke Rats. *Stem Cells Trans Med* (2016) 5:1656–67. doi: 10.5966/sctm.2015-0349
76. AbuBakr N, Haggag T, Sabry D, Salem ZA. Functional and Histological Evaluation of Bone Marrow Stem Cell-Derived Exosomes Therapy on the Submandibular Salivary Gland of Diabetic Albino Rats Through Tgfβ/Smad3 Signaling Pathway. *Heliyon* (2020) 6:e03789. doi: 10.1016/j.heliyon.2020.e03789
77. Lin Y, Zhang F, Lian XF, Peng WQ, Yin CY. Mesenchymal Stem Cell-Derived Exosomes Improve Diabetes Mellitus-Induced Myocardial Injury and Fibrosis *Via* Inhibition of TGF-β1/Smad2 Signaling Pathway. *Cell Mol Biol (Noisy-le-Grand France)* (2019) 65:123–6. doi: 10.14715/cmb/2019.65.7.21
78. Baltzis D, Eleftheriadou I, Veves A. Pathogenesis and Treatment of Impaired Wound Healing in Diabetes Mellitus: New Insights. *Adv Ther* (2014) 31:817–36. doi: 10.1007/s12325-014-0140-x
79. Singh N, Armstrong DG, Lipsky BA. Preventing Foot Ulcers in Patients With Diabetes. *JAMA* (2005) 293:217–28. doi: 10.1001/jama.293.2.217
80. Ramsey SD, Newton K, Blough D, McCulloch DK, Sandhu N, Reiber GE, et al. Incidence, Outcomes, and Cost of Foot Ulcers in Patients With Diabetes. *Diabetes Care* (1999) 22:382–7. doi: 10.2337/diacare.22.3.382
81. Brem H, Sheehan P, Rosenberg HJ, Schneider JS, Boulton AJ. Evidence-Based Protocol for Diabetic Foot Ulcers. *Plast Reconstr Surg* (2006) 117:193S–209S; discussion 210S-211S. doi: 10.1097/01.prs.0000225459.93750.29
82. Verreck FA, de Boer T, Langenberg DM, Hoeve MA, Kramer M, Vaisberg E, et al. Human IL-23-producing Type 1 Macrophages Promote But IL-10-producing Type 2 Macrophages Subvert Immunity to (Myco)Bacteria. *Proc Natl Acad Sci United States America* (2004) 101:4560–5. doi: 10.1073/pnas.0400983101
83. Gordon S, Martinez FO. Alternative Activation of Macrophages: Mechanism and Functions. *Immunity* (2010) 32:593–604. doi: 10.1016/j.immuni.2010.05.007
84. Mantovani A, Sozzani S, Locati M, Allavena P, Sica A. Macrophage Polarization: Tumor-Associated Macrophages as a Paradigm for Polarized M2 Mononuclear Phagocytes. *Trends Immunol* (2002) 23:549–55. doi: 10.1016/S1471-4906(02)02302-5
85. Pradhan Nabzdyk L, Kuchibhotla S, Guthrie P, Chun M, Auster ME, Nabzdyk C, et al. Expression of Neuropeptides and Cytokines in a Rabbit Model of Diabetic Neuroischemic Wound Healing. *J Vasc Surg* (2013) 58:766–775.e712. doi: 10.1016/j.jvs.2012.11.095
86. Mi Q, Rivière B, Clermont G, Steed DL, Vodovotz Y. Agent-Based Model of Inflammation and Wound Healing: Insights Into Diabetic Foot Ulcer Pathology and the Role of Transforming Growth Factor-Beta1. *Wound Repair Regen: Off Publ Wound Heal Soc Eur Tissue Repair Soc* (2007) 15:671–82. doi: 10.1111/j.1524-475X.2007.00271.x
87. Kraakman MJ, Murphy AJ, Jandeleit-Dahm K, Kammoun HL. Macrophage Polarization in Obesity and Type 2 Diabetes: Weighing Down Our Understanding of Macrophage Function? *Front Immunol* (2014) 5:470. doi: 10.3389/fimmu.2014.00470
88. Ochoa O, Torres FM, Shireman PK. Chemokines and Diabetic Wound Healing. *Vascular* (2007) 15:350–5. doi: 10.2310/6670.2007.00056
89. Tonnesen MG, Feng X, Clark RA. Angiogenesis in Wound Healing. *J Invest Dermatol Symp Proc* (2000) 5:40–6. doi: 10.1046/j.1087-0024.2000.00014.x
90. Altavilla D, Saitta A, Cucinotta D, Galeano M, Deodato B, Colonna M, et al. Inhibition of Lipid Peroxidation Restores Impaired Vascular Endothelial Growth Factor Expression and Stimulates Wound Healing and Angiogenesis in the Genetically Diabetic Mouse. *Diabetes* (2001) 50:667–74. doi: 10.2337/diabetes.50.3.667
91. Rackov G, Garcia-Romero N, Esteban-Rubio S, Carrión-Navarro J, Belda-Iniesta C, Ayuso-Sacido A. Vesicle-Mediated Control of Cell Function: The Role of Extracellular Matrix and Microenvironment. *Front Physiol* (2018) 9:651. doi: 10.3389/fphys.2018.00651
92. Merino-González C, Zuñiga FA, Escudero C, Ormazabal V, Reyes C, Nova-Lamperti E, et al. Mesenchymal Stem Cell-Derived Extracellular Vesicles Promote Angiogenesis: Potencial Clinical Application. *Front Physiol* (2016) 7:24. doi: 10.3389/fphys.2016.00024
93. Espinel E, Agraz I, Ibernon M, Ramos N, Fort J, Serón D. Renal Biopsy in Type 2 Diabetic Patients. *J Clin Med* (2015) 4:998–1009. doi: 10.3390/jcm4050998
94. Doshi SM, Friedman AN. Diagnosis and Management of Type 2 Diabetic Kidney Disease. *Clin J Am Soc Nephrol: CJASN* (2017) 12:1366–73. doi: 10.2215/CJN.11111016
95. Zhou LL, Hou FF, Wang GB, Yang F, Xie D, Wang YP, et al. Accumulation of Advanced Oxidation Protein Products Induces Podocyte Apoptosis and Deletion Through NADPH-dependent Mechanisms. *Kidney Int* (2009) 76:1148–60. doi: 10.1038/ki.2009.322

96. Siddiqi FS, Advani A. Endothelial-Podocyte Crosstalk: The Missing Link Between Endothelial Dysfunction and Albuminuria in Diabetes. *Diabetes* (2013) 62:3647–55. doi: 10.2337/db13-0795
97. Sawada K, Toyoda M, Kaneyama N, Shiraiwa S, Moriya H, Miyatake H, et al. Upregulation of α3β1-Integrin in Podocytes in Early-Stage Diabetic Nephropathy. *J Diabetes Res* (2016) 2016:9265074. doi: 10.1155/2016/9265074
98. Zhu F, Chong Lee Shin OLS, Pei G, Hu Z, Yang J, Zhu H, et al. Adipose-Derived Mesenchymal Stem Cells Employed Exosomes to Attenuate AKI-CKD Transition Through Tubular Epithelial Cell Dependent Sox9 Activation. *Oncotarget* (2017) 8:70707–26. doi: 10.18632/oncotarget.19979
99. Lin KC, Yip HK, Shao PL, Wu SC, Chen KH, Chen YT, et al. Combination of Adipose-Derived Mesenchymal Stem Cells (ADMSC) and ADMSC-derived Exosomes for Protecting Kidney From Acute Ischemia-Reperfusion Injury. *Int J Cardiol* (2016) 216:173–85. doi: 10.1016/j.ijcard.2016.04.061
100. Simonson MS. Phenotypic Transitions and Fibrosis in Diabetic Nephropathy. *Kidney Int* (2007) 71:846–54. doi: 10.1038/sj.ki.5002180
101. Essawy M, Soylemezoglu O, Muchaneta-Kubara EC, Shortland J, Brown CB, l Nahas AM. Myofibroblasts and the Progression of Diabetic Nephropathy. *Nephrol Dial Transplant: Off Publ Eur Dialysis Transplant Assoc Eur Renal Assoc* (1997) 12:43–50. doi: 10.1093/ndt/12.1.43
102. Hills CE, Squires PE. Tgf-beta1-induced Epithelial-to-Mesenchymal Transition and Therapeutic Intervention in Diabetic Nephropathy. *Am J Nephrol* (2010) 31:68–74. doi: 10.1159/000256659
103. Nicholson L, Patrao NV, Ramu J, Vazquez-Alfageme C, Muwas M, Rajendram R, et al. Influence of Baseline Diabetic Retinopathy Status on Initial Anatomical Response of Intravitreal Ranibizumab Therapy for Diabetic Macular Oedema. *Eye (London England)* (2017) 31:1358–64. doi: 10.1038/eye.2017.69
104. Choi JA, Kim HW, Kwon JW, Shim YS, Jee DH, Yun JS, et al. Early Inner Retinal Thinning and Cardiovascular Autonomic Dysfunction in Type 2 Diabetes. *PloS One* (2017) 12:e0174377. doi: 10.1371/journal.pone.0174377
105. Ezquer M, Urzua CA, Montecino S, Leal K, Conget P, Ezquer F. Intravitreal Administration of Multipotent Mesenchymal Stromal Cells Triggers a Cytoprotective Microenvironment in the Retina of Diabetic Mice. *Stem Cell Res Ther* (2016) 7:42. doi: 10.1186/s13287-016-0299-y
106. Mishra M, Flaga J, Kowluru RA. Molecular Mechanism of Transcriptional Regulation of Matrix Metalloproteinase-9 in Diabetic Retinopathy. *J Cell Physiol* (2016) 231:1709–18. doi: 10.1002/jcp.25268
107. Sun C, Alkhoury K, Wang YI, Foster GA, Radecke CE, Tam K, et al. IRF-1 and miRNA126 Modulate VCAM-1 Expression in Response to a High-Fat Meal. *Circ Res* (2012) 111:1054–64. doi: 10.1161/CIRCRESAHA.112.270314
108. Lizza EF, Rosen RC. Definition and Classification of Erectile Dysfunction: Report of the Nomenclature Committee of the International Society of Impotence Research. *Int J Impot Res* (1999) 11:141–3. doi: 10.1038/sj.ijir.3900396
109. Fedele D. Therapy Insight: Sexual and Bladder Dysfunction Associated With Diabetes Mellitus. *Nat Clin Pract Urol* (2005) 2:282–90; quiz 309. doi: 10.1038/ncpuro0211
110. Musicki B, Burnett AL. Endothelial Dysfunction in Diabetic Erectile Dysfunction. *Int J Impot Res* (2007) 19:129–38. doi: 10.1038/sj.ijir.3901494
111. Moore CR, Wang R. Pathophysiology and Treatment of Diabetic Erectile Dysfunction. *Asian J Androl* (2006) 8:675–84. doi: 10.1111/j.1745-7262.2006.00223.x
112. Alwaal A, Zaid UB, Lin CS, Lue TF. Stem Cell Treatment of Erectile Dysfunction. *Adv Drug Deliv Rev* (2015) 82-83:137–44. doi: 10.1016/j.addr.2014.11.012
113. Dillmann WH. Diabetic Cardiomyopathy. *Circ Res* (2019) 124:1160–2. doi: 10.1161/CIRCRESAHA.118.314665
114. Jia G, DeMarco VG, Sowers JR. Insulin Resistance and Hyperinsulinaemia in Diabetic Cardiomyopathy. *Nat Rev Endocrinol* (2016) 12:144–53. doi: 10.1038/nrendo.2015.216
115. Ni J, Sun Y, Liu Z. The Potential of Stem Cells and Stem Cell-Derived Exosomes in Treating Cardiovascular Diseases. *J Cardiovasc Trans Res* (2019) 12:51–61. doi: 10.1007/s12265-018-9799-8
116. Yu B, Kim HW, Gong M, Wang J, Millard RW, Wang Y, et al. Exosomes Secreted From GATA-4 Overexpressing Mesenchymal Stem Cells Serve as a Reservoir of Anti-Apoptotic microRNAs for Cardioprotection. *Int J Cardiol* (2015) 182:349–60. doi: 10.1016/j.ijcard.2014.12.043
117. Arslan F, Lai RC, Smeets MB, Akeroyd L, Choo A, Aguor EN, et al. Mesenchymal Stem Cell-Derived Exosomes Increase ATP Levels, Decrease Oxidative Stress and Activate PI3K/Akt Pathway to Enhance Myocardial Viability and Prevent Adverse Remodeling After Myocardial Ischemia/Reperfusion Injury. *Stem Cell Res* (2013) 10:301–12. doi: 10.1016/j.scr.2013.01.002

Permissions

All chapters in this book were first published Frontiers; hereby published with permission under the Creative Commons Attribution License or equivalent. Every chapter published in this book has been scrutinized by our experts. Their significance has been extensively debated. The topics covered herein carry significant findings which will fuel the growth of the discipline. They may even be implemented as practical applications or may be referred to as a beginning point for another development.

The contributors of this book come from diverse backgrounds, making this book a truly international effort. This book will bring forth new frontiers with its revolutionizing research information and detailed analysis of the nascent developments around the world.

We would like to thank all the contributing authors for lending their expertise to make the book truly unique. They have played a crucial role in the development of this book. Without their invaluable contributions this book wouldn't have been possible. They have made vital efforts to compile up to date information on the varied aspects of this subject to make this book a valuable addition to the collection of many professionals and students.

This book was conceptualized with the vision of imparting up-to-date information and advanced data in this field. To ensure the same, a matchless editorial board was set up. Every individual on the board went through rigorous rounds of assessment to prove their worth. After which they invested a large part of their time researching and compiling the most relevant data for our readers.

The editorial board has been involved in producing this book since its inception. They have spent rigorous hours researching and exploring the diverse topics which have resulted in the successful publishing of this book. They have passed on their knowledge of decades through this book. To expedite this challenging task, the publisher supported the team at every step. A small team of assistant editors was also appointed to further simplify the editing procedure and attain best results for the readers.

Apart from the editorial board, the designing team has also invested a significant amount of their time in understanding the subject and creating the most relevant covers. They scrutinized every image to scout for the most suitable representation of the subject and create an appropriate cover for the book.

The publishing team has been an ardent support to the editorial, designing and production team. Their endless efforts to recruit the best for this project, has resulted in the accomplishment of this book. They are a veteran in the field of academics and their pool of knowledge is as vast as their experience in printing. Their expertise and guidance has proved useful at every step. Their uncompromising quality standards have made this book an exceptional effort. Their encouragement from time to time has been an inspiration for everyone.

The publisher and the editorial board hope that this book will prove to be a valuable piece of knowledge for researchers, students, practitioners and scholars across the globe.

List of Contributors

Kriti Joshi
Department of Endocrinology and Metabolism, All India Institute of Medical Sciences Rishikesh, Uttarakhand, India
Department of Molecular Medicine & Biotechnology, Sanjay Gandhi Post Graduate Institute of Medical Sciences, Lucknow, India
Department of Cell Biology, Murdoch Children's Research Institute, Parkville, Vic, Australia

Swasti Tiwari
Department of Molecular Medicine & Biotechnology, Sanjay Gandhi Post Graduate Institute of Medical Sciences, Lucknow, India

Fergus Cameron
Department of Cell Biology, Murdoch Children's Research Institute, Parkville, Vic, Australia
Department of Endocrinology and Diabetes, The Royal Children's Hospital, Parkville, Vic, Australia
Department of Paediatrics, University of Melbourne, Parkville, Vic, Australia

Stuart I. Mannering
Immunology and Diabetes Unit, St. Vincent's Institute of Medical Research, Fitzroy, Vic, Australia

Andrew G. Elefanty and Edouard G. Stanley
Department of Cell Biology, Murdoch Children's Research Institute, Parkville, Vic, Australia
Department of Paediatrics, University of Melbourne, Parkville, Vic, Australia
Department of Anatomy and Developmental Biology, Monash University, Clayton, Vic, Australia

Carmel Braverman-Gross and Nissim Benvenisty
The Azrieli Center for Stem Cells and Genetic Research, Department of Genetics, The Alexander Silberman Institute of Life Sciences, The Hebrew University of Jerusalem, Jerusalem, Israel

Matthew N. George
Center for Cellular and Molecular Therapeutics, The Children's Hospital of Philadelphia, Philadelphia, PA, United States

Karla F. Leavens
Center for Cellular and Molecular Therapeutics, The Children's Hospital of Philadelphia, Philadelphia, PA, United States
Department of Pediatrics, Perelman School of Medicine, University of Pennsylvania, Philadelphia, PA, United States
Division of Endocrinology and Diabetes, The Children's Hospital of Philadelphia, Philadelphia, PA, United States

Paul Gadue
Center for Cellular and Molecular Therapeutics, The Children's Hospital of Philadelphia, Philadelphia, PA, United States
Department of Pathology and Laboratory Medicine, The Children's Hospital of Philadelphia, Philadelphia, PA, United States

Eleonora de Klerk and Matthias Hebrok
Diabetes Center, University of California San Francisco, San Francisco, CA, United States

Jared P. Taylor
Department of Pathology, Immunology, and Laboratory Medicine, College of Medicine, University of Florida, Gainesville, FL, United States

Lucas H. Armitage, Scott E. Stimpson, Similoluwa Ogundare, Brittney N. Newby, Mollie K. Huber, Daniel J. Perry, Todd M. Brusko and Jing Chen
Department of Pathology, Immunology, and Laboratory Medicine, College of Medicine, University of Florida, Gainesville, FL, United States
University of Florida Diabetes Institute, University of Florida, Gainesville, FL, United States

Katherine E. Santostefano
Department of Pathology, Immunology, and Laboratory Medicine, College of Medicine, University of Florida, Gainesville, FL, United States
Center for Cellular Reprogramming, College of Medicine, University of Florida, Gainesville, FL, United States
Century Therapeutics, iPSC Biology, Philadelphia, PA, United States

Lina Sui and Dieter Egli
Department of Pediatrics, Naomi Berrie Diabetes Center, Columbia Stem Cell Initiative, Columbia University, New York, United States

Prerana Sharma, Ilian A. Radichev and Alexei Y. Savinov
Children's Health Research Center, Sanford Research, Sioux Falls, SD, United States

Natalie C. Fredette and Naohiro Terada
Department of Pathology, Immunology, and Laboratory Medicine, College of Medicine, University of Florida, Gainesville, FL, United States
Center for Cellular Reprogramming, College of Medicine, University of Florida, Gainesville, FL, United States

Mark A. Wallet
Department of Pathology, Immunology, and Laboratory Medicine, College of Medicine, University of Florida, Gainesville, FL, United States
University of Florida Diabetes Institute, University of Florida, Gainesville, FL, United States
Century Therapeutics, Immunology, Philadelphia, PA, United States

Clayton E. Mathews
Department of Pathology, Immunology, and Laboratory Medicine, College of Medicine, University of Florida, Gainesville, FL, United States
University of Florida Diabetes Institute, University of Florida, Gainesville, FL, United States
Center for Cellular Reprogramming, College of Medicine, University of Florida, Gainesville, FL, United States

Kfir Molakandov, Denise A. Berti, Avital Beck, Karina Yavriyants, Michal Zimerman, Ella Volman, Itzik Toledo, Anna Erukhimovich, Alon M. Levy, Arik Hasson, Joseph Itskovitz-Eldor and Judith Chebath
Kadimastem Ltd., Weizmann Science Park, Ness Ziona, Israel

Ofer Elhanani, Michael D. Walker and Yoav Soen
Department of Biomolecular Sciences, Weizmann Institute of Science, Rehovot, Israel

Michel Revel
Kadimastem Ltd., Weizmann Science Park, Ness Ziona, Israel
Department of Molecular Genetics (emeritus), Weizmann Institute of Science, Rehovot, Israel

Erin M. MacFarlane and Jennifer E. Bruin
Department of Biology & Institute of Biochemistry, Carleton University, Ottawa, ON, Canada

Juan Ignacio Burgos and Santiago A. Rodríguez-Seguí
Departamento de Fisiología, Biología Molecular y Celular, Facultad de Ciencias Exactas y Naturales, Universidad de Buenos Aires and Instituto de Fisiología, Biología Molecular y Neurociencias (IFIBYNE), CONICET-Universidad de Buenos Aires, Ciudad Universitaria, Buenos Aires, Argentina

Ludovic Vallier
Wellcome-Medical Research Council Cambridge Stem Cell Institute and Department of Surgery, University of Cambridge, Cambridge, United Kingdom

Diego Balboa
Regulatory Genomics and Diabetes, Centre for Genomic Regulation, The Barcelona Institute of Science and Technology, Barcelona, Spain

Diepiriye G. Iworima
Department of Cellular and Physiological Sciences, University of British Columbia, Vancouver, BC, Canada
School of Biomedical Engineering, The University of British Columbia, Vancouver, BC, Canada

Timothy J. Kieffer
Department of Cellular and Physiological Sciences, University of British Columbia, Vancouver, BC, Canada

School of Biomedical Engineering, The University of British Columbia, Vancouver, BC, Canada
Department of Surgery, University of British Columbia, Vancouver, BC, Canada

Taylor M. Triolo
The Barbara Davis Center for Diabetes, School of Medicine, University of Colorado Denver, Aurora, CO, United States

Melena D. Bellin
Department of Pediatrics, School of Medicine, University of Minnesota, Minneapolis, MN, United States

Shimon Efrat
Department of Human Molecular Genetics and Biochemistry, Sackler School of Medicine, Tel Aviv University, Tel Aviv, Israel

Sandra Heller, Ninel Azoitei and Alexander Kleger
Department of Internal Medicine I, Ulm University Hospital, Ulm, Germany

Michael Karl Melzer
Department of Internal Medicine I, Ulm University Hospital, Ulm, Germany
Department of Urology, Ulm University Hospital, Ulm, Germany

Cécile Julier
Université de Paris, Institut Cochin, INSERM U1016, CNRS UMR-8104, Paris, France

Amanda Oakie
McEwen Stem Cell Institute, University Health Network, Toronto, ON, Canada

Maria Cristina Nostro
McEwen Stem Cell Institute, University Health Network, Toronto, ON, Canada
Department of Physiology, University of Toronto, Toronto, ON, Canada

Roberto Castro-Gutierrez, Aimon Alkanani, Aaron Michels and Holger A. Russ
Barbara-Davis Center for Diabetes, University of Colorado Anschutz, Aurora, CO, United States

Clayton E. Mathews
Department of Pathology, Immunology, and Laboratory Medicine, College of Medicine, University of Florida, Gainesville, FL, United States

Naomi Koehler and Carmen Gonelle-Gispert
Surgical Research Unit, Faculty of Science and Medicine, University of Fribourg, Fribourg, Switzerland

Leo Buhler
Department of Surgery, Cantonal Hospital Fribourg, Fribourg, Switzerland

Bernhard Egger
Surgical Research Unit, Faculty of Science and Medicine, University of Fribourg, Fribourg, Switzerland
Department of Surgery, Cantonal Hospital Fribourg, Fribourg, Switzerland

Jaquellyne G. Penaforte-Saboia and Livia A. Batista
Post Graduate Program in Medical Sciences, Federal University of Ceará, Ceará, Brazil

Renan M. Montenegro Jr.
Post Graduate Program in Medical Sciences, Federal University of Ceará, Ceará, Brazil
Federal University of Ceará, Ceará, Brazil

Ana Paula D. R. Montenegro and Virginia O. Fernandes
Federal University of Ceará, Ceará, Brazil

Hussain Akhtar
Federal University of Ceará, Ceará, Brazil
University of Oslo, UIO, Oslo, Noruega

Carlos A. Negrato
Brazilian Society of Diabetes, São Paulo, Brazil

Kelen Cristina Ribeiro Malmegrim
Center for Cell-Based Therapy, Regional Blood Center of Ribeirão Preto, Ribeirão Preto Medical School, University of São Paulo, Ribeirão Preto, Brazil

Daniela Aparecida Moraes, Juliana B. E. Dias, Belinda P. Simões, Maria Carolina Oliveira, Carlos E. Couri, Carlos Eduardo Barra Couri and Juliana Bernardes Elias Dias
Center for Cell-Based Therapy, Regional Blood Center of Ribeirão Preto, Ribeirão Preto Medical School, University of São Paulo, Ribeirão Preto, Brazil
Department of Internal Medicine, Ribeirão Preto Medical School, University of São Paulo, Ribeirão Preto, Brazil

Marilia Brito Gomes
State University of Rio de Janeiro, Rio de Janeiro, Brazil

Jaquellyne Gurgel Penaforte-Saboia
Postgraduate Program in Medical Sciences, Department of Clinical Medicine, Federal University of Ceará, Fortaleza, Brazil
Clinical Research Unit, Walter Cantidio University Hospital, Federal University of Ceará/EBSERH, Fortaleza, Brazil
Faculty of Health Sciences, Nord University, Bodø, Norway
Centre for Global Health Research, Diabetic Association of Bangladesh, Dhaka, Bangladesh

Virginia Oliveira Fernandes, Ana Paula Dias Rangel Montenegro and Renan Magalhães Montenegro Jr.
Postgraduate Program in Medical Sciences, Department of Clinical Medicine, Federal University of Ceará, Fortaleza, Brazil
Clinical Research Unit, Walter Cantidio University Hospital, Federal University of Ceará/EBSERH, Fortaleza, Brazil
Postgraduate Program in Public Health, Department of Community Health, Federal University of Ceará, Fortaleza, Brazil

Lívia Aline De Araújo Batista
Postgraduate Program in Medical Sciences, Department of Clinical Medicine, Federal University of Ceará, Fortaleza, Brazil
Postgraduate Program in Public Health, Department of Community Health, Federal University of Ceará, Fortaleza, Brazil

Lenita Zajdenverg
University Hospital, Federal University of Rio de Janeiro, Rio de Janeiro, Brazil

Carlos Antonio Negrato
Brazilian Society of Diabetes, São Paulo, Brazil

Kelen Cristina Ribeiro Malmegrim
Center for Cell-Based Therapy, Regional Blood Center of Ribeirão Preto, Ribeirão Preto Medical School, University of São Paulo Ribeirão Preto, Brazil

Akhtar Hussain
Postgraduate Program in Medical Sciences, Department of Clinical Medicine, Federal University of Ceará, Fortaleza, Brazil
Centre for Global Health Research, Diabetic Association of Bangladesh, Dhaka, Bangladesh
Faculty of Health Sciences, Nord University, Bodø, Norway

Marilia Brito Gomes
Diabetes Unit, Department of Internal Medicine, State University of Rio de Janeiro, Rio de Janeiro, Brazil

Xin-Xing Wan
Department of Endocrinology, Third Xiangya Hospital, Central South University, Changsha, China

Dan-Yi Zhang and Sheng-Yuan Zheng
Clinical Medicine Eight-Year Program, Xiangya School of Medicine, Central South University, Changsha, China

Md. Asaduzzaman Khan
The Research Centre for Preclinical Medicine, Southwest Medical University, Luzhou, China

Xi-Min Hu and Qi Zhang
Department of Anatomy and Neurobiology, School of Basic Medical Science, Central South University, Changsha, China

Rong-Hua Yang
Department of Burn and Plastic Surgery, Guangzhou First People's Hospital, School of Medicine, South China University of Technology, Guangzhou, China

Kun Xiong
Department of Anatomy and Neurobiology, School of Basic Medical Science, Central South University, Changsha, China
Hunan Key Laboratory of Ophthalmology, Changsha, China
Key Laboratory of Emergency and Trauma, Ministry of Education, College of Emergency and Trauma, Hainan Medical University, Haikou, China

Jiachao Xiong, Hao Hu, Rong Guo, Hui Wang and Hua Jiang
Department of Plastic Surgery, Shanghai East Hospital, Tongji University School of Medicine, Shanghai, China

Index

A

Alloimmunity, 50, 135-136

Allotransplantation, 42, 51, 133-137, 182

Antigen Presenting Cells, 1-3, 5, 7, 10

Autoimmunity, 2, 4, 6-7, 10, 12-13, 15-16, 42, 54-56, 67, 108, 135-137, 139, 169, 179, 193, 195, 205, 208

B

Beta Cell Mass, 2, 4, 11, 89, 124, 131, 135, 138, 184

C

Cell Differentiation, 7, 12-13, 23-25, 28, 32, 37, 50, 67, 81, 88, 93, 100, 102, 104-106, 108, 110, 112, 119, 121-122, 130, 132-133, 140, 143, 149, 152, 156, 159, 161, 163-165, 167, 173

Cell Survival, 42-43, 47, 49-50, 85, 153, 169, 184, 205

Chemotaxis, 9, 61

Chronic Pancreatitis, 51, 134-135, 137-138, 147-148, 157

Clinical Trial, 41-42, 45, 207-208, 212, 215

D

Diabetes Mellitus, 2, 9, 13, 15-16, 23, 26, 35-36, 39-41, 45-46, 51-54, 89, 92, 95, 101, 109, 111-112, 114, 126-128, 133, 137, 144-145, 148, 155-160, 168, 178, 185, 188-189, 194-197, 201-204, 211-214, 216, 218, 223-225

Diabetes Subtypes, 26-28, 32-33, 35, 126

Diabetic Encephalopathy, 216-218

Diabetic Foot, 204, 216-218, 223-224

Direct Differentiation, 169, 175

Disease Modeling, 14, 17, 22-23, 25, 37, 85, 88, 96, 106, 109, 114, 120-121, 125, 140, 142, 146

Disease Pathogenesis, 1-2, 6, 9-12, 89

Donor Pancreas, 133

E

Embryonic Stem Cells, 8, 14-15, 17-18, 21, 25, 28, 36, 40, 44, 48, 52-53, 55, 66-67, 80, 85, 87, 93, 97, 110, 115, 119, 129-131, 134-135, 138-140, 144, 146, 149, 156-157, 165, 167, 179, 195, 203, 205, 210-211, 214

Encapsulation, 41, 43, 49-50, 53, 69, 74, 79, 124, 135-136, 151, 169, 179-184, 186, 209-210, 213, 215

Endocrine Cells, 6-7, 12, 14, 28, 36, 48-49, 79-80, 82, 84, 86-87, 89, 97-98, 102-104, 106, 108, 110, 114-115, 119, 122, 124, 129, 131, 138-139, 149-151, 153-154, 156, 162, 164, 166, 183, 187, 208

Endocrine Progenitors, 105, 107, 111, 122, 142, 149, 163-165

Endothelium, 15, 30, 55, 57, 61, 66

Environmental Toxins, 84, 87-88

Epigenetic Aberrations, 17, 22-23, 25

Exogenous Insulin, 41-43, 48, 55, 133-134, 180, 189, 194, 197, 204

Exosomes, 47-48, 53, 205, 212, 216-217, 220-225

G

Genetic Defects, 114-118, 121

Genetic Variants, 1, 40, 54, 65, 114-118, 120-122, 125, 147

Genome Editing, 13, 24, 26-30, 32-35, 37-38, 50, 53, 67, 87-88, 93, 101-103, 105, 109, 112, 115, 120-121, 125, 127, 130, 136, 157, 159, 179, 213

Genome Engineering, 37, 67, 130, 136, 139, 168, 170, 174

Genotype, 33, 56, 66-67, 95, 128, 148-149, 157-158

Glucose Tolerance, 27, 44, 47, 69, 71, 76, 79, 91-92, 99, 124, 209

Glycemic Control, 14, 51, 53, 80, 84, 104, 125, 131, 133, 135, 179, 188-189, 191-193, 197, 201, 212

H

Haploinsufficiency, 20-22, 24-25, 27, 36, 38, 98, 105-107, 111-112, 130, 154, 157-159

Hypoglycemia, 19, 38, 42-43, 45, 49, 51, 126-127, 134, 137, 168, 193, 196-200, 202, 204, 208

Hypoglycemic Events, 41-42, 79, 197-200, 204

I

Immune Interactions, 4, 135

Immune Rejection, 41, 50, 135, 151

Immune Response, 2, 9, 23, 44-45, 55, 133, 135-136, 194, 205-206, 212, 216

Immune System, 2, 5, 11, 29, 35, 48-50, 59, 67, 81, 145-146, 169, 181, 193, 205, 207-208

Immunofluorescence, 107, 170, 172-175, 178

Immunogenicity, 67, 136, 139, 177, 179-180, 183, 216-217

Immunosuppression, 42-43, 49, 134-137, 151, 180, 184, 203-204, 212-213, 215

Immunotolerance, 203, 205-206

Inflammatory Reaction, 10, 42

Insulin Dose, 188, 190-194, 197-198, 202, 204, 211

Insulin Independence, 41-42, 49, 133-135, 137-138, 156, 180, 194, 201

Insulin Resistance, 17, 27, 83-84, 86, 90, 92, 95, 100, 109, 112, 114, 134, 137, 145, 155, 166, 203, 217-218, 223, 225

Insulin Secretion, 2, 7, 20, 27, 35, 38, 47, 50, 60, 69, 74, 77, 80-81, 83-89, 91-92, 99-100, 103, 105, 107, 112, 114-129, 132-133, 135, 137, 146-147, 150-154, 158-159, 164-165, 180, 182, 184, 186, 192-193, 197, 201-204, 210, 213, 215-217

Insulin Therapy, 41, 43-44, 48, 51, 148, 168, 189, 193, 198, 208, 216

Ischemia, 135, 138, 212, 217, 221-222, 225

Islet Cells, 16, 35, 53, 63-64, 68-69, 114-116, 118-119, 122-125, 130-131, 139, 142, 162, 179-180, 182, 186, 207-208

Islet Mass, 12, 42, 49, 134-135, 186

Islet Transplantation, 18, 41-43, 49, 51-53, 80, 132-135, 137, 139, 146, 156, 167, 178, 180-181, 184, 186, 204, 214

Isogenic Lines, 23, 28, 30-31, 103

K

Kidney Capsule, 162, 177, 184

Kidney Injury, 216-218, 220

Kidney Transplantation, 42, 51, 136

M

Macrophages, 1-2, 4-5, 7-10, 13-15, 48, 54-57, 61-62, 64, 66, 217-219, 223-224

Mesenchymal Stem Cells, 11, 41, 44-46, 52-53, 89, 151, 181, 184-186, 203, 206, 210-211, 213-218, 220, 222-225

Methylation, 22, 25, 88, 111, 123, 131, 141-143

Microvascular Complications, 134, 188-194, 196-199, 201, 216, 218

Mitochondrial Respiration, 21, 33, 123

Monocytes, 7-8, 14, 57, 61-62, 64

Mortality, 26, 35, 51, 89, 134, 204, 212, 222

N

Neonatal Porcine Islets, 180, 215

O

Organoids, 8, 23, 41, 43-44, 47, 49-50, 65, 112, 131, 156, 179, 205, 208

P

Pancreas, 3-4, 11-12, 21, 24-25, 27, 29, 34, 36, 38, 42-43, 49, 51, 55, 76-77, 81-82, 84-87, 89, 91-93, 95-113, 115, 118-122, 124, 126, 128-131, 133-140, 142-144, 147-149, 151, 153-154, 156-160, 164, 166-168, 183, 186, 204, 206, 208-209, 212-213, 215

Pancreatectomy, 42, 51, 133-135, 138

Pancreatic Agenesis, 20-21, 27, 36, 38, 100-101, 105-107, 112-113, 118, 148, 150, 154, 158, 160

Pancreatic Differentiation, 14, 20-21, 37, 88, 93, 95, 97, 100, 104, 107, 110-111, 113, 120, 129-130, 143, 151, 153-154, 159, 166, 215

Pancreatic Progenitors, 14, 20, 22, 24, 28, 44, 47-50, 52-53, 66, 80, 86, 93, 97-98, 101-102, 104, 106-107, 110-111, 119-122, 124, 129-130, 138-139, 143, 149, 151, 156, 159, 161, 163-167, 179, 208, 210, 215

Pancreatic Specimens, 9

Pancreatic Tissue, 2-3, 9, 55, 97, 102, 107-108, 162, 217

Pathomechanisms, 149, 152-153, 155

Peripheral Blood, 5, 7-10, 14, 40, 56-57, 136, 141, 154, 177, 182, 186, 189, 197-198, 207

Phenotype, 4-5, 7, 21, 27, 30-31, 33, 38, 40, 67, 76-77, 95-96, 98-106, 108, 111, 118, 121, 124, 128, 135, 141-142, 147-151, 153-155, 157-158, 162, 166, 181-182, 194, 208, 217

Progenitors, 14, 20, 22, 24, 28, 44, 47-50, 52-53, 57, 61, 65-66, 73, 77, 79-80, 86, 93, 97-98, 101-102, 104-107, 110-112, 119-122, 124, 129-131, 138-139, 142-143, 149, 151, 156, 158-159, 161, 163-167, 179, 185, 208, 210, 215

S

Stem Cells, 2, 4, 7-8, 11, 13-18, 21, 24-26, 28, 36-41, 44-46, 48, 50-56, 58, 66-69, 76, 80, 82, 84-85, 87-89, 93-98, 109-112, 114-115, 119, 126-127, 129-131, 133-136, 138-141, 143-146, 149, 151, 154-157, 159, 162-163, 189, 193, 195, 198, 203-218, 220, 222-225

Streptozotocin, 18, 47, 52, 69, 71, 74, 119, 124, 131, 153, 195, 208-209, 215, 224

T

Therapeutic Strategies, 12-13, 98, 136, 193-194

Toxicology, 82-84, 86-88, 90-91, 93

Transplanted Islet, 136, 186

V

Viscosity, 74, 79